P9-DTX-562

Chemical Industries
Information Sources

OTHER BOOKS IN THE
MANAGEMENT INFORMATION GUIDE SERIES
Write for Complete List

ELECTRONIC INDUSTRIES INFORMATION SOURCES—Edited by Gretchen R. Randle, 1968 (MIG No. 13)

INTERNATIONAL BUSINESS AND FOREIGN TRADE INFORMATION SOURCES—Edited by Lora Jeanne Wheeler, 1968 (MIG No. 14)

COMPUTERS AND DATA PROCESSING INFORMATION SOURCES—Edited by Chester Morrill, Jr., 1969 (MIG No. 15)

FOOD AND BEVERAGE INDUSTRIES—Edited by Albert C. Vara, 1970 (MIG No. 16)

COMMERCIAL LAW INFORMATION SOURCES—Edited by Julius J. Marke and Edward J. Bander, 1970 (MIG No. 17)

ACCOUNTING INFORMATION SOURCES—Edited by Rosemary R. Demarest, 1970 (MIG No. 18)

INVESTMENT INFORMATION—Edited by James B. Woy, 1970 (MIG No. 19)

RESEARCH IN TRANSPORTATION—Edited by Kenneth U. Flood, 1970 (MIG No. 20)

ETHICS IN BUSINESS CONDUCT—Edited by Portia Christian with Richard Hicks, 1970 (MIG No. 21)

PUBLIC RELATIONS INFORMATION SOURCES—Edited by Alice Norton, 1970 (MIG No. 22)

AMERICAN ECONOMIC AND BUSINESS HISTORY INFORMATION SOURCES—Edited by Robert W. Lovett, 1971 (MIG No. 23)

INSURANCE INFORMATION SOURCES—Edited by Roy Edwin Thomas, 1971 (MIG No. 24)

COMMUNICATION IN ORGANIZATIONS—Edited by Robert M. Carter, 1972 (MIG No. 25)

PUBLIC AND BUSINESS PLANNING IN THE UNITED STATES—Edited by Martha B. Lightwood, 1972 (MIG No. 26)

NATIONAL SECURITY AFFAIRS—Edited by Arthur D. Larson, 1973 (MIG No. 27)

OCCUPATIONAL SAFETY AND HEALTH—Edited by Theodore P. Peck, 1974 (MIG No. 28)

PURCHASING INFORMATION SOURCES—Edited by Douglas C. Basil, Emma Jean Gillis, and Walter R. Davis, 1977 (MIG No. 30)

EXECUTIVE AND MANAGEMENT DEVELOPMENT FOR BUSINESS AND GOVERNMENT—Edited by Agnes O. Hanson, 1976 (MIG No. 31)

MANAGEMENT PRINCIPLES AND PRACTICE—Edited by K.G.B. Bakewell, 1977 (MIG No. 32)

MANAGEMENT AND ECONOMICS JOURNALS—Edited by Vasile Tega, 1977 (MIG No. 33)

AGRICULTURAL ENTERPRISES MANAGEMENT IN AN URBAN-INDUSTRIAL SOCIETY—Edited by Portia Christian (MIG No. 34)

DEVELOPING ORGANIZATIONS—Edited by Jerome L. Franklin, 1978 (MIG No. 35)

HEALTH SCIENCES AND SERVICES—Edited by Lois F. Lunin, 1979* (MIG No. 36)

EMPLOYEE COUNSELING IN INDUSTRY AND GOVERNMENT—Edited by Theodore P. Peck, 1979* (MIG No. 37)

NEW PRODUCT PLANNING—Edited by Sarojini Balachandran, 1979* (MIG No. 38)

*in preparation

MANAGEMENT
INFORMATION
GUIDE : : 29

Chemical Industries Information Sources

Theodore P. Peck

Head of Public Service
St. Paul Campus Library System
University of Minnesota
St. Paul

GALE RESEARCH COMPANY · BOOK TOWER · DETROIT, MICHIGAN

Library of Congress Cataloging in Publication Data

Peck, Theodore P
    Chemical industries.

    (Management information guide; v. 29)
    1. Chemical engineering—Information services.
2. Chemical industries—Information services.
3. Chemical engineering—Documentation. 4. Chemical
industries—Documentation. I. Title. II. Series.
TP155.P4   660'07   76-6891
ISBN 0-8103-0829-0

VITA

Theodore P. Peck is currently head of public service at the St. Paul Campus Library System, University of Minnesota, St. Paul. Peck has written OCCUPATIONAL SAFETY AND HEALTH: A GUIDE TO INFORMATION SOURCES for the Management Information Guide Series and is presently working on EMPLOYEE COUNSELING IN INDUSTRY AND GOVERNMENT: A GUIDE TO INFORMATION SOURCES for the Series.

CONTENTS

Contents

Contents

Contents

Contents

Contents

Contents

INTRODUCTION

This guide attempts to draw together in a resource publication details on information sources useful in the chemical and related industries. It lists government agencies, professional societies, trade associations, and independent research institutes. It also includes a selected bibliography of information tools which seem to be pertinent to chemical and industrial information. Liberal use has been made of government publications and society publications in reproducing the descriptive material contained in this Guide.

The special subject areas include agriculture, bioengineering, food engineering, materials science, nuclear engineering, paper and pulp technology, and petroleum engineering. Pollution and safety aspects are covered in the bibliographical sections.

This guide should prove helpful to chemical engineers in learning of the broad range of data and information available today on a cross-disciplinary basis.

It is arranged in sections which are alphabetical by title of section with the exception of Section 1, General Information on Chemical Engineering and Chemical Industries. Within each Section there are two main divisions: A, Organizations and Associations; and B, Literature. Each of these main divisions are subdivided into parts which group the type of information source described as the subject of the literature cited into a logical pattern. A final grouping listed as addenda gives details on large libraries with collections of chemical process information and the names and addresses of publishing firms in the Guide.

Section I

GENERAL INFORMATION ON

CHEMICAL ENGINEERING AND CHEMICAL INDUSTRIES

Section 1

GENERAL INFORMATION ON
CHEMICAL ENGINEERING AND CHEMICAL INDUSTRIES

A. ORGANIZATIONS AND ASSOCIATIONS

1. Governmental Agencies

Government agencies at the federal, state, and local level originate and distribute a prodigious amount of data and detail on a wide spectrum of subjects, perhaps exceeding the output from all the professional societies and organizations which exist today. A certain percentage of the output is concerned with official transactions and results from the activities of governing the country as a whole or a geographical region of the country. However, among the documents, reports, bibliographies, hearings, statistical compendia, maps, charts, graphs, handbooks, and circulars issued by the various official and quasi-official agencies are numerous items of practical information use to chemical engineers.

Federal government documentation is the most prolific and covers the widest range of topics, having generally the greatest impact.

State government documents naturally are of special significance to a particular state government agency such as the laws and codes of that governmental unit. State government agencies also produce other items of value such as maps, codes and regulations, directories, guides, and descriptive booklets on a variety of subjects.

On the other hand, documents issued by county, city, and other governmental units are frequently limited to ordinances, planning reports, annual reports of agencies, and similar publications. The information contained within these official releases will, on occasion, be of use in the chemical engineer's activities, especially in plant design and layout, and in becoming aware of regulations concerning industrial activities in a specific geographical area.

Governmental agencies have established public and specialized information offices where inquiries can be received and answered or provisions made whereby the inquiries can be referred to the other departments who can provide the specific details being sought.

a. FEDERAL

The subject of chemical manufacture and chemical process industries would fit into the programs of many different government agencies on federal, state, and municipal government levels. There are direct relationships such as in the areas of industrial safety in plants which manufacture chemicals and in product standards in drug and cosmetic industries. More obvious are the functions of government which appear in the areas of agriculture and nuclear energy, etc.

In this guide, an attempt is made to highlight governmental involvement in support and regulation of various branches of chemical engineering and process industries. The plan for inclusion of these agencies to list first in Section 1 - A those that pertain in a general way to the broad field of chemical processing industries and then in the following sections to present details on government departments and programs which seem to be closely associated with individual branches of chemical engineering, such as agricultural engineering, nuclear engineering, petroleum engineering, etc.

For more detailed information concerning these governmental programs, users of this guide are referred to the U.S. GOVERNMENT ORGANIZATION MANUAL and to other resource tools which describe the responsibilities and functions of the federal government.

THE CONGRESS

The United States Congress is one of the prime producers of information in the federal government. The legislation it enacts is in itself a major information resource and of particular importance are the information materials prepared for Congress as a backup for its investigations and considerations. These are issued as "Hearings" of the Senate and House committees.

To find out about legislation and hearings, the following are useful guides:

1. MONTHLY CATALOG

   U.S. Superintendent of Documents. Lists committee hearings, reports on public bills, and House and Senate documents.

2. DIGEST OF PUBLIC GENERAL BILLS AND RESOLUTIONS

3. CONGRESSIONAL RECORD

4. FEDERAL REGISTER

5. CODE OF FEDERAL REGULATIONS

6. UNITED STATES CODE

   Further details on the availability, frequency, and cost of the above are given in the MONTHLY CATALOG of the Superintendent of Documents.

7. National newspapers list major legislation and announce hearings. THE NEW YORK TIMES INDEX, The New York Times Company, New York, New York 10036, is useful for locating such detail.

8. THE PUBLIC AFFAIRS INFORMATION SERVICE BULLETIN, 11 West 40th Street, New York, New York 10018.

This is another index which provides current coverage of congressional hearings and activities. Document depository libraries collect copies of hearings and legislation as well as the official lists of these information forms issued by the federal government. One of these is the MONTHLY CATALOG of the U.S. Superintendent of Documents, Government Printing Office, Washington, D.C. 20402, however a primary information guide to legislation is the DIGEST OF PUBLIC BILLS which is also available from the Government Printing Office.

9. CONGRESSIONAL QUARTERLY SERVICE, 1735 K Street N.W., Washington, D.C. 20006.

An annotated service publication which explains the phases that a piece of legislation experiences from the time it is first introduced until it dies in committee or ends up as a public law. Details on sponsors and other interesting materials concerning each bill are included. This publication is subscribed to by all large libraries and is either shelved in the main reference collection or in the History/Documents collection. Other commercial services do exist. This is one of the more outstanding ones.

10. CONGRESSIONAL INFORMATION SERVICE INDEX TO PUBLICATIONS OF THE U.S. CONGRESS, 500 Montgomery Building, Washington, D.C. 20014.

This publication is not to be confused with the CONGRESSIONAL QUARTERLY SERVICE. THE CONGRESSIONAL INFORMATION SERVICE INDEX is a new indexing service for documents issued or prepared at the request of Congress. Its primary use is as an index; whereas the former publication is a valuable source of information on the passage of legislation through Congress.

ONE MAY ALSO CONTACT HIS SENATOR OR REPRESENTATIVE

The offices of the congressional representatives in the House or Senate will provide information on legislation as well as texts of hearings, bills, and laws. They can also provide valuable assistance in keeping one informed on legislation moving through Congress and can be an important referral source.

DIRECTORIES OF CONGRESS

OFFICIAL CONGRESSIONAL DIRECTORY (annual), Government Printing Office, Washington, D.C. 20402.

CONGRESSIONAL STAFF DIRECTORY (annual), The Congressional Staff Directory Publishing Company, 300 New Jersey Avenue S.E., Washington, D.C. 20003.

The above directories, the first, an official guide of the federal government, the other prepared by a commercial organization, give biographical information, committee assignments and other details on the Congress and its members.

OTHER USEFUL DIRECTORIES

U.S. GOVERNMENT ORGANIZATION MANUAL, Government Printing Office, Washington, D.C. 20402.

TAYLOR'S ENCYCLOPEDIA OF GOVERNMENT OFFICIALS: FEDERAL AND STATE (annual). Edited by J. Clements. Political Research Inc., 510 Continental Building, Dallas, Texas.

The following journal article on federal information is useful to chemical engineers:

E. Biggert, "Federal Government Information Sources." CHEMICAL ENGINEERING, May 15, 1972, pp. 103-14.

Congressional committees with interests and responsibilities pertinent to chemical industries.

A. House Committee on Public Works-Rayburn House Office Building, Room 2165, South Capitol Street and Independence Avenue S.W., Washington, D.C. 20515.

Areas of interest include: Economic development programs; public buildings and grounds; water resources; water pollution control and abatement; sewage treatment facilities; rivers and harbors; environmental quality; and other areas.

Information services: Answers inquiries concerning activities in the Committee's areas of interest.

B. Senate Committee on Commerce-New Senate Office Building, Room 5202, Constitution Avenue and 1st Street N.E., Washington, D.C. 20510, Tel: (202) 225-5115.

Areas of interest include: Interstate commerce generally, including consumer protection; foreign commerce generally; transportation generally; maritime matters; domestic surface transportation, including pipelines and highway safety; federal power matters; and other matters.

Publications: Reports, transcripts of hearings.

Information services: Answers inquiries.

C. Senate Committee on Interior and Insular Affairs, New-Senate Office Building, Room 3106, Constitution Avenue and 1st Street N.E., Washington, D.C. 20510.

Areas of interest include: Nationwide water resource and hydroelectric power development; desalination; river basin and regional water research and planning programs; irrigation and water quality control; public lands, including the Outer Continental Shelf; mines and mining; environmental studies; national land use policies, etc.

Publications: Reports, transcripts of hearings, special studies.

Information services: Answers inquiries; provides reference services.

D. Senate Committee on Public Works, New-Senate Office Building, Room 4204, Constitution Avenue and 1st Street N.E., Washington, D.C. 20510.

Areas of interest include: Public works legislation relating to water resources conservation and development, rivers and harbors, hydro-electric power, pollution abatement in air and water, highway construction, public buildings, and regional economic development.

Publications: Reports, transcripts of Committee hearings.

Information services: Answers inquiries; makes referrals.

DEPARTMENT OF COMMERCE

Address: 14th Street between Constitution Avenue and E Street N.E., Washington, D.C. 20230.

"The mission of the department is to foster, serve and promote the nation's economic development and technological advancement." The many programs of this federal department are of concern to chemical engineers. A particularly important branch of this organization is the National Bureau of Standards.

A. NATIONAL BUREAU OF STANDARDS - Washington, D.C. 20234

"Conducts research and provides testing and other services to assure maximum application of physical and engineering sciences to the advancement of technology in commerce and industry." These functions are assigned to various divisions and groups.

CENTERS AND PROGRAMS IN CHEMICAL ENGINEERING - GENERAL

1. Chemical Kinetics Information Center - Gaithersburg, Maryland.

    Areas of interest: Chemical kinetics; photochemistry; energy transfer; elastic scattering. Excludes heterogeneous catalysis.

    Publications: Bibliographies, tables, and rate data.

    Information services: Answers brief inquiries; provides data and bibliographic information on published research on rates of chemical reactions. Services are available to all scientists and technical personnel with professional interests.

2. Chemical Thermodynamics Data Center - Physical Chemistry Division, Institute for Materials Research, Gaithersburg, Maryland.

    Area of interest: Thermochemical and thermophysical properties of pure chemical substances in gas, liquid, and solid phases, and their aqueous solutions.

    Holdings: About 125,000 abstract data cards of thermodynamic information from about 60,000 books, reports, theses, papers, and periodicals.

    Publications: SELECTED VALUES OF CHEMICAL THERMODYNAMIC PROPERTIES (NBS Circular 500, out of print; revisions in NBS Technical Notes 270-1 and 270-2); THERMAL PROPERTIES OF AQUEOUS UNI-UNIVALENT ELECTROLYTES (Monograph NSRDS-NBS 2); INORGANIC SUBSTANCE-PROPERTY INDEX TO THE BULLETIN OF THERMODYNAMICS AND THERMOCHEMISTRY (annual).

Information services include: Answers to inquiries and some consulting services.

3. Cryogenic Data Center - National Bureau of Standards, Boulder, Colorado 80302.

Areas of interest: Low-temperature data on properties of materials, thermodynamic properties of gases, thermal conductivity and expansion.

Publications: Bibliographies; selected entropy Mollier diagrams of cryogenic fluids; technical information sheets, including graphs of thermal conductivity and data on thermal expansion; weekly announcement bulletins.

Information services: Answers inquiries and provides consulting services.

4. Heat Division - National Bureau of Standards, Gaithersburg, Maryland.

Areas of interest: Cryogenic physics, temperature, radiation thermometry, and pressure measurements.

Publications: Specifications and standards.

Information services: Provides consulting services on the physics of heat without cost or on a contract basis to other governmental agencies and to the technical community.

5. Low Temperature Specific Heats Data Center - National Bureau of Standards, Gaithersburg, Maryland.

Areas of interest: Heat-capacity data in the range 0-300 K.

Information services: Answers inquiries; provides tables of closely spaced values of heat capacities from 0-300 degrees K and information on calorimeter design and equipment for making heat measurements.

6. Office of Engineering Standards Services - Information Section, National Bureau of Standards, State Road 124 adjacent to Rte. 705, Gaithersburg, Maryland.

Areas of interest: Engineering standards, specifications, test methods, and recommended practices covering most products and subjects except food and drugs.

Publications: Key-Word-In-Context (KWIC) Index of the 19,000 standards in the collection.

Information services: Answers inquiries; provides reference services; makes referrals; permits onsite use of collection of standards.

7. Office of Invention and Innovation - National Bureau of Standards, Gaithersburg, Maryland.

Areas of interest: Promotes programs for use of inventions in government agencies and in industry.

Publications: Pamphlets, including the following: HOW TO SUBMIT IDEAS & INVENTIONS TO THE UNITED STATES GOVERN-

MENT; STATE INVENTION EXPOSITIONS; and DEVELOPING A STATE INVENTION EXPOSITION.

Information services: Answers inquiries; provides consulting services; makes referrals.

8. Office of Standard Reference Data - National Bureau of Standards, Gaithersburg, Maryland.

Areas of interest: Physical and chemical properties of well characterized substances; thermodynamics and transport data; chemical kinetics; nuclear data; solid state data; colloid and surface chemistry; atomic and molecular data; computer processing of data.

Information services: Answers inquiries; provides consulting services.

B. NATIONAL INDUSTRIAL POLLUTION CONTROL COUNCIL - Washington, D.C. 20230.

The Council, consisting of 63 industry leaders, has 30 subcouncils dealing with industrial pollution activities and an International Environmental Advisory Committee.

Publications: More than 35 reports have been issued by the Council and are available to the public at various prices. A publications list may be obtained from the National Industrial Pollution Control Council, Department of Commerce, Washington, D.C. 20230.

C. PATENT OFFICE - Washington, D.C. 20231

The Patent Office was established to administer the patent laws enacted by Congress in accordance with Article I, section 8, of the Constitution. The Patent Office also administers the federal trademark law enacted July 5, 1946 (60 Stat. 427; 15 U.S.C. 1051), as amended.

Functions and activities: The Patent Office examines patent applications to ascertain whether applicants are entitled to patents under the law, and grants the patents when they are so entitled; publishes and disseminates patent information, records the assignment of patents; maintains a Scientific Library and a search file of U.S. and foreign patents, and general reference literature for public use; and supplies copies of patents and official records of the Patent Office. Similar functions are performed in carrying out the statutory provisions for the registration of trademarks.

For further information, contact the Commissioner of Patents, Washington, D.C. 20231.

1. Office of Research and Development - U.S. Patent Office, Crystal City Plaza, Building 2, Room 6DO7, 2110 Jefferson Davis Highway, Arlington, Virginia.

Areas of interest: Mechanical information retrieval systems for conducting novelty searches of the prior art; automata theory; pattern recognition; cybernetics; artificial intelligence.

Publications: PATENT OFFICE RESEARCH AND DEVELOPMENT REPORTS.

Information services: Answers inquiries; makes referrals; permits onsite reference to literature collection.

2. Office of Information Services – U.S. Patent Office, Crystal City Plaza, Building 3, Room 1D01, 2110 Jefferson Davis Highway, Arlington, Virginia.

Publications: OFFICIAL GAZETTE OF THE UNITED STATES PATENT OFFICE (weekly; Decisions leaflet and Trademark Section sold separately); PATENT LAWS; RULES OF PRACTICE IN PATENT CASES; TRADEMARK RULES OF PRACTICE OF THE PATENT OFFICE WITH FORMS AND STATUTES; MANUAL OF PATENT EXAMINING PROCEDURE; MANUAL OF CLASSIFICATION OF PATENTS; INDEX OF PATENTS and INDEX OF TRADEMARKS (both annual); DIRECTORY OF REGISTERED PATENT ATTORNEYS AND AGENTS; ROSTER OF ATTORNEYS AND AGENTS REGISTERED TO PRACTICE BEFORE THE U.S. PATENT OFFICE; PATENTS AND INVENTIONS, an information aid to inventors; GENERAL INFORMATION CONCERNING PATENTS and GENERAL INFORMATION CONCERNING TRADEMARKS; information circulars and pamphlets on patents and trademarks.

Information services: Answers general information inquiries, including those concerning the use of Patent Office research facilities; makes referrals; furnishes lists of registered patent attorneys and agents in a specified area, etc.

3. Search Room and Record Room – U.S. Patent Office, Crystal City Plaza, 2021 Jefferson Davis Highway, Arlington, Virginia.

Collections: U.S. patents granted since 1836, arranged according to the Patent Office classification; a bound set of U.S. patents arranged in numerical order; a complete set of the OFFICIAL GAZETTE.

Information services: The public may search and examine patents in the Search Room. In the adjoining Record Room, records and files of issued patents and other public records may be inspected, and applicants and their attorneys or agents may examine their own cases.

4. Trademark Search Room – Crystal City Plaza, Building 2, 2110 Jefferson Davis Highway, Arlington, Virginia, Tel: (703) 557-3277.

Collection: Register of all trademarks (over 800,000) registered under federal statutes.

Information services: Answers inquiries; identifies commercial origin of goods sold or services rendered under known marks; provides duplication services. The collection is accessible for onsite use; Trademark Search Room employees will show visitors how to search, but will make searches for government agencies only.

D. WATER RESOURCES AND ENGINEERING SERVICES DIVISION – Bureau of Domestic Commerce, U.S. Department of Commerce, 14th Street and Constitution Avenue N.W., Washington, D.C. 20230.

Areas of interest: Industrial water use (trends and projections for market research and river basin planning purposes); industrial pollution of water; materials and equipment for water supply and sewage systems; water works and sewage plant construction; water and sewage utilities (manpower and distribution of costs); capital investment values in water resources; power and municipal water use.

Information services: Answers inquiries; makes referrals; provides assistance and advice to U.S. consulting engineers operating overseas.

DEPARTMENT OF DEFENSE

A. DEFENSE ADVANCED RESEARCH PROJECTS AGENCY - 1400 Wilson Boulevard, Arlington, Virginia 22209, Director: Stephen J. Lukasik.

The Advanced Research Projects Agency, established in 1959 as an operating agency under the Director of Defense Research and Engineering, was redesignated as the Defense Advanced Research Projects Agency (DARPA) on March 23, 1972, and made a separate agency of the Department of Defense under the direction, authority, and control of the Secretary of Defense.

Activities: DARPA provides for the conduct of basic and applied research and development for such advanced projects as may be designated by the Secretary of Defense. In the performance of its projects, the Agency utilizes the services of the military departments, other government agencies, private industrial and public entities, individuals and educational or research institutions. The Agency's objective is to carry advanced programs to feasibility demonstration and then transfer them to an appropriate military service.

B. U.S. AIR FORCE - Aerospace Research Laboratories, Building 450, Wright-Patterson Air Force Base, Ohio 45433.

Areas of interest: Research in the physical and engineering sciences, particularly in areas of hypersonic aerodynamics; fluid dynamic facilities; energy conversion processes; thermomechanics, including heat transfer and electric arc phenomena; metallurgy and certamic, including fracture mechanics and electrical properties; chemistry, including organosicon and metal chelate chemistry, combustion, and surface chemistry; etc.

Publications: An average of 150 journal articles per year and a similar number of ARL reports, both internal and contractor prepared; biannual response on research; annual lists of publications; summary of coupling activities; other staff reports.

Information services: Requests for specific information should be directed to the ARL Information Office at the above address.

C. U.S. ARMY - Technical Support Directorate, Army Chemical Information and Data System Project, Technical Support Directorate, Edgewood Arsenal, Maryland 21010.

Areas of interest: Chemical structures of toxic compounds; storage of chemical structures with accessible nonstructural information.

Holdings: 260,000 chemical structures on magnetic tapes.

Publications: Reports.

Information services: Answers inquiries; makes referrals, provides reference and literature-searching services. Services are available only to organizations that have a working arrangement with the Project, but inquiries regarding participation are solicited.

DEPARTMENT OF HEALTH, EDUCATION, AND WELFARE

Address: 330 Independence Avenue S.W., Washington, D.C. 20201.

Charged with the responsibility of promoting improvement and general welfare of society in these broad areas. Like all the departments of the federal government, HEW covers many points which are also of importance in chemical engineering endeavors. Examples of activity in this department are the following:

A. BUREAU OF OCCUPATIONAL SAFETY AND HEALTH - Scientific Reference Service, U.S. Public Health Service, 1014 Broadway, Cincinnati, Ohio 45202.

Areas of interest: Occupational health; industrial hygiene; toxicology and hazards of industrial materials and conditions; chemical engineering, nursing, medicinal, and physiological aspects of occupational health hazards; prevention and treatment of injurious effects on health; poisonous gases, vapors, mist, and dust; radiation; heat and humidity; noise; effects of pesticides.

Information services: Answers inquiries; makes referrals; provides consulting, reference, literature-searching, and abstracting services; permits onsite use of collections.

B. BUREAU OF WATER HYGIENE - U.S. Public Health Service, 222 East Central Parkway, Cincinnati, Ohio 45202.

Areas of interest: Public drinking water supply (source, storage, protection, treatment, distribution, plumbing); water quality; toxicological evaluations of chemicals and materials used in or with water; health aspects of water pollution and water resources.

Publications: PUBLIC HEALTH SERVICE DRINKING WATER STANDARDS; manuals of recommended practice such as PUBLIC DRINKING WATER SUPPLY EVALUATION, EVALUATION OF WATER LABORATORIES, and WATER SUPPLY AND PLUMBING CROSS-CONNECTION; technical reports, data compilations, list of approved materials for water treatment, i.e., coagulant aids.

Information services: Answers inquiries; provides consulting services; analyzes samples from selected public water supplies for government agencies and state departments of health.

C. CHEMICAL/BIOLOGICAL INFORMATION-HANDLING REVIEW COMMITTEE - Biotechnology Resources Branch, Division of Research Resources, National Institutes of Health, Bethesda, Maryland 20014.

Program: Committee reviews program objectives and contract proposals in specific research and development efforts in the area of handling chemical/biological information and advises the Director, Division of Research Resources, on both the scientific merit and the general program relevance of individual proposals.

D. NATIONAL AIR POLLUTION CONTROL ADMINISTRATION - Office of Technical Information and Publications, Air Pollution Technical Information Center, 1030 Wade Avenue, Raleigh, North Carolina 27605.

Area of interest: Data on chemical, physical and biological effects of varying air quality and other information pertaining to air pollution and its prevention and control.

Publishes: APCA ABSTRACTS and APTIC ACCESSIONS BULLETIN.

Information services: Answers requests for technical information on air pollution and its control. Provides details in form of literature searches, and extracts from files and also distributes publications free to wide range of interests.

E. NATIONAL INSTITUTE FOR OCCUPATIONAL SAFETY AND HEALTH - Health, Education, and Welfare Department, Rockville, Maryland 20852.

The Institute plans, directs, and coordinates the national program effort to develop and establish recommended occupational safety and health standards and to conduct research, training, and related activities to assure safe and healthful working conditions for every working man and woman.

DEPARTMENT OF INTERIOR

Address: C Street between 18th and 19th Streets N.W., Washington, D.C. 20240.

Responsible for the protection and development of the public lands and natural resources. Pollution control is a major program of this department and a prime information source is the following:

A. FEDERAL WATER QUALITY ADMINISTRATION - Division of Technical Support, Washington, D.C. 20242.

Provides technical assistance in water pollution control by answering inquiries and making referrals, field and laboratory studies, and supplies water quality and waste source data.

Branches of this agency are:

1. Division of Technical Support - Federal Water Quality Administration, U.S. Department of the Interior, Washington, D.C. 20242.

Areas of interest: Pollution from municipal, industrial, and agricultural sources; fishkills; sources of mineral pollutants; mathematical models of river basins and estuaries; stream flow regulation; methods and costs of waste water treatment related to quality management in receiving waters; water quality data and waste sources inventories.

Information services: Answers inquiries, makes referrals, and provides assistance in research studies.

2. Technical Advisory and Investigations Branch - Federal Water Quality Administration, 5555 Ridge Avenue, Cincinnati, Ohio 45213.

Information services: Answers inquiries, provides consulting services.

3. National Water Quality Laboratory - Federal Water Quality Administration, 6201 Congdon Boulevard, Duluth, Minnesota 55804.

Areas of interest: Development of limits of contamination (water quality requirements) for all water uses, including municipal, industrial, agricultural, and aquatic life, with current emphasis on aquatic life and chronic toxicity testing of fish and invertebrates.

Information services: Answers inquiries; provides consulting services to public agencies.

4. Office of Research and Development - Federal Water Quality Administration, 633 Indiana Avenue N.W., Washington, D.C. 20242.

Areas of interest: Treatment of water supplies and waste water; sources, fate, and identification (physical, chemical, biological) of pollutants; eutrophication; stream pollution; sewage wastes and treatment of nutrients, organics, colloidal solids, inorganics, and microorganisms; water quality control; water resources planning, renovation, and reuse; water conservation; salt water intrusion; mine drainage; thermal pollution; oil pollution; waterborne wastes from municipalities, industries, agriculture, mining, ships, dredging, recreation, and other sources.

Information services: Answers inquiries; provides consulting and limited reference services; provides bibliographic citations and abstracts; makes referals; makes interlibrary loans.

5. Pacific Northwest Water Laboratory - Federal Water Quality Administration, 200 Southwest 35th Street, Corvallis, Oregon 97330.

Areas of interest: Water pollution control research activities, national and regional in scope. The national research and development programs relate to eutrophication, coastal pollution, thermal pollution, and to industrial waste treatment in the areas of food, paper and allied products, and forestry and logging.

Information services: Makes interlibrary loans; permits onsite use of collection. Training courses are available to technical and administrative personnel to communicate the latest information in general and specialized areas of water pollution control technology.

B.   OFFICE OF SALINE WATER - U.S. Department of the Interior, 18th and
C Streets N.W., Washington, D.C. 20240.

Areas of interest: Chemical, physical, engineering, and economic
feasibility of saline water conversion; conversion processes, including distillation, membrane processes, crystalization processes, solvent extraction, and removal of ions; scale-formation, heat-transfer,
pre-treatment, materials, and corrosion problems relating to conversion problems; design, construction, and operation of pilot and
test-bed plants.

Publications: SALINE WATER CONVERSION REPORT (annual).

Information services: Answers inquiries; makes referrals; permits
onsite use of collection.

C.   OFFICE OF WATER RESOURCES RESEARCH - U.S. Department of the Interior, 18th and C Streets N.W., Washington, D.C. 20240.

Areas of interest: Research, training, and information exchanges
in the water resources field, as authorized by Public Law 88-379
as amended.

Publications: THE WATER RESOURCES RESEARCH CATALOG (listing
and describing current research projects); COOPERATIVE WATER
RESOURCES RESEARCH AND TRAINING -- 1967 ANNUAL REPORT;
WATER RESOURCES THESAURUS; RESEARCH REPORTS SUPPORTED
BY OFFICE OF WATER RESOURCES RESEARCH (quarterly cumulative
listing); SELECTED WATER RESOURCES ABSTRACTS; state-of-the-art
reports on selected water problem areas.

Information services: Answers inquiries on Office research programs
and on procedures for applying for contracts and grants for water
resources research under P. L. 88-379 as amended.

The Office manages a Water Resources Scientific Information Center
which, through the acquisition, storage, selective dissemination, and
retrieval of information pertaining to water resources scientific and
technical literature, is designed to be responsive to the needs of
federal water resources organizations.

DEPARTMENT OF LABOR

A.   OCCUPATIONAL SAFETY AND HEALTH ADMINISTRATION - Washington,
D.C. 20210.

The Assistant Secretary for Occupational Safety and Health has responsibility
for the occupational safety and health activities in the Department of Labor.

The Occupational Safety and Health Administration develops and promulgates
occupational safety and health standards; develops and issues regulations; conducts investigations and inspections to determine the status of compliance with
safety and health standards and regulations; and issues citations for noncompli-
ance with safety and health standards and regulations.

The Occupational Safety and Health Administration has regional offices established in 10 areas throughout the United States.

Regional Offices -- Occupational Safety and Health Administration

Region	Regional Administrator
I. Boston, Mass. 02203 John F. Kennedy Federal Bldg.	Donald E. MacKenzie
II. New York, N.Y. 10001 341 Ninth Ave.	Alfred Barden
III. Philadelphia, Pa. 19107 1317 Filbert St.	David H. Rhone, Acting
IV. Atlanta, Ga. 30309 1375 Peachtree St. N.E.	Basil A. Needham, Jr.
V. Chicago, Ill. 60606 300 S. Wacker Dr.	Edward E. Estkowski
VI. Dallas, Tex. 75201 1512 Commerce St.	John K. Barto
VII. Kansas City, Mo. 64106 823 Walnut St.	Joseph A. Reidinger
VIII. Denver, Colo. 80202 Federal Bldg., P.O. Box 3588	Howard J. Schulte
IX. San Francisco, Calif. 94102 Federal Bldg., P.O. Box 36017	Warren H. Fuller
X. Seattle, Wash. 98104 506 2nd Ave.	James W. Lake

Contact individual offices for current details on occupational Safety and Health Act provisions and regulations.

DEPARTMENT OF TRANSPORTATION

A. CHEMICAL TRANSPORTATION INDUSTRY ADVISORY COMMITTEE – Marine Safety Council, Coast Guard Headquarters (CMC/82), U.S. Department of Transportation, 400 Seventh Street, S.W., Washington, D.C. 20590.

The Committee is a subcommittee of the Marine Safety Council and as such it is a public advisory committee which functions within the jurisdiction of the Coast Guard, U.S. Department of Transportation.

> Program: Committee provides advice and consultation with respect to the water transportation of chemicals or hazardous materials.

B. NATIONAL TRANSPORTATION SAFETY BOARD – 800 Independence Avenue S.W., Washington, D.C. 20591.

The mission of the National Transportation Safety Board is to promote safety in all modes of transportation.

> Functions: Aviation Accident Cause Determination and Safety Promotion: The Board investigates U.S. civil aviation accidents, except investigations which are delegated by the Board to the Federal Aviation Administration; determines accident cause and reports the facts and circumstances in all aviation accidents; and conducts special studies and makes recommendations on matters pertaining

to aviation safety and aviation accident prevention.

Surface Transportation Accident Cause Determination and Safety Promotion: This activity covers the fields of railroad, highway, pipeline, and marine safety. The Safety Board delegates accident cause determinations of most accidents to the Administrations within the Department of Transportation, but reserves the right to investigate, determine cause, and report the facts and circumstances of all surface transportation accidents which it declares to be major. The Safety Board conducts special studies and makes recommendations on matters pertaining to surface transportation safety promotion and surface transportation accident prevention.

Information services: Answers inquiries; makes referrals; and allows use of its extensive library holdings.

C. MATERIALS TRANSPORTATION BUREAU - 2100 Second Street S.W., Washington, D.C. 20590.

This bureau was formed in 1974 to coordinate the Department of Transportation's overall responsibilities for transportation of hazardous materials and pipeline safety.

Areas of interest: These include transport of hazardous materials, regulation and safety, and related research.

Publications: Prepares annual reports on transport of hazardous materials and pipeline safety, contributes to compilation of federal standards, and issues monthly newsletters.

Information services: Answers inquiries and makes referrals to appropriate agencies.

ENERGY RESEARCH AND DEVELOPMENT ADMINISTRATION

Address: 20 Massachusetts Avenue N.W., Washington, D.C. 20545.

The Energy Research and Development Administration was established by the Energy Reorganization Act of 1974 (replaces former Atomic Energy Commission).

Purpose: The Energy Research and Development Administration (ERDA) aims to reorganize and consolidate federal activities relating to research and development on the various sources of energy in order to develop and increase the efficiency and reliability of the use of all energy sources; to increase the productivity of the national economy and strengthen its position in regard to international trade; to make the nation self-sufficient in energy; to advance the goals of restoring, protecting, and enhancing environmental quality; and to assure public health and safety.

Areas of interest: Coal research, nuclear energy, energy research centers, wind power, solar heating and cooling, geothermal energy and development, demonstration of commercial feasibility and practical applications of extraction conversion, storage, transmission, and utilization of energy forms.

Publications: RESEARCH AND DEVELOPMENT REPORTS, STAN-
DARDS, CRITICAL REVIEW SERIES, Standard safety analysis reports,
NUCLEAR SCIENCE ABSTRACTS.

Information services: Answers inquiries, makes referrals, maintains
a speakers' bureau, produces and distributes films through its Office
of Public Affairs. A public document room is available at 20 Massachusetts Avenue N.W., Washington, D.C. 20545.

Energy Research and Development Administration has the following energy research
centers.

Bartlesville Energy Research Center - P.O. Box 1398, Bartlesville, Okla.
74003.

Grand Forks Energy Research Center - Box 8213, University Station, Grand
Forks, N. Dak. 58202.

Laramie Energy Research Center - P.O. Box 3395, University Station, Laramie,
Wyo. 82070.

Morgantown Energy Research Center - P.O. Box 880, Morgantown, W. Va.
26505.

Pittsburgh Energy Research Center - 4800 Forbes Ave., Pittsburgh, Pa. 15213.

EXECUTIVE OFFICE OF THE PRESIDENT

A.   ENVIRONMENTAL PROTECTION AGENCY - 401 M Street S.W., Washington, D.C. 20460.

Activities: Air and Water Programs: The functions of this program
area include (1) air and water quality standards development; (2)
program policy development and evaluation; (3) technical direction,
support, and evaluation of regional air and water activities; (4)
development of programs for technical assistance and technology
transfer; and (5) selected demonstration programs. Standards are
developed and set against major types of air pollutants to protect
the public health and to protect against effects on soil, water,
vegetation, materials, animals, weather, visibility, and personal
comfort and well being. Through the technology transfer program,
pollution control information gained from research and demonstration projects is freely offered to potential users so that it can be
utilized. These programs are under the direction of the Assistant
Administrator for Air and Water Programs.

Pesticides, Solid Waste, and Radiation: The Assistant Administrator for Categorical Programs is responsible for the pesticides, solid
waste management, and radiation programs of the Agency. These
activities include (1) environmental and pollution source standards

development; (2) pesticides registration activities designed to assure compliance of present and proposed products with requirements for safety and effectiveness; (3) selected demonstration programs, such as activities related to grants to municipalities to demonstrate existing or advanced technology in waste disposal; and (4) selected technical assessment and assistance programs, such as the provision of assistance to states and other agencies having radiation programs.

Research and Monitoring: The principal science adviser to the Administrator is the Assistant Administrator for Research and Monitoring, who is reponsible for a national research program in pursuit of technological controls of all forms of pollution. The Office provides direct supervision to the activities of Agency laboratories engaged in national or basic research, and technical policy direction to those laboratories engaged in operations in support of the responsibilities of the Agency Regional Administrators. Close coordination of the various research programs is designed to yield a synthesis of knowledge from the biological, physical, and social sciences which can be interpreted in terms of total human and environmental needs. Management of selected demonstration programs, planning for Agency environmental quality monitoring programs, and coordination of Agency monitoring efforts with those of other federal agencies, the states, and other public bodies are general functions of this Office.

Research Centers and Laboratories: The primary areas of responsibility of the National Environmental Research Centers are health effects research (Research Triangle Park, N.C.), pollution control technology and engineering research (Cincinnati, Ohio), and ecological systems research (Corvallis, Oreg.).

Region	Regional Administrator	Address
I	John A. S. McGlennon	John F. Kennedy Federal Bldg., Boston, Mass. 02203
II	Gerald M. Hansler	26 Federal Plaza, New York, N.Y. 10007
III	Edward W. Furia, Jr.	Curtis Bldg., 6th and Walnut Sts., Philadelphia, Pa. 19106
IV	Jack E. Ravan	1421 Peachtree St. NE, Atlanta, Ga. 30309
V	Francis T. Mayo	1 N. Wacker Dr., Chicago, Ill. 60606
VI	Arthur W. Busch	1600 Patterson St., Dallas, Tex. 75201
VII	Jerome H. Svore	1735 Baltimore Ave., Kansas City, Mo. 64108
VIII	John A. Green	1860 Lincoln St., Denver, Colo. 80203
IX	Paul DeFalco, Jr.	100 California St., San Francisco, Calif. 94111

X   James L. Agee   1200 6th Ave., Seattle, Wash.
                                    98101

The Western Environmental Research Laboratory at Las Vegas, Nevada, conducts
radiation effects research.

B. OFFICE OF SCIENCE AND TECHNOLOGY - Executive Office Building,
Washington, D.C. 20506.

The Director of the Office of Science and Technology provides advice and
assistance to the President with respect to developing policies and evaluating
and coordinating programs to assure that science and technology are used most
effectively in the interests of national security and general welfare.

Activities: Functions of the Office include:

1. Evaluation of major policies, plans, and programs of science
and technology of the various agencies of the federal government,
giving appropriate emphasis to the relationship of science and technology to foreign policy, and measures for furthering science and
technology in the nation;
2. Assessment of selected scientific and technical developments
and programs in relation to their impact on national policies;
3. Review, integration, and coordination of major federal activities
in science and technology giving due consideration to the effects
of such activities as non-federal resources and institutions;
4. Assuring that good and close relations exist with the nation's
scientific and engineering communities so as to further in every
appropriate way their participation in strengthening science and
technology in the United States and the free world; and
5. Such other matters consonant with law as may be assigned by
the President to the Office.

The Director of the Office of Science and Technology serves as
Chairman of the Federal Council for Science and Technology. The
Council secretariat is provided by the Office.

For further information, contact the Executive Officer, Office of
Science and Technology, Executive Office Building, Washington,
D.C. 20506.

C. FEDERAL COUNCIL FOR SCIENCE AND TECHNOLOGY - 216 Executive
Office Building, Washington, D.C. 20506.

"Established to promote closer cooperation among federal agencies, to facilitate
resolution of common problems and to improve planning and management in science and technology, and to advise and assist the President regarding federal
programs affecting more than one agency."

Programs are carried out through a committee structure. They include: Council committees on Atmospheric Sciences, Materials Re-

search and Development, High Energy Physics, Water Resources Research, Behavioral Sciences, Patent Policy, Long Range Planning, International Programs, Science Information, Academic Science and Engineering, Federal Laboratories, Environmental Quality, and Solid Earth Sciences.

The Committee on Scientific and Technical Information (COSATI) of the Federal Council for Science and Technology has the goal of establishing a National Information System. In this pursuit they have organized a Panel on Informational Analysis Centers which has been examining the problems with the formation and development of Information Analysis Centers. An Information Analysis Center is a specialized scientific and technical information resource outlet. The institution's raw data and detail are gathered, analyzed, synthesized and issued as a package or customized information program. There are over 100 of these centers in operation. The Nuclear Safety Information Center described in the section on Atomic Energy Commission activities is illustrative of the valuable services these agencies provide to scientists, engineers, and others.

The Committee on Scientific and Technical Information has prepared a DIRECTORY OF FEDERALLY SUPPORTED INFORMATION ANALYSIS CENTERS, available from National Technical Information Service.

D.   FEDERAL SAFETY COUNCIL - Room 411, Railway Labor Building, First and D Streets N.W., Washington, D.C. 20210.

The Council was reestablished by Executive Order 10990 of February 2, 1962, to advise the Secretary of Labor with respect to development and maintenance of safety organizations and programs in federal departments and agencies and with respect to criteria, standards, and procedures designed to eliminate work hazards and health risks and to prevent injuries and accidents in federal employment.

GENERAL SERVICES ADMINISTRATION

Address: 18th and F Streets N.W., Washington, D.C. 20405.

This agency is concerned with the management of federal government property and records, and is an important information source for doing business with the federal government. Details on procurement programs, bids, specifications, contract forms, and other matters of this nature are made available. Contact Office of Information, at the above address.

Regional offices in Boston, New York, Washington, D.C., Atlanta, Chicago, Kansas City, Fort Worth, Denver, Los Angeles, and Seattle may be also contacted for information or assistance.

General Services Administration issues the federal civilian specifications. These specifications carry an identifying code made up of single, double or triple letters and numbers and are listed in the INDEX OF FEDERAL SPECIFICATIONS AND STANDARDS, Government Printing Office, Washington, D.C. 20402, printed periodically to update previous issues. This has a two-part listing of necessary documents. The first arranges them alphabetically and gives the price. In the second part, the specifications and standards are listed by procurement groups. Copies of federal specifications and standards can be obtained from the General Services Administration Business Centers in various cities in the U.S. See the local telephone directory or consult the U.S. Government Organization Manual for addresses.

NATIONAL ACADEMY OF SCIENCES

Address: Office of Information, 2101 Constitution Avenue N.W., Washington, D.C. 20418.

A. NATIONAL ACADEMY OF ENGINEERING

B. NATIONAL RESEARCH COUNCIL

The National Academy of Sciences is an organization of distinguished scientists and engineers who may be called upon to act in an advisory capacity to the federal government, and who encourage research by suggesting areas where a particular need is apparent. The National Academy of Engineering and The National Research Council are supporting organizations to The National Academy of Sciences and make possible broad participation in research programs. The Office of Information is the official entry point to services of these organizations. It maintains awareness of their varied activities in the pure and applied sciences and refers inquirers to the office suitable for each information request.

The National Research Council has established a number of committees to promote and implement research and to promulgate the results of investigative and exploratory scientific and technological undertakings.

National Research Council Committees include:

1. Committee on Chemical Information - Division of Chemistry and Chemical Technology.

Areas of interest: Nonconventional methods for handling chemical information; chemical notation; mechanical processing of chemical information.

Publications: SURVEY OF CHEMICAL NOTATION SYSTEMS; PROCEEDINGS OF A CONFERENCE ON MECHANICAL PROCESSING OF CHEMICAL INFORMATION (1964); SURVEY OF EUROPEAN NON-CONVENTIONAL CHEMICAL NOTATION SYSTEMS; CHEMICAL STRUCTURE INFORMATION HANDLING, A REVIEW OF THE LITERATURE 1962-1968.

Information services: Answers inquiries; makes referrals.

2. Committee on Colloid and Surface Chemistry.

Areas of interest: Development and standardization of symbols, definitions, nomenclature, and methods of measurement in the field of colloid and surface chemistry; physical constants in this field.

Information services: The Division of Chemistry and Chemical

Technology refers inquiries to members of the Committee.

3. Committee on Fundamental Constants.

Areas of interest: Fundamental physical constants.

Publications: GENERAL PHYSICAL CONSTANTS RECOMMENDED BY NAS-NRC (NBS Misc. Publ. 253).

4. Committee on High-Temperature Chemical Phenomena.

Areas of interest: Activities concerned with high-temperature phenomena and related aspects of materials sciences.

Publications: HIGH-TEMPERATURE CHEMISTRY: CURRENT AND FUTURE PROBLEMS.

Information services: The Division of Chemistry and Chemical Technology refers inquiries to members of the Committee.

5. Committee on Kinetics of Chemical Reactions.

Areas of interest: Chemical kinetics.

Publications: TABLES OF CHEMICAL KINETICS: HOMOGENEOUS REACTIONS (NBS Circular 510 and Supplement I, and Monograph 34).

Information services: The Division of Chemistry and Chemical Technology refers inquiries to members of the Committee.

6. Committee on Macromolecular Chemistry.

Areas of interest: National and international nomenclature and standards in macromolecular chemistry; polymer science.

Publications: UNSOLVED PROBLEMS IN POLYMER SCIENCE, A COMPILATION OF ESSAYS WITH A BIBLIOGRAPHY OF REVIEWS.

7. Committee on Nomenclature.

Areas of interest: Chemical nomenclature; standardization of chemical nomenclature.

Information services: The Division of Chemistry and Chemical Technology refers inquiries to members of the Committee.

8. Committee on Symbols, Units, and Terminology – Office of Critical Tables, National Research Council, 2101 Constitution Avenue N.W., Washington, D.C. 20418, Tel: (202) 961-1388.

Areas of interest: Symbols, units, and terminology in all areas of the physical sciences.

Information services: Answers inquiries; makes referrals.

C. NATIONAL COMMITTEE FOR THE INTERNATIONAL UNION OF PURE AND APPLIED CHEMISTRY – National Academy of Sciences--National Research Council, 2101 Constitution Avenue N.W., Washington, D.C. 20418, Tel: (202) 961-1257.

Areas of interest: Activities of the International Union of Pure and Applied Chemistry.

Holdings: Reports of biennial IUPAC conferences; IUPAC periodical, PURE AND APPLIED CHEMISTRY; IUPAC INFORMATION BULLETIN.

Publications: Reports of delegations to IUPAC conferences.

Information services: Answers inquiries.

NATIONAL AERONAUTICS AND SPACE ADMINISTRATION

Address: Washington, D.C. 20546.

Conducts research into problems of flight within and without the earth's atmosphere by testing and exploration; cooperates in peaceful applications; and disseminates research results. The primary information source for this agency is the Scientific and Technical Information Division.

A. SCIENTIFIC AND TECHNICAL INFORMATION DIVISION - 300 7th Street S.W., Washington, D.C. 20546, Tel: (202) 962-3648.

The Scientific and Technical Information Division serves as the agency's primary technical information resource in answering inquiries and making referrals. Services include the following: assistance with inquiries concerning NASA's research or programs; computer generated bibliographies for NASA contractors.

Publications: Several categories of scientific and technical report literature are regularly issued resulting from NASA research. These are Technical Reports (TP-prefix), Technical Memorandums (TM-prefix), Technical Notes (TN-prefix), Technical Translations (TT-prefix), Special Publications (books, manuals, data compilations, bibliographies, etc.)

These are indexed in: SCIENTIFIC AND TECHNICAL AEROSPACE REPORTS (STAR). STAR also contains abstracts of other literature related to space research throughout the world.

NASA report literature is also indexed in the GOVERNMENT REPORT LITERATURE and in GOVERNMENT REPORTS INDEX.

NASA REGIONAL DISSEMINATION CENTERS: The seven centers provide machine literature searches and other engineering and marketing data on a fee basis. They offer a means for non-NASA contractors to obtain research information which could be applied to industrial purposes.

NATIONAL SCIENCE FOUNDATION

Address: 1800 G Street N.W., Washington, D.C. 20550.

The fundamental purpose of the National Science Foundation is to strengthen research and education in the sciences in the United States.

Activities: The activities of the Foundation are described as follows: Grants and contracts are awarded to universities, nonprofit, and other research organizations to support fundamental research in all

the science disciplines. Efforts are supported in research on un-
resolved scientific questions concerning fundamental life processes,
natural laws and phenomena, fundamental processes influencing
man's environment, and the forces impacting on man as a member
of society as well as on the behavior of his society. Research
aimed at stimulating economic growth and productivity, improving
environmental quality, and enhancing our capabilities for dealing
effectively with social issues is supported.

Publications: Detailed brochures are issued announcing and describing new programs and critical dates and application procedures for
competitions, and may be obtained from the Distribution Section,
National Science Foundation, 1800 G Street N.W., Washington,
D.C. 20550. Other publications include MOSAIC, the NSF
quarterly magazine, and GUIDE TO PROGRAMS, available at the
Superintendent of Documents, Government Printing Office, Washington, D.C. 20402.

OCCUPATIONAL SAFETY AND HEALTH REVIEW COMMISSION

Address: 1825 K Street N.W., Washington, D.C. 20006.

The Occupational Safety and Health Review Commission is an independent executive agency established by section 12 of the Occupational Safety and Health
Act of 1970 (84 Stat. 1590; 29 U.S.C. 651). The Commission was created to
insure just and equitable enforcement of occupational safety and health standards promulgated by the Secretary of Labor under the said act which are contested by employers, employees, or representatives of employees.

Organization: The Commission, composed of three members, has
its central and principal offices in Washington, D.C. Each of
the members is appointed by the President with the advice and
consent of the Senate for a term of six years. One member is
designated by the President to serve as Chairman of the Commission. The Chairman is responsible on behalf of the Commission
for the administrative operations of the Commission and appoints
such hearings examiners and other employees as he deems necessary
to assist in the performance of the Commission's functions. He
exercises full and final authority over the Office of the Executive
Secretary, the Office of General Counsel, the Office of Administration and, subject to the provisions of the Administrative Procedure Act and section 12 of the Occupational Safety and Health
Act of 1970, all Review Commission Hearings Examiners.

Each Commission member exercises full and final authority over a
staff of legal assistants, each staff being under the supervision of
the Special Counsel of the respective Commission member.

Functions and activities: The Commission has three principal functions under the Occupational Safety and Health Act of 1970: (1)
adjudicating alleged violations of occupational safety and health
standards cited by the Secretary of Labor which are contested by
employers, (2) determining the period of time for correction of

alleged occupational safety and health violations cited by the Secretary of Labor where the reasonableness thereof is contested by any employer, employee or representative of employees, and (3) assessing all civil penalties for occupational safety and health violations under section 17 of this act.

For further information, contact the Office of the Executive Secretary, 1825 K Street N.W., Washington, D.C. 20006.

TENNESSEE VALLEY AUTHORITY

Address: Division of Environmental Research and Development, 720 Edney Building, 11th and Market Streets, Chattanooga, Tennessee 37401.

Areas of interest: Water and water research, including aquatic biology, reservoir ecology, control of aquatic weeds, eutrophication, limnology, phreatophyte control, water pollution, waste water disposal, water quality, water analysis, water chemistry, water economics, water microbiology, effects of impoundment on water quality, and thermal pollution; air and air research (related to coal-fired and nuclear steam-electric plants), including plume rise, stack height, plume dispersion, ground level concentration of stack emissions, environmental effects of stack emissions, meteorology, removal of sulphur from stack gas in usable form, control of sulphur oxide emissions, control of particulate emissions, air monitoring instrumentation, cooling tower emissions, air chemistry, and stack gas chemistry; solid waste research, including compost processing, utilization of compost, solid waste collection systems, and solid waste disposal; industrial and radiological hazards control; reservoir mosquito control.

Information services: Answers technical inquiries through correspondence, technical material, or discussion and consultation.

WATER RESOURCES COUNCIL

Address: 2120 L Street N.W., Washington, D.C. 20037.

The Council was established to maintain a continuing study of the adequacy of supplies of water necessary to meet the requirements in each water resource region in the United States of the relation of regional or river basin plans and programs to the requirements of larger regions of the nation, and of the adequacy of administrative and statutory means for the coordination of the water and related land resources policies and programs of the several federal agencies. The Council also reviews the plans of the river basin commissions and transmits these plans with its recommendations to the President for his review and transmittal by him to Congress. It administers a program of federal financial grants to states to aid them in comprehensive water and related land resource planning (79 Stat. 244; 42 U.S.C. 1962a).

SELECTED FEDERAL GOVERNMENT DATA BANKS AND INFORMATION OUTLETS

U.S. SUPERINTENDENT OF DOCUMENTS

Address: Government Printing Office, Washington, D.C. 20402.

Documents from the federal government have been made available to the public for many years through the Superintendent of Documents, who supervises the operation of the Government Printing Office in Washington, D.C.

The MONTHLY CATALOG lists a major portion of the materials developed and issued by the departments and agencies of the national government. Some materials are not listed in the MONTHLY CATALOG because they are not printed in the Government Printing Office. Also, some of the items recorded in the MONTHLY CATALOG are restricted to departmental use and are usually unavailable to the public.

Many of the documents whose titles appear in the MONTHLY CATALOG are "depository" items, which means that copies of these are sent to libraries throughout the U.S. which have been designated as official U.S. Government document depositories.

The depository libraries receive, classify and arrange the documents for use by the public, though often these materials are for reference use only. Depository items in the MONTHLY CATALOG are indicated by a solid black dot. Other symbols indicate limitations on availability of document.

Documents issued by the federal government are very inexpensive and most may be purchased through the Superintendent of Documents, Government Printing Office, Washington, D.C. 20402, or through the U.S. Department of Commerce field offices located on a regional basis in the larger cities. Purchase of publications from the Government Printing Office is facilitated for those who frequently use these items by use of coupons issued from the Superintendent of Documents and by means of deposit accounts with the Government Printing Office against which purchases may be charged. Complete details are given in the MONTHLY BULLETIN.

DEFENSE DOCUMENTATION CENTER

Address: Defense Supply Agency, U.S. Department of Defense, Cameron Station, Alexandria, Virginia 22314.

The Defense Documentation Center (DDC) is the central depository for technical reports generated by the research, development, test, and evaluation activities of the Department of Defense and its contractors, subcontractors, and grantees. Technical reports are received from other U.S. Government agencies and from several foreign nations on a reciprocal or voluntary basis. DDC operates, for organizations registered for service, the Research and Technology Work Unit Information System, a fully automated data bank containing work unit information descriptions of current Defense-sponsored and NASA-sponsored efforts in research and technology. Data banks concerning Contractor Performance Evaluation Reports and Contractor Cost Reduction Reports are maintained for use by selected military procurement organizations. Other data banks are maintained for special projects directed by the Director, Defense Research and Engineering. In additon, the Center is responsible for the development of long-range concepts and requirements for new or improved Defense documentation and information systems, services, and products.

Areas of interest: The fields of science and technology are subdivided into the following categories for announcement and distribution control: aeronautics; agriculture; astronomy and astrophysics; atmospheric sciences; behavioral and social sciences; biological and medical sciences; chemistry; earth sciences and oceanography; electronics and electrical engineering; energy conversion (nonpropulsive); materials; mathematical sciences; mechanical, industrial, civil, and marine engineering; methods and equipment; military sciences; missile technology; navigation, communications, detection, and countermeasures; nuclear science and technology; ordnance; physics; propulsion and fuels; and space technology.

Holdings: The technical report collection totals more than a million titles with approximately 45,000 titles added each year. About 20,000 summaries of management and technical information concerning on-going DoD and NASA research and development projects are maintained in the R&T Work Unit Information System.

Publications: TECHNICAL ABSTRACT BULLETIN, a classified (Confidential) publication, appears twice monthly and contains announcement descriptions of classified documents and of unclassified reports having release limitations. TECHNICAL ABSTRACT BULLETIN INDEXES, unclassified, is distributed to all registered user organizations. This index volume is issued twice monthly to coincide with release of the announcement publication and is cumulated quarterly and annually. Other publications include scheduled bibliographies, bibliographies tailor-made to individual user requests, etc.

Information services: DDC products and services are available to U.S. Government agencies, contractors, subcontractors and grantees, and to Defense potential contractors formally associated with a military department awareness program. To receive service, an organization must be officially registered with the Center. Technical reports are supplied to users in microform without cost; a service charge is imposed for paper copies. Bibliographic searches are made; all of the AD collection is computerized for rapid searching. Scheduled bibliographies, which concern subjects of high interest, are published as part of the technical report collection.

THE NATIONAL REFERRAL CENTER FOR SCIENCE AND TECHNOLOGY

Address: Science and Technology Division, Library of Congress, Washington, D.C. 20540.

A service of the Science and Technology Division of the Library of Congress offering directory assistance by subject, and connecting those in need of information with individuals or all other information outlets throughout the United

States. This is a free service. For assistance, write to the above address or call (202) 426-5670.

Publications: In order to expedite the referral services, several directories have been issued by this agency and are periodically updated.

1. A DIRECTORY OF INFORMATION RESOURCES IN THE UNITED STATES: PHYSICAL SCIENCES, ENGINEERING, 1972.
2. A DIRECTORY OF INFORMATION RESOURCES IN THE UNITED STATES: SOCIAL SCIENCES, 1967.
3. A DIRECTORY OF INFORMATION RESOURCES IN THE UNITED STATES: WATER, 1966.
4. A DIRECTORY OF INFORMATION RESOURCES IN THE UNITED STATES: BIOLOGICAL SCIENCES.
5. A DIRECTORY OF INFORMATION RESOURCES IN THE UNITED STATES: FEDERAL GOVERNMENT, 1967.
6. A DIRECTORY OF INFORMATION RESOURCES IN THE UNITED STATES: GENERAL TOXICOLOGY, 1969.

These may be purchased from: U.S. Superintendent of Documents, U.S. Government Printing Office, Washington, D.C. 20402.

NATIONAL TECHNICAL INFORMATION SERVICE (NTIS)

Address: Springfield, Virginia 22151.

NTIS is a filtering center for report literature, both from federal agencies and for reports generated under federal research contracts. Prior to 1970, this information outlet was called the Clearinghouse for Federal Scientific and Technical Information. NTIS receives copies of report material, microfilms them, and makes copies in microfilm or paper form available to the public for a fee. Most reports are freely available, but some are restricted to those who have contracts or other "need to know."

Announcement Services of NTIS:

GOVERNMENT REPORTS ANNOUNCEMENTS
A semi-monthly journal providing an abstract of technical reports received by NTIS. The index publication.

GOVERNMENT REPORTS INDEX
Has listings by subject, personal author, corporate author (company or institution issuing report), and contract number. An accession/ report number for each entry coincides with the abstract of the report, U.S. GOVERNMENT REPORTS ANNOUNCEMENTS.

FAST ANNOUNCEMENT SERVICE
Brief abstracts of reports recently received at NTIS are given in this service. Subscribers may choose one or more categories of interest for notification by the FAST Announcement Service. These are not restricted to science and technology, but cover social sciences as well.

CAST-CLEARINGHOUSE ANNOUNCEMENTS IN SCIENCE AND TECHNOLOGY

Announces to subscribers in brief abstract form the existence of new report literature in science-technology in forty-six broad subject categories.

FAST and CAST services are a practical means of learning of new developments, research or literature in a field without having to scan numerous publications or to build up files of indexing journals. These services are available in paper format or on microfiche.

SCIM is a new subscription service which enables one to obtain microfiche copies of technical reports in any of 90 subject categories. Microfiche are sent automatically once categories desired are registered with the NTIS.

SCIENCE INFORMATION EXCHANGE

Address: 1730 M Street N.W., Washington, D.C. 20036.

The Exchange is a clearinghouse for information on scientific research in progress from all available sources--government, industry, and private. It complements the services of technical libraries and documentation centers by providing information about research in progress between the time a project is proposed or started and the time results are made available in published form.

Areas of interest: Current and proposed research projects in all areas of science, including the life, physical, medical, engineering, agricultural, behavioral, and educational sciences; geographical distribution of research grants; automatic computer indexing and retrieval; comparative indexing techniques; computer dictionaries.

Publications: WATER RESOURCES RESEARCH CATALOG, WATER RESOURCES THESAURUS, and OUTDOOR RECREATION RESEARCH (prepared for Department of the Interior); ABSTRACTS OF RESEARCH AND DEMONSTRATION PROJECTS IN SOCIAL WELFARE AND RELATED FIELDS (for U.S. Department of Health, Education, and Welfare); other compilations of research records in such fields as urban research, environmental sciences, pesticides, and oceanography.

Information services: The Exchange answers questions from individual investigators, such as: Who is currently working in specific fields? Where? Under whose support? It provides program administrators with information that can be used to avoid duplication of research effort and to determine the advantageous distribution of research funds, and with information on general trends over a period of years (patterns of research activity, changes in support and personnel, and changes in topical interest). The system is capable of compiling all research records related to a specific topic or any combination of items to answer questions relating to program management or investigator interests. Service charges cover only the output cost (fee schedules are available).

b. STATE

The chemical engineer will, no doubt, find a different kind of information need satisfied by departments of the state government. The types of service offered and the extent to which the state government can and will assist in research or industrial information problems varies. Important agencies to contact for engineering or technical information and assistance are the Department of Commerce or Economic Development (as it is known in some states), the State Planning Agency, and the Pollution Control or State Environmental Agency. Office locations for these are listed in state capital city phone directories and in state government manuals or guides which are in the reference collections of most libraries.

Each state has an agency similar to the U.S. Superintendent of Documents which prints and issues state regulations, publications, maps, and brochures. An annual catalog describing items for sale is one approach in finding materials such as boiler regulations, industrial safety laws, pollution control legislation, and building codes, as well as other useful and practical informational materials.

Libraries in the states which collect federal documents also attempt to build collections of state publications. These libraries usually subscribe to the MONTHLY CHECKLIST OF STATE PUBLICATIONS, prepared by the Library of Congress as a guide to official documents of each of the state governments.

Examples of directories are the following:

BOOK OF THE STATES
The Council of State Governments, Lexington, Kentucky 40405.

TAYLOR'S ENCYCLOPEDIA OF GOVERNMENT OFFICIALS:
FEDERAL AND STATE
Ed. by J. Clements. Political Research Inc., 510 Continental Building, Dallas, Texas 75201.

In addition, each state issues a legislative manual or guide which has names of state agencies and officials who are potential sources of information.

c. MUNICIPAL

County, city, and other governing bodies for geographical areas smaller than the state are potential sources of data and knowledge for research and engineering needs. In county government, the county commissioner's office is a good first contact. Individual departments such as the Highway Department can also supply specific details related to their work. Telephone directories are the principal guide for connecting with such agencies. To retrieve official documentation is difficult because the distribution system is not as large as that on higher levels of government. A direct request to the commissioner's office or a county department is the best solution.

Especially useful in this regard is the article, "State and City Information Sources" by Elizabeth Biggert, CHEMICAL ENGINEERING, August 21, 1972, pp. 94-103.

In city government, the city engineer's office would be a useful contact for the chemical engineer. This department of municipal government is informed

on ordinances and codes and such matters as industrial waste disposal and pollution control. City government publications are annual reports, regulatory
documents, and financial records and statements. Commissions of the city such
as those for planning are concerned with industry and would be knowledgable
about these subjects. They also issue reports concerning industrial sites and
similar items which contain charts, statistics, and maps showing transportation
access to various city areas and industrial parks.

Libraries attempt to acquire these annual and special reports from local government, but often direct application to an official city agency is the only way
to learn about or to obtain copies of these. A Chamber of Commerce office
can be helpful in identifying the appropriate city departments and in locating
copies of publications.

2. Professional Societies, Trade Associations, and
Other Scientific-Technological Related Organizations

The role of the professional society, trade association, and scientific-technolog-
ical related organization, as providers of information in all areas of commercial
and educational endeavor, is extensive and inextricably interwoven with the
activities of engineers, scientists, and businessmen. Though each group has
unique purposes and goals, similarity occurs in their efforts to expand the amount
of knowledge available in their particular area of concern and in the services
they provide for the exchange of data and information. Thus the lines of distinction which outline the goals and programs of the scientific, technical, and
research societies and those organizations whose membership represents a trade
or commercial enterprise are becoming less clear today than previously has been
the case.

Engineers and scientists recognize the benefits of belonging to at least several
professional societies and have an acquaintance with other such groups in their
field. The information servicing functions of the professional and trade associations, however, are generally less well understood. One of the aims of this
guide is to highlight these available resources. Though only national organizations are singled out, local and student sections are also able to serve the
interests of their members in specialized ways.

a. SCIENTIFIC, ENGINEERING, AND RESEARCH
SOCIETIES AND INSTITUTES

Services: The transfer of information from research to user is an important
function carried on by these groups. This is accomplished through special publications and meetings. Services usually include national scientific and technical meetings and conferences for presentations and discussion of papers covering
reviews, research, development, design and operating experiences. In special
areas, joint sponsorship by societies is arranged. These professional programs
provide exposure to a wide range of research developments and in combination
with these and with onsite visits to industrial plants, the chemical engineer is
given an opportunity to keep informed on new systems and to trade his own
observations with others in his specialty.

Preprints of papers written for these programs are an important part of the total scientific and technical information transfer system. They are useful in the communication network among technical people because they make details of new concepts or reviews of experimentation in an area available long before they are formally published as journal articles or in books. Preprints can be purchased in advance of meetings and are listed in announcements by the sponsoring organizations.

Selected papers read at conferences may be collected and published some months after the meeting in the form of proceedings, and these are advertised in journals and in news releases from the professional societies. Some are sold through regular publishing firms and are described in their catalogs. One can be made aware of these by asking professional societies and publishers of Sci-Tech (science and technical) literature to add your name to their mailing list for announcements.

The dissemination of news and information through the publication of journals, books and other materials is a major contribution of the societies. The professional journal provides an essential link between the source and the user of information, and a variety of journals serve the needs of the specialists as well as of the general practitioners. In addition to the more formal subject and research presentations, journals also contain news items, book and literature reviews, advertisements of equipment and products and services. Additional details on journal literature are provided in the parts entitled "Libraries" and "General Chemical Engineering Literature."

Societies compile research data and special subject information in book format. These appear as handbooks, guides, bibliographies, state of the art reviews, yearbooks and monographs. They are announced in professional society journals, in flyers mailed out by the society, and are also cited in standard book trade publications such as the PUBLISHERS WEEKLY, available from R.R. Bowker Company, 1180 Avenue of the Americas, New York, N.Y. 10036.

The individual society should be contacted for its lists of publications. Limited services are provided for responding to specific inquiries. The holdings of libraries of the societies may consist of not only the back issues of the journals and of the agencies' special publications, but also extensive literature and information in special interest areas. Items in these libraries include non-print materials such as tapes, slides, films and microforms as well as the more familiar publications in book and journal format.

The societies function as a clearinghouse for developments in their subject fields, for the purposes of public relations as well as for the continuing benefit of the profession. The society, by being alert to developments and problems in the field, can keep the membership, the government, and the public as a whole informed.

b. TRADE ASSOCIATIONS

(In this context, associations of individuals or groups who have formed together a voluntary, non-profit organization to benefit and promote a trade, enterprise, or industry.)

Services: In presenting to the public the story of an industry, the association also provides information specific to a kind of processing or manufacture of a product. It interprets and explains the uses of the "product" and acts in an advisory capacity to those with questions about specifications, operating procedures or testing of a certain item or material. To this end, manuals, detailed drawings, and even books are made available. The public information office of the trade association therefore operates as a specialized information resource and should not be overlooked by engineers.

The association may also seek sponsorship of research programs which lead to new or improved products or utilization of byproducts resulting from manufacturing processes or improvements thereof. Furthermore, the experimentation and investigation supported by trade associations often produce significant industrial standards.

The creation of standards in parts design and in their manufacturing is another service of the trade association. Standards and specifications are publicized in trade journals and in special announcements. Most industrial standards are cataloged and issued through the American National Standards Institute, 1430 Broadway, New York, New York 10018, an umbrella organization to which trade associations belong.

Some trade organizations also provide educational programs for members, similar to those of the professional societies. These take the form of courses and training sessions, meetings and lectures, publication of manuals and literature, and production of audio-visual material.

AIR-CONDITIONING AND REFRIGERATION INSTITUTE - 1815 North Fort Myer Drive, Arlington, Virginia 22209.

> Areas of interest: Engineering activities and other functions involved with air-conditioning and refrigeration equipment and components.
>
> Publications: Standards and technical publications in the area of air-conditioning and cooling.
>
> Information services: Answers inquiries; makes referrals.

AIR POLLUTION CONTROL ASSOCIATION (Environmental Quality) - 4400 Fifth Avenue, Pittsburgh, Pennsylvania 15213.

An association of industrialists, researchers, equipment manufacturers, governmental control personnel, and others concerned with the problems of air pollution.

> Committees: Chemical; Petroleum; Coal; Public Utilities; Steel; Ferrous and Nonferrous Foundries; Nonferrous Smelting and Refining; Waste Disposal, Incinerator, Oil Burner Equipment; Dust, Fumes, Mists and Fog Collectors; Coal Utilization; Odor and Gas Treatment; Meteorological; Power Boiler Equipment Information; Vehicular Exhaust; Air Pollution Measurements; Economic Effects; Radiation; Ambient Air Standards; Atmospheric Reaction; Cleaner Air Week; Cement, Pulp and Paper.
>
> Publications: (1) JOURNAL OF THE AIR POLLUTION CONTROL ASSOCIATION (monthly); (2) DIRECTORY OF GOVERNMENTAL AIR POLLUTION AGENCIES (in cooperation with U.S. Public Health Service).

Information services: Answers inquiries; makes referrals.

AMERICAN ACADEMY OF ENVIRONMENTAL ENGINEERS (AAEE) - P.O. Box 1278, Rockville, Maryland 20850.

An organization of environmental engineers licensed to practice engineering in the United States or Canada.

Areas of interest: Air pollution control, industrial hygiene, sanitary engineering, and radiation and hazard control. Conducts research to improve undergraduate engineering education; publishes a list of graduate programs in environmental engineering accredited by the Engineers Council for Professional Development with the participation of the Engineering Education Committee of the Environmental Engineering Intersociety Board.

Publications: (1) THE DIPLOMATE (quarterly); (2) AAEE ROSTER (annual).

Information services: Answers inquiries; makes referrals.

AMERICAN ASSOCIATION FOR CONTAMINATION CONTROL - 6 Beacon Street, Boston, Massachusetts 02108.

Purpose: To (1) promote the technology of contamination control in closed systems of gases and liquids, and engage in research in related fields of contamination control; (2) promote the exchange of ideas and information with allied technologists and technical engineering societies; etc.

Professional activities: Research, technical lecture series, and seminars on contamination control and the promulgation of research to the scientific world through publications.

Publications: JOURNAL OF AMERICAN ASSOCIATION FOR CONTAMINATION CONTROL (quarterly); CONTAMINATION CONTROL (bimonthly).

AMERICAN ASSOCIATION FOR THE ADVANCEMENT OF SCIENCE - 1515 Massachusetts Avenue, N.W., Washington, D.C. 20005.

An organization which touches upon all aspects of science in its efforts to bring about a continually improved environment for scientific activity. It attempts to do this through meetings, publications and research designed to increase the understanding of science and to ensure the application of scientific progress to human welfare. The association is of general interest to the chemical engineer. Particularly informative are two of its publications: SCIENCE, a weekly journal that covers a variety of topics, and SCIENCE BOOKS, a quarterly review which lists and reviews new sci-tech publications reaching the market.

AMERICAN ASSOCIATION OF COST ENGINEERS - P.O. Box 368, Bellaire, Texas 77401.

Purpose: To (1) provide forums and media through which experiences with the principles and techniques of cost engineering may be reported, discussed, and published in furtherance of the public interest; (2) promote standardization of terminology in cost engineering and so far as practicable develop standard methods.

Publications: AACE BULLETIN (bimonthly); TRANSACTIONS (annual).

AMERICAN CARBON COMMITTEE - 101 Mineral Industries Building, Pennsylvania State University, University Park, Pennsylvania 16802.

Committee membership includes: physicists, chemists, technologists, and other scientific personnel from throughout the world who are interested in carbon specialties and research. The conferences and publications of the Committee cover the physics and chemistry and other scientific aspects of a class of materials including organic crystals and polymers through chars and carbons and graphite. Its conferences, which are held in the United States, are supported also by governmental agencies and by corporations.

Publications: (1) CARBON (bimonthly); (2) CONFERENCE REGISTRATION LIST (biennial); (3) CONFERENCE PROGRAM; and ABSTRACTS of Conference Papers.

AMERICAN CHEMICAL SOCIETY - 1155 16th Street N.W., Washington, D.C. 20006.

The organization works for the advancement and promotion of the discipline of the pure and applied chemical sciences. The American Chemical Society is particularly prolific in its information activities. These activities are illustrated foremost by its CHEMICAL ABSTRACTS SERVICE and in addition by the many practical publications issued under its sponsorship.

Publications: ANALYTICAL CHEMISTRY (monthly); JOURNAL OF THE AMERICAN CHEMICAL SOCIETY (biweekly); BIOCHEMISTRY (biweekly); JOURNAL OF MEDICINAL CHEMISTRY (monthly); INORGANIC CHEMISTRY (monthly); JOURNAL OF ORGANIC CHEMISTRY (biweekly); JOURNAL OF AGRICULTURAL AND FOOD CHEMISTRY (bimonthly); JOURNAL OF CHEMICAL AND ENGINEERING DATA (quarterly); MACROMOLECULES (bimonthly); ACCOUNTS OF CHEMICAL RESEARCH (monthly); CHEMICAL REVIEWS (bimonthly); JOURNAL OF PHYSICAL CHEMISTRY (biweekly); JOURNAL OF CHEMICAL DOCUMENTATION (quarterly); INDUSTRIAL AND ENGINEERING CHEMISTRY--FUNDAMENTALS (quarterly); INDUSTRIAL AND ENGINEERING CHEMISTRY--PROCESS DESIGN AND DEVELOPMENT (quarterly); INDUSTRIAL AND ENGINEERING CHEMISTRY--PRODUCT RESEARCH AND DEVELOPMENT (quarterly); ENVIRONMENTAL SCIENCE AND TECHNOLOGY (monthly); CHEMICAL AND ENGINEERING NEWS (weekly); CHEMISTRY (11 issues per year); CHEMICAL TECHNOLOGY (monthly); CHEMICAL ABSTRACTS (weekly); CHEMICAL-BIOLOGICAL ACTIVITIES (biweekly); CHEMICAL TITLES (biweekly); PLASTICS INDUSTRY NOTES (weekly); POLYMER SCIENCE AND TECHNOLOGY--JOURNALS (biweekly); POLYMER SCIENCE AND TECHNOLOGY--PATENTS (biweekly).

Monographs: Advancement in Chemistry Series. These cover a range of subject interests and include bibliographies, guides and reports of research. Some examples of this series are:

#10 LITERATURE RESOURCES FOR CHEMICAL PROCESS
INDUSTRIES. 1954.

#46 PATENTS FOR CHEMICAL INVENTIONS. 1964.

AMERICAN COKE AND COAL CHEMICALS INSTITUTE (ACCCI) - 1010 16th
Street N.W., Washington, D.C. 20036.

Producers of oven coke and coal chemicals, metallurgical coal for distillers,
and producers of chemicals.

Committees: Chemicals; Coke; Manufacturing and Environment;
Traffic.

Publications: (1) Monthly newsletter, plus numerous other newsletters; (2) DIRECTORY of ACCCI.

Information services: Answers and refers inquiries.

AMERICAN CONFERENCE OF GOVERNMENTAL INDUSTRIAL HYGIENISTS -
1014 Broadway, Cincinnati, Ohio 45202.

Areas of interest: Industrial hygiene and occupational health, with
emphasis on Threshold Limit Values for air contaminants, industrial
ventilation, air sampling instruments, health services in small industries, and uniform industrial hygiene codes or regulations.

Publications: AIR POLLUTION CONTROL PROCESS FLOW SHEETS;
AIR POLLUTION REFERENCE LIBRARY; AIR SAMPLING INSTRU-
MENTS MANUAL; ANALYTICAL METHODS MANUAL; BIBLIOG-
RAPHY OF RADIATION PROTECTION ORGANIZATIONS; GUIDE
TO RECORDS FOR HEALTH SERVICES IN SMALL INDUSTRIES;
GUIDE TO UNIFORM INDUSTRIAL HYGIENE CODES OR REGU-
LATIONS; STANDARD REPORTS OF OCCUPATIONAL HEALTH
ACTIVITIES AMONG STATE AND LOCAL HEALTH AGENCIES;
THRESHOLD LIMIT VALUES; GUIDE TO RECORDS AND REPORTS
FOR EVALUATING ENVIRONMENTAL CONDITIONS IN INDUS-
TRY; TRANSACTIONS OF ANNUAL MEETINGS. A publications
list is available.

Information services: Answers inquiries; makes referrals.

AMERICAN DYE MANUFACTURERS INSTITUTE - 74 Trinity Place, New York,
New York 10006.

Producers of dyes in the United States. To promote the proper end use of dyes
through advertising and educational programs; to develop testing and quality
control procedures to promote the use of dyes.

Information services: Answers inquiries; makes referrals.

AMERICAN INSTITUTE OF CHEMICAL ENGINEERS (AICHE) - 345 East 47th
Street, New York, New York 10017.

Areas of interest: All aspects of chemical engineering and chemical
industries including environmental, food, pharmaceutical, bioengineering, forest products, fuels and petrochemicals, heat transfer, nuclear

engineering, and materials sciences.

Publications: AICHE JOURNAL (bimonthly); CHEMICAL ENGINEER-
ING PROGRESS (monthly); INTERNATIONAL CHEMICAL ENGINEER-
ING (quarterly). Also publishes a symposium series on topics of concern in chemical engineering and chemical industries as well as pamphlets on chemical engineering problems, standards, and educational
and career information.

Information services: Provides lists of publications, answers inquiries
from nonmembers that come within scope of the association, and makes
referrals.

AICHE MONOGRAPH AND SYMPOSIUM SERIES: These cover a
broad spectrum in chemical engineering disciplines including: Biochemical Engineering, Heat Transfer and Energy Conversion, Nuclear Engineering, Processing Technology, Aerospace, Mathematics
and Ion Exchange. The publications consist of papers read at symposia and/or individual lectures. The titles reflect a particular
aspect of the general topical heading with which they are associated. Each symposium topic is identified by the letter "S" and a
number. Other categories are:

T (series)   AMMONIA PLANT SAFETY SERIES
X-5   CHEMICAL ENGINEERING THESAURUS
X-6   PUMP MANUAL
E (series)   EQUIPMENT TESTING PROCEDURES
A (series)   CONTINUING EDUCATION SERIES OF
             MOTION PICTURES, SLIDES AND TAPES
             STANDARDS
W (series)   AICHE WORKSHOP VOLUMES

Informational materials about the publications, programs, meetings
and other activities are available from the society.

AMERICAN INSTITUTE OF CHEMISTS (AIC) - 79 Madison Avenue, New York,
New York 10016.

Membership includes chemists and chemical engineers.

Purpose: "To elevate the professional and economic status of
chemists and chemical engineers."

Committees: Ethics; Interprofessional Relations; Legislation; Professional Accreditation; Professional Education and Training; Professional Employment.

Publications: THE CHEMIST (monthly).

Information services: Answers inquiries.

AMERICAN LEATHER CHEMISTS ASSOCIATION - University of Cincinnati,
Cincinnati, Ohio 45221.

Association membership includes chemists, leather technologists and educators
concerned with the tanning and leather industry.

Program: Development of methods for the analysis and testing of leathers and materials used in leather manufacture; to promote communications within industry and to foster research applications to problems confronting the leather industry.

Publications: JOURNAL OF ALCA (monthly); DIRECTORY (annual).

Information services: Answers inquiries; makes referrals.

AMERICAN NATIONAL STANDARDS INSTITUTE - 1430 Broadway, New York, New York 10018.

Areas of interest: The Institute acts as the national coordinating institution for voluntary standardization through which interested organizations may cooperate in establishing, recognizing, and improving voluntary standards of the United States. It also represents United States interests in international standards work.

Publications: AMERICAN NATIONAL STANDARDS (new and revised); Standards Institute REPORTER (biweekly); STANDARDS ACTION (biweekly); AMERICAN NATIONAL STANDARDS INSTITUTE CATALOG (annual).

Information services: The Institute's Department of Public Information answers inquiries and makes referrals.

AMERICAN OIL CHEMISTS' SOCIETY - 35 East Wacker Drive, Chicago, Illinois 60601.

Areas of interest: Industrial technology of fats, detergents, and protective coatings, including synthesis and separation of industrial fat products and other areas of research in fats and lipids.

Publications: THE JOURNAL OF THE AMERICAN OIL CHEMISTS' SOCIETY (monthly, industrially oriented); LIPIDS (monthly, basic research oriented); SHORT COURSE LECTURES (annual); OFFICIAL AND TENTATIVE METHODS OF ANALYSIS (revised annually); MEMBERSHIP DIRECTORY (annual).

Information services: Answers inquiries; makes referrals; dispenses official standard supplies; conducts two national technical meetings annually; conducts short courses; certifies referee chemists.

AMERICAN PHARMACEUTICAL ASSOCIATION - 2215 Constitution Avenue N.W., Washington, D.C. 20037.

Purpose: To (1) aid in improving, promoting, and safeguarding the public health and welfare in every practical manner and by all practical means; (2) foster and encourage interprofessional relations; (3) improve the art and science of pharmacy for the general welfare of the public by fostering the publication of scientific information, etc.

Professional activities: Research interests include drug interactions evaluation program; pharmaceutical services in poverty areas (Office of Economic Opportunity grant); drug standards laboratory research on pharmaceutical analysis and standards; slide talk on drug interactions; handbook on non-prescription drugs.

Publications: JOURNAL OF THE AMERICAN PHARMACEUTICAL

ASSOCIATION (monthly); APhA NEWSLETTER (biweekly); JOURNAL OF PHARMACEUTICAL SCIENCES (monthly).

AMERICAN PUBLIC WORKS ASSOCIATION - 1313 East 60th Street, Chicago, Illinois 60637.

Areas of interest: Planning, design, construction, operation, maintenance, and administration of all public works, including water treatment and distribution facilities, sewer systems, sewage treatment facilities, refuse collection and disposal facilities, etc.

Publications: APWA REPORTER (monthly magazine); annual directory; newsletters of the Association's Institute for Solid Wastes and Institute for Municipal Engineering (quarterly); special reports, standards and specifications, handbooks. A publications list is available.

Information services: Answers inquiries.

AMERICAN SOCIETY FOR ENGINEERING EDUCATION - 1 Dupont Circle, Washington, D.C. 20036.

A society formed to promote and advance high standards of engineering education. This organization is active in its endeavors to bring together those concerned with the training and development of engineers and it accomplishes this through annual conferences, regional meetings, and through its publications.

The transfer of information among engineers and students of the engineering sciences is a particular concern of the Engineering School Libraries Division (one of the subject specialty divisions within the parent organization).

Publications: ENGINEERING EDUCATION (monthly); CHEMICAL ENGINEERING EDUCATION (quarterly); ENGINEERING ECONOMIST (quarterly); ENGINEERING RESEARCH AND METHODS (quarterly); ENGINEERING DESIGN GRAPHICS JOURNAL; MECHANICAL ENGINEERING NEWS (quarterly); DIRECTORY OF ENGINEERING COLLEGE RESEARCH AND GRADUATE STUDY (annually).

Bibliographies and guides to literature: AEROSPACE ENGINEERING; AGRICULTURAL ENGINEERING; CHEMICAL ENGINEERING; COMPUTERS ENGINEERING; ELECTRICAL & ELECTRONIC ENGINEERING; ENVIRONMENTAL ENGINEERING; INDUSTRIAL ENGINEERING; MECHANICAL ENGINEERING; METALS & METALLURGY ENGINEERING; TRANSPORTATION ENGINEERING.

AMERICAN SOCIETY FOR INFORMATION SCIENCE (ASIS) - 1140 Connecticut Avenue N.W., Washington, D.C. 20036.

Information sepcialists, scientists, librarians, administrators, social scientists and others interested in the use, organization, storage, retrieval, and dissemination of recorded specialized information. Seeks to improve the information transfer process through research, development, application and education. Provides a forum for the discussion, publication, and critical analysis of work dealing with the theory, practice, research, and development of elements involved in communication of information.

Publications: (1) NEWSLETTER (bimonthly); (2) JOURNAL OF THE ASIS (bimonthly); (3) INFORMATION SCIENCE ABSTRACTS (bimonthly); (4) ASIS Membership Handbook and Directory (annual); (5) Annual Meeting Proceedings; (6) Annual Review of Information Science and Technology.

AMERICAN SOCIETY FOR QUALITY CONTROL, INC. - 161 West Wisconsin Avenue, Milwaukee, Wisconsin 53203.

Purpose: To create, promote, and stimulate interest in the advancement and diffusion of knowledge of the science of control and its application to industrial administrative and other processes, especially to the quality of industrial products.

Publications: JOURNAL OF QUALITY TECHNOLOGY (quarterly); QUALITY PROGRESS (bimonthly).

Information services: Answers inquiries; makes referrals.

AMERICAN SOCIETY OF HEATING, REFRIGERATING, AND AIR-CONDITION-ING ENGINEERS, INC. - 345 East 47th Street, New York, New York 10017.

Areas of interest: Heating; refrigeration; air conditioning; ventilation; food preservation; physiology and human comfort; plant and animal physiology; odors; sound and vibration; lubrication; corrosion; water treatment; thermal insulation; cryogenics; ice making plants; solar energy utilization; urban environment.

Publications: ASHRAE JOURNAL (monthly); ASHRAE TRANSACTIONS (annual, in 2 parts); ASHRAE GUIDE AND DATA BOOK (in 3 volumes); ASHRAE HANDBOOK OF FUNDAMENTALS; standards, bulletins, abstracts, circulars, psychrometric charts.

Information services: Answers inquiries. Other services, including reference, literature-searching, and duplication services, are available from the Engineering Societies Library, some on a fee basis.

AMERICAN SOCIETY OF LUBRICATION ENGINEERS (ASLE) - 838 Busse Highway, Park Ridge, Illinois 60068.

Engineers and others engaged in the technological development and use of lubricants, lubricating equipment and systems, etc.

Activities: Sponsors joint committees and councils with industry to advance the science of lubrication; sponsors annual short course in lubrication engineering.

Technical committees: Bearings and Bearing Lubrication; Engine Lubrication; Gears and Gear Lubrication; Hydraulics; Lubrication Equipment and Practices; Lubrication Fundamentals; Metalworking Fluids; Pollution Control; Properties of Lubricants; Seals; Solid Lubricants. Industry Councils: Aerospace; Cleaners and Solvents; Machine Tool; Mining; National Railroad; Non-Ferrous Metals; Petroleum and Chemicals; Power Generation; Steel.

Publications: (1) LUBRICATION ENGINEERING (monthly); (2) ASLE TRANSACTIONS (quarterly); (3) MEMBERSHIP ROSTER (annual); also publishes handbooks, Glossary of Seal Terms, Recent Developments in Seal Technology.

Information services: Answers inquiries; makes referrals.

AMERICAN SOCIETY OF SAFETY ENGINEERS - 850 Busse Highway, Park Ridge, Illinois 60068.

Areas of interest: All aspects of safety engineering, accident prevention, traffic safety, and physical and environmental hazards; development of educational activities in response to the increased demand for professionals in the accident and loss prevention fields.

Publications: ASSE JOURNAL (monthly, available to nonmembers by subscription); A SELECTED BIBLIOGRAPHY OF REFERENCE MATERIALS IN SAFETY ENGINEERING AND RELATED FIELDS.

Information services: In handling requests for information, the Society in most cases attempts to provide the inquirer with a list of specialized sources of the information desired. When technical questions require considerable research and knowledge of a specialized area, the inquirer is referred to safety engineering consultants.

AMERICAN SOCIETY OF SANITARY ENGINEERING - 228 Standard Building, Cleveland, Ohio 44113.

Areas of interest: All aspects of sanitary engineering, including plumbing; gas and water supplies; drainage; sanitation codes; lake, river, ocean, and underground water pollution; radioactive waste disposal; sewage disposal; water purification; health as related to sanitation.

Publications: ASSE NEWSLETTER (monthly); YEARBOOK OF THE AMERICAN SOCIETY OF SANITARY ENGINEERING.

Information services: Answers inquiries; makes referrals; provides reference services.

AMERICAN WATER RESOURCES ASSOCIATION - P.O. Box 434, Urbana, Illinois 61801; Location: 130 North Race Street, Urbana, Illinois.

Areas of interest: All aspects of water resources, including water pollution, irrigation, ground water, surface water, meteorology, and drainage.

Publications: WATER RESOURCES BULLETIN JOURNAL, which includes the WATER RESOURCES NEWSLETTER (bimonthly); HYDATA (monthly review and index of journals on water resources); WATER RESOURCES ABSTRACTS (monthly); PROCEEDINGS OF THE AMERICAN WATER SYMPOSIUM (annual).

Information services: Makes referrals to specialists in the water resources field.

AMERICAN WATER WORKS ASSOCIATION, INC. - 2 Park Avenue, New York, New York 10016.

Areas of interest: Management, operation, and maintenance of public water supply utilities; water quality; water treatment techniques and facilities; distribution of public water supplies; water rates; water storage.

Publications: JOURNAL AMERICAN WATER WORKS ASSOCIATION (monthly); WILLING WATER (semimonthly bulletin, for members and public libraries only); books, pamphlets, manuals, standards. A list of publications and a bibliography entitled THE WATER SUPPLY LIBRARY are available on request.

Information services: Answers brief inquiries; makes referrals to consultants, usually Association members; permits onsite use of collection; collects and publishes statistical data on water resources, management of public water utilities, water treatment, and water distribution.

ASSOCIATION OF CONSULTING CHEMISTS AND CHEMICAL ENGINEERS, INC. - 50 East 41st Street, New York, New York 10017.

Areas of interest: Chemistry; chemical engineering.

Publications: A directory, CONSULTING SERVICES (available for purchase), published in looseleaf form and revised biennially, contains a classified index of member specialists and descriptive data on each member, as well as alphabetical and geographical indexes.

Information services: Provides, without charge, the names and addresses of members specializing in particular fields of chemistry and chemical engineering.

ASSOCIATION OF STATE AND INTERSTATE WATER POLLUTION CONTROL ADMINISTRATORS (ASIWPCA) - c/o NEIWPCC, 607 Boylston Street, Boston, Massachusetts 02116.

Organization consists of administrators of state and interstate governmental agencies legally responsible for prevention, abatement and control of water pollution.

Purpose: "To improve the water quality objectives of the states." Promotes coordination among agency programs.

Publications: (1) Membership Directory (annual); (2) Proceedings (irregular).

CHEMICAL COATERS ASSOCIATION - Industrial Finishing Magazine, Hitchcock Publishing Co., 25W550 Geneva Road, Wheaton, Illinois 60187.

Industrial users of organic finishing systems; suppliers of chemicals, equipment and paints.

Purpose: Works toward improvement of decorative, functional and performance standards of chemical coatings. Encourages members to continue improvements in application technology. Provides coating industry with representation to public authorities and government agencies.

Activities: Sponsors programs and other activities toward control of environmental pollution.

Information services: Will compile marketing data and other general information of value to the industry.

CHEMICAL MARKETING RESEARCH ASSOCIATION - 100 Church Street, New York, New York 10007.

Individuals engaged in chemical market research for chemical companies and chemical process industries in such fields as rubber, glass, soap, textiles, plastics, steel, non-ferrous metals, petroleum, drugs, and food. Keeps members informed on developments in the field, particularly industrial market research. Sponsors market research. Also conducts seminars, business schools and short courses in marketing research and allied disciplines.

CHEMICAL SPECIALTIES MANUFACTURERS ASSOCIATION, INC. - 50 East 41st Street, New York, New York 10017.

Areas of interest: Aerosol spray products of all types, as well as conventional packed automotive chemicals, detergents and cleaning compounds, disinfectants and sanitizers, insecticides, waxes, polishes, and floor finishes.

Publications: COMPILATION OF ECONOMIC POISON LAWS; COMPILATION OF LABELING LAWS AND REGULATIONS FOR HAZARDOUS SUBSTANCES; COMPILATION OF OFFICIAL INSEC- TICIDE BIO-ASSAY METHODS AND CHEMICAL ANALYSIS; AGEN- CIES AND REGULATIONS OF INTEREST TO THE AEROSOL IN- DUSTRY; VENDORS TO THE TRADE; AEROSOL GUIDE; COMPIL- ATION OF ANTI-FREEZE & BRAKE FLUID LAWS; COMPILATION OF WEIGHTS & MEASURES LAWS; WAX MANUAL OF TESTING METHODS; STATE REGISTRATION OF PESTICIDES; COMPOSITE INDEX OF PROCEEDINGS.

Information services: Answers inquiries; makes referrals; provides legislative reporting service; develops specifications and testing methods; performs market surveys; conducts meetings and seminars. Some services are provided to members only.

CHLORINE INSTITUTE, INC. - 342 Madison Avenue, New York, New York 10017.

Areas of interest: All problems relating to the handling and use of chlorine, such as safety, transportation, regulations and legislation, equipment, containers, statistical data, and engineering and design; disinfection and other chlorine applications in water; sewage and indus- trial waste treatment processes; state regulations related to disinfection.

Publications: THE CHLORINE MANUAL; pamphlets, engineering and design recommendations, specifications and drawings.

Information services: Makes referrals; provides bibliographic, con- sulting, and duplication services.

THE COMBUSTION INSTITUTE - Union Trust Building, Pittsburgh, Pennsylvania 15219.

Areas of interest: All aspects of combustion, flame, and explosion phenomena.

Publications: COMBUSTION AND FLAME (bimonthly); SYMPOSIUM

(INTERNATIONAL) ON COMBUSTION (biennial).

Information services: Answers inquiries; makes referrals. The Institute's primary purpose is to organize a biennial symposium, with subsequent publication of contributed papers and attendant discussion in a bound volume of 1,200-1,400 pages.

COMMITTEE FOR ENVIRONMENTAL INFORMATION - 438 North Skinker Boulevard, St. Louis, Missouri 63130.

Areas of interest: Air and water pollution; pesticides; biological effects of radiation; nuclear power reactors; nuclear testing; nuclear war and civil defense; peaceful uses of nuclear explosives; chemical and biological warfare; supersonic transport; noise; weather modification.

Publications: ENVIRONMENT (10 issues a year).

Information services: Answers inquiries; provides consulting and reference services; permits onsite reference.

CONFERENCE OF STATE SANITARY ENGINEERS (CSSE) - c/o Environmental Health Service, West Virginia State Department of Health, 1800 Washington Street, East, Charleston, West Virginia 25305.

Membership includes sanitary engineering officials from departments of health in the United States, and administrative officers of state water pollution control agencies.

Purpose: Coordinates the various public health engineering activities of official state and territorial health organizations, acts as clearinghouse for exchange of information in this area, makes studies and compiles information on sewage and water treatment problems.

Committees: Air; Environmental Planning and Program Development; General Environmental Health; Indian Health; Milk and Food Protection; Occupational Health; Personnel Management; Radiological Health; Water and Waste Water.

Publications: (1) Committee Reports, and (2) Proceedings, both annual.

CONSULTING ENGINEERS COUNCIL OF THE UNITED STATES - 1155 15th Street, N.W., Washington, D.C. 20005.

Areas of interest: All branches of engineering are represented in the programs of this organization. Also concerned with project feasibility and economic studies, reports, planning, design, and construction management services for buildings, industrial plants, transportation systems, housing projects, site development, water supply, sanitation, solid waste disposal, air pollution control, water pollution control, power generation and transmission, irrigation, flood control, waterfront development, air and marine navigation, bridges, tunnels, etc.; valuation; rate studies; product processing, handling, and storage; structural evaluations.

Publications: NEWSLETTER (monthly); DIRECTORY OF MEMBERSHIP; INTERNATIONAL ENGINEERING DIRECTORY; MANUAL OF PRINCIPLES AND PERFORMANCE FOR CONSULTING ENGINEERS; reports, bulletins, standard forms of agreement.

Information services: Answers inquiries; makes referrals; provides consulting and reference services; provides information to students and others interested in career opportunities in the field; conducts seminars and conferences.

COOLING TOWER INSTITUTE - 3003 Yale Street, Suite 107, Houston, Texas 77018.

Areas of interest: Technology, design, performance, and maintenance of water cooling towers, including water treating practices, corrosion of tower hardware, prevention of decay of tower lumber, use of biocides, and influences of wood preservatives on the corrosion of metal components.

Publications: CTI NEWS (quarterly); CTI BOOK OF PERFORMANCE CURVES; standards and specifications, research reports.

Information services: Provides consulting services for fee; tests cooling towers and analyzes the results; provides speakers for meetings.

COSMETIC INDUSTRY BUYERS AND SUPPLIERS (CIBS) - c/o James Feigin, Almay Cosmetics, 562 Fifth Avenue, New York, New York 10036.
Organization made up of buyers and suppliers of essential oil, chemicals, packaging, finished goods, etc.

Purpose: To enhance growth and promote prosperity and protection of the American cosmetic industry. Provides a forum for the exchange of ideas and experiences within industry.

Information services: Answers inquiries; makes referrals.

COSMETIC, TOILETRY AND FRAGRANCE ASSOCIATION (CTFA) - 1625 Eye Street, N.W., Washington, D.C. 20006.
Membership includes manufacturers and distributors of finished cosmetics and toilet preparations; suppliers of raw materials and services. Services to members are: advertising copy review service, trade mark records, and a small library; conducts color safety research, and scientific, management, and marketing conferences, and compiles statistics on wages and industry sales.

Publications: (1) INSIDE NEWS FROM CTFA (semimonthly); (2) TRADE-MARK BULLETIN (semimonthly); (3) EXECUTIVE NEWSLETTER (monthly); (4) CTFA COSMETIC JOURNAL (includes directory and scientific conference proceedings) (quarterly); (5) LEGISLATIVE BULLETIN (irregular).

Information services: Answering of inquiries and directory referrals.

COUNCIL ON ENGINEERING LAWS (CEL) - 280 Madison Avenue, New York, New York 10016.

Organization consists of corporations, partnerships, and firms interested in furthering the practice of engineering under adequate and reasonable safeguards for public health, safety, and welfare.

Purpose: To promote and maintain high professional standards in the practice of engineering by corporations. Maintains library of laws, opinions, and information from various states regarding the corporate practice of engineering.

Information services: Answers inquiries and makes referrals.

CRYOGENIC SOCIETY OF AMERICA (CSA) - P.O. Box 1147, Huntington Beach, California 92647.

Membership includes individuals, firms, associations, and institutions engaged in cryogenic work.

Purpose: To encourage the dissemination of knowledge of low temperature processes and techniques.

Activities: Promotes research and development through meetings, professional contacts, and publications. Undertakes information gathering in response to industrial and other needs.

Publication: CRYOGENIC TECHNOLOGY (bimonthly).

Information services: Answers inquiries; makes referrals.

CRYOGENICS ENGINEERING CONFERENCE (CEC) - Public Information Office, National Bureau of Standards, Boulder, Colorado 80302.

Membership includes academic, industrial, and governmental researchers and managers involved in basic and applied work in cryogenics. Provides a forum for an annual three-day presentation of papers and seminars concerning recent advances in the science and technology of cryogenics in such areas as super-conductivity, heat transfer, insulation, cryogenic cables, cryo-health-services, cryo-biology, LNG, and power generation.

Publication: ADVANCES IN CRYOGENIC ENGINEERING (Proceedings) (biennial).

DRUG, CHEMICAL AND ALLIED TRADES ASSOCIATION (DCAT) - 350 Fifth Avenue, Suite 3014, New York, New York 10001.

Membership includes manufacturers and distributors of drugs, chemicals and related products.

Publication: BULLETIN (biweekly).

Information services: Answers inquiries; makes referrals; acts as clearinghouse on related trade matters.

ELECTROCHEMICAL SOCIETY (ECS) - P.O. Box 2071, Princeton, New Jersey 08540.

A professional society of electrochemists, chemists, chemical and electrochemical engineers, metallurgists and metallurgical engineers, physical chemists, physicists, electrical engineers, research engineers, teachers, technical sales representatives, and patent attorneys.

Purpose: To advance the science and technology of electrochemistry, electronics, electrothermics, electrometallurgy, and applied subjects.

Publications: (1) JOURNAL OF THE ECS (monthly); (2) EXTENDED ABSTRACTS AND SYMPOSIA VOLUMES; (3) Membership Directory.

Information services: Answers inquiries; makes referrals.

ENGINEERING FOUNDATION - 345 East 47th Street, New York, New York 10017.

Purpose: To promote research in science and engineering. To advance the profession of engineering.

Activities: Research related to fatigue in fiber-reinforced composite materials; fundamental combustion study for the disposal of industrial and municipal wastes by incineration; updating transportation planning surveys; equipment selection and evaluation for vertical transportation systems in construction; development of a formal approach to full utilization of the engineer's capability in resolution of health problems; study of non-ferrous reaction rates and mechanisms with the hot filament microscope.

ENGINEERS' COUNCIL FOR PROFESSIONAL DEVELOPMENT, INC. - 345 East 47th Street, New York, New York 10017.

Areas of interest: Engineering education; engineering technician education; creditation by curricula through first professional degree for engineering; accreditation by curricula for 2-year and 4-year programs of engineering technology; guidance for student grades K through 12 and for engineering students; professional development of engineers; engineering ethics; continuing education for practicing engineers.

Holdings: Literature and reports on accreditation and guidance activities.

Publications: Pamphlets, reprints, books, annual report. A publication list is available.

Information services: Answers inquiries; permits onsite reference to nonconfidential information.

ENGINEERS JOINT COUNCIL - 345 East 47th Street, New York, New York 10017.

Areas of interest: Learning resources for engineers, scientists, and managers; engineering societies; engineering manpower supply and demand data; engineering salaries; selective service policies; standardized vocabulary reference for engineering and scientific terms.

Publications: ENGINEER (bimonthly magazine); LEARNING RESOURCES (directory issued 3 times a year); THESAURUS OF ENGINEERING AND SCIENTIFIC TERMS (1967); PROFESSIONAL INCOME OF ENGINEERS, SALARIES AND INCOME OF ENGINEERING TECHNICIANS, DEMAND FOR ENGINEERS AND TECHNICIANS (all biennial); other directories, reports, survey results, bulletins, handbooks, and guidebooks. A publication catalog is available.

Information services: Answers inquiries; provides consulting and reference services.

THE FATTY ACID PRODUCERS' COUNCIL - 485 Madison Avenue, New York, New York 10022.

Areas of interest: Fatty acids, including production and raw materials.

Publications: Abstracts.

Information services: Answers inquiries.

FEDERATION OF AMERICAN SCIENTISTS (FAS) - 203 C Street, N.E., Washington, D.C. 20002.

National and social scientists and engineers and others concerned with problems of science and society. "To act on public issues where the opinions of scientists are relevant, those which affect science or in which the experience or perspective of scientists is a needed guide." Functions through testimony to Congress and government agencies, public statements, articles, etc.

Publication: FAS NEWSLETTER (9/year).

FEDERATION OF SOCIETIES FOR PAINT TECHNOLOGY (FSPT) - 121 South Broad Street, Philadelphia, Pennsylvania 19107.

A society of chemists, chemical engineers, technologists, and supervisory production personnel in the decorative and protective coatings industry and allied industries.

Purpose: To develop and provide practical and technical facts, data, and standards fundamental to the manufacturing and use of paints, varnishes, lacquer, related protective coatings, and printing inks. Promotes research related to this field. Works to develop standards and methods of testing for raw materials and finished products.

Publications: (1) JOURNAL OF PAINT TECHNOLOGY (monthly); (2) Yearbook (annual directory).

Information activities: In addition to above, answers inquiries and makes referrals.

FLUID POWER SOCIETY - P.O. Box 905, Elm Grove, Wisconsin 53122.

Purpose: To further the art, science, and technology of fluid power.

Publications: HYDRAULICS AND PNEUMATICS (monthly); FLUID POWER SOCIETY DIRECTORY (annual); FPS NEWS (quarterly).

Information services: Answers inquiries, makes referrals.

INDUSTRIAL SAFETY EQUIPMENT ASSOCIATION, INC. - 60 East 42nd Street, Suite 1914, New York, New York 10017.

Areas of interest: Safety equipment and wearing apparel for protection from industrial hazards; personal protective equipment for head, eyes, respiratory system, body, and skin.

Information services: Answers inquiries regarding standards, product performance, or other broad industry matters.

INSTITUTE OF ENVIRONMENTAL SCIENCES - 940 East Northwest Highway, Mt. Prospect, Illinois 60056.

Areas of interest: Environmental science, environmental engineering, and related technologies; acoustics; corrosion; cryogenics; data acquisition; electromagnetics; environmental pollution; nuclear radiation; shock and vibration; solar radiation; ultra high temperatures; space vacuum, marine environments; instrument metrology.

Publications: JOURNAL OF ENVIRONMENTAL SCIENCES (bi-monthly); IES MEETINGS PROCEEDINGS (annual); special publications, newsletter, brochure. A publications list is available.

Information services: Answers inquiries; sponsors tutorial sessions, technical equipment exhibits, and symposia.

INSTRUMENT SOCIETY OF AMERICA - 400 Stanwix Street, Pittsburgh, Pennsylvania 15222.

Areas of interest: Application of all aspects of instrumentation to industrial, laboratory, biophysical, marine, and space environments; instrumentation for analysis, automatic control, data handling and computation, telemetry, metrology, physical and mechanical measurement, and information display.

Publications: INSTRUMENTATION TECHNOLOGY (monthly); ISA TRANSACTIONS (quarterly); ISA INSTRUMENTATION INDEX (quarterly); ISA TRANSDUCER COMPENDIUM; STANDARDS AND PRACTICES FOR INSTRUMENTATION; English translations of Russian instrumentation journals (monthly and bimonthly); proceedings, preprints, and abstracts of ISA sponsored and cosponsored symposia and conferences; standards and specifications; bibliographies, state-of-the-art reviews, educational aids, references, monographs.

Information service: Answers inquiries; makes referrals; permits onsite use of collection.

INTERNATIONAL ASSOCIATION ON WATER POLLUTION RESEARCH - c/o Mr. Robert Canham, Secretary, U.S. National Committee, IAWPR, 3900 Wisconsin Avenue, N.W., Washington, D.C. 20016.

Areas of interest: All aspects of water quality management; effects of pollution on aquatic and marine biota; municipal and industrial waste treatment; stream pollution; applied limnology; quantitative ecology and marine water quality, including waste dispersal systems; waste water reclamation and reuse.

Publications: WATER RESEARCH (monthly journal); PROCEEDINGS OF INTERNATIONAL CONFERENCES ON WATER POLLUTION RESEARCH.

Information services: Answers inquiries; makes referrals; sponsors international conferences on water pollution research and supports national symposia; provides consulting services to international agencies.

INTERSTATE SANITATION COMMISSION - 10 Columbus Circle, New York, New York 10019.

Areas of interest: Water pollution abatement and control in the New York Metropolitan Area of New Jersey, New York, and Connecticut and certain activities pertaining to air pollution problems among New York, New Jersey, and Connecticut, including coordination and planning activities; tracing sources of pollutants by water and air samplings; continuous monitoring of the Arthur Kill and East River waterways by means of electronic water quality analysis system; polluting effects of sulfur dioxide in the air; interstate air pollution complaints; coordination of alert system; effect of industrial pollution on fish life; planktonic water pollution; sampling treatment plants; water area surveys.

Publications: Technical and annual reports.

Information services: Answers inquiries; makes referrals; permits onsite reference.

JOINT INDUSTRIAL COUNCIL - 7901 Westpark Drive, McLean, Virginia 22101.

Membership includes producers and users of machine tools and industrial machinery or equipment.

Purpose: "To provide direction, coordination, and continuity of effort in the development, advancement, and referral to the American National Standards Institute of standards which will encourage the safe and reliable application of controls to machines and equipment used in industrial applications."

Information services: Answers inquiries on standardization; makes referrals.

MANUFACTURING CHEMISTS ASSOCIATION - 1825 Connecticut Avenue N.W., Washington, D.C. 20009.

Membership made up of manufacturers of basic chemicals who sell a substantial portion of their production to others.

Areas of interest: Air and water abatement; environmental control; solid wastes management; safe chemicals; chemical engineering; chemical industry information.

Activities: (1) Sponsoring research; (2) operation of the Chemical Transportation Emergency Center (CHEMTREC) for guidance to emergency services on handling transportation accidents involving chemicals; (3) promoting interest in chemistry among educators and students.

Publications: CHEMICAL STATISTICS HANDBOOK; manuals of recommended practice relating to safe transportation, packaging, handling, and storage of chemicals; chemical safety data sheets and guides; Guide to the Precautionary Labeling of Hazardous Chemicals; environmental health publications relating to water resources, air quality, and safety; teaching and guidance materials on chemistry.

Information services: In addition to those listed above, answers inquiries; makes referrals; permits onsite reference use of collection by researchers.

METRIC ASSOCIATION, INC. - c/o Mr. Louis F. Sokol, Secretary, 624 North Drury Lane, Arlington Heights, Illinois 60004.

Areas of interest: Promotion of the use of the metric system of measurement.

Publications: METRIC ASSOCIATION NEWSLETTER (quarterly); METRIC SUPPLEMENT TO MATHEMATICS (1967); BIBLIOGRAPHY OF THE METRIC SYSTEM (1968); METRIC UNITS OF MEASURE (1968).

Information services: Answers brief inquiries free; makes referrals; provides copies of reference materials in files for fee.

NATIONAL BOARD OF BOILER AND PRESSURE VESSEL INSPECTORS - 1155 North High Street, Columbus, Ohio 43201.

Areas of interest: Mechanical engineering and industrial engineering as applied to safety of boilers, unfired pressure vessels, and nuclear vessels.

Holdings: 5,000,000 manufacturers' data reports of boilers and unfired pressure vessels on microfilm, plus 2,500,000 manufacturers' data reports in original form; boiler and pressure vessel laws of states and Canadian provinces.

Publications: NATIONAL BOARD BULLETIN (quarterly); MANUFACTURERS DIRECTORY; PROCEEDINGS OF ANNUAL MEETING; NATIONAL BOARD INSPECTION CODE; RULES FOR BOILER BLOW-OFF EQUIPMENT; RELIEVING CAPACITIES OF SAFETY VALVES AND RELIEF VALVES for steam, air, and gas service.

Information services: Answers inquiries; makes referrals; provides copies of manufacturers' data reports for boilers and pressure vessels upon request.

NATIONAL CONFERENCE OF STANDARDS LABORATORIES (NCSL) - c/o National Bureau of Standards, Washington, D.C. 20234.

An independent organization sponsored by National Bureau of Standards. Membership includes representatives of measurements standards and calibration laboratories and organizations with related interests.

Purpose: To find solutions of problems, both technical and administrative, involved in measurement activities, in the physical sciences, engineering, and technology.

Activities: Conducts conferences, workshops, seminars, and meetings for presentation of papers and discussions pertaining to technical and managerial problems, operating practices, and policies for measurement standards laboratories. Compiles statistics on Calibration Intervals and Measurement Agreement Comparisons.

Committees: Calibration Procedures Distribution; Calibration Systems Management; Education and Training; Information; Long Range Planning; Measurement, Comparison and Statistical Procedures; Metric System Conversion; National Measurement Requirements; Recommended Practices; Specifications.

Publications: (1) NCSL NEWSLETTER (quarterly); (2) DIRECTORY OF STANDARDS LABORATORIES (biennial, with annual supplement); (3) CONFERENCE PROCEEDINGS.

Information services: Answers inquiries; makes referrals.

NATIONAL ENVIRONMENTAL HEALTH ASSOCIATION - 1600 Pennsylvania, Denver, Colorado 80203.

A professional society of persons engaged in environmental health for governmental public health agencies, public health education, or inspection services for private employers. Seeks to standardize methods of law enforcement and general public health practices among governmental agencies. Activities are carried on by a series of committees.

Publications: (1) JOURNAL OF ENVIRONMENTAL HEALTH (bimonthly); (2) BEACON NEWSLETTER (quarterly).

Information services: In addition to the above, answering inquiries; making referrals.

NATIONAL FIRE PROTECTION ASSOCIATION (NFPA) - 60 Batterymarch Street, Boston, Massachusetts 02110.

Membership includes representatives of business and industry, both individual and corporate public safety officials, including state and municipal fire authorities and building inspectors; fire insurance executives and engineers; colleges, hospitals, libraries, and others interested in the protection of life and property against loss by fire. Prime purpose is to serve as clearinghouse for information on fire.

Activities: (1) Through some 130 technical committees, develops and publishes advisory standards on virtually every aspect of fire protection and prevention; (2) provides field service of specialist engineers to promote electrical fire safety through wider application of National Electrical Code and to solve fire problems associated with storage, handling, and use of flammable liquids and gases; (3) sponsors Fire Prevention Week in October, Spring Clean-Up Campaign, and Sparky the Fire Dog; (4) conducts annual contest for fire prevention programs of municipalities, industrial plants, the military, and nonmilitary government establishments. Specialized information collection contains 1000 volumes on fire prevention, extinguishing, casualties, etc., and files of data analyzing more than 200,000 fires; (5) compiles annual statistics on fires and fire losses classified, large loss fires, and multiple death fires.

Publications: (1) FIRE NEWS (monthly); (2) FIRE COMMAND (monthly); (3) FIRE JOURNAL (bimonthly); (4) FIRE TECHNOLOGY (quarterly); (5) ADVANCE REPORTS (annual); (6) PROCEEDINGS (annual); (7) YEARBOOK and COMMITTEE LIST; (8) NATIONAL

FIRE CODES (annual); (9) NFPA FIRE PROTECTION HANDBOOK;
also publishes reference books; standards, laws, and ordinances;
educational pamphlets; fire records and reports; folders, posters,
and signs.

NATIONAL FLUID POWER ASSOCIATION - P.O. Box 49, Thiensville,
Wisconsin 53092; Location: 227 South Main Street, Thiensville, Wisconsin
53092.

Areas of interest: Fluid power and hydraulics; terminology; technical standards; accumulators; cylinders; filters; lubricators; regula-
tors; hydraulic valves, pumps, motors, and fluid conditioners;
pneumatic valves, pumps, and motors; sealing devices; reservoirs;
fluidics; fluids; sales statistics for the industry; education; fluid
power safety.

Publications: NATIONAL FLUID POWER ASSOCIATION DIRECTORY;
NFPA REPORTER (monthly); FLUID POWERGRAM; FLUID POWER
INDEX (monthly); bulletins, preprints, reprints, graphs, tables,
standards and specifications, pamphlets.

Information services: Answers inquiries; makes referrals; permits
onsite use of collection.

NATIONAL LIME ASSOCIATION - 4000 Brandywine Street N.W., Washington,
D.C. 20016.

Areas of interest: All aspects of lime technology, including manufacture, method of tests, specifications, and all uses in chemical,
structural, highway, and agricultural fields; lime-soil and lime-
silica reactions; water purification and softening with lime; chemical treatment of sewage and industrial wastes.

Holdings: Books, periodicals, abstracts, technical reports, graphs,
tables, specifications, reviews, newsletters, reprints, bibliographies,
publication lists, pamphlets, maps and charts, display materials,
films, photos, slides, press releases, clippings, data compilations,
samples of soils, limes, mortars, and slags.

Publications: CHEMICAL LIME FACTS (bulletin); WATER SUPPLY
AND TREATMENT (book); CHEMICAL TREATMENT OF SEWAGE
AND INDUSTRIAL WASTE (book); films.

Information services: Answers inquiries; makes referrals; provides
consulting services; provides analytical and testing services to
member companies (lime manufacturers) and certain government
agencies.

NATIONAL PAINT AND COATINGS ASSOCIATION (NPCA) - 1500 Rhode
Island Avenue N.W., Washington, D.C. 20005.

Membership includes manufacturers of paints and chemical coatings and suppliers
of raw materials and equipment used in the manufacture of paints and chemical
coatings.

Service program: Conducts statistical surveys, research, government and public relations programs, management information and

management and technician development programs through a series of technical committees.

Publications: (1) COATINGS (newsletter); (2) NPCA Member/ Services Directory; (3) Annual Report; also publishes "how-to" consumer pamphlets; Technical, Legislative and Legal, Statistical, Traffic and Safety, and Health and Community Services Bulletins; Trademark Directory; Guide to U.S. Government Paint Specifications; Raw Materials Indexes; Abstract Review; Scientific Circulars; Paint Industry Labeling Laws and Regulations; Compensation, Operating Cost, Sales, Bad Debt, and Credits and Collections Surveys; Credit Manual.

Information services: Include answering inquiries and making referrals.

NATIONAL PAINT, VARNISH, AND LACQUER ASSOCIATION - 1500 Rhode Island Avenue N.W., Washington, D.C. 20005.

Areas of interest: Paint, varnishes, lacquers, organic coatings, pigments, printing inks, resins, oils, solvents, plasticizers, and other products of the industry; wallpapering; decorating; color.

Publications: ABSTRACT REVIEW (monthly), containing abstracts from English, Russian, Italian, Spanish, French, and German sources arranged by subject; GUIDE TO U.S. GOVERNMENT PAINT SPECIFICATIONS; COATINGS (semimonthly newsletter); NPVLA TRADEMARK DIRECTORY (annual); RAW MATERIALS INDEXES; reports, pamphlets.

Information services: Provides reference, consulting, and duplication services; permits onsite reference.

NATIONAL SOCIETY OF PROFESSIONAL ENGINEERS - 2029 K Street N.W., Washington, D.C. 20006.

Areas of interest: Vocational guidance for engineering as a career; social, professional, ethical, and economic considerations of engineering as a profession.

Publications: PROFESSIONAL ENGINEER (monthly); ENGINEERS IN GOVERNMENT NEWSLETTER (monthly); PROFESSIONAL ENGINEER IN INDUSTRY NEWSLETTER (monthly); LEGISLATIVE BULLETIN (monthly); PRIVATE PRACTICE NEWS (monthly); ROSTER OF PROFESSIONAL ENGINEERS IN PRIVATE PRACTICE (indexed according to specialization); ENGINEER IN GOVERNMENT NEWSLETTER (quarterly); pamphlets, reports. A publications list is available.

Information services: Answers inquiries; makes referrals.

NATIONAL WATER INSTITUTE - 744 Broad Street, Room 3405, Newark, New Jersey 07102.

"The Institute is the public relations arm of the Water and Wastewater Equipment Manufacturers Association; its activities are conducted chiefly for the layman, the press, radio, and television."

Areas of interest: Equipment for water and wastewater fields.

Publications: Booklets, pamphlets.

Information services: Answers inquiries; makes referrals; distributes motion pictures, pamphlets, and booklets.

OFFICE OF INTERGOVERNMENTAL SCIENCE AND RESEARCH UTILIZATION PROGRAMS - National Science Foundation, Washington, D.C. 20550.

Sponsored by the National Science Foundation to help state and local governments develop improved programs and institutions for applying science and technology to governmental problems, and for implementing recommendations or utilizing information resulting from NSF programs. The office funds mainly pilot efforts and information exchange activities, giving preference to innovative approaches to develop models for governmental use of science and technology.

PAINT RESEARCH INSTITUTE - Dr. Raymond R. Myers, Research Director, Paint Research Institute, c/o Kent State University, Kent, Ohio 44240.

Areas of interest: Basic research on properties, application, manufacture, and use of protective and decorative coatings (paints, varnishes, lacquers, adhesives).

Information services: Answers inquiries; makes referrals.

SAFETY RESEARCH INFORMATION SERVICE - National Safety Council, 425 North Michigan Avenue, Chicago, Illinois 60611.

Areas of interest: Safety; accidents; injuries; hazards; errors; protective devices.

Holdings: About 3,000 abstracted research reports, of which 80 percent are related to traffic safety and 20 percent to industrial, home, public, school, and farm safety research; 100 periodical titles; 200 reference books.

Publications: Abstracts in the JOURNAL OF SAFETY RESEARCH (quarterly) and in NATIONAL SAFETY NEWS (monthly), both publications of the National Safety Council.

Information services: Answers inquiries; makes referrals; provides advisory technical reference and literature-searching services and limited duplication services. Arrangements for formal exchange of information may be made.

SOAP AND DETERGENT ASSOCIATION - 485 Madison Avenue, New York, New York 10022.

Areas of interest: All aspects of the soap and detergent industry, including problems of sewage treatment and water supply attributed to the presence of synthetic detergents; glycerine; fatty acids; household chemicals; cleaning agents; surface-active agents.

Publications: Division newsletters (available to members only); WATER IN THE NEWS (monthly, free to technical and semitechnical publics in the field of water); SCIENTIFIC AND TECHNICAL REPORTS (series on water-related subjects); the following motion pictures: "It's Your Decision--Clean Water" (cosponsored by the Association and the League of Women Voters of the United States) and "The Pursuit of Cleanliness." Loan prints of both films are available free on request.

Information services: Answers inquiries; makes referrals; provides limited literature-searching and bibliographic services; permits on-site use of collection.

SOCIETY OF AMERICAN VALUE ENGINEERS - 2550 Hargrove Road, L-205, Smyrna, Georgia 30080.

Purpose: To promote advancement of value engineering and value analysis and its application to the research, design, development, test, evaluation, engineering, production, purchasing and distribution phases in government, private industry and commerce.

Publications: SAVE COMMUNICATIONS AND JOURNAL OF VALUE ENGINEERING (monthly).

SOCIETY OF COSMETIC CHEMISTS - 50 East 41st Street, New York, New York 10017.

A professional society of cosmetic and perfume chemists. Programs are carried on through series of committees which include International Affairs, Library, Special Literature Award.

Publication: JOURNAL OF THE SCC (13/year).

Information services: Answers inquiries; makes referrals.

SOCIETY OF ENGINEERING SCIENCE (SES) - c/o Prof. E. Rodin, Department of Applied Mathematics, Washington University, St. Louis, Missouri 63130.

Individuals with at least a baccalaureate degree who are engaged in any aspect of engineering science or in other pursuits which contribute to the advancement of engineering science. Corporate members are business concerns and other organizations actively engaged in engineering and science.

Purpose: To foster and promote interchange of ideas and information among the various fields of engineering science and between engineering science and the fields of theoretical and applied physics, chemistry, and mathematics.

Publication: ENGINEERING SCIENCE BULLETIN (irregular).

Information services: Answers inquiries; makes referrals.

SPECIAL LIBRARIES ASSOCIATION (SLA) - 235 Park Avenue, South, New York, New York 10003.

An international organization of professional librarians and information experts who serve manufacturing concerns, banks, corporations, law firms, newspapers, advertising and insurance agencies, transportation companies, research organizations, museums, hospitals, business branches and other departments of public and university libraries, federal, state and municipal government bureaus, associations and other organizations in the fields of business, medicine, science, technology and the social sciences.

Purpose: To promote the collection, organization and dissemination of information in specialized fields and to improve the usefulness of special libraries and information services. Activities are carried on through committees.

Committees: Consultation Service; Copyright; International Relations; Standards; Publisher Relations; Recruitment; Scholarship. Divisions of association are: Advertising and Marketing; Aerospace; Biological Sciences; Business and Finance; Chemistry; Documentation; Engineering; Food; Geography and Map; Insurance; Metals/Materials; Military Librarians; Museums, Arts and Humanities; Natural Resources; Newspaper; Nuclear Science; Petroleum; Pharmaceutical; Picture; Publishing; Science-Technology; Social Science; Transportation.

Publications: (1) SPECIAL LIBRARIES (10/year); (2) TECHNICAL BOOK REVIEW INDEX (10/year); (3) SCIENTIFIC MEETINGS (4/year).

STANDARDS ENGINEERS SOCIETY (SES) - P.O. Box 7507, Philadelphia, Pennsylvania 19101.

A society of engineers, teachers, executives and scholars interested in practicing standardization.

Purpose: "To further standardization as a means of enhancing general welfare and to promote knowledge and use of approved standards issued by regularly constituted standardizing bodies. The development or issuance of standards is expressly excluded from the intent of the Society."

Publications: (1) STANDARDS ENGINEERING (bimonthly); (2) PROCEEDINGS (annual).

Information services: Answers inquiries; makes referrals.

SULPHUR INSTITUTE - 1725 K Street N.W., Washington, D.C. 20006.

Areas of interest: Sulphur; chemistry; engineering; agronomy; soil science; animal nutrition; fertilizer technology.

Holdings: 500 books, 2,200 journals; 500 reports; 2,400 patents; 13,000 published documents; films, photos, data.

Publications: SULPHUR INSTITUTE JOURNAL (quarterly); technical bulletins, brochures.

Information services: Provides reference services; permits onsite use of collections.

SYNTHETIC ORGANIC CHEMICAL MANUFACTURERS ASSOCIATION (SOCMA) - 1075 Central Park Avenue, Scarsdale, New York 10583.

Manufacturers of synthetic organic chemicals, including intermediates, dyes, flavor and perfume materials, medicinals, plasticizers, synthetic resins, and rubber processing chemicals. Promotes interest and research in the field of synthetic organic chemicals through a series of committees.

Publications: (1) NEWSLETTER; (2) SOCMA STATISTICS (monthly).

Information services: Answers inquiries; makes referrals.

WATER AND WASTEWATER EQUIPMENT MANUFACTURERS ASSOCIATION, INC. - 744 Broad Street, Newark, New Jersey 07102.

Areas of interest: Water and waste water treatment and related equipment.

Information services: Answers inquiries about equipment and makes referrals to member companies and other organizations; sponsors educational exhibits of water and waste water equipment at conferences of American Water Works Association and Water Pollution Control Federation.

WATER CONDITIONING ASSOCIATION INTERNATIONAL - P.O. Box 651, Wheaton, Illinois 60187; Location: 325 West Wesley Street, Wheaton, Illinois, Tel: (312) 668-8892.

An association composed of retail dealers in water conditioning equipment.

Areas of interest: Water quality; water conditioning equipment.

Publications: Proceedings.

Information services: Answers inquiries; makes referrals; provides reference and consulting services; permits onsite reference.

WATER CONDITIONING FOUNDATION - 1780 Maple Street, Northfield, Illinois 60093.

Areas of interest: Domestic, commercial, and industrial water-conditioning equipment and materials; purification chemicals; water softeners; reverse osmosis deionizers; demineralizers; filters; feeders; mechanical equipment for conversion of water at the point of use.

Publications: WCFI training course on Water Treatment Fundamentals (available at nominal cost); technical, educational, and promotional bulletins and pamphlets.

Information services: Answers inquiries; makes referrals; provides consulting services, water laboratory analyses, and testing of water treatment equipment for fee.

WATER CONDITIONING RESEARCH COUNCIL - P.O. Box 651, Wheaton, Illinois 60187; Location: 325 West Wesley Street, Wheaton, Illinois.

This Research Council is cosponsored by the Water Conditioning Association International and the Association of Water Conditioning Equipment Manufacturers.

Areas of interest: Water resources; water quality; water conditioning.

Information services: Answers inquiries; makes referrals.

WATER CONDITIONING FOUNDATION (WCF) - 1780 Maple Street, North-field, Illinois 60093.

Manufacturers of water conditioning equipment. Associate members are suppliers of water conditioning equipment.

>   Publication: WATER CONDITIONING REPORTER (monthly).

>   Information services: Answers inquiries; makes referrals.

WATER CONVERSION INSTITUTE - 1435 G Street N.W., Washington, D.C. 20005.

>   Areas of interest: Conversion of salt and brackish waters into potable water; ground water; municipal and industrial water treat-ment; water economics.

>   Information services: Answers technical inquiries; makes referrals; provides consulting services.

WATER POLLUTION CONTROL FEDERATION - 3900 Wisconsin Avenue N.W., Washington, D.C. 20016.

>   Areas of interest: All aspects of waste water handling and control of water pollution.

>   Publications: JOURNAL: WATER POLLUTION CONTROL FEDER-ATION (monthly, with quarterly research supplement); HIGHLIGHTS (monthly newsletter); DEEDS AND DATA (monthly educational news-letter for operating personnel); manuals of technical practice in water pollution control, including safety, sewer use ordinances, chlorination, aeration, units of mathematical expression, sewer maintenance, sewer design, waste water treatment plant design and operation, accounting procedures, laboratory procedures, an-aerobic sludge digestion, paints and protective coatings for waste water treatment facilities, and sludge dewatering; waste water treatment plant operator training courses and visual aids; public relations handbook; safety instructional materials.

>   Information services: Answers inquiries; makes referrals; provides reference services; permits onsite use of literature collection by appointment.

WATER QUALITY RESEARCH COUNCIL (WQRC) - 330 South Naperville Street, Suite 403, Wheaton, Illinois 60187.

Research arm of the Water Conditioning Association International and the Water Conditioning Foundation. Publishes materials for schools on water conditioning and project ideas related to water.

WATER RESOURCES COUNCIL - 1025 Vermont Avenue N.W., Washington, D.C. 20005.

The Council maintains a continuing study of U.S. water resources and approves grants to states for comprehensive water and related land resources planning. Its members are the Secretaries of Interior, Agriculture, Army, Health, Educa-tion, and Welfare, and Transportation, and the Chairman of the Federal Power

Commission; associate members are the Secretaries of Commerce and Housing and Urban Development; observers are the Attorney General and the Director of the Bureau of the Budget.

Areas of interest: Water resources (federal planning and action programs); standards and procedures for water resources planning and project evaluation; river basin planning.

Publications: PROCEEDINGS OF NATIONAL CONFERENCE OF STATE AND FEDERAL WATER OFFICIALS, SALT LAKE CITY, UTAH (1969); PROCEDURES FOR EVALUATION OF WATER AND RELATED LAND RESOURCES PROJECTS (1969); PROPOSED FLOOD HAZARD EVALUATION GUIDELINES FOR FEDERAL EXECUTIVE AGENCIES (1969).

Information services: Answers inquiries regarding federal water resources policy and planning; makes referrals.

c. DIRECTORIES

Libraries maintain a variety of directories of associations and organizations. The librarian can also suggest guidebooks which list agencies and information sources in different subject fields. The following is a selective list of directories:

NATIONAL TRADE AND PROFESSIONAL ASSOCIATIONS OF THE UNITED STATES (annual). Columbia Books, Inc., Suite 300, 917 15th Street N.W., Washington, D.C. 20005.

Provides title and address, phone number, names of executive secretary and other principal officers, annual budget, number of members, titles of regular publications, and convention dates for current and next year. Key word index facilitates locating listings in directory and also a roster of executives is given.

DIRECTORY OF ENGINEERING SOCIETIES AND RELATED ORGANIZATIONS. 1968. Engineers Joint Council, 345 East 47th Street, New York, New York 10017.

Contains information about 300 or more national, state, and local organizations which have engineering or engineering related activities. Details given are name, address and phone number of organization principal officers, membership details and qualification, objectives and brief general data on services or programs.

Federal Council for Science and Technology. Committee on Domestic Technology Transfer. FEDERAL TECHNOLOGY TRANSFER DIRECTORY OF PROGRAMS, RESOURCES, CONTACT POINTS. Washington, D.C.: Government Printing Office, 1975.

ENCYCLOPEDIA OF ASSOCIATIONS. 3 vols. 12th ed. Edited by Mary Wilson Pair. Detroit: Gale Research Company, 1978.

Extensive listing of non-profit national and special interest organizations in the United States not limited to scientific or engineering associations, but covering a very diverse group of programs and associations.

SCIENTIFIC, TECHNICAL AND RELATED SOCIETIES OF THE U.S. 9th ed. 1971. National Academy of Sciences, 2101 Constitution Avenue N.W., Washington, D.C. 20418.

3. Research Centers

Basic research programs are fundamental to the continuing unfoldment of new scientific and technological development. This experimental and exploratory effort is carried on in a variety of settings on academic campuses, industrial laboratories at military installations and in independent research institutes.

a. RESEARCH INSTITUTES AND INDUSTRIAL LABORATORIES

These agencies perform studies and experimentation on scientific and technical matters on a contract basis. Individual staff and the agencies themselves offer assistance on specific inquiries or are able to act in a referral role for requests which are outside the scope of their programmed activities. Their libraries, which are especially developed resource collections in a discipline, are also valuable information sources. They can be contacted directly, or through the nearest library.

A SELECTIVE LIST OF RESEARCH INSTITUTES AND INDUSTRIAL LABORATORIES

BATELLE MEMORIAL INSTITUTE - 505 King Avenue, Columbus, Ohio 43201.

Offers research assistance and specialized information in many disciplines through its divisions and centers. Maintains an extensive library. Publishes: BATTELLE RESEARCH OUTLOOK, a quarterly.

Chemical Engineering related divisions:

1. Ecological Information and Analysis Center - Batelle Memorial Institute, Columbus Laboratories, 505 King Avenue, Columbus, Ohio 43201.

Areas of interest: Systems ecology; food-chain studies; bioenergetics; mathematical modeling of ecosystems; thermal affluent effects; population dynamics; radionuclide cycling in the environment.

Information services: Provides technical advisory services and information on environmental problems on an individual project basis. Information files are accessible for onsite use.

2. Measurement Laboratory Information Service - Batelle Memorial Institute

Columbus Laboratories, 505 King Avenue Columbus, Ohio 43201.

Areas of interest: Physical measurements; instrumentation; measurement standards related to the National Bureau of Standards.

Holdings: 210 books; 420 reports; 5,000 manufacturers' literature; 350 standards and specifications; periodicals.

Information services: Answers inquiries; makes referrals; provides consulting, reference, and bibliographic services for fee; permits onsite reference.

CARNEGIE-MELLON UNIVERSITY

1. API Standard Reference Materials - Schenley Park, Pittsburgh, Pennsylvania 15213.

Provides a fee basis standard reference material of hydrocardons, sulfur and nitrogen compounds of very high purity.

2. Mellon Institute Library - 4400 Fifth Avenue, Pittsburgh, Pennsylvania 15213.

Areas of interest: Chemistry, metallurgy, industrial hygiene, air and water pollution.

THE FRANKLIN INSTITUTE RESEARCH LABORATORIES

Science Information Services - 20th and Race Streets, Philadelphia, Pennsylvania 19103.

Areas of interest: Materials science and engineering, including metallurgy, solid state physics, and ceramics; mechanical and nuclear engineering, including applied mechanics, heat and fluid mechanics, and friction and lubrication; chemistry, including colloids and polymers, bio-instrumentation; systems science, including operations research, environmental resources, transportation and safety, science information services, including bibliographic activities, technical writing, technical data compilations, R&D in scientific documentation, state-of-the-art surveys, technical translations, review writing, handbook writing, indexing, and abstracting.

Publications: THE JOURNAL OF THE FRANKLIN INSTITUTE (monthly); OZONE CHEMISTRY AND TECHNOLOGY (semimonthly); various one-shot literature surveys on such subjects as transportation, gas lubricated bearings, heat transfer, braking systems, and CODEN (ASTM/FI).

Information services: The following services are provided on a contract basis or for a fee, with a minimum charge of one day's effort: retrospective bibliographies, selective dissemination on a current awareness basis, state-of-the-art surveys, development of indexing systems, design of information centers, development of thesauri, and technical translations.

MIDWEST RESEARCH INSTITUTE - 425 Volker Boulevard, Kansas City, Missouri 64110.

Areas of interest: Chemical engineering; combustion research; energy conversion research; chemistry, including inorganic, organic, physical, polymer, solid state, surface, industrial, and biochemistry; advanced technology; market research; water and air pollution; applied research on materials, including metal physics, yield and fracture of materials, fatigue of metals, composite materials, stress corrosion cracking, dispersion-hardened alloy systems, metallography, fractography, aerospace science and technology; reliability; biomedical engineering; research on analytical methods for pesticides, in formulations and as residues in all types of samples, including foodstuffs, soils, air, water, vegetation, and tissues.

Publications: Books, reports, bibliographies, periodicals, abstracts, indexes, directories.

Information services: Answers brief inquiries; makes referrals; provides consulting, reference, and literature-searching services; permits onsite reference; makes interlibrary loans.

The Institute has extensive laboratory facilities.

NORTH STAR RESEARCH AND DEVELOPMENT INSTITUTE - 3100 38th Avenue South, Minneapolis, Minnesota 55406.

A research institute operating on an independent, non-profit basis. Areas of concern are materials, science, pollution control, chemical engineering, food technology, marketing of new systems and products, and other subjects and disciplines. Research and investigation done under contract.

PAINT RESEARCH INSTITUTE - Chemistry Department, Kent State University, Kent, Ohio 44240.

Separately incorporated nonprofit research organization affiliated with Federation of Societies for Paint Technology.

Principal field of research: Chemistry, particularly colloid, surface and polymer chemistry. Sponsors research programs through grants to universities in the United States, Canada and Great Britain.

Research results published in professional journals.

SOUTHERN RESEARCH INSTITUTE - 2000 Ninth Avenue South, Birmingham, Alabama 35205.

Areas of interest: Chemistry, including organic, physical, analytical, and radiochemistry; chemical, mechanical, and electrical engineering; physics, specifically solid state devices; metallurgy; biological sciences; pharmacology; foods; flavors; coal and coal products; plastics; polymers; fibers and textiles; water desalination; cement; paint; inks; fuels; combustion; wood; paper; fertilizers; air pollution; instrumentation; infrared technology; ceramics; corrosion; cancer and virus chemotherapy; toxicology; waste treatment; industrial hygiene.

Publications: Scientific articles, quarterly bulletin, annual report.

Information services: Answers inquiries; makes interlibrary loans; provides reference services; performs contract research; permits on-site use of collection.

Facilities include animal laboratories, analytical chemistry laboratories, machine shop, polygraphs, spectrophotometers, radioactive counting equipment, electron microscope, computer services, microcard reader, microfiche reader, 36mm film reader, and copier.

SOUTHWEST RESEARCH INSTITUTE LIBRARY - P.O. Drawer 28510, San Antonio, Texas 78228; Location: 8500 Culebra Road, San Antonio, Texas.

Areas of interest: Primary research fields include chemistry and chemical engineering, biochemistry, materials engineering, and bioengineering. In addition, the Library has special facilities for research in air and water pollution and engine exhaust emissions, analytical chemistry, artificial organs, corrosion, metal fatigue, nondestructive testing, ecology, electric power, explosive and impact effects, gas technology, fire research, flavors, fluid mechanics and hydrodynamics, fuels, lubricants, lubrication, gas dynamics, highway engineering, oceanography, metallurgy, mineralogy, nuclear reactors, polymers, radiochemistry, safety engineering, stress analysis, telemetry, testing, transducers, waste disposal, water resources, water treatment, and welding.

Holdings: 22,000 books; 25,000 volumes of journals, with current subscriptions to 1,000 journals in English, French, Dutch, Spanish, German, Russian, Polish, Hungarian, Italian, the Scandinavian languages, Czech, Bulgarian, Yugoslavian, Arabic, Hindi, Japanese, and Turkish; 60,000 reports and documents.

Publications: Accession lists. Other divisions of the Institute issue TOMORROW THROUGH RESEARCH (bimonthly), technical reports, state-of-the-art reviews, standards and specifications, data handbooks, bibliographies, proceedings, and translations.

Information services: Answers inquiries; makes referrals; makes interlibrary loans; permits onsite use of collection. Xerographic and microfilm copying is provided for a fee.

STANFORD RESEARCH INSTITUTE - 333 Ravenswood Avenue, Menlo Park, California 94025.

An independent research institute which has as its principal areas of interest the life and physical sciences, engineering, materials, technology and space related investigation. Performs research on contact; maintains an extensive library collection in the areas of research interest.

b. EDUCATIONAL INSTITUTIONS

Educational institutions are major centers of research activity and along with the independent research institutes continually feed into information networks new data and knowledge on a wide range of science-related subjects. This material reaches the engineering community on a formal basis through published articles in scientific and technical journals, books or by way of conferences.

Informal channels are a much faster means of obtaining recent research information, but finding the most productive sources can be difficult. Engineering experiment stations, laboratories, and research centers are often the places from which specialized research papers emanate. Abstracting services such as ENGINEERING INDEX pick up some of these individual items. Certain of these agencies publish annually a list of research projects and papers and these can be obtained by contacting the engineering and science libraries or the public relations office of various institutions.

DIRECTORIES AND GUIDES OF RESEARCH PROGRAMS AND OF INDIVIDUALS KNOWLEDGEABLE IN SPECIFIC FIELDS OF INQUIRY

AMERICAN MEN AND WOMEN OF SCIENCE: A BIOGRAPHICAL DIRECTORY. R.R. Bowker Co., 1180 Avenue of the Americas, New York, New York 10036.

American Chemical Society. DIRECTORY OF GRADUATE RESEARCH. Biennial. 1155 16th Street N.W., Washington, D.C. 20036.

AMERICAN UNIVERSITIES AND COLLEGES. American Council on Education, One Dupont Circle, Washington, D.C. 20036.

ANNUAL DIRECTORY OF ENGINEERING COLLEGE AND GRADUATE STUDY. American Society for Engineering Education, Suite 400, One Dupont Circle, Washington, D.C. 20036.

ANNUAL GUIDES TO GRADUATE STUDY BOOK VII. ENGINEERING AND APPLIED SCIENCES. Peterson's Guides, Inc., Alexander Street, Princeton, New Jersey 08540.

DICTIONARY OF SCIENTIFIC BIBLIOGRAPHY. Charles Scribner, 597 Fifth Avenue, New York, New York 10017.

DIRECTORY OF INFORMATION RESOURCES IN THE UNITED STATES PHYSICAL SCIENCES AND ENGINEERING (1971). U.S. Superintendent of Documents, Government Printing Office, Washington, D.C. 20402.

Engineers Joint Council. ENGINEERS OF DISTINCTION INCLUDING SCIENTISTS IN RELATED FIELDS. 345 East 47th Street, New York, New York 10017.

GUIDES TO RESEARCH PROGRAMS. Annual. American Institute of Chemical Engineers, Chemical Engineering Faculties, 345 East 47th Street, New York, New York 10017.

INDUSTRIAL RESEARCH LABORATORIES OF THE UNITED STATES. 13th ed. R.R. Bowker, New York, New York 10036.

INTERNATIONAL CHEMISTRY DIRECTORY (1969/70). W. A. Benjamin, Inc., 2 Park Avenue, New York, New York 10016.

LEADERS IN AMERICAN SCIENCE. Who's Who in American Education, Nashville, Tennessee.

McGRAW-HILL MODERN MEN OF SCIENCE. McGraw-Hill Book Co., 330 West 42nd Street, New York, New York 10036.

NATIONAL FACULTY DIRECTORY, 1977. Gale Research Co., Book Tower, Detroit, Michigan 48226.

RESEARCH CENTERS DIRECTORY. 5th ed. (1975). Gale Research Co., Book Tower, Detroit, Michigan 48226.

> (New Research Centers -- a looseleaf updating service of the Gale Research Company which describes newly established centers.) See also ENCYCLOPEDIA OF ASSOCIATIONS.

SCIENCE CITATION INDEX. Institute for Scientific Information, 325 Chestnut Street, Philadelphia, Pennsylvania 19106.

> A guide to cited articles which are frequently referred to in footnotes and bibliographies and a tool to learn of authors who publish in science and technology frequently.

WHO IS PUBLISHING IN SCIENCE (1971). Institute for Scientific Information, 325 Chestnut Street, Philadelphia, Pennsylvania 19106.

> Another type of source listing of individuals writing in the scientific disciplines. The content is arranged by author, organization and geographically, and is useful for obtaining addresses of authors in order to get reprints and to find locations of specific organizations.

WORLD WHO'S WHO IN SCIENCE. A. N. Marquis Co., 200 East Ohio Street, Chicago, Illinois 60611.

4. International Organizations

Science and technology have taken on a global significance since World War II. Scientists and engineers the world over are able to reach one another quickly through improved communication systems and the flow and exchange of specialized knowledge is increasing between countries. There is a need for greater encouragement of these activities as the world's distances shrink before advances in communications and transportation. The oceans and space that separate countries have now become the common research ground linking together scientists and technologists in joint efforts to find solutions to major problems such as hunger, diminishing resources, and duplication of research effort.

International organizations such as those listed here perform this essential role of promoting understanding and research on common problems among many countries and act in the valuable support function of clearinghouses for new research details data and information.

UNISIST

A significant development for improvement of scientific information exchange has been the three-year cooperative effort between UNESCO and the International Council of Scientific Unions in studying the feasibility of a World Science Information System called UNISIST. A review of the program and specific recommendations for implementation of UNISIST appear in INFORMATION, May-June 1971, pages 113-134.

COMMITTEE ON DATA FOR SCIENCE AND TECHNOLOGY (CODATA) - Westendstrasse 19, 6 Frankfurt/Main, Germany FR.T.74.80.44.

Purpose: The promotion and furtherance of knowledge of distribution of selected numerical values of properties of substances of importance and interest to science and industry on a worldwide basis, and to compile and promulgate widely numerical and data listings and standards.

Publications: CO-DATA BULLETIN; CODATA NEWSLETTER. INT COMPENDIUM OF NUMERICAL DATA PROJECTS (1969).

COMMONWEALTH ENGINEERING CONFERENCE - c/o Council of Engineering Institutions, 2 Little Smith Street, London SW1P 3DL, England.

Purpose: To bring about cooperation with other international organizations in support of the engineering field and to encourage engineering education and training in the Commonwealth.

EUROPEAN COMMITTEE OF AIR HANDLING AND AIR CONDITIONING EQUIPMENT MANUFACTURERS - 10 avenue Hoche, 75-Paris 8E, France.

Purpose: To foster cooperation and information exchange among air handling equipment and manufacturers in European countries in order to unify standards and rules for testing materials.

EUROPEAN FEDERATION OF CHEMICAL ENGINEERING - c/o DECHEMA, Theodor Heuss-Allee 25, 6 Frankfurt/Main, Germany FR.T.77.04.81.

Purpose: Promote cooperation in Europe on advancement of chemical engineering, and to further development of new processes in manufacturing in chemical industries.

EUROPEAN LEAD DEVELOPMENT COMMITTEE - Lead Development Association, 34 Berkeley Square, London W1X 6AJ, England.

Purpose: Stimulate exchange of information on development and use of lead throughout Europe.

Publications: Conference proceedings.

EUROPEAN OIL HYDRAULIC AND PNEUMATIC COMMITTEE - VDMA, Postfach 71 109, 6000 Frankfurt/Main, Germany FR.T.

Purpose: To investigate and review the problems of the oil hydraulic and pneumatic industries.

Activities: Makes recommendations and publishes results of studies.

EUROPEAN ORGANIZATION FOR QUALITY CONTROL - Weena 734, Rotterdam, Netherlands.

Purpose: To foster the application of quality control in industry in order to bring about improved quality and reliability of goods and services, reduce costs and increase productivity. This is done by means of committees, meetings, and publications.

Publications: QUALITY (quarterly); NEWSLETTER (monthly). Conference proceedings. GLOSSARY OF TERMS USED IN QUALITY CONTROL in fifteen languages; publications on sampling procedures and others.

INTER-AMERICAN ASSOCIATION OF SANITARY ENGINEERING - 2526 Trophy Lane, Reston, Virginia 22070.

Made up of sanitary engineers in government, private business, and educational institutions throughout the Western Hemisphere. Fosters study of sanitary problems in the Americas and elsewhere; establishes uniform standards for permanent protection of health of all the inhabitants of the Western Hemisphere.

Publication: SANITARIA (quarterly).

INTERNATIONAL ASSOCIATION OF THE SOAP AND DETERGENT INDUSTRY - 49 square Marie-Louise, 1040 Brussels, Belgium.

Purpose: Promote the interests of industries concerned with soap-making, detergents and allied products.

INTERNATIONAL ASSOCIATION ON WATER POLLUTION RESEARCH - P.O. Box 395, Pretoria, South Africa.

Purpose: To promote communication, exchange of information on water pollution research and water quality management. To sponsor regular international meetings and conferences for dissemination of important research on pollution, and to provide a source for publication of research reports and activities in this field.

Publications: WATER RESEARCH (monthly); conference proceedings.

INTERNATIONAL BUREAU OF WEIGHTS AND MEASURES - Pavillon de Breteuil, 92-Sevres, France, T. Paris 027.00-51.

Purpose: Establishment of international standards and of measurement scales of physical quantities; verification of standards and measurement scales, and to maintain precision determination of fundamental physical constants, which is accomplished by means of seven consultative committees (electricity, ionising radiations, metre, photometry, second, thermometry, units).

Publications: Comptes Rendus des Conferences Generales des Poids et Mesures, Proces-verbaux des seances du Comite International des Poids et Mesures, Sessions des Comites consultatifs, Recueil de Travaux du Bureau International des Poids et Mesures.

INTERNATIONAL FEDERATION OF CONSULTING ENGINEERS - Javastraat 44, The Hague T. (070), Netherlands.

Purpose: To encourage relations between the Associations of Consulting Engineers, promote study of their professional interests, and encourage the setting up of standards and rules of professional conduct in conformity with the responsibility and dignity of the profession. Acts as an information source for this field on an international scale.

INTERNATIONAL FEDERATION OF PATENT AGENTS - 25 rue de Leningrad, 75-Paris 8e, France.

Purpose: To stimulate relations among patent agents in different countries, facilitate professional information exchange, collect and disseminate useful professional information, and carry out research on administrative and legislative reforms for improvement of international patent system. Publishes a bulletin of its activities.

INTERNATIONAL INFORMATION CENTRE FOR TERMINOLOGY - INFOTERM, Austrian Standards Institute, Osterreichischer Normungsinstitut, Leopoldgasse 4, A-1020, Wien, Austria.

This center has been established to act as a clearinghouse for scientific and technical terminology as used in all parts of the world. Center personnel have the responsibility of collecting terminological publications, gathering details on special collections and libraries with this type of material, and announcing the findings to the world at large. The center will work in close cooperation with UNISIST.

INTERNATIONAL INSTITUTE OF REFRIGERATION (IIR) - 177 boulevard Malesherbes, 75-Paris 7e, France.

Purpose: Further the development of scientific research in the field of refrigeration; collect scientific, technical and economic information in this area; publish such studies and documents which may be important; further the development of the uses of refrigeration in food and agricultural industries as well as in the area of health and hygiene.

Publications: TECHNICAL BULLETIN (6 a year); REFRIGERATION SCIENCE AND TECHNOLOGY; proceedings of meetings, congresses, and symposia.

INTERNATIONAL MEASUREMENT CONFEDERATION (MEKO) - POB 457, Budapest 5.T., Hungary.

Purpose: Encourage the international exchange of scientific and technical information relating to measuring techniques, instrument design and manufacture as well as the application of instrumentation in scientific research and in industry.

Publications: ACTA IMEKO (Congress proceedings); proceedings of each symposium.

INTERNATIONAL ORGANIZATION FOR STANDARDIZATION (ISO) - 1 rue de Varembe, 1211 Geneva 20.T.34.12.40., Switzerland.

Purpose: Promote development of standards in the world in order to bring about uniformity in measurement and standardization to facilitate international exchange of goods and services.

Publications: ISO BULLETIN (monthly); ISO MEMENTO (annual); ISO Catalogue (annual); ISO Annual REVIEW.

Members: National standard bodies in 59 countries.

INTERNATIONAL PATENT INSTITUTE - P. van Wassbergen, Technical Director, 97 Nieuwe Parklaan, The Hague. T.51.22.31., Netherlands.

Purpose: To provide technical reports on inventions and to further patent and invention systems by undertaking documentary research in all fields of science and technology as concerns the patenting process.

Publications: IIB DOCUMENTARY SEARCH CENTER (for USA); INSTITUT INTERNATIONAL DES BREVETS, in English, French, German.

Members: Governments of eight countries.

INTERNATIONAL UNION OF INDEPENDENT LABORATORIES - Ashbourne House, Alberon Gardens, London NW 11, England.

Purpose: To promote information exchange and to represent the role of independent laboratories in the world of science as well as in the economy.

Publication: UILI DIRECTORY.

INTERNATIONAL UNION OF LEATHER CHEMISTS SOCIETIES (IULCS) - S. Wolstenholme, Hon. Sec., The Procter, Department of Food and Leather Science, University of Leeds, Leeds LS2 9JT, U.K. T. 31751.

Purpose: Promote international collaboration in the field of leather chemistry through meetings and conferences.

INTERNATIONAL UNION OF PURE AND APPLIED CHEMISTRY (IUPAC) - Dr. M. Williams, Exec. Sec., Bank Court Chambers, 2-3 Pound Way, Cowley Centre, Oxford OX4 3YF, U.K.

Purpose: To foster co-operation among the chemists of the member countries; study subjects of international scope in pure and applied chemistry which need regulation, standardisation or codification, and to co-operate with other international organisations which deal with topics of a chemical nature.

Publications: INFORMATION BULLETIN, JOURNAL OF PURE AND APPLIED CHEMISTRY (monthly).

OIL AND COLOUR CHEMISTS' ASSOCIATION (GREAT BRITAIN AND COMMONWEALTH) (OCCA) - R. H. Hamblin, Director, Wax Chandlers' Hall, Gresham Street, London EC2. T. 606-1439, England.

Purpose: To further the development of industries concerned with the manufacture of paint, printing inks, pigments, varnishes, drying oils, resins, lacquers, soaps, linoleum, etc., and plant equipment and materials used in their production.

Publications: JOURNAL OF THE OIL AND COLOUR CHEMISTS' ASSOCIATION (monthly); PAINT TECHNOLOGY MANUALS (7 vols).

ORGANIZATION FOR ECONOMIC CO-OPERATION AND DEVELOPMENT (OECD) - 2 rue Andre-Pascal, Paris 16e, France.

Purpose: "Promote policies designed to: achieve the highest sustainable economic growth and employment and a rising standard of living in Member countries, while maintaining financial stability, and thus to contribute to the development of the world economy; contribute to sound economic expansion in Member as well as non-Member countries in the process of economic development; contribute to expansion of world trade on a multilateral, non-discriminatory basis in accordance with international obligations."

Activities of this organization touch upon various areas of common interest to chemical industries including agriculture, nuclear power, energy and environment, science policies and international trade and commerce.

Publications: Annual and monthly reports; OECD OBSERVER (bimonthly); ACTIVITIES (monthly); information, technical and statistical bulletins; numerous specialized books and reports on all aspects of the economy.

Many of these publications would be of general interest to chemical engineers.

ADDITIONAL INTERNATIONAL ORGANIZATIONS

COMMITTEE FOR INTERNATIONAL COOPERATION IN INFORMATION RETRIEVAL AMONG PATENT OFFICES - c/o BIRRI, 32 Chemis des Colombettes, 1211 Geneva 20 Switzerland.

COMMITTEE ON SCIENCE AND TECHNOLOGY IN DEVELOPING COUNTRIES - ICSU Secretariat, 7 Via Cornello Celso, 00161 Rome, Italy.

INTERNATIONAL ASSOCIATION OF SCIENTIFIC HYDROLOGY - 61 Braemstraat, 9001 Gentbrugge, Belgium.

INTERNATIONAL FEDERATION OF DOCUMENTATION - 7 Hofweg, The Hague, Netherlands.

INTERNATIONAL ASSOCIATION FOR THE PHYSICAL SCIENCES OF THE OCEAN - c/o Naval Undersea Research and Development Center, San Diego, California 92132.

INTERNATIONAL COMMMITTEE OF ELECTRO-CHEMICAL THERMODYNAMICS AND KINETICS - Institute Battelle, 7 Route de Drize, 1227 Carauge-Geneva, Switzerland.

INTERNATIONAL CONFERENCE ON THE PROPERTIES OF STEAM - American Society of Mechanical Engineers, 345 East 47th Street, New York, New York 10017.

INTERNATIONAL FEDERATION FOR INFORMATION PROCESSING - 32 rue de l'Athenee, 1206 Geneva, Switzerland.

INTERNATIONAL INSTITUTE OF PHYSICS AND CHEMISTRY - Universite Libre de Bruxelles, 50 Ar Franklin Roosevelt, 1050 Brussels, Belgium.

INTERNATIONAL OCEANOGRAPHIC FOUNDATION - 10 Rickenbacker Causeway, Virginia Key, Florida 33149.

INTERNATIONAL ORGANIZATION FOR THE ADVANCEMENT OF HIGH PRESSURE RESEARCH - Maison des Industries Chimiques, 49 Square Marie-Louise, 1040 Brussels, Belgium.

INTERNATIONAL UNION OF PURE AND APPLIED PHYSICS - The Nuffield Foundation, Nuffield Lodge, Regents Park, London NW1, England 4RS.

INTERNATIONAL UNION OF THEORETICAL AND APPLIED MECHANICS - Frithiof Niordson, Rigensgade 13, 1316 Copenhagen, Denmark.

ORGANIZATION FOR ECONOMIC COOPERATION AND DEVELOPMENT - 2 rue Andre-Pascal, 75 Paris 16e, France, (has interest in scientific and technical information).

SCIENTIFIC COMMITTEE ON OCEANIC RESEARCH - Institute fur Meerskunde, Warnemunde 253, Germany, Federal Republic.

UNITED NATIONS EDUCATION SCIENTIFIC AND CULTURAL ORGANIZATION (UNESCO) - New York Office, Director, United Nations, New York, New York 10017.

WORLD ACADEMY OF ART AND SCIENCE - American Division, 630 Fifth Avenue, Suite 627, New York, New York 10020.

Further details concerning these organizations may be found in the directories cited:

EUROPEAN RESEARCH INDEX. Francis Hodgson Limited, 1969, 2 vol., P.O. Box 74, Guernsey, Channel Islands, England.

INTERNATIONAL SCIENTIFIC ORGANIZATIONS. OECD Publications Office, 1965, 2 rue Andre-Pascal, 75 Paris 16e, France.

INTERNATIONAL COMPENDIUM OF NUMERICAL DATA PROJECTS: A SURVEY AND ANALYSIS. International Council of Scientific Unions, Committee on Data and Technology, 7 Via Cornello Celso, 00161 Rome, Italy.

U.S. LIBRARY OF CONGRESS. INTERNATIONAL ORGANIZATION SECTION. A guide to their library, documentation, and information services. Government Printing Office, Washington, D.C. 1962.

YEARBOOK OF INTERNATIONAL ORGANIZATIONS. Union of International Associations, rue aux Laines 1, 1000 Brussels, Belgium.

WORLD GUIDE TO TECHNICAL INFORMATION & DOCUMENTATION SERVICES. UNESCO, Unipub, Inc., 650 - 1st Avenue, New York, New York 10016.

5. Other Information Sources and Programs

a. SPECIALIZED INFORMATION SERVICES

Data are obtainable from computerized data banks through government and commercial services. Data in this form are read directly into the computer through organized reference item collection programs which have been established to make available such stores of specialized information for the scientific community. Professional scientific and engineering societies, federal government agencies and research and education institutions are active in developing these services. A brief list of some of these services is provided here. For additional details on these and other computerized services, the directories listed here are helpful. Inquirers should also contact the national professional societies in their specialized field.

AMERICAN INSTITUTE OF PHYSICS - 335 East 45th Street, New York, New York 10017.

> Searchable Physics Information service. Includes bibliographic details and abstracts of over 2000 article records per month from 60 physics journals. Current awareness and bibliographic searches available in this system.

AMERICAN NATIONAL STANDARDS INSTITUTE - 1430 Broadway, New York, New York 10018.

> Service to chemical firms and others known as "Technical Help to Exporters" in which marketing information on a worldwide basis is provided. This program incorporates details gathered by the British Standards Institution on exporting to other countries.

AMERICAN SOCIETY OF HOSPITAL PHARMACISTS - 4630 Montgomery Avenue, Washington, D.C. 20014.

> A drug information service which is computer based is available from this agency. Current awareness service and research details in drug related fields are offered and should be of importance to scientists in the drug industries. Contact the above listed agency for further information.

BIO-SCIENCES INFORMATION SERVICE OF BIOLOGICAL ABSTRACTS - 2100 Arch Street, Philadelphia, Pennsylvania 19103.

> Has available ABSTRACTS ON HEALTH EFFECTS OF ENVIRONMENTAL POLLUTANTS which is both in print and machine readable tape format. The substance of the data base consists of information on occupational health and industrial medicine, chemicals and materials in the environment which have a toxic effect on man. The data base in magnetic tape format may be obtained on a lease basis.

CHEMICAL ABSTRACTS SUBJECT INDEX - Marketing Department, Chemical Abstracts Services, Columbus, Ohio 43210.

> Available as a computer readable information service. Based on subject indexing of CA material. Complete details such as molecular

formulas, general subject entries, etc., all included on tape.
Frequency is bi-weekly.

CHEMICAL HORIZONS, INC. - 274 Madison Avenue, New York, New York 10016.

A division of Predicasts, Inc., of Cleveland, Ohio.

Services include: "Chemical Horizons Intelligence File," made up of 3 x 5 file cards updated periodically which provide details on chemical market research, corporate mergers, developments, plans and statistics included.

SDI and data collection available through Predicasts, Inc.

CHEMICAL SYSTEMS, INC.

COMPUTERIZED STRUCTURAL GROUP INDEX OF COMMERCIAL ORGANIC CHEMICALS - P.O. Box 5523, Southfield Station, Shreveport, Louisiana 71105.

Provides computerized literature searches, reference information and manufacture/trade name details on organic chemicals. This is available on a contract basis.

COMPENDEX (COMPUTERIZED ENGINEERING INDEX)

Monthly tape service of ENGINEERING INDEX data base providing worldwide coverage of engineering input.

Engineering Index Inc., 345 East 47th Street, New York, New York 10017.

FRANKLIN INSTITUTE RESEARCH LABORATORIES SCIENCE INFORMATION SERVICES DEPARTMENT (SIS) - 20th and Benjamin Franklin Parkway, Philadelphia, Pennsylvania 19103.

Science Information Services Department offers an extensive program including gathering, analyzing, and processing of information according to user requirements and specifications. Services include literature searching, special data compilations, state-of-the-art reports, and translation.

GEORGE WASHINGTON UNIVERSITY PATENT, TRADEMARK, AND COPYRIGHT INSTITUTE (PTC) - Washington, D.C. 20006.

An organization with the primary purpose of carrying on research and education in fields of invention and patenting and trademarks. This is carried on by means of serial and monographic publications and through varied information transfer programs. Also maintains a sizable library with materials related to patents, trademarks, copyrights, etc.

INSTITUTE FOR SCIENTIFIC INFORMATION (ISI) - 325 Chestnut Street, Philadelphia, Pennsylvania 19106.

Offers information retrieval and dissemination services. Current awareness service as well as retrospective searches are provided. Source tapes and citation tapes provide data from approximately 5,500 items each week published in 1,800 journals. ICRS (Index Chemicus Registry System) tapes include all the bibliographic, use-profile, and analytical instrumentation information published in INDEX CHEMICUS, plus indexing terms. Tapes are issued monthly, cumulated annually, and are available on subscription basis.

NATIONAL SAFETY COUNCIL - 425 North Michigan Avenue, Chicago, Illinois 60611.

Offers a "Safety Research Information Service" which collects, classifies and distributes safety information, a major portion of which is of pertinence to chemical industries. Services include indexing, searching literature, reference and referral service.

PLENUM PUBLISHING CORPORATION; IFI/PLENUM DATA CORPORATION - 1000 Connecticut Avenue, N.W., Washington, D.C. 20036.

Provides coverage of all U.S. chemical and chemically related patents. Uses a Uniterm Index to analyze, edit, classify, and cross index all domestic patents issued from January 1950 to date. Indexing is done from the text of the patent which is updated bimonthly and annually, and is available on a subscription or a lease program in different formats and as data bases.

SADTLER RESEARCH LABORATORIES, INC. - 3316 Spring Garden Street, Philadelphia, Pennsylvania 19104.

Issues the Sadtler Research Laboratories Data Program, in infrared, ultraviolet, and NMR Spectroscopy in looseleaf binders with indexing by name of compound, molecular formula, chemical class and serial number. Also supplies the IRIS (Infrared Information System) data base which users may search via their own hardware. Contact the Sadtler Research Labs for further details.

STANFORD RESEARCH INSTITUTE (SRI) CHEMICAL INFORMATION SERVICES - 333 Ravenswood Avenue, Menlo Park, California 94025.

Chemical Information services consist of research programs in chemical economics presented in print and tape format.

U.S. DEPARTMENT OF COMMERCE

A. NATIONAL BUREAU OF STANDARDS CRYOGENIC DATA CENTER - Boulder, Colorado 80302.

Provides literature searches, data collection, SDI programs, and state-of-the-art reports besides reference and referral assistance.

B. NATIONAL TECHNICAL INFORMATION SERVICE - 5285 Port Royal Road, Springfield, Virginia 22151.

Provides computer literature searches, data collection, SDI programs, and state-of-the-art reports besides reference and referral assistance.

U.S. ENVIRONMENTAL PROTECTION AGENCY

Air Pollution Control Office, Office of Technical Information and Publications, Air Pollution Technical Information Center - P.O. Box 12055, Research Triangle Park, North Carolina 27709.

Provides computer literature searches, data collection, SDI programs, and state-of-the-art reports besides reference and referral assistance.

b. DIRECTORIES

COMPUTER PROGRAMS DIRECTORY. New York: C.C.M. Information Corp., 1971.

"Computer Programs for Chemical Engineers." CHEMICAL ENGINEERING, July 12, 1971, pp. 66-86.

DIRECTORY OF COMPUTERIZED DATA FILES, SOFTWARE AND RELATED TECHNICAL REPORTS. Springfield, Virginia: National Technical Information Service, 1976.

1972 EDP INDUSTRY DIRECTORY, A CATALOG OF EDP PRODUCTS AND SERVICES. Barrington, Illinois: Technical Publishing Co., 1972.

ENCYCLOPEDIA OF INFORMATION SYSTEMS AND SERVICES. 2nd ed. Ann Arbor, Michigan: Edwards Brothers, 1974.

"More Computer Programs for Chemical Engineers." CHEMICAL ENGINEERING, December 27, 1971, pp. 63-73.

Robinson, P. "Survey of Computer Systems for Chemists." CHEMISTRY AND INDUSTRY, December 20, 1975, pp. 1032-37.

WORLDWIDE DIRECTORY OF COMPUTER COMPANIES. Annual. Orange, New Jersey: Academic Press.

B. LITERATURE

This part of the Guide contains selected literature related to chemical process industries. It includes print material on broad engineering topics as well as specific information tools on chemical engineering. Literature having to do with particular parts of the total chemical engineering industry, such as food engineering, nuclear engineering, etc., are listed separately in the individual sections with these subject descriptions. Items which are broad in scope are filed in

Section I, even though they may also contain details on other engineering subjects (food, agriculture, etc.) which are arranged separately. It is hoped therefore that the listings in Section 1 of this guide will provide a useful representation of the literature of chemical engineering and an overview of information sources which support these information needs. The literature with a more well defined scope in a particular field will be found in the sections following.

1. Guides and Handbooks

a. GENERAL

American Chemical Society. Committee on Analytical Reagents. REAGENT CHEMICALS: AMERICAN CHEMICAL SOCIETY SPECIFICATIONS. 4th ed. Washington, D.C.: American Chemical Society, 1968.

Blackadder, D. A HANDBOOK OF UNIT OPERATION. New York: Academic, 1971.

> "Textbook of the basics of the unit operations. Covers binary distillation (and rectification), solvent extraction, gas absorption, multicomponent separations, heat exchangers, drying, evaporators and filtration as entirely separate and distinct topics."

British Valve Manufacturer's Association. TECHNICAL REFERENCE BOOK ON VALVES FOR THE CONTROL OF FLUIDS. London: 1972.

> "The first 35 pages are devoted to detailed descriptions and the characteristics of all types of valve... Useful tables are given on fluid flow, steam, and materials including a corrosion guide... The second half of the book is a directory and catalogue of the products produced by individual members of the Association."

CHEMICAL ENGINEER'S HANDBOOK. Edited by Robert H. Perry. 5th ed. New York: McGraw-Hill, 1973.

> "Completely modernized and reorganized to incorporate the latest innovations in theory and practice."

CHEMICAL TECHNICIANS' READY REFERENCE HANDBOOK. By G. Shugar, et al. New York: McGraw-Hill, 1973.

> "For the staffs in today's chemical laboratories. It is designed as an omnibook for the chemical technician, the student taking chemical laboratory courses, and the graduate student who needs a ready source of information which meaningfully relates to laboratory practice."

CHEMICAL TECHNOLOGY HANDBOOK. Washington, D.C.: American Chemical Society, 1975.

COLOUR INDEX. 3rd ed. 5 vols. Bradford, Yorkshire, England: Society of Dyers and Colourists, 1971.

Cremer, H. CHEMICAL ENGINEERING PRACTICE. 12 vols. London: Butterworth, 1965.

Cutting, W. HANDBOOK OF PHARMACOLOGY. Edgewood Cliffs, New Jersey: Prentice Hall, 1972.

Devon, T. HANDBOOK OF NATURALLY OCCURING COMPOUNDS. New York: Academic Press, 1972.

Dow Chemical Co. DESALINATION MATERIALS MANUAL. By P.F. George, et al. Report no. PB252043. Springfield, Virginia: National Technical Information Service, 1976.

Evans, F. EQUIPMENT DESIGN HANDBOOK FOR REFINERIES & CHEMICAL PLANTS. 2 vols. Houston: Gulf Publishing Co., 1972.

Gladstone, J. MECHANICAL ESTIMATING GUIDEBOOK. 4th ed. New York: McGraw-Hill, 1971.

> "Contains an updated and comprehensive discussion of the methods and procedures for estimating the costs of a variety of mechanical equipment in many different types of installations."

Gordon, A. THE CHEMIST'S COMPANION; A HANDBOOK OF PRACTICAL DATA, TECHNIQUES, AND REFERENCES. New York: Interscience, Division of Wiley, 1973.

> "Properties of molecular systems. Properties of atoms and bonds. Kinetics and energetics. Spectroscopy. Photochemistry. Chromatography. Experimental techniques. Mathematical and numerical information. Miscellaneous. Suppliers index. Subject index."

Graham, A. ELECTROPLATING ENGINEERING HANDBOOK. 2nd ed. New York: Reinhold, 1962.

> "Chapters are written by specialists and directed to the designer, manufacturer, supplier, and purchaser of electroplating equipment."

HANDBOOK OF CHEMISTRY AND PHYSICS. Annual. Cleveland, Ohio: Chemical Rubber Publishing Co.

HANDBOOK OF CORROSION RESISTANT PIPING. By P. Schweitzer. New York: Industrial Press, 1969.

HANDBOOK OF PRECISION ENGINEERING. Edited by A. Davidson. 12 vols. New York: McGraw-Hill, 1972.

Hilor, W., ed. HANDBOOK ON CORROSION TESTING AND EVALUATION. New York: Wiley, 1971.

"Based on the work of 50 or so authors, mostly American, presented at the four-day symposium on 'Corrosion Testing' held in Toronto in June 1970."

King, H. HANDBOOK OF HYDRAULICS FOR THE SOLUTION OF HYDRO-STATIC AND FLUID FLOW PROBLEMS. 5th ed. New York: McGraw-Hill, 1963.

LANGE'S HANDBOOK OF CHEMISTRY. Edited by J. Deane. 11th ed. New York: McGraw-Hill, 1973.

"New edition contains vast, authoritative compilations of facts and data covering every aspect of chemistry. Extensively updated, re-organized, and improved by the editor."

Liptak, B., ed. INSTRUMENT ENGINEERS' HANDBOOK. VOLUME I: PRO-CESS MEASUREMENT; VOLUME II: PROCESS CONTROL. Philadelphia: Chilton, 1969-70.

"Volume I describes ...Techniques of measurement, processing of these measurements and regulating mechanisms which utilize the results..."
"Volume II treats such major subject matters as process control systems, panels and panel instruments, control theory, control valves, and computers."

_____. INSTRUMENT ENGINEERS' HANDBOOK. Supplement 1. Philadelphia: Chilton, 1972.

"First supplement adds new topics to subjects covered in Volumes I and II."

McDermott, J., ed. DESALINIZATION BY DISTILLATION; RECENT DEVELOP-MENTS. Park Ridge, New Jersey: Noyes Data Corp., 1971.

Lists U.S. patents on desalinization by distillation.

_____. DESALINIZATION BY FREEZE CONCENTRATION. Park Ridge, New Jersey: Noyes Data Corp., 1971.

A guide to U.S. patents on freeze concentration.

Mellan, I. INDUSTRIAL SOLVENTS HANDBOOK 1970. Park Ridge, New Jersey: Noyes Data Corp., 1970.

"A handbook with complete, up-to-date, pertinent data regarding industrial solvents."

Morrow, L. MAINTENANCE ENGINEERING HANDBOOK. 2nd ed. New York: McGraw-Hill, 1966.

ORGANICUM. Practical Handbook of Organic Chemistry Translated from the 5th Edition of the German Book "Organikum" and Revised During Translation to Correspond to the 11th German Edition Published in 1971. Revised and translated by B. J. Hazzard, et al. Edited by P. A. Ongley. Reading, Massachusetts: Addison-Wesley, 1973.

Perry, J. ENGINEERING MANUAL; A PRACTICAL REFERENCE OF DATA AND METHODS IN ARCHITECTURAL, CHEMICAL, CIVIL, ELECTRICAL, MECHANICAL, AND NUCLEAR ENGINEERING. New York: McGraw-Hill, 1959.

PIPING HANDBOOK. Edited by S. Crocker. 5th ed. New York: McGraw-Hill, 1967.

"Data and procedures on piping in industrial, municipal, and building piping systems."

PLANT ENGINEERING HANDBOOK. Edited by W. Staniar. New York: McGraw-Hill, 1959.

Rabald, E. CORROSION GUIDE. New York: Elsevier, 1968.

Rohsenow, W. HANDBOOK OF HEAT TRANSFER. New York: McGraw-Hill, 1972.

"A collection of basics, calculations, tables, charts, and design parameters."

Rosen, M. SYSTEMATIC ANALYSIS OF SURFACE-ACTIVE AGENTS. 2nd ed. New York: Wiley-Interscience, 1972.

A guide to the field of surface active agents.

Synthetic Organic Chemical Manufacturers Association. SOCMA HANDBOOK; COMMERCIAL ORGANIC CHEMICAL NAMES. Washington, D.C.: American Chemical Society, 1965.

Lists 6300 industrial compounds.

U.S. Department of Interior. DESALINATION PLANT LABORATORY PROCEDURES MANUAL (with list of references and bibliography). By R. M. Burd, et al. Washington, D.C.: Government Printing Office, 1973.

U.S. Department of Interior, Office of Saline Water. DESALINATION PLANT OPERATOR CHEMICAL CONTROL GUIDE. By W. R. Greenway. Washington, D.C.: Government Printing Office, 1973.

Walker, W. GUIDE TO INDUSTRIAL HYDRAULICS. Levittown, New York: Transatlantic, 1973.

Wiberg, K., ed. ORGANIC SYNTHESES; AN ANNUAL PUBLICATION OF SATISFACTORY METHODS FOR THE PREPARATION OF ORGANIC CHEMICALS. New York: Wiley, 1969.

b. TABLES, DATA, AND SCIENTIFIC UNITS

American Society of Heating, Refrigerating, and Air Conditioning Engineering, Inc. ASHRAE GUIDE AND DATA BOOK. HANDBOOK OF FUNDAMENTALS. Annual. New York: Author.

American Society of Mechanical Engineers. ASME THERMODYNAMIC AND TRANSPORT PROPERTIES OF STEAM. New York: The Society, 1969.

American Society of Mechanical Engineers. ASME HANDBOOK: ENGINEER-ING TABLES. New York: The Society, 1956.

Aylward, G. SI CHEMICAL DATA. New York: Wiley, 1971.

Barin, I., et al. THERMOCHEMICAL PROPERTIES OF INORGANIC SUB-STANCES. New York: Springer-Verlag, 1973.

Blackman, D. S.I. UNITS IN ENGINEERING. New York: Macmillan, 1969.

> "Introduction to the SI system of units aimed basically at the engineer."

Bolz, R. HANDBOOK OF TABLES FOR APPLIED ENGINEERING SCIENCE. Cleveland: Chemical Rubber, 1970.

> "Reference book for practicing engineering students. Data on various processes in the different fields of engineering are presented. The emphasis is on the mechanical, thermal, electrical and radiation properties and phenomena."

CHEMICAL ENGINEERING DATA BOOK. Irregular. Edited by T. Ross. London: Leonard Hill (Books) Ltd.

Chiswell, B. SI UNITS. New York: Wiley, 1971.

> "A compilation of tables of the International System of Units (SI Units) as developed by the Conference General des Poides et Mesures."

Dow Chemical Company, Thermal Research Laboratory. JANAF THERMO-
CHEMICAL TABLES. 2nd ed. Washington, D.C.: Government Printing
Office, 1971.

Erskine, M. CHEMICAL CONVERSION FACTORS AND YIELDS. Menlo Park,
California: Chemical Information Services, Stanford Research Institute, 1969.

    "Provides commercial yield data for calculation of output and raw
    material requirements."

Gieck, K. ENGINEERING FORMULAS. New York: McGraw-Hill, 1971.

    "Guide to the more important mathematical and technical formulas
    and definitions."

Green, M. INTERNATIONAL AND METRIC UNITS OF MEASUREMENT. New
York: Chemical Publishing Co., 1973.

HANDBOOK OF CHEMISTRY AND PHYSICS. Annual. Cleveland: Chemical
Rubber Publishing Co.

HANDBOOK OF TABLES FOR APPLIED ENGINEERING SCIENCE. 2nd ed.
Cleveland: CRC Press, 1973.

Henley, E. MATERIAL AND ENERGY BALANCE COMPUTATION. New York:
Wiley, 1969.

Hicks, T. STANDARD HANDBOOK OF ENGINEERING CALCULATIONS.
New York: McGraw-Hill, 1972.

    "Presents more than 2,000 step-by-step calculation procedures for
    solving problems in 12 different engineering disciplines: civil,
    architectural, mechanical, electrical, electronics, chemical, con-
    trol, aeronautical and astronautical, marine, nuclear, sanitary and
    engineering economics."

Himmelbau, D. BASIC PRINCIPLES AND CALCULATIONS IN CHEMICAL
ENGINEERING. 2nd ed. Edgewood Cliffs, New Jersey: Prentice Hall,
1967.

INTERNATIONAL COMPENDIUM OF NUMERICAL DATA PROJECTS: A SUR-
VEY AND ANALYSIS. International Council of Scientific Unions Committee
on Data and Technology, Rome, Italy, 1969.

INTERNATIONAL CRITICAL TABLES OF NUMERICAL DATA. PHYSICS.
CHEMISTRY AND TECHNOLOGY. Washington, D.C.: National Research
Council, National Science Foundation, 1933.

Jerrard, H. A DICTIONARY OF SCIENTIFIC UNITS; INCLUDING DIMEN-
SIONLESS NUMBERS AND SCALES. 3rd ed. New York: Harper & Row,
1972.

    Physical constants, conversion tables, etc.

Kaufman, H. HANDBOOK OF ORGANOMETALLIC COMPOUNDS. Princeton, New Jersey: Van Nostrand, 1961.

"Gives tables for each element within a periodic group, and a bibliography."

Kaye, G. TABLES OF PHYSICAL AND CHEMICAL CONSTANTS AND SOME MATHEMATICAL FUNCTIONS. 12th ed. New York: Longmans, 1959.

Keenan, J. GAS TABLES: THERMODYNAMICS. New York: Wiley, 1948.

Properties of air products of combustion and component gases. Compressible flow functions.

Keller, R., ed. BASIC TABLES IN CHEMISTRY. New York: McGraw-Hill, 1967.

Kuong, J. APPLIED NOMOGRAPHY. 4 vols. Houston: Gulf Publishing, 1965.

LANDOLT-BORNSTEIN: NUMERICAL DATA AND FUNCTIONAL RELATIONSHIPS IN SCIENCE AND TECHNOLOGY. New Series. Edited by K. H. Hellwege. Berlin: Springer, 1969-70.

Updating of standard data handbook of physical sciences.

MATHEMATICAL HANDBOOK FOR SCIENTISTS AND ENGINEERS. Edited by Granino and Theresa Korn. New York: McGraw-Hill, 1968.

Parsons, R. HANDBOOK OF ELECTROCHEMICAL CONSTANTS. London: Butterworth, 1959.

PSYCHROMETRIC TABLES AND CHARTS. Dover, New Hampshire: Industrial Research Service, 1964.

Ramaswamy, G. SI UNITS: A SOURCE BOOK. New York: McGraw-Hill, 1973.

Raznjevic, R. HANDBOOK OF THERMODYNAMIC TABLES AND CHARTS. New York: McGraw-Hill, 1976.

SADTLER STANDARD SPECTRA (INFRARED). Continually updated. Philadelphia, Pennsylvania: Sadtler Research Laboratories.

"Looseleaf services including spectra information by alphabetical arrangement, by chemical classes and a numerical index."

Seidell, A. SOLUBILITIES OF INORGANIC AND METAL ORGANIC COMPOUNDS: A COMPILATION OF QUALITATIVE SOLUABILITY DATA FROM THE PERIODICAL LITERATURE. 4th ed. New York: Van Nostrand-Reinhold, 1958.

Spencer, R., et al. THEORETICAL STEAM RATE TABLES COMPATIBLE WITH THE 1967 ASME STEAM TABLES. New York: American Society of Mechanical Engineers, 1969.

Stephen, H., ed. SOLUBILITIES OF INORGANIC AND ORGANIC COMPOUNDS. 2 vols. New York: Pergamon, 1963.

U.S. National Bureau of Standards. LIQUID DENSITIES OF OXYGEN, NITROGEN, ARGON, AND PARAHYDROGEN WITH LIST OF REFERENCES. By H. M. Roder, R. D. McCarty, V. J. Johnson. U.S. National Bureau of Standards, Gaithersburg, Maryland, 1972.

"Revises and updates issue of January 31, 1968."

_____. CIRCULAR 500. SELECTED VALUES OF CHEMICAL THERMODYNAMIC PROPERTIES. Washington, D.C.: Government Printing Office, 1952.

_____. SELECTED VALUES OF CHEMICAL THERMODYNAMIC PROPERTIES. In standard order of arrangement; by D. Wagman, et al. Physical Chemistry Division, Institute for Material Research. Washington, D.C.: Government Printing Office, 1971.

_____. SUPPLEMENTARY BIBLIOGRAPHY OF KINETIC DATA ON GAS PHASE REACTIONS OF NITROGEN, OXYGEN, AND NITROGEN OXIDES. By F. Westley. Chemical Kinetics Information Center. Washington, D.C.: Government Printing Office, 1973.

This supplement covers period 1900–January 1972.

_____. APPLIED MATHEMATICS SERIES HANDBOOK 55. HANDBOOK OF MATHEMATICAL FUNCTIONS WITH FORMULAS, GRAPHS AND MATHEMATICAL TABLES. Washington, D.C.: Government Printing Office, 1964.

Utermark, W. MELTING POINT TABLES OF ORGANIC COMPOUNDS. 2nd revised and supplemented ed. New York: Interscience, 1963.

Zimmerman, O. INDUSTRIAL RESEARCH SERVICE'S CONVERSION FACTORS AND TABLES. 3rd ed. Dover, New Hampshire: Industrial Research Service, 1961.

"Designed to provide an accurate source of fundamental physical relationships and thousands of useful constants for the conversion of units."

2. Dictionaries

a. TECHNICAL

Ballentyne, D. A DICTIONARY OF NAMED EFFECTS AND LAWS IN CHEMISTRY, PHYSICS AND MATHEMATICS. 3rd ed. New York: Barnes & Noble, 1971.

A revision of earlier work.

Bennett, H., ed. CONCISE CHEMICAL AND TECHNICAL DICTIONARY. 3rd ed. Cleveland: Chemical Publishing Co., 1974.

Booth, K. DICTIONARY OF REFRIGERATION AND AIR CONDITIONING. New York: Elsevier, 1970.

"Covers terminology in all English-speaking countries."

Clason, W. ELSEVIER'S DICTIONARY OF CHEMICAL ENGINEERING. Vol. 1, Chemical engineering and laboratory equipment. Vol. 2, Chemical engineering; processes and products. New York: Elsevier, 1968.

"Includes chemical engineering terms, with exception of purely chemical terms and processes derived from a proper noun."

Collcott, T., ed. DICTIONARY OF SCIENCE AND TECHNOLOGY. New York: Harper & Row, 1972.

CONDENSED CHEMICAL DICTIONARY. 7th ed. Completely revised and enlarged by A. Rose. New York: Reinhold, 1966.

Dettner, H. DICTIONARY OF METAL FINISHING AND CORROSION. New York: Elsevier, 1972.

"1650 words and phrases commonly used in the metal finishing field are listed in a basic table in English with foreign language equivalents."

Dreyfuss, H. SYMBOL SOURCEBOOK; AN AUTHORITATIVE GUIDE TO INTERNATIONAL GRAPHIC SYMBOLS. New York: McGraw-Hill, 1972.

"Useful sourcebook of symbols. Has an arrangement by discipline and then by graphic form. The section on the meaning and use of color in various applications and countries makes fascinating reading."

Engineers Joint Council. THESAURUS OF ENGINEERING TERMS. 1st ed. New York: The Council, 1964.

"A list of engineering terms and their relationships for use in vocabulary control in indexing and retrieving engineering information."

A GLOSSARY OF URETHANE INDUSTRY TERMS. Louisville, Kentucky: Martin Sweets Company, 1973.

"Comprehensive glossary of urethane terms plus nomographs."

Hackh, I. CHEMICAL DICTIONARY, AMERICAN AND BRITISH USAGE; CONTAINING THE WORDS GENERALLY USED IN CHEMISTRY, AND MANY OF THE TERMS USED IN THE RELATED SCIENCES OF PHYSICS, ASTROPHYSICS, MINERALOGY, PHARMACY, AGRICULTURE, BIOLOGY, MEDICINE, ENGI-

NEERING, ETC. BASED ON RECENT CHEMICAL LITERATURE. 4th ed. Completely revised and ed. by Julius Grant. New York: McGraw-Hill, 1969.

Haynes, W. CHEMICAL TRADE NAMES AND COMMERCIAL SYNONYMS: A DICTIONARY OF AMERICAN USAGE. 2nd ed., revised and enlarged. New York: Van Nostrand, 1955.

"Dictionary of chemical trade names, with explanation of usage."

INTERNATIONAL DICTIONARY OF PHYSICS AND ELECTRONICS. W. Michels, ed. in chief. 2nd ed. New York: Van Nostrand, 1961.

"Designed for the use of students, professional physicists, and others whose primary activities lie in a different field of science (e.g., chemistry, engineering, biology, etc.)."

James, A. DICTIONARY OF THERMODYNAMICS. New York: Halstead Press, 1976.

McGRAW-HILL DICTIONARY OF SCIENTIFIC AND TECHNICAL TERMS. New York: McGraw-Hill, 1974.

MERCK INDEX; AN ENCYCLOPEDIA OF CHEMICALS AND DRUGS. 8th ed. Rahway, New Jersey: Merck, 1968.

"This new edition includes nearly 10,000 descriptions of individual substances, more than 42,000 names of chemicals and drugs."

NEW INTERNATIONAL DICTIONARY OF REFRIGERATION. Paris: Institut International du Froid, 1975.

Snell, F. DICTIONARY OF COMMERCIAL CHEMICALS. 3rd ed. New York: Van Nostrand, 1962.

VAN NOSTRAND'S CHEMIST'S DICTIONARY. Edited by J. Hornig. New York: Van Nostrand-Reinhold, 1953.

b. FOREIGN LANGUAGE

Alford, M. RUSSIAN-ENGLISH SCIENTIFIC AND TECHNICAL DICTIONARY. 2 vols. New York: Pergamon, 1970.

De Vries, L. DICTIONARY OF CHEMISTRY AND CHEMICAL ENGINEERING. 2 vols. New York: Academic, 1972.

"A comprehensive English-German dictionary for students, scientists, engineers and chemists. Covers chemistry, chemical engineering and many other related fields. Many words recently appearing in the technical literature have been included. Contains also temperature conversion chart."

Denti, R. ITALIAN-ENGLISH, ENGLISH-ITALIAN TECHNICAL DICTIONARY. 7th ed. New York: Ernst Heineman, 1970.

DICTIONARY OF CHEMISTRY AND CHEMICAL TECHNOLOGY IN SIX LAN-GUAGES: ENGLISH, GERMAN, SPANISH, FRENCH, POLISH, RUSSIAN. Edited by Z. Sobecka. New York: Pergamon, 1966.

Dorian, A. DICTIONARY OF SCIENCE AND TECHNOLOGY, ENGLISH-GERMAN. New York: Elsevier, 1967.

"English-German technical and engineering dictionary.."

_____ DICTIONARY OF SCIENCE AND TECHNOLOGY: GERMAN-ENGLISH. New York: Elsevier, 1970.

East China School of Chemical Technology. RUSSIAN-CHINESE-ENGLISH CHEMICAL AND TECHNICAL DICTIONARY. New York: Transatlantic, 1968.

ELSEVIER'S DICTIONARY OF CHEMICAL ENGINEERING. In six languages: English/American, French, Spanish, Italian, Dutch, German. Compiled by W. E. Clason. Amsterdam: Elsevier, 1968.

Ernst, R. DICTIONARY OF INDUSTRIAL TECHNICS, INCLUDING RELATED FIELDS OF SCIENCE AND CIVIL ENGINEERING. 15th ed. Wiesbaden: Brandstetter-Verlag., 1970. Distributed in New York, Ernst Heineman.

c. BIBLIOGRAPHIES

INTERNATIONAL BIBLIOGRAPHIES OF DICTIONARIES. 5th ed. New York: Bowker, 1972.

7000 dictionaries in more than 100 languages in areas of science, technology and economics are included.

Rechenbach, C. A BIBLIOGRAPHY OF SCIENTIFIC, TECHNICAL, AND SPECIALIZED DICTIONARIES; POLYGLOT, BILINGUAL, UNILINGUAL. Washington, D.C.: Catholic University of America Press, 1969.

UNESCO. BIBLIOGRAPHY OF INTERLINGUAL SCIENTIFIC AND TECHNICAL DICTIONARIES. 5th ed. New York: UNIPUB, 1970 (c1969).

Extensive listing of dictionaries in many languages.

3. Directories

a. GENERAL

Brown, R. HOW TO FIND OUT ABOUT THE CHEMICAL INDUSTRY. New York: Pergamon, 1969.

An overall guide to sources and literature, particularly British and European.

Klein, B., ed. GUIDE TO AMERICAN SCIENTIFIC AND TECHNICAL DIRECTORIES. Rye, New York: Todd Publishers, 1975.

RESEARCH GRANTS INDEX. Annual. 3 vols. Washington, D.C.: Government Printing Office.

Annual directory of grants for scientific research from Department of Health, Education and Welfare.

b. BIOGRAPHY

AMERICAN CHEMISTS AND CHEMICAL ENGINEERS. Washington, D.C.: American Chemical Society, 1976.

AMERICAN MEN AND WOMEN OF SCIENCE; A BIOGRAPHICAL DIRECTORY. 12th ed. New York: Bowker, 1973.

"The physical and biological sciences. Volumes 1-6. Over 137,500 entries in the six volumes."

BIOGRAPHICAL INFORMATION SERVICES

BIOGRAPHY INDEX. Monthly. New York: H.W. Wilson Co.

CHEMICAL INDUSTRY DIRECTORY AND WHO'S WHO ANNUAL. London: Benn Publishing Co.

CURRENT BIOGRAPHY. Monthly. New York: H.W. Wilson Co.

DICTIONARY OF SCIENTIFIC BIOGRAPHY. 6 volumes to date (1972). In process. New York: Scribners.

Engineers Joint Council. ENGINEERS OF DISTINCTION INCLUDING SCIENTISTS IN RELATED FIELDS. 1st ed. New York: The Council, 1970.

Kain, W. WHO IS WHO IN THE FRENCH CHEMICAL INDUSTRY. New York: International Publications Service, 1972.

_____. WHO IS WHO IN THE GERMAN CHEMICAL INDUSTRY. New York: International Publications Service, 1972.

LEADERS IN AMERICAN SCIENCE. Nashville, Tennessee: Who's Who in American Education, 1969.

Biographical directory of 20,000 eminent American and Canadian leaders in the industrial, governmental, and educational science and research fields.

NATIONAL FACULTY DIRECTORY, 1977. 7th ed. 2 vols. Detroit: Gale Research Co., 1977.

> Directory, names, departments, and addresses of about 450,000 full and part-time faculty members.

THE NEW YORK TIMES BIOGRAPHICAL EDITION MONTHLY SERVICE. New York Times Biographical Edition, Arno Press, 330 Madison Avenue, New York, New York 10017.

Turkevich, J. PROMINENT SCIENTISTS OF CONTINENTAL EUROPE. New York: Elsevier, 1968.

> Brief biographical detail about each individual entered.

WHO'S WHO IN SCIENCE IN EUROPE. 2nd ed. 4 vols. Guernsey, British Isles: Hodgson, 1972.

WHO'S WHO OF BRITISH ENGINEERS, 1968. 2nd ed. London: Maclaren, 1968.

WHO'S WHO OF BRITISH SCIENTISTS, 1971-72. London: Longman, 1972.

WORLD WHO'S WHO IN SCIENCE; A BIOGRAPHICAL DICTIONARY OF NOTABLE SCIENTISTS FROM ANTIQUITY TO THE PRESENT. 1st ed. Edited by Allen G. Dubus. Chicago: Marquis, 1968.

> Brief details on over 30,000 persons.

c. PRODUCTS AND MANUFACTURERS

i. United States

AMERICAN CHEMICAL SOCIETY LABORATORY GUIDE TO INSTRUMENTS, EQUIPMENT, AND CHEMICALS, 1971-72. Washington, D.C.: American Chemical Society, 1971.

"Blue Book: Reference and Buyer's Guide." SOAP/COSMETICS/CHEMICAL SPECIALTIES. McNair-Dorland Co., 101 West 31st Street, New York, New York 10001.

> An annual section of SOAP/COSMETICS/CHEMICAL SPECIALTIES.

CHEM SOURCES U.S.A. Annual. Directories Publishing Company, P.O. Box 422, Flemington, New Jersey 08822.

> Chemical buyer's guide for chemical products produced in the United States.

CHEMICAL ENGINEERING CATALOG. Annual. New York: Rheinhold.

> Contains information for purchase of equipment, materials, and specialized services.

CHEMICAL GUIDE TO THE U.S. Biennial. Park Ridge, New Jersey: Noyes Data Corp.

CHEMICAL ORIGINS AND MARKETS. Flow charts and tables. 4th ed. Chemical Information Services, Stanford Research Institute, Menlo Park, California 94025, 1972.

> Desk reference volume with product flow charts, tables of major organics and inorganics.

CHEMICAL PURCHASING BUYER'S DIRECTORY. Annual. Myers Publishing Company, Inc., 381 Park Avenue S., New York, New York 10016.

CHEMICAL WEEK CHEMICAL BUYERS' GUIDE ISSUE. Annual. New York: McGraw-Hill.

> "Lists chemical producers and distributers. Trade names, products, descriptions, company names."

CHEMICAL WEEK EQUIPMENT/PACKAGING BUYERS' GUIDE ISSUE. Annual. New York: McGraw-Hill.

CONOVER-MAST PURCHASING DIRECTORY. Revised and published annually. Conover-Mast Purchasing Directory, 95 East Putnam Avenue, Greenwich, Connecticut 06830.

> "Contains a grouping of listings consisting of entries on approximately 33,000 American manufacturers."

"Construction Alert." CHEMICAL ENGINEERING. Semiannual. McGraw-Hill, Inc., 1221 Avenue of the Americas, New York, New York 10020.

> An inventory of new plants and facilities planned or under construction. This semiannual directory usually appears in the April and October issues.

CORPORATE DIAGRAMS AND ADMINISTRATIVE PERSONNEL OF THE CHEMICAL INDUSTRY. Annual. Chemical Economic Services, 92B Nassau Street, Princeton, New Jersey 08540.

> "Publication presents in looseleaf, indexed form, 7,000 top- and middle-management executives in 200 companies."

Croft, A. CRYOGENIC LABORATORY EQUIPMENT. New York: Plenum, 1970.

> "Deals with the equipment essential to a low temperature laboratory, their sources of supply, and modus operandi."

Crowley, Ellen T., ed. TRADE NAMES DICTIONARY. 2 vols. Detroit: Gale Research Co., 1978.

"Directory and Planning Aids." CHEMICAL AND ENGINEERING NEWS, Vol. 50, December 2, 1972, pp. 24-25.

DIRECTORY OF CHEMICAL PRODUCERS. Annual with quarterly supplements. Chemical Information Services, Stanford Research Institute, Menlo Park, California 94025.

   Lists company, products and regions.

DIRECTORY OF THE CHEMICAL MARKETING RESEARCH ASSOCIATION. Annual. Chemical Marketing Research Association, 100 Church Street, New York, New York 10007.

"Equipment Buyers Guide." CHEMICAL ENGINEERING. Annual. McGraw-Hill, Inc., 1221 Avenue of the Americas, New York, New York 10020.

Gardner, W. CHEMICAL SYNONYMS AND TRADE NAMES. 7th ed. London: Technical Press, 1971.

Godel, J. GUIDE TO SOURCES OF CONSTRUCTION INFORMATION. New York: Construction Publishing Co., 1975.

GUIDE TO SCIENTIFIC INSTRUMENTS. Annual. Instrument Society of America, 400 Stanwix Street, Pittsburgh, Pennsylvania 15222.

   "Directory of scientific instruments and their manufacturers."

HYDROCARBON PROCESSING CATALOG. Annual. Houston, Texas: Gulf Publishing Company.

   "Specialized hydrocarbon processing industry; oil, petrochemical and natural gas processing."

INDUSTRIAL CONSTRUCTION CATALOG FILE, PLANT ENGINEERING CATALOG FILE and other series are available from: General Directories of Equipment and Supplies Sweets Division, McGraw-Hill Information Systems Company, 1221 Avenue of the Americas, New York, New York 10020.

THE KLINE GUIDE TO PAINT INDUSTRY. 3rd ed. Fairfield, New Jersey: C.H. Kline & Co., 1972.

KLINE GUIDE TO THE CHEMICAL INDUSTRY. 2nd ed. Fairfield, New Jersey: C.H. Kline & Co., 1974.

MACRAE'S BLUE BOOK. Annual. MacRae's Blue Book, 100 Shore Drive, Hinsdale, Illinois 60521.

   A directory of products and manufacturers.

MARKETING GUIDE: U.S. SYNTHETIC ORGANIC CHEMICALS INDUSTRY. Westport, Connecticut: Technomic Publishing Company, 1966.

   "An analysis of organic chemicals marketing methods, based on interviews with marketing executives."

MARKETING GUIDE TO THE COSMETICS INDUSTRY. Westport, Connecticut: Technomic Publishing Company, July 1973.

> "Contents: Economic Environment·Industry Structure Markets for: hair preparations, make-up preparations, oral products, men's toiletries, deodorants, fragrances, toilet soaps, face creams, hand preparations, and other products·Marketing Channels·Information Sources·Major Companies·Directory of Companies."

MUNICIPAL INDEX. Annual. Buttenheim Publishing Corporation, Berkshire Common, Pittsfield, Massachusetts 01201.

> Purchasing guide for municipal offices.

"New Plants and Facilities: CE Construction Alert." CHEMICAL ENGINEER-ING. Vol. 82, March 31, 1975, pp. 110-20.

"New Processes and Technology Alert." CHEMICAL ENGINEERING. Semi-annual. McGraw-Hill, Inc., 1221 Avenue of the Americas, New York, New York 10020.

> This directory usually appears in the February and July issues. It includes inorganic chemicals, metals, organic chemicals, petro-leum and fuels, plastics, pollution control, pulp and paper, tech-nology, and licenses.

"1972-73 Lab Guide." ANALYTICAL CHEMISTRY, Vol. 44, No. 10, August 1972, pp. 3LG-440LG.

> A special section of ANALYTICAL CHEMISTRY for August 1972.

OPD CHEMICAL BUYER'S DIRECTORY. Annual. Issued annually to subscribers of CHEMICAL MARKETING REPORTER. Schnell Publishing Company, 100 Church Street, New York, New York 10007.

> Annual buyer's guide for oil, paint, drug and chemical industry.

"Semi-Annual Inventory of New Plants and Facilities." CHEMICAL ENGINEER-ING, Vol. 70, December 2, 1972, pp. 72-78.

SOAP/COSMETICS/CHEMICAL SPECIALTIES. Blue Book Reference and Buyer's Guide Issue. Annual. MacNair-Dorland Company, Inc., 101 West 31st Street, New York, New York 10001.

> Supply information and data for chemical specialties field.

SOUTHERN CHEMICAL PRODUCERS, 1971. Chemical Information Services, Stanford Research Institute, 333 Ravenswood Avenue, Menlo Park, California 94025.

THOMAS REGISTER OF MANUFACTURERS. Annual. Thomas Publishing Company, 461 Eighth Avenue, New York, New York 10001.

> A comprehensive listing of products, trade names and manufacturers.

U.S. SYNTHETIC ORGANIC CHEMICALS INDUSTRY MARKETING GUIDE. Annual. Stamford, Connecticut: Technomic Publishing Company.

WORLD SPACE DIRECTORY. Ziff-Davis Publishing Company, 1156-15th Street, N.W., Washington, D.C. 20005.

"Includes oceanography, issued twice per year."

ii. International

ABC EDITION, EUROP PRODUCTION. 13th ed. 2 vols. Darmstadt: Europ Export Edition GMBH, 1972. Available from International Publications Service, New York.

A directory of European manufacturers.

BUYER'S GUIDE OF CHEMICALS. Annual. Society of Chemical Industry, 14 Belgrave Square, London, SW1, England.

"Buyers Guide to Process Equipment Controls and Instrumentation." CANADIAN CHEMICAL PROCESSING. Southern Business Publications, Ltd., 1450 Don Mills Road, Don Mills, Ontario, Canada.

"An annual section of Canadian chemical processing periodicals."

Canada. Department of Industry, Trade and Commerce. Chemicals branch. CANADIAN CHEMICAL REGISTER, 1971. Ottawa, Ontario, Canada.

CHEM SOURCES, EUROPE. Annual. Directories Publishing Company, P.O. Box 422, Flemington, New Jersey 08822.

Chemical buyer's guide for chemical products produced in Europe.

CHEMICAL BUYER'S GUIDE. Annual. Southam Business Publications Ltd., 1450 Don Mills Road, Don Mills, Ontario, Canada.

CHEMICAL INDUSTRIES OF WESTERN EUROPE. Annual. London: ECN Chemical Data Services.

CHEMICAL INDUSTRY DIRECTORY. Annual. Benn Brothers, Ltd., 159 Fleet Street, London E.C. 4, England.

CHEMICAL INDUSTRY OF THE AMERICAS. Annual. London: ECN Chemical Data Services.

CHEMICALS 1976. New York: International Publications Service, 1976.

CHEMISTRY AND INDUSTRY BUYERS' GUIDE. Annual. Society of Chemical Industry, 14 Belgrave Square, London, SW1, England.

Derz, F., ed. CHEMBUYDIRECT: INTERNATIONAL CHEMICAL BUYERS DIRECTORY. New York: De Gruyter, 1976.

DIRECTORY OF CHEMICAL PLANT CONTRACTORS. London: IPC Business Press, 1975.

EUROPEAN DIRECTORY OF ECONOMIC AND CORPORATE PLANNING, 1973-74. Epping, England: Gower Press, 1973.

INTERNATIONAL CHEMISTRY DIRECTORY. Annual. New York: W. A. Benjamin.

> "Offers a directory of academic departments and faculties (with faculty and geographical indexes), laboratories, societies, meetings, grants, fellowships, and awards. Also a section of journals and a listing of chemistry books in print."

JAPAN CHEMICAL DIRECTORY. Annual. (Text in English) Chemical Daily Company, Ltd., 19-16, 3-chome, Shibaura, Minato-ku, Tokyo, Japan.

LLOYD'S CANADIAN CHEMICAL PHARMACEUTICAL AND PRODUCT DIRECTORY. Irregular. West Hill, Ontario: Lloyd Publications of Canada.

> Revised periodically.

MAJOR CHEMICAL COMPANIES OF THE WORLD. Annual. London: ECN Chemical Services.

Rojana, C. CHEMICAL AND PETROCHEMICAL INDUSTRIES OF RUSSIA AND EASTERN EUROPE, 1960-1980. New York: Praeger, 1975.

SOAP, PERFUMERY AND COSMETICS YEAR BOOK AND BUYERS GUIDE. Annual. United Trade Press, Ltd., 9 Gough Square, London, E.C.4, England.

United Nations. Industrial Development Organization. INFORMATION SOURCES ON THE PAINT AND VARNISH INDUSTRY. New York: United Nations, 1975.

WHERE TO BUY CHEMICALS AND CHEMICAL PLANT. Semiannual. London: John Adam House Publishers.

WORLD GUIDE TO SCIENTIFIC ASSOCIATIONS. Munchen: Verlag Dokumentation Pullach, 1974.

Wunder, D. DIRECTORY OF THE WEST-GERMAN CHEMICAL INDUSTRY 1973-1975. 8th ed. New York: International Publications Service, 1975.

iii. Journals and Serial Publications
Related to Products and Manufacturers

CHEMICAL AND ENGINEERING NEWS. Weekly. Patrick P. McCurdy.
American Chemical Society, 1155 16th Street, N.W., Washington, D.C. 20036.

CHEMICAL EQUIPMENT. Monthly. Gordon Publications Inc., 20 Community
Place, Morristown, New Jersey 07960.

Deals with new products for chemical processing fields.

CHEMICAL DIGEST. Quarterly. Foster D. Snell, Inc., Hanover Road, Florham
Park, New Jersey 07932.

CHEMICAL HORIZONS WEEKLY REPORTS; AN INTERNATIONAL WEEKLY SUM-
MARY FOR CHEMICAL MARKETING & SALES EXECUTIVES. North American
and overseas editions. Chemical Horizons Inc., 274 Madison Avenue, New York,
New York 10016.

CHEMICAL INDUSTRY NOTES. American Chemical Society, Chemical Abstracts
Service, Accounting Department, Box 3012, Ohio State University, Columbus
43210.

CHEMICAL INDUSTRY REPORT. Quarterly. U.S. Department of Commerce,
Business and Defense Services Administration. Superintendent of Documents,
Government Printing Office, Washington, D.C. 20402.

CHEMICAL MARKET ABSTRACTS. Monthly. Foster D. Snell, Inc., 29 W.
15th Street, New York, New York 10011.

CHEMICAL MARKETING REPORTER. Weekly. Schnell Publishing Company,
Inc., 100 Church Street, New York, New York 10007.

Covers chemical and process materials industries.

CHEMICAL PEDDLER. Annual. Salesmen's Association of the American Chemical Industry, Inc., 79 Madison Avenue, New York, New York 10016.

CHEMICAL PROCESSING (U.S.). Monthly. Putnam Publishing Company,
111 E. Delaware Place, Chicago, Illinois 60611.

CHEMICAL PURCHASING. Monthly. Myers Publishing Company, Inc., 381
Park Avenue South, New York, New York 10016.

Wide coverage of purchasing in chemical process industries.

CHEMICAL SPOTLIGHT. Weekly. Corporate Intelligence, 25 Broad Street,
New York, New York 10004.

CHEMICAL WEEK. Weekly. McGraw-Hill Book Co., 1221 Avenue of the
Americas, New York, New York 10036.

CHEMICALS, QUARTERLY INDUSTRY REPORT. Commerce Department, Domestic Commerce Bureau, Chemicals Division. Washington, D.C.: Government Printing Office.

FACTORY EQUIPMENT NEWS. Fortnightly. Production Publications (London), Ltd., 10-16 Elm Street, London W.C.1, England.

INDUSTRIAL EQUIPMENT, MATERIALS & SERVICES. Monthly. Envoy Journals Ltd., 67 Clerkenwell Road, London E.C.1, England.

INDUSTRIAL EQUIPMENT NEWS (U.S.); WHAT'S NEW IN EQUIPMENT PARTS, MATERIALS AND LITERATURE AND CATALOGS. Monthly. Thomas Publishing Co., 461 Eighth Avenue, New York, New York 10001.

INDUSTRIAL GASES, 1971. Prepared in Industry Division, U.S. Department of Commerce, Washington, D.C. 20230, November 1972.

INTERNATIONAL CHEMICAL REGISTER; A MONTHLY BUYERS' GUIDE. Economic Documentation Office, Graaf Florislaan 30A, Box 505, Hilversum, Netherlands.

MODERN CHEMICALS. Bimonthly. Modern Chemicals Publishing Company, Box 810, Red Bank, New Jersey 07701.

NEW EQUIPMENT DIGEST; EQUIPMENT, MATERIALS, PROCESSES, DESIGNS, LITERATURE. Monthly. Penton Publishing Company, Penton Building, Cleveland, Ohio 44113.

NEW EQUIPMENT NEWS (Canada). Monthly. Canadian Engineering Publications Ltd., 46 St. Clair Avenue, East, Toronto 7, Canada.

NEW EQUIPMENT NEWS (South Africa). Monthly. Viking Publications (Pty.) Ltd., Box 47062, Parklands, Transvaal, South Africa.

WHAT'S NEW IN CHEMICAL PROCESSING EQUIPMENT. Quarterly. Putnam Publishing Company, 111 E. Dearborn Street, Chicago, Illinois 60611.

d. LABORATORIES AND RESEARCH CENTERS

American Society for Testing and Materials. DIRECTORY OF TESTING LABORATORIES, COMMERCIAL-INSTITUTIONAL. Philadelphia: ASTM Special Technical Publication, 33B, 1971.

DIRECTORY OF EUROPEAN ASSOCIATIONS. Vol. 1: NATIONAL INDUSTRIAL, TRADE AND PROFESSIONAL ASSOCIATIONS. Vol. 2: NATIONAL LEARNED, SCIENTIFIC AND TECHNICAL SOCIETIES. Detroit: Gale Research Co., vol. 1 1971; vol. 2, 1975.

DIRECTORY OF TESTING LABORATORIES. Philadelphia: American Society for Testing and Materials, 1973.

INDUSTRIAL RESEARCH LABORATORIES OF THE UNITED STATES. 14th ed. New York: R.R. Bowker Company, 1975.

> Provides information on more than 5,000 laboratories in over 3,000 organizations.

U.S. Federal Energy Administration. DIRECTORY OF FEDERAL AGENCIES ENGAGED IN ENERGY RELATED ACTIVITIES. Washington, D.C.: Government Printing Office, 1975.

U.S. National Bureau of Standards. DIRECTORY OF STANDARDS LABORATORIES. Prepared by the National Conference Standards Laboratories. Gaithersburg, Maryland: 1971.

4. Encyclopedias

BEILSTEINS' HANDBUCH DER ORGANISCHEN CHEMIE. 4th ed. New York: Springer-Verlag, 1972.

BROCKHAUS ABC CHEMIE. 2 vols. Leipzig: Brockhaus, 1965.

> "An encyclopedia of chemical terms. Some of the longer articles carry bibliographies."

CHEMICAL TECHNOLOGY: AN ENCYCLOPEDIC TREATMENT; THE ECONOMIC APPLICATION OF MODERN TECHNOLOGICAL DEVELOPMENTS. New York: Barnes and Noble, 1969-1975.

> "The economic application of modern technological developments. 8 volume work in progress. Covers world's raw materials from chemical standpoint."

Considine, D., ed. ENCYCLOPEDIA OF INSTRUMENTATION AND CONTROL. New York: McGraw-Hill, 1972.

DICTIONARY OF ORGANIC COMPOUNDS. 4th ed., completely revised, enlarged and reset. 5 vols. New York: Oxford University Press, 1965.

> Updated by supplements. Supplement 7, 1971.

ELSEVIER'S ENCYCLOPAEDIA OF ORGANIC CHEMISTRY. Edited by F. Radt. Berlin: Springer-Verlag, 1965.

> Updated by supplements.

ENCYCLOPAEDIC DICTIONARY OF PHYSICS: GENERAL, NUCLEAR, SOLID STATE, MOLECULAR, CHEMICAL, METAL AND VACUUM PHYSICS, ASTRONOMY, GEOPHYSICS, BIOPHYSICS AND RELATED SUBJECTS. J. Thewlis, editor-in-chief. 9 vols. New York: Pergamon, 1961-64.

ENCYCLOPEDIA OF CHEMICAL TECHNOLOGY. 2nd revised ed. 22 vols. 1st supplement. New York: Wiley-Interscience, 1972.

ENCYCLOPEDIA OF CHEMISTRY. George L. Clark, editor-in-chief. Gessner G. Hawley, managing editor. 2nd ed. New York: Reinhold, 1966.

ENCYCLOPEDIA OF INDUSTRIAL CHEMICAL ANALYSIS. Edited by Foster Snell. 17 vols. New York: Interscience, 1966-73.

THE ENCYCLOPEDIA OF THE CHEMICAL ELEMENTS. New York: Reinhold, 1968.

Hampel, C.A. THE ENCYCLOPEDIA OF ELECTROCHEMISTRY. New York: Reinhold, 1964.

Hampel, C.A., et al., eds. THE ENCYCLOPEDIA OF CHEMISTRY. 3rd ed., revised and enlarged. New York: Van Nostrand-Reinhold, 1973.

KINGZETT'S CHEMICAL ENCYCLOPAEDIA; A DIGEST OF CHEMISTRY AND ITS INDUSTRIAL APPLICATIONS. D.H. Hey, general ed. 9th ed. New York: Van Nostrand, 1968.

Mead, W. THE ENCYCLOPEDIA OF CHEMICAL PROCESS EQUIPMENT. New York: Reinhold, 1964.

> Describes types of equipment used in industry.

RODD'S CHEMISTRY OF CARBON COMPOUNDS. 12 vols. to date. New York: Elsevier, 1973.

Sittig, M. INORGANIC CHEMICAL AND METALLURGICAL PROCESS ENCYCLOPEDIA. Park Ridge, New Jersey: Noyes Data Corp., 1968.

_____. ORGANIC CHEMICAL PROCESS ENCYCLOPEDIA. 2nd ed. Park Ridge, New Jersey: Noyes Data Corp., 1969.

Thorpe, J. THORPE'S DICTIONARY OF APPLIED CHEMISTRY. 4th ed., revised and enlarged. 12 vols. New York: Longmans, 1956.

> Standard encyclopedia for applied chemistry field.

Todd, D., ed. THE WATER ENCYCLOPEDIA. New York: Water Information Center, 1970.

> Provides much detailed information on water and water resources.

VAN NOSTRAND'S, SCIENTIFIC ENCYCLOPEDIA. 4th ed. New York: Van Nostrand, 1968.

5. Bibliographies and Guides to the Literature

Includes general guides to scientific and technical literature and bibliographies on chemicals and chemical industry.

Bibliographies on particular topics such as Nuclear Engineering are included in those sections of this Guide.

a. GENERAL

Auger, C. ed. USE OF REPORTS LITERATURE. Hamden, Connecticut: Archon Books, 1975.

BIBLIOGRAPHIC GUIDE TO TECHNOLOGY. Annual. Boston: G.K. Hall Co.

BOOKS IN PRINT. Annual. New York: Bowker.

> Annual listing of books currently in print from American publishers. Author and subject volumes.

CUMULATIVE BOOK INDEX, A WORLD LIST OF BOOKS IN THE ENGLISH LANGUAGE. 11/year (quarterly cumulations). New York: H. W. Wilson.

ENGINEERING SOCIETIES LIBRARY. Classed subject catalog of the Engineering Societies Library. 8th supplement. Boston: G. K. Hall, 1972.

ENGINEERING SOCIETIES LIBRARY. Reference collection of the Engineering Societies Library. New York: Engineering Societies Library, 1970.

> "A list of 600 titles (arranged in 31 classes) available in the Societies' Library Reference Department reading room. An excellent checking copy for technology collections."

Grogan, D. SCIENCE AND TECHNOLOGY; AN INTRODUCTION TO THE LITERATURE. Hamden, Connecticut: Shoe String Press, 1970.

> "A detailed study of the various types of literature available in the fields of science and technology."

Herner, S. A BRIEF GUIDE TO SOURCES OF SCIENTIFIC AND TECHNICAL INFORMATION. Washington, D.C.: Information Resources Press, 1969.

Hittle, D. SOURCEBOOK FOR CHEMISTRY AND PHYSICS. New York: Macmillan, 1973.

Japan Information Center of Science and Technology. CURRENT BIBLIOGRAPHY ON SCIENCE AND TECHNOLOGY; CHEMISTRY AND CHEMICAL INDUSTRY (Japan). 3 per month. Tokyo: Japan Information Center of Science and Technology.

The Johns Hopkins University Applied Physics Laboratory. GUIDE TO THE LITERATURE OF ENGINEERING, MATHEMATICS AND THE PHYSICAL SCIENCES. 2nd ed. Edited by Sylvia Goldman. Silver Spring, Maryland: 1964.

Kyed, J. SCIENTIFIC, ENGINEERING AND MEDICAL SOCIETIES PUBLICA-
TIONS IN PRINT. Annual. New York: Bowker.

McGRAW-HILL BASIC BIBLIOGRAPHY OF SCIENCE AND TECHNOLOGY.
New York: McGraw-Hill, 1966.

    A bibliographical supplement to the MCGRAW-HILL ENCYCLO-
    PEDIA OF SCIENCE AND TECHNOLOGY.

Malinowsky, H. SCIENCE AND ENGINEERING REFERENCE SOURCES. 2nd
ed. New York: Libraries Unlimited, 1976.

Mount, E. GUIDE TO BASIC INFORMATION SOURCES--IN ENGINEERING.
New York: Halstead, 1976.

NEW TECHNICAL BOOKS. Monthly. New York: New York Public Library.

Pavlic, V. GOVERNMENT PUBLICATIONS: A GUIDE TO BIBLIOGRAPHIC
TOOLS. 4th ed. Washington, D.C.: Library of Congress, 1975.

SCIENCE BOOKS; A QUARTERLY REVIEW. Washington, D.C.: American
Association for the Advancement of Science.

    "An annotated, classified listing of new books in the pure and
    applied sciences."

SCIENCE REFERENCE SOURCES. Edited by F. Jenkins. 5th ed. Cambridge,
Massachusetts: MIT Press, 1969.

SCIENTIFIC AND TECHNICAL BOOKS IN PRINT. Annual. New York:
Bowker.

    Listing of 52,000 titles and publishers. Does not include medicine,
    psychology, juvenile books.

Sheehy, E. GUIDE TO REFERENCE BOOKS. 9th ed. Chicago: American
Library Association, 1976.

SUBJECT GUIDE TO BOOKS IN PRINT. Annual. New York: Bowker.

SUBJECT GUIDE TO FORTHCOMING BOOKS. Bimonthly. Supplement to:
SUBJECT GUIDE TO BOOKS IN PRINT. New York: Bowker.

TECHNICAL BOOK REVIEW INDEX. Monthly. Special Libraries Association,
235 Park Avenue, New York, New York 10003.

TECHNICAL BOOKS IN PRINT; A REFERENCE CATALOGUE OF . . .BOOKS IN
PRINT AND ON SALE IN GREAT BRITAIN. Biennial. London: Whitaker.

U.S. National Bureau of Standards. PUBLICATIONS OF THE BUREAU OF
STANDARDS. Annual. Washington, D.C.: Government Printing Office.

Includes brief abstracts. Supplements bring list up to date from first listing in 1901.

U.S. National Technical Information Service. FEDERAL GOVERNMENT'S FIRST PUBLIC SALE OF 500 COMPUTER SEARCHES INTO ADVANCED TECHNOLOGY FOR BUSINESSMEN, SCIENTISTS AND SPECIALISTS IN URBAN, ECONOMIC, AND SOCIAL AFFAIRS. Washington, D.C.: Government Printing Office, 1975.

A list of published searches available.

Walford, A.J. GUIDE TO REFERENCE MATERIAL. Vol 1: SCIENCE AND TECHNOLOGY. 3rd ed. London: Library Association, 1973.

Woodburn, H. USING THE CHEMICAL LITERATURE: A PRACTICAL GUIDE. New York: Marcel Dekker, 1974.

Woodward, A., comp. DIRECTORY OF REVIEW SERIALS IN SCIENCE AND TECHNOLOGY, 1970-1973. London: ASLIB, 1975.

WORLD GUIDE TO TECHNICAL INFORMATION AND DOCUMENTATION SERVICES. 2nd ed. Paris: UNESCO, 1975.

b. CHEMICAL INDUSTRIES—RELATED MATERIALS

American Chemical Society. Division of Chemical Literature. LITERATURE OF CHEMICAL TECHNOLOGY. Advances in Chemistry Series #78. Washington, D.C.: American Chemical Society, 1968.

American Chemical Society. Division of Chemical Literature. LITERATURE RESOURCES FOR CHEMICAL PROCESS INDUSTRIES. Advances in Chemistry Series #10. Washington, D.C.: American Chemical Society, 1954.

American Society for Engineering Education. Engineering. Engineering Libraries Division. GUIDE TO LITERATURE OF CHEMICAL ENGINEERING. Washington, D.C.: The Society.

BIBLIOGRAPHY ON HIGH PRESSURE RESEARCH IN CHEMISTRY AND PHYSICS. Bimonthly. High Pressure Data Center, Brigham Young University, 574 JRCL, Provo, Utah 84601.

Bottle, R. USE OF THE CHEMICAL LITERATURE. London: Butterworth, 1962.

Bourton, K. CHEMICAL AND PROCESS ENGINEERING - UNIT OPERATIONS; A BIBLIOGRAPHICAL GUIDE. New York: IFI/Plenum, 1968.

"Presents a selection from the chemical engineering literature which will be a guide to the available aids to searching and to those works which are likely to serve as the most productive sources for retrieval."

Burman, C. HOW TO FIND OUT IN CHEMISTRY. 2nd ed. New York: Pergamon, 1967.

Cahn, R. SURVEY OF CHEMICAL PUBLICATIONS AND REPORT TO THE CHEMICAL SOCIETY. London: Chemical Society, 1965.

CHEMBOOKS; A BIBLIOGRAPHY OF NEW AND FORTHCOMING BOOKS. Annual. Edited by Peter G. Isler. Karger Libri AG, Petergraben 15, Basel 11, CH-4011, Switzerland.

Text in English, German, French, Italian.

Codlin, E. CRYOGENICS AND REFRIGERATION; A BIBLIOGRAPHICAL GUIDE. 2 vols. New York: Plenum, 1970.

Connolly, T. "Information Sources for Ultrapurification and Characterization." In ULTRAPURITY; METHODS AND TECHNIQUES, edited by Morris Zief and Robert Speights. New York: Dekker, 1972.

Crane, E. A GUIDE TO THE LITERATURE OF CHEMISTRY. 2nd ed. New York: Wiley, 1957.

"Treatment and discussion of the procedures of literature searching, with listings and descriptions of the standard reference books, periodicals, and organizations."

Davis, D., ed. CHEMICAL PROCESSING NOMOGRAPHS. 2 vols. New York: Chemical Publishing, vol. 1, 1960; vol. 2, 1970.

Each volume contains charts and formulas as well as literature references.

HEAT PIPE TECHNOLOGY, A CONTINUING BIBLIOGRAPHY WITH AB-STRACTS. Quarterly. Prepared by University of New Mexico, Technology Application Center, Albuquerque. Available from: National Technical Information Service, Springfield, Virginia 22161.

HYDROGEN PRODUCTION. By D. Cavagnaro. Rept. no. NTIS/PS-76/0459. Springfield, Virginia: National Technical Information Service, 1976.

A bibliography with abstracts. Covers 1967-76.

HYDROGEN STORAGE. By D. Cavagnaro. Part 1: Storage as a Gas or Liquid. Rept. no. NTIS/PS-76/0460. Part 2: Hydrogen as a Hydride. Rept. no. NTIS/PS-76/0461. Springfield, Virginia: National Technical Information Service, 1976.

A bibliography with abstracts. Covers 1974-76.

International Institute of Refrigeration, Paris. BIBLIOGRAPHIC GUIDE TO REFRIGERATION, 1953-1968. 3 vols. Oxford: Pergamon, 1962-69.

"Lists all documents abstracted in the Bulletin of the institute during the 1953-68 period."

Mellon, M. CHEMICAL PUBLICATIONS, THEIR NATURE AND USE. 4th ed. New York: McGraw-Hill, 1965.

Describes literature and reference resource material for field.

OSMOTIC DESALTING MEMBRANES. By D. Cavagnaro. Vol. 1: 1970-73. Rept. no. NTIS/PS-76/0085. Vol. 2: 1974-75. Rept. no. NTIS/PS-76/0086. Springfield, Virginia: National Technical Information Service, 1976.

Citations from ENGINEERING INDEX.

OSMOTIC DESALTING MEMBRANES. By K. Warner. Vol. 1: 1964-72. Rept. no. NTIS/PS-76/0083. Vol. 2: 1973-76. Rept. no. NTIS/PS-76/0084. Springfield, Virginia: National Technical Information Service, 1976.

Citations from NTIS data base.

Poole, J. SOLID-LIQUID SEPARATION: A REVIEW AND A BIBLIOGRAPHY. New York: Chemical Publishing, 1968.

Schamus, J. BIBLIOGRAPHY OF SALINE WATER CONVERSION LITERATURE. Washington, D.C.: U.S. Department of the Interior, 1965. Distributed by U.S. Government Printing Office.

"U.S. Office of Saline Water. Research and development progress report no. 146."

SEA WATER CORROSION. By M. Smith. Rept. no. NTIS/PS-76/0058. Springfield, Virginia: National Technical Information Service, 1976.

A bibliography with abstracts.

WASTE HEAT UTILIZATION. By A. Hunderman. Rept. no. NTIS/PS-76/0276. Springfield, Virginia: National Technical Information Service, 1976.

Citations from NTIS data base.

WASTE HEAT UTILIZATION. By A. Hunderman. Vol. 1: 1970-74. Rept. no. NTIS/PS-76/0277. Vol. 2: 1975-76. Rept. no. NTIS/PS-76/0278. Springfield, Virginia: National Technical Information Service, 1976.

Taken from ENGINEERING INDEX data base.

c. RELATED BIBLIOGRAPHIES IN CURRENT JOURNALS

The following journals frequently contain book reviews and/or bibliographies. See the heading Journals in this section for addresses.

AICHE (journal)
AMERICAN OIL CHEMISTS' SOCIETY (journal)
AMERICAN SOCIETY OF SAFETY ENGINEERS (journal)
AUSTRALIAN CHEMICAL ENGINEERING
AUSTRALIAN CHEMICAL PROCESSING

BRITISH CHEMICAL ENGINEERING
CANADIAN CHEMICAL PROCESSING
CANADIAN JOURNAL OF CHEMISTRY
CHEMICAL AGE
CHEMICAL AND PROCESS ENGINEER-ING
CHEMICAL ENGINEER AND TRANSAC-

TIONS OF THE INSTITUTION OF CHEMICAL ENGINEERS
CHEMICAL ENGINEERING
CHEMICAL ENGINEERING JOURNAL
CHEMICAL ENGINEERING SCIENCE
CHEMICAL ENGINEERING WORLD
CHEMICAL INSTRUMENTATION
CHEMICAL PROCESSING
CHEMICAL PROCESSING AND ENGINEERING
CHEMICAL SOCIETY. QUARTERLY REVIEWS
CHEMICAL TECHNOLOGY
CHEMISTRY IN BRITAIN
CHEMISTRY IN CANADA
CLEAN AIR AND WATER NEWS
COLOR ENGINEERING
COMPUTER ABSTRACTS
COMPUTING REVIEWS
CORROSION SCIENCE
CRYOGENIC TECHNOLOGY
DATA MANAGEMENT
DETERGENTS AND SPECIALTIES
EFFLUENT AND WATER TREATMENT JOURNAL
ENVIRONMENTAL CONTROL AND SAFETY MANAGEMENT
ENVIRONMENTAL SCIENCE AND TECHNOLOGY
FIRE RESEARCH ABSTRACTS AND REVIEWS
FLUID POWER INTERNATIONAL
FRANKLIN INSTITUTE JOURNAL
FUTURES
IDEA
INDUSTRIAL AND ENGINEERING CHEMISTRY
INDUSTRIAL AND ENGINEERING CHEMISTRY - FUNDAMENTALS
INDUSTRIAL AND ENGINEERING CHEMISTRY PROCESS DESIGN AND DEVELOPMENT
INDUSTRIAL MANAGEMENT
INSTITUTION OF CHEMICAL ENGINEERS TRANSACTIONS

INTERNATIONAL CHEMICAL ENGINEERING
INTERNATIONAL JOURNAL OF ENGINEERING SCIENCE
INTERNATIONAL JOURNAL OF ENVIRONMENTAL STUDIES
JOURNAL OF APPLIED CHEMISTRY
JOURNAL OF CHEMICAL DOCUMENTATION
JOURNAL OF ENVIRONMENTAL SCIENCES
JOURNAL OF HUMAN ECOLOGY
JOURNAL OF LUBRICATION TECHNOLOGY
MANAGEMENT ABSTRACTS
MANAGEMENT REVIEW
MANAGEMENT SCIENCE
MANAGEMENT TODAY
MARKETING INFORMATION GUIDE
NON-DESTRUCTIVE TESTING
PATENT LICENSING GAZETTE
POLLUTION ENGINEERING
POLLUTION EQUIPMENT NEWS
PROCESS ENGINEERING, PLANT AND CONTROL
PRODUCT DESIGN AND DEVELOPMENT
PROFESSIONAL ENGINEER
QUALITY ASSURANCE
RECORD OF CHEMICAL PROGRESS
SAFETY JOURNAL
SCI-TECH NEWS
SOUTH AFRICAN CHEMICAL AND ENGINEERING AGE
SOUTH AFRICAN CHEMICAL PROCESSING CHEMICAL REVIEWS
STANDARDS ENGINEERING
THEORETICAL CHEMICAL ENGINEERING ABSTRACTS
VERFAHRENSTECHNISCHE BERICHTE
WATER AND AIR POLLUTION NEWSLETTER
WATER POLLUTION CONTROL
WATER POLLUTION CONTROL FEDERATION (journal)

6. Monographs, Reports, and Symposia Proceedings

A selection of recent literature which seems pertinent to this field. The titles have been chosen randomly from the large amount of literature published and are presented here as an indication of the variety of information available in this format.

Additional titles can be located in the bibliographies and reference tools which list new books and in the journal and serial publications included in the different Sections of the Guide.

ADVANCES IN CHEMICAL ENGINEERING. Vol. 8. New York: American Institute of Chemical Engineers, 1970.

ADVANCES IN DESALINATION. 7th National Symposium on Desalination, Ayyelet Hashabar, Israel, April 1970. Hakirya, Jerusalem, Israel: National Council for Research and Development, 1972.

American Institute of Chemical Engineers. AICHE CONTINUING EDUCATION SERIES. Vol. 4. New York: 1970.

_____. AICHE MONOGRAPH SERIES. Vol. 8. New York: 1974.

_____. AICHE SYMPOSIUM SERIES. Irregular. New York: American Institute of Chemical Engineers.

_____. ANNUAL MEETING. PROCEEDINGS. Annual. New York: American Institute of Chemical Engineers.

_____. CHEMICAL ENGINEERING PROGRESS SYMPOSIUM. Vol. 87. New York: 1968.

_____. 80TH NATIONAL MEETING. Boston, Massachusetts, September 1975. New York: 1976.

_____. 68TH ANNUAL MEETING. Los Angeles, California, November 1975. New York: 1976.

Bailey, C., ed. ADVANCED CRYOGENICS. New York: Plenum, 1971.

Book is based on a course of lectures given to Oxford by authorities in field. Provides overview of the subject field.

Bain, A. THE HYDRAULIC TRANSPORT OF SOLIDS BY PIPELINE. New York: Pergamon, 1971.

A practical guide to fluid conveying.

Bakish, R., ed. PRACTICE OF DESALINATION. Park Ridge, New Jersey: Noyes Data Corp., 1973.

"This volume is based upon 17 instructional papers presented as a special topics course by various academic and industrial experts. The course was held during December 1971 and was sponsored jointly by Fairleigh Dickinson University and the Association of Caribbean Desalination Plants Owners and Operators."

Banks, R. FLUOROCARBONS AND THEIR DERIVATIVES. 2nd ed. New York: Elsevier, 1971.

"Advanced, updated monograph on fluorocarbon compounds and their derivatives."

Bard, A., ed. ELECTROANALYTICAL CHEMISTRY. 6 vols. New York: Dekker, 1973.

"Presents authoritative reviews of topics of current interest in this field and background and problems in areas currently investigated."

Batelle Columbus Laboratories. MOLECULAR SIEVE CONTROL PROCESS SULFURIC ACID PLANTS. By D. Hissong. Rept. no. PB 249 563. Springfield, Virginia: National Technical Information Service, 1975.

_____. MOLECULAR SIEVE MERCURY CONTROL PROCESS IN CHLOR-AKLALI PLANTS. By M. Anastas. Rept. no. PB 251 203. Springfield, Virginia: National Technical Information Service, 1976.

Bechtel Corp. FEASIBILITY STUDY: LIQUID HYDROGEN PLANT, 30 TONS PER DAY. Rept. no. N76-13312. Springfield, Virginia: National Technical Information Service, 1975.

Bennett, J. EDITING FOR ENGINEERS. New York: Interscience, Division of Wiley, 1970.

Bosich, J. CORROSION PREVENTION FOR PRACTICING ENGINEERS. New York: Barnes & Noble, 1970.

"An aid in identifying various types of corrosion and how to control some corrosion problems."

Catalytic, Inc. SULFURIC ACID PLANT EMISSIONS DURING STARTUP, SHUTDOWN, AND MALFUNCTION. By E. Calvin et al. Rept. no. PB 249 508. Springfield, Virginia: National Technical Information Service, 1976.

Chedaille, J. MEASUREMENTS IN FLAMES. 3 vols. London: Edward Arnold, 1972.

"Deals with measurements, techniques, and instruments which have been developed to study large industrial flames."

CHEMICAL ENGINEERING CONGRESS, 1ST PACIFIC. Kyoto, Japan, October 1972. 3 vols. New York: American Institute of Chemical Engineers, 1973.

CHEMICAL ENGINEERING DEVELOPMENTS IN MINERAL PROCESSING. Symposium, London, March 1972. London: Institution of Chemical Engineers, 1973.

CHEMICAL ENGINEERING, SCANDINAVIAN CONGRESS. Copenhagen, January 1974. Copenhagen: Bella Centret, 1975.

THE CHEMICAL INDUSTRY. Irregular. Paris: Organization for Economic Cooperation and Development.

Revised periodically.

CHEMICAL REACTION ENGINEERING. 1st International Symposium on Chemical Reaction Engineering, Washington, D.C. Advances in Chemistry Series 109. Washington, D.C.: American Chemical Society, 1972.

Cockayne, B., ed. MODERN OXIDE MATERIALS. New York: Academic, 1972.

"Reviews the increasing role of oxide materials throughout materials science and technology."

Conference on Industrial Carbons and Graphite, 3rd, London, 1970. PROCEED-INGS. Edited by J. Gregory. London: Society of Chemical Industry, 1971.

"Collection of 90 papers presented at eight sessions covering graph-ite intercalation compounds, oxidation of carbons and graphite, carbon fibres, carbon as a material in mechanical, electrical and chemical engineering, adsorptive properties and pore structure, carbon formation and graphitization, nuclear graphite, and crystal structure, texture and physical properties."

Cook, T. CHEMICAL PLANT AND ITS OPERATION. New York: Pergamon, 1969.

"A practical survey of new operating techniques, plant and equip-ment operations for increased productivity, improved safety stan-dards, and reduced manufacturing costs."

COOLING TOWERS, Symposium at 68th National Meeting. Houston, Texas, February 1971. New York: American Institute of Chemical Engineers, 1972.

CORROSION ENGINEERING, Engineering Summer Conference, Ann Arbor, Michigan, June 1972. Ann Arbor, Michigan: University of Michigan, 1972.

Coulson, J. CHEMICAL ENGINEERING. 3 vols. New York: Pergamon, 1971.

Overview of entire field.

Crowe, C. CHEMICAL PLANT SIMULATION: AN INTRODUCTION TO COMPUTER-AIDED STEADY-STATE PROCESS ANALYSIS. Englewood Cliffs, New Jersey: Prentice Hall, 1971.

Cutler, W., ed. DETERGENCY; THEORY AND TEST METHODS. 2 Pts., Pt. 1. New York: Dekker, 1972.

Surfactant science series.

DEVELOPMENT & INNOVATION FOR AUSTRALIAN PROCESS INDUSTRIES. Chemical Engineering Conference, Australian, 1972. Sydney, Australia: Institution of Engineers, 1972.

Dow Chemical Co. ENERGY CONSUMPTION: FUEL UTILIZATION AND CONSERVATION IN INDUSTRY. Rept. no. PB 246 888. Springfield, Virginia: National Technical Information Service, 1975.

_____. EVALUATION OF NEW ENERGY SOURCES FOR PROCESS HEAT. Rept. no. PB 245 604. Springfield, Virginia: National Technical Information Service, 1975.

Gisser, P. LAUNCHING THE NEW INDUSTRIAL PRODUCT. New York: American Management Association, 1972.

Gosman, A. HEAT AND MASS TRANSFER IN RECIRCULATING FLOWS. New York: Academic, 1969.

"The purpose...is to explain the nature and justification of a general method of predicting those two dimensional flows which require elliptic differential equations for their mathematical description."

Govier, G., ed. FLOW OF COMPLEX MIXTURES IN PIPES. New York: Van Nostrand-Reinhold, 1972.

Gregory, J., ed. THIRD CONFERENCE ON INDUSTRIAL CARBONS AND GRAPHITE. New York: Academic, 1971.

"Contains about a hundred papers, all of which are reports of original research and are accompanied by transcripts of the discussions."

Guthrie, K. PROCESS ESTIMATING, EVALUATION AND CONTROL. Los Angeles: Craftsman Books, 1974.

Haber, L. THE CHEMICAL INDUSTRY, 1900-1930; INTERNATIONAL GROWTH AND TECHNOLOGICAL CHANGE. New York: Oxford University Press, 1971.

"A thorough survey of the United States and European chemical industry from 1900 to 1930."

Hawk, M., ed. BULK MATERIALS HANDLING, VOL. 1. Pittsburgh, Pennsylvania: University of Pittsburgh, 1972.

"Most of the papers are 'How To' types that deal with both actual and imagined problems from a practical standpoint...Can serve as a reference for the operating engineer or provide a good overview of distribution for those in management responsible for cutting the significant cost of bulk-materials handling."

Hicks, T. PUMP APPLICATION ENGINEERING. New York: McGraw-Hill, 1971.

Practical guide to selection of pumps.

Institution of Chemical Engineers. HIGH TEMPERATURE CHEMICAL REACTION ENGINEERING. London: 1972.

"Based on a survey carried out by a multi-disciplinary task force under the sponsorship of the Institution of Chemical Engineers, London. This book is a source of reference material for high temperature processes."

Jones, J. DESIGN METHODS; SEEDS OF HUMAN FUTURES. New York: Interscience, Division of Wiley, 1971.

Provides review of the subject and related topics such as management and marketing.

Lewis, C. SCIENTIFIC INVENTORY CONTROL. New York: Elsevier, 1971.

An introduction to inventory control with explanation of mathematical principles involved.

Nichols, R., ed. PRESSURE VESSEL ENGINEERING TECHNOLOGY. New York: Elsevier, 1971.

McCabe, W., et al. UNIT OPERATIONS IN CHEMICAL ENGINEERING. 3rd ed. New York: McGraw-Hill, 1976.

PERFORMANCE CONCEPT IN BUILDINGS. Symposium, International Union of Testing and Research Laboratories for Materials and Structures (RILEM), American Society for Testing of Materials, and International Council for Building Research and Documentation (CLB), Philadelphia, Pennsylvania, 1972. National Bureau of Standards Special Publication no. 361. Washington, D.C.: Government Printing Office, 1972.

PROBLEMS & PROSPECTUS OF THE CHEMICAL INDUSTRIES IN THE LESS DEVELOPED COUNTRIES. Symposium of Division of Chemical Marketing and Economics at 158th Meeting. New York: American Chemical Society, 1969.

PROGRESS IN REFRIGERATION SCIENCE & TECHNOLOGY. 13th International Congress of Refrigeration, sponsored by International Institute of Refrigeration, Washington, D.C., 1973. 4 vols. Westport, Connecticut: Avi, 1973.

Purchas, D. INDUSTRIAL FILTRATION OF LIQUIDS. Columbus, Ohio: Chemical Rubber Co., 1972.

Ranney, M. COATINGS: RECENT DEVELOPMENTS. Park Ridge, New Jersey: Noyes Data Corporation, 1976.

Reuben, B. THE CHEMICAL ECONOMY. New York: Longmans, 1973.

"Describes the development of the chemical industry in terms of the chemical, social, economic, & engineering factors which have shaped it."

Rosenstock, H. THE ROLE OF CHEMICAL KINETICS IN ENERGY CONSERVATION. Springfield, Virginia: National Technical Information Service, 1975.

Searcy, A., ed. CHEMICAL AND MECHANICAL BEHAVIOR OF INORGANIC MATERIALS. Eleventh course of the Guido Donegani Foundation, directed by Alan W. Searcy, in Tremezzo (Lake Como), Italy, September 8-20, 1968. New York: Interscience, Division of Wiley, 1970.

Concerned with ceramic materials.

Siegel, R. THERMAL RADIATION HEAT TRANSFER. New York: McGraw-Hill, 1972.

> Serves as a guide to engineers working on thermal radiation heat transfer.

Sittig, M. AROMATICS MANUFACTURE AND DERIVATIVES, 1968. Park Ridge, New Jersey: Noyes Data Corp.

> "Contents: Introduction, Production of Aromatics, Separation of Aromatics, Purification of Aromatics, Reactions Giving Hydrocarbon Products, Other Reactions, Phenol Production, Styrene Manufacture and Derivatives, Future Trends."

Smith, J. CHEMICAL ENGINEERING KINETICS. 2nd ed. New York: McGraw-Hill, 1970.

> "Kinetics, homogeneous reactions, heterogeneous catalytic and non-catalytic reactors, and residence time distribution effects are treated in detail in this book."

Solomons, G. MATERIALS AND METHODS IN FERMENTATION. New York: Academic, 1969.

> Covers engineering aspects of fermentation and includes a bibliography of literature.

SOLVENT EXTRACTION CONFERENCE. International (ISEC 71), sponsored by Society of Chemical Industry, Royal Netherlands Chemical Society, and European Federation of Chemical Engineering (101st event), The Hague, Netherlands. New York: Academic.

SONOCHEMICAL ENGINEERING. Published symposium, sponsored by American Institute of Chemical Engineers. Chemical Engineering Progress Symposium Series, vol. 67, no. 109. New York: American Institute of Chemical Engineers, 1971.

SPECIALTY CHEMICALS. New York: Chemical Marketing and Economics Division, American Chemical Society, 1970.

> "Symposium sponsored by the Division of Chemical Marketing and Economics at the 160th meeting of the American Chemical Society, Chicago, Illinois, September 14-18, 1970."

Stepanoff, A. GRAVITY FLOW OF BULK SOLIDS AND TRANSPORTATION OF SOLIDS IN SUSPENSION. New York: Wiley, 1969.

> Covers gravity flow of solids from hoppers. A general review of techniques.

STORAGE & FLOW OF SOLIDS. 2nd Symposium, sponsored by American Society of Mechanical Engineers, Materials Handling Engineering Division, Chicago, Illinois. New York: American Society of Mechanical Engineers.

Uhlig, H. H. CORROSION AND CORROSION CONTROL. 2nd ed. New York: Wiley, 1971.

Standard monograph on corrosion.

UNESCO. UNISIST: STUDY REPORT ON THE FEASIBILITY OF A WORLD SCIENCE INFORMATION SYSTEM. New York: UNIPUB, 1971.

Report on plan for a worldwide scientific information network.

Williams, A. PAINT AND VARNISH REMOVERS, 1972. Park Ridge, New Jersey: Noyes Data Corp.

"This book, based on 163 U.S. patents, gives several hundred formulas from simple alcohols and ketones thru dimethyl sulfoxide to today's elaborate high performance compositions for the removal of epoxies, urethanes, and silicones."

Williams, A. PRINTING INKS, 1972. Park Ridge, New Jersey: Noyes Data Corp.

Based on details from reviews of recent patent literature.

Wolf, F. SEPARATION METHODS IN ORGANIC CHEMISTRY AND BIOCHEMISTRY. New York: Academic, 1969.

7. Abstracting and Indexing Services

a. GENERAL

These publications cover the areas of science and technology but do include specific references to chemical engineering.

APPLIED MECHANICS REVIEWS. Monthly. American Society of Mechanical Engineers, 345 East 47th Street, New York, New York 10017.

Covers applied mechanics and related engineering fields.

APPLIED SCIENCE & TECHNOLOGY INDEX. Monthly. New York: Wilson.

AUSTRALIAN SCIENCE INDEX. Monthly. Commonwealth Industrial Research Organization, Box 89, East Melbourne, Victoria 3002, Australia.

"An index of articles published in Australian scientific and technical periodicals."

BRITISH TECHNOLOGY INDEX. Monthly. Library Assn., 7 Ridgmount Street Street, Store Street, London WCIE 7AE, England.

"A current subject-guide to articles in British technical journals."

CANADIAN PERIODICAL INDEX. Monthly. Canadian Library Association, 63 Sparks Street, Ottawa 4, Canada.

CSIR RESEARCH REVIEW. Semiannual. Council for Scientific and Industrial Research, Box 395, Pretoria, South Africa.

ENERGY ABSTRACTS FOR POLICY ANALYSIS. Monthly. Government Printing Office, Washington, D.C. 20042.

ENGINEERING INDEX. Monthly with annual cumulation. Engineering Index, Inc., 345 East 47th Street, New York, New York 10017.

Abstracts world's engineering and technical literature.

ENGINEERING INDEX CARD SERVICE. Weekly. Engineering Index, Inc., 345 E. 47th Street, New York, New York 10017.

"Annotated bibliography of current worldwide engineering literature. 249 divisions of fields of interest. Divisions can be purchased separately. Listing appears on 3 x 5 card and includes title, author, citation and abstract in addition to subject heading. 1200 per week."

HUNGARIAN TECHNICAL ABSTRACTS. Quarterly. Hungarian Central Technical Library and Centre for Documentation. Text in English, German and Russian. Kultura, Box 149, Budapest 62, Hungary.

INTERNATIONAL AEROSPACE ABSTRACTS. AIAA Technical Information Service, 750 Third Avenue, New York, New York 10017.

Lists abstracts of journal literature.

PANDEX. Quarterly microfiche ed. with annual cumulation. New York: Pandex, Inc., published by CCM Information Sciences, New York.

PHYSICS ABSTRACTS. SCIENCE ABSTRACTS: Section A. Monthly with annual index. Institute of Electrical Engineers, Savoy Place, London W2CR, OBL, England.

Indexing service for physics and related literature.

RESEARCH INDEX. Fortnightly. Business Surveys Ltd., The Mead, Wallington, Suney, England.

SCIENCE CITATION INDEX. Quarterly with annual cumulation. Institute for Scientific Information, 325 Chestnut Street, Philadelphia, Pennsylvania 19106.

"An international interdisciplinary index to the literature of science, medicine, agriculture and technology. A computer produced index, which provides access to related articles by indicating sources in which a known article by a given author has been cited."

WEEKLY GOVERNMENT ABSTRACTS. ENERGY. Government Printing Office, Washington, D.C. 20042.

b. SPECIFIC TO CHEMICAL INDUSTRY

ABSTRACT REVIEW. Monthly. National Paint, Varnish & Lacquer Association, 1500 Rhode Island Avenue, N.W., Washington, D.C. 20005.

American Chemical Society. ABSTRACTS OF PAPERS AT THE ANNUAL MEET-ING. Biannual. American Chemical Society, Special Issues Sales, 1155 16th Street, N.W., Washington, D.C. 20036.

_____. ABSTRACTS OF PAPERS AT THE REGIONAL MEETINGS. Irregular. American Chemical Society, Special Issues Sales, 1155 16th Street, N.W., Washington, D.C. 20036.

_____. ACS SINGLE ARTICLE ANNOUNCEMENTS. Semimonthly. American Chemical Society, 1155 16th Street, N.W., Washington, D.C. 20036.

ANALYTICAL ABSTRACTS. Monthly. Society for Analytical Chemistry, 10 Savile Row, London WIX 1AF, England.

BASIC JOURNAL ABSTRACTS / B J A. A computer based chemical information service. Chemical Abstracts Service, Ohio State University, Columbus, Ohio 43210.

Computer tape format.

BULLETIN SIGNALETIQUE. Part 170: Chimie. Chimie generale, chimie physique, chimie minerale, chimie analytique, chimie organique. Quarterly. Centre National de la Recherche Scientifique, Centre de Documentation du C.N.R.S., 15 Quai Anatole-France, Paris (7c), France.

CARBON BLACK ABSTRACTS. Quarterly. Cabot Corp., 125 High Street, Boston, Massachusetts 02110.

CHEMICAL ABSTRACTS. Weekly. Abstracts Service, American Chemical Society, Ohio State University, Columbus, Ohio 43210.

CHEMICAL ABSTRACTS - APPLIED CHEMISTRY SECTIONS. Fortnightly. American Chemical Society, 1155 16th Street, N.W., Washington, D.C. 20036.

CHEMICAL ABSTRACTS SERVICE SOURCE INDEX (formerly ACCESS). Chemical Abstracts Service, Columbus, Ohio.

A service of chemical abstracts which describes source literature from the "abstracts" and lists library holdings and locations.

CHEMICAL MARKET ABSTRACTS. Monthly. Chemical Horizons, Inc., 274 Madison Avenue, New York, New York 10016.

CHEMICAL SUBSTRUCTURE INDEX / C.S.I. MONTHLY. Institute for Scientific Information, 325 Chestnut Street, Philadelphia, Pennsylvania 19106.

"A permuted listing of Wiswesser Line Notations that indexes new compounds reported in CURRENT ABSTRACTS OF CHEMISTRY AND INDEX CHEMICUS."

CHEMICAL TITLES: CURRENT AUTHOR AND KEYWORD INDEXES FROM SELECTED CHEMICAL JOURNALS. Biweekly. Abstracts Service, American Chemical Society, Easton, Pennsylvania.

"A concordance, produced by electronic computers, to chemical research papers. Titles are selected from some 700 journals of pure and applied chemistry."

CORROSION ABSTRACTS. Bimonthly. National Association of Corrosion Engineers, 2400 W. Loop South, Houston, Texas 77027.

Cryogenic Data Center. CURRENT AWARENESS SERVICE. Weekly. Commerce Department, National Bureau of Standards, Institute for Basic Standards, Boulder, Colorado. Cryogenic Data Center National Bureau of Standards, Boulder, Colorado 80302.

CURRENT ABSTRACTS OF CHEMISTRY AND INDEX CHEMICUS / C A C & I C. Weekly. Formerly INDEX CHEMICUS. Institute for Scientific Information, 325 Chestnut Street, Philadelphia, Pennsylvania 19106.

CURRENT CONTENTS, PHYSICAL & CHEMICAL SCIENCES / C C-P CS. Weekly. Institute for Scientific Information, 325 Chestnut Street, Philadelphia, Pennsylvania 19106.

"Reproduction of the contents pages of approximately 800 international journals."

DESALINATION ABSTRACTS. Quarterly. Center of Scientific and Technological Information, 84 Hachashmonaim Street, Tel Aviv, Israel.

ENCYCLOPAEDIA CHIMICA INTERNATIONALIS. Annual. Institute for Scientific Information, 325 Chestnut Street, Philadelphia, Pennsylvania 19106.

The annual cumulation of CURRENT ABSTRACTS OF CHEMISTRY AND INDEX CHEMICUS.

INDEX TO REVIEWS, SYMPOSIA VOLUMES AND MONOGRAPHS IN ORGANIC CHEMISTRY FOR THE PERIOD 1940-1960. New York: Pergamon, 1962.

Japan Information Center of Science and Technology. CURRENT BIBLIOGRAPHY ON SCIENCE AND TECHNOLOGY: COMPLETE CHEMICAL ABSTRACTS OF JAPAN (Japan). Monthly. Japan Information Center of Science and Technology, 2-5-2 Nagata-cho, Chiyoda-ku, Tokyo, Japan.

JOURNAL OF APPLIED CHEMISTRY. Abstracting. Society of Chemical Industry, 14 Belgrave Square, London, S.W. 1, England.

LEAD ABSTRACTS. Monthly. Lead Development Association, 34 Berkeley Square, London, W. 1, England.

A review journal of recent technical literature on the use of lead and its products.

Muszaki Lapszemle Kemia. VEGYIPAR/TECHNICAL ABSTRACTS CHEMICAL IN-DUSTRY. Orszagos Muszaki Konyvtar es Dokumentacios, Kozpont, Reviczky u. 6, Budapest 8, Hungary.

PREDICASTS ABSTRACTING SERVICES. Quarterly. Predicasts, 200 University Circle Research Center, 11001 Cedar Avenue, Cleveland, Ohio 44106.

Economic information on chemicals.

SEARCH. Abstracting Service – Monthly. Compendium Publishers International Corp., 2175 Lemoine Avenue, Fort Lee, New Jersey 07024.

Has following divisions in general chemical related industry:
SEARCH: INORGANIC CHEMICALS DIVISION
SEARCH: METALS DIVISION
SEARCH: NON-METALLIC MINERALS DIVISION
SEARCH: OILS, FATS & WAXES DIVISION
SEARCH: ORGANIC CHEMICALS DIVISION
SEARCH: DYES, PIGMENTS & COATINGS DIVISION
SEARCH: ESSENTIAL OILS, SOAPS & TOILETRIES DIVISION

SOLID-LIQUID FLOW ABSTRACTS. Monthly. Gordon and Breach Science Publishers, 12 Bloomsbury Way, London, England.

THEORETICAL CHEMICAL ENGINEERING ABSTRACTS. Bimonthly. Chemical Engineering Abstracts, Box 146, Liverpool, L69 2BL, England.

THERMAL ABSTRACTS. Bimonthly. Heating & Ventilating Research Association, Bracknell, Berks., England.

WEEKLY GOVERNMENT ABSTRACTS. CHEMISTRY. Government Printing Office, Washington, D.C. 20042.

WORLD SURFACE COATINGS ABSTRACTS. Monthly. Charley & Pickersgill, Ltd., Amberley House, Norfolk Street, Strand, London W.C. 2, England.

URETHANE ABSTRACTS. Monthly. Prepared by the Franklin Institute. Technomic Publishing Co., Inc., 265 W. State Street, Westport, Connecticut 06880.

"Each month URETHANE ABSTRACTS reviews current articles, papers, patents, government reports, and other literature on urethanes – both from the U.S. and around the world."

VERFAHRENSTECHNISCHE BERICHTE / CHEMICAL AND PROCESS ENGINEER-ING ABSTRACTS. Bimonthly. Verlag Chemie GmbH, Pappelallee 3, Postfach 129, 6940, Weinheim/Bergstrasse, West Germany.

ZINC ABSTRACTS. Zinc Development Association, 34 Berkeley Square, London W. 1, England.

c. RELATED ABSTRACTS IN CURRENT JOURNALS

The following journals and serials frequently contain abstracts of related literature. See the Journals in this section for additional details as to publisher's
address, etc.

AICHE. Journal

AMERICAN OIL CHEMISTS' SOCIETY JOURNAL

BIBLIOGRAPHY OF PUBLICATIONS OF UNIVERSITY BUREAUS OF BUSINESS
AND ECONOMIC RESEARCH. Boulder, Colorado: Business Research
Division, University of Colorado.

BRITISH CHEMICAL ENGINEERING

CHEMICAL ENGINEER AND TRANSACTIONS OF THE INSTITUTION OF
CHEMICAL ENGINEERS

CHEMICAL ENGINEERING

CHEMICAL ENGINEERING PROGRESS

CHEMICAL ENGINEERING SCIENCE

CHEMICAL PROCESSING AND ENGINEERING

CHEMICAL TECHNOLOGY

CLEAN AIR AND WATER NEWS

COMPUTER AND INFORMATION SYSTEMS

COMPUTING REVIEWS

CORROSION SCIENCE

CRYOGENIC TECHNOLOGY

DESIGN NEWS

ENVIRONMENTAL SCIENCE AND TECHNOLOGY

IDEA

INDUSTRIAL AND ENGINEERING CHEMISTRY

INDUSTRIAL MANAGEMENT

INSTITUTION OF CHEMICAL ENGINEERS TRANSACTIONS

INTERNATIONAL CHEMICAL ENGINEERING

JOURNAL OF ENVIRONMENTAL SCIENCES

JOURNAL OF SAFETY RESEARCH

NEW EQUIPMENT NEWS (Canada)

NEW EQUIPMENT NEWS (South Africa)

NON-DESTRUCTIVE TESTING

PATENT LICENSING GAZETTE

SANITARY ENGINEERING RESEARCH LABORATORY NEWS QUARTERLY

WATER AND AIR POLLUTION NEWSLETTER

d. GUIDES TO ABSTRACTING AND INDEXING SERVICES
FOR CHEMICAL INDUSTRIES

ABSTRACTING SERVICES. Vol. 1: Science and Technology. Hague: International Federation for Documentation, 1969.

Sheehy, E. GUIDE TO REFERENCE BOOKS. 9th ed. Chicago: American Library Association, 1976.

ULRICH'S INTERNATIONAL PERIODICAL DIRECTORY. Annual. New York: Bowker.

U.S. Library of Congress. Science and Technology Division. NATIONAL FEDERATION OF SCIENCE ABSTRACTING AND INDEXING SERVICES. Report no. 102. Washington, D.C.: 1963.

> A guide to the world's abstracting and indexing services in science and technology.

8. Journals and Serial Publications

This part contains a majority of the titles of journals, periodicals, newsletters, bulletins, and serial publications which pertain to chemical engineering industries.

Some publications whose scope is general are included, such as the management journals, because these may be of interest to chemical engineers who are in administrative positions.

The remaining titles listed here pertain to the chemical manufacturing and process type industries.

Titles of journals in the other varied fields of chemical engineering, such as Petroleum Engineering, Food Engineering, are to be found in other sections of this Guide.

a. DIRECTORIES AND LISTS

ACCESS. Bimonthly. Environment Information Center, Inc., 124 E. 39th Street, New York, New York 10016.

> Abstracting service for all phases of environmental effects.

ANNOTATED DIRECTORY OF PRODUCT AND INDUSTRY JOURNALS. Unipub, Inc., Box 433, New York, New York 10018.

> "Lists 800 journals and includes content analysis and ordering information for each."

AYER'S DIRECTORY OF PUBLICATIONS. Annual. Ayer Press, Westminster Square, Philadelphia, Pennsylvania 19106.

DIRECTORY OF CANADIAN SCIENTIFIC AND TECHNICAL PERIODICALS. National Science Library, Ottawa, Canada.

GUIDES TO SCIENTIFIC PERIODICALS. Edited by M. Fowler. The Library Association, London, 1966.

> Annotated bibliography.

Public Documents Department. GOVERNMENT PERIODICALS AND SUBSCRIPTION SERVICES, 147th ed. Washington, D.C.: Government Printing Office, February 1973.

STANDARD PERIODICAL DIRECTORY. Oxbridge Publishing Co., New York, New York.

ULRICH'S INTERNATIONAL PERIODICAL DIRECTORY. 2 vols. Annual. New York: Bowker.

b. JOURNAL AND SERIAL TITLES

ABSTRACT REVIEW. Monthly. National Paint, Varnish & Lacquer Association, 1500 Rhode Island Avenue, N.W., Washington, D.C. 20005.

ABSTRACTS FROM TECHNICAL AND PATENT PUBLICATIONS. Weekly. British Internal Combustion Engine Research Institute, 111/112 Buckingham Avenue, Slough, Bucks, England.

ABSTRACTS OF AIR AND WATER CONSERVATION LITERATURE AND PATENTS. Weekly. American Petroleum Institute, Central Abstracting and Indexing Service, 1271 Avenue of the Americas, New York, New York 10020.

> Available in microfilm.

ABSTRACTS ON HEALTH EFFECTS OF ENVIRONMENTAL POLLUTANTS. Monthly. Biosciences Information Service of Biological Abstracts, 2100 Arch Street, Philadelphia, Pennsylvania 19103.

ACCESS. Bimonthly. Environment Information Service, 124 East 39th Street, New York, New York 10016.

> Computerized information service for environmental area literature searches, special studies on contract bases; copy abstract service as well as microfiche of governmental reports on subscription.

ADVANCES IN CHEMICAL ENGINEERING. Edited by Thomas B. Drew et al. New York: Academic Press, 1970.

> "Critical reviews bringing standard topics up to date and to summarize new developments for chemical engineers."

ADVANCES IN CORROSION SCIENCE AND TECHNOLOGY. Vol. 4.
Edited by Mars G. Fontana and Roger W. Staehle. New York: Plenum, 1974.

ADVANCES IN CRYOGENIC ENGINEERING. Annual. New York: Plenum.

Proceedings of cryogenic engineering conferences.

ADVANCES IN ECOLOGICAL RESEARCH. Vol. 6. New York: Academic
Press, 1969.

ADVANCES IN ELECTROCHEMISTRY AND ELECTROCHEMICAL ENGINEERING.
Vol. 9. Edited by Paul Delahay. New York: Wiley, 1973.

Reviews latest developments on the subject.

ADVANCES IN ENVIRONMENTAL SCIENCES AND TECHNOLOGY. Vol. 4.
New York: Wiley, 1973.

ADVANCES IN HEAT TRANSFER. Vol. 9. New York: Academic Press,
1973.

ADVANCES IN HIGH PRESSURE RESEARCH. Edited by R.S. Bradley. New
York: Academic Press, 1970 (c1969).

"Reviews recent work on topics of current interest to research scientists and engineers."

ADVANCES IN HYDROSCIENCE. Vol. 9. New York: Academic Press, 1973.

Continuing series on hydrology and wave motion in fluids.

ADVANCES IN WATER POLLUTION RESEARCH. Biennial. New York:
Pergamon.

AICHE JOURNAL. Bimonthly. American Institute of Chemical Engineers,
345 East 47th Street, New York, New York 10017.

AIR POLLUTION ABSTRACTS. Monthly. U.S. Government Printing Office,
Washington, D.C . 20402.

A bulletin of the Environmental Protection Agency.

AIR POLLUTION ABSTRACTS. Monthly. United Kingdom, Department of
Trade and Industry. Warren Spring Laboratory, Gunnels Wood Road, Stevenage,
Herts., SG1 2BX, England.

AIR POLLUTION CONTROL ASSOCIATION. JOURNAL. Monthly. Air
Pollution Control Association, 4400 Fifth Avenue, Pittsburgh, Pennsylvania
15213.

AIR POLLUTION TITLES. Bimonthly. Center for Environment Studies.
Pennsylvania State University, 226 Chemical Engineering II, University Park,
Pennsylvania 16802.

AIR/WATER POLLUTION REPORT. Weekly. Business Publications, Inc., P.O. Box 1067, Silver Spring, Maryland 20910.

    The pioneer business newsletter in the environmental field--since 1963--"Covering significant aspects of environmental problems."

American Chemical Society. ABSTRACTS OF PAPERS AT THE NATIONAL MEETING. Biannual. American Chemical Society, Special Issues Sales, 1155 16th Street, N.W., Washington, D.C. 20036.

    _____. ABSTRACTS OF PAPERS AT THE REGIONAL MEETINGS. Irregular. American Chemical Society, Special Issues Sales, 1155 16th Street, N.W., Washington, D.C. 20036.

    _____. ACS SINGLE ARTICLE ANNOUNCEMENTS. Semimonthly. American Chemical Society, 1155 16th Street, N.W., Washington, D.C. 20036.

AMERICAN OIL CHEMISTS' SOCIETY JOURNAL. Monthly. 35 E. Wacker Drive, Chicago, Illinois 60601.

AMERICAN SOCIETY OF SAFETY ENGINEERS JOURNAL. Monthly. American Society of Safety Engineers, 850 Busse Highway, Park Ridge, Illinois 60068.

American Water Resources Association. PROCEEDINGS. Annual. American Water Resources Association, 206 E. University Avenue, Urbana, Illinois 61801.

AMERICAN WATER WORKS ASSOCIATION JOURNAL. Monthly. American Water Works Association, 2 Park Avenue, New York, New York 10012.

ANALYTICAL ABSTRACTS. Monthly. Society for Analytical Chemistry, 9-10 Savile Row, London, W1X 1AF, England.

ANNUAL REPORTS ON PROGRESS OF CHEMISTRY. Chemical Society, Burlington House, Piccadilly, London, W1V 0BN, England.

ANNUAL REVIEW OF FLUID MECHANICS. Annual. Palo Alto, California: Annual Reviews.

ANNUAL REVIEW OF PHYSICAL CHEMISTRY. Annual. Palo Alto, California: Annual Reviews.

ANSI REPORTER/STANDARDS ACTION. Fortnightly. Formerly: STANDARDS INSTITUTE REPORTER. American National Standards Institute, 1430 Broadway, New York, New York 10018.

APCA ABSTRACTS. Monthly. Air Pollution Control Association, 4400 Fifth Avenue, Pittsburgh, Pennsylvania 15213.

APPLIED MECHANICS REVIEWS. Monthly. American Society of Mechanical Engineers.

    Covers applied mechanics and related engineering fields.

APPLIED SCIENCE AND TECHNOLOG INDEX. Monthly. New York: Wilson.

ATMOSPHERIC ENVIRONMENT. Quarterly. New York: Pergamon.

AUSTRALIAN CHEMICAL ENGINEERING; A TECHNICAL JOURNAL FOR CHEM-ICAL ENGINEERS. Monthly. Box 250, North Sydney, N.S.W., Australia.

AUSTRALIAN CHEMICAL PROCESSING AND ENGINEERING. Monthly. Law-son Publications Ltd., 49 Clarence Street, Sydney, Australia.

AUSTRALIAN SCIENCE INDEX. Monthly. Industrial Research Organization, Box 89, East Melbourne, Victoria 3002, Australia.

"Index of articles published in Australian scientific and technical periodicals."

BASIC JOURNAL ABSTRACTS/B J A. Fortnightly. Chemical Abstracts Service, Ohio State University, Columbus, Ohio 43210.

"A computer based chemical information service."

BRITISH CHEMICAL ENGINEERING. Monthly. 33-40 Bowling Green Lane, London E.C.1, England.

BRITISH PATENTS ABSTRACTS. Weekly. Derwent Publications Ltd., Rochdale House, 128 Theobalds Road, London, WC1X8RP, England.

BRITISH TECHNOLOGY INDEX. Monthly. Library Association, 7 Ridgmount Street, Store Street, London, WC1E7AE, England.

"A current subject-guide to articles in British technical journals."

BULLETIN SIGNALETIQUE. Part 170: Chemistry. Quarterly. Centre National de la Recherche Scientifique. 1961. m. 500 F. Centre de Documentation du C.N.R.S., 15 Quai Anatole-France, Paris (7e), France.

CALLS FOR PAPERS. Weekly. CCM Information Corp., 909 Third Avenue, New York, New York 10022.

Lists upcoming meetings to alert scientists and engineers.

CANADIAN CHEMICAL PROCESSING. Monthly. Southam Business Publica-tions Ltd., 1450 Don Mills Road, Don Mills, Ontario, Canada.

CANADIAN JOURNAL OF CHEMICAL ENGINEERING. Monthly. Canadian Society for Chemical Engineering. Chemical Institute of Canada, 151 Slater Street, Ottawa 4, Canada.

CANADIAN JOURNAL OF CHEMISTRY. Semimonthly. National Research Council of Canada, Ottawa 7, Ontario.

Text in English and French.

CANADIAN OCCUPATIONAL SAFETY. Bimonthly. Parkins, 1215 Greene Avenue, Montreal 215, Quebec, Canada.

CANADIAN PERIODICAL INDEX. Monthly. Canadian Library Association, 63 Sparks Street, Ottawa 4, Canada.

CARBON BLACK ABSTRACTS. Quarterly. Cabot Corp., 125 High Street, Boston, Massachusetts 02110.

CHEMICAL ABSTRACTS. 1907. Weekly. Chemical Abstracts Service, American Chemical Society, Ohio State University, Columbus, Ohio 43210.

CHEMICAL ABSTRACTS - APPLIED CHEMISTRY SECTIONS. Fortnightly. American Chemical Society, 1155 16th Street, N.W., Washington, D.C. 20036.

CHEMICAL ABSTRACTS SERVICE SOURCE INDEX. Quarterly. Formerly ACESS. Chemical Abstracts Service, American Chemical Society, Ohio State University, Columbus, Ohio 43210.

> A service of Chemical Abstracts which describes source literature from the "Abstracts" and lists library holdings and locations.

CHEMICAL AGE. Weekly. Benn Brothers Ltd., Bouverie House, Fleet Street, London E.C.4, England.

CHEMICAL AND ENGINEERING NEWS. Weekly. American Chemical Society, 1155 16th Street, N.W., Washington, D.C. 20036.

CHEMICAL AND PROCESS ENGINEERING. Monthly. 28 Essex Street, Strand, London, W.C.2, England.

CHEMICAL COMMUNICATIONS. Semimonthly. Chemical Society, Burlington House, London W1V OBN, England.

> "A journal for urgent preliminary accounts of important chemical research."

CHEMICAL DIGEST. Quarterly. Foster D. Snell, Inc., Hanover Road, Florham Park, New Jersey 07932.

CHEMICAL ENGINEER AND TRANSACTIONS OF THE INSTITUTION OF CHEMICAL ENGINEERS. Monthly. Belgrave Square, London S.W.1, England.

CHEMICAL-ENGINEERING. Monthly. McGraw-Hill Book Co., 1221 Avenue of the Americas, New York, New York 10020.

CHEMICAL ENGINEERING PROGRESS. Monthly. American Institute of Chemical Engineers, 345 East 47th Street, New York, New York 10017.

CHEMICAL ENGINEERING SCIENCE. Monthly. Editor-in-Chief, Professor A. W. Hendry. Elmsford, New York: Pergamon.

> "Publishes original papers dealing with the application of chemical engineering to chemistry, physics and mathematics."

CHEMICAL ENGINEERING WORLD. Monthly. Industrial Publications, A-3 Jeevan Jyot, 18/20, Cowasji Patel Street, Bombay 1, India.

CHEMICAL EQUIPMENT. Monthly. Gordon Publications, Inc., 20 Community Place, Morristown, New Jersey 07960.

Deals with new products for chemical processing fields.

CHEMICAL HORIZONS WEEKLY REPORTS. Weekly. North American and overseas editions. Chemical Horizons, Inc., 274 Madison Avenue, New York, New York 10016.

"An international weekly summary for chemical marketing and sales executives."

CHEMICAL INDUSTRY NOTES. Weekly. American Chemical Society, Chemical Abstracts Service, Accounting Department, Box 3012, Ohio State University, Columbus, Ohio 43210.

CHEMICAL INSTRUMENTATION. Quarterly. Marcel Dekker, Inc., 95 Madison Avenue, New York, New York 10016.

CHEMICAL MARKET ABSTRACTS. Monthly. Predicasts Inc., 200 University Circle, 1101 Cedar Avenue, Cleveland 44106.

CHEMICAL MARKETING REPORTER. Weekly. Schnell Publishing Company, Inc., 100 Church Street, New York, New York 10007.

Covers chemical and process materials industries.

CHEMICAL NEWSLETTER. Quarterly. National Safety Council, 425 North Michigan Avenue, Chicago, Illinois 60611.

CHEMICAL PEDDLER. Annual. Salesmen's Association of the American Chemical Industry, Inc., 79 Madison Avenue, New York, New York 10016.

CHEMICAL PROCESSING (ENG.). Monthly. Engineering, Chemical and Marine Press Ltd., 33-39 Bowling Green Lane, Box 42, London EC1P 1 AH, England.

CHEMICAL PROCESSING (U.S.). Monthly. Putnam Publishing Co., 111 East Delaware Place, Chicago, Illinois 60611.

CHEMICAL PROCESSING & ENGINEERING. Monthly. Colour Publications Private Ltd., 126-A, Dhuruwadi, Off Dr. Nariman Road, Bombay 25 D D, India.

CHEMICAL PURCHASING. Monthly. Myers Publishing Co., 381 Park Avenue South, New York, New York 10016.

Wide coverage of purchasing in chemical process industries.

CHEMICAL REVIEWS. Bimonthly. American Chemical Society, 1155 16th Street, N.W., Washington, D.C. 20036.

Chemical Society, London. ANNUAL REPORTS ON THE PROGRESS OF CHEM-
ISTRY. Section A: General, Physical, and Inorganic Chemistry. Chemical
Society, Burlington House, Piccadilly, London W1V 0BN, England.

_____. QUARTERLY REVIEWS. Chemical Society, Burlington House, Picca-
dilly, London W1V 0BN, England.

CHEMICAL SPOTLIGHT. Weekly. Corporate Intelligence, 25 Broad Street,
New York, New York 10004.

CHEMICAL SUBSTRUCTURE INDEX/C D I. Monthly. Institute for Scientific
Information, 325 Chestnut Street, Philadelphia, Pennsylvania 19106.

CHEMICAL TECHNOLOGY. Monthly. American Chemical Society, 1155 16th
Street, N.W., Washington, D.C. 20036.

CHEMICAL TECHNOLOGY REVIEW. Annual. Noyes Data Corp., Mill Road,
Park Ridge, New Jersey 07656.

CHEMICAL TITLES. Fortnightly. Chemical Abstracts Service, American Chemical
Society, Ohio State University, 1775 S. College Road, Columbus, Ohio 43210.

> "Current author and keyword indexes from selected chemical journals."

CHEMICAL WEEK. Weekly. McGraw-Hill Book Co., 1221 Avenue of the Americas,
New York, New York 10020.

CHEMISTRY IN BRITAIN. Monthly. Royal Institute of Chemistry, Burlington
House, Piccadilly, London W1N 0BN, England.

CHEMISTRY IN CANADA. Monthly. Chemical Institute of Canada, 151 Slater
Street, Ottawa 4, Canada.

> Text in English and French.

CHEMOSPHERE. Bimonthly. Elmsford, New York: Pergamon.

> "Chemosphere offers rapid dissemination of original articles related
> to environmental affairs and reports scientific investigations in the
> fields of chemistry, physics, and biology."

CIS INDEX. Monthly. Congressional Information Service, 500 Montgomery
Building, Washington, D.C. 20014.

> Looseleaf service.

CLEAN AIR AND WATER NEWS. Weekly. Commerce Clearing House, Inc.,
4025 W. Peterson Avenue, Chicago, Illinois 60646.

> "News about air and water pollution, treatment, waste and disposal."

CLEAN WATER REPORT. Biweekly. Business Publishers Inc., P.O. Box 1067,
Silver Spring, Maryland 20910.

"Market opportunities in water pollution abatement...government controls - water research programs - demonstration results...and reuse."

CLEARINGHOUSE ANNOUNCEMENTS IN SCIENCE & TECHNOLOGY. U.S. Department of Commerce, National Technical Information Service, Springfield, Virginia 22151.

Abstracts of report literature.

COLOR ENGINEERING (U.S.). Bimonthly. Chromatic Communications, Inc., 799 Roosevelt Road, Building 4, Suite 300, Glen Ellyn, Illinois 60137.

Committee for International Cooperation in Information Retrieval Among Examining Patent Offices. PROCEEDINGS OF ANNUAL MEETINGS. Committee for International Cooperation in Information Retrieval Among Examining Patent Offices, 32 Chemin des Colombelles, 1211 Geneva 10, Switzerland.

COMPUTER ABSTRACTS. Monthly. Fancourt Technical Information Co., Ltd., Box 59, St. Helier, Jersey, British Channel Islands.

COMPUTER AND INFORMATION SYSTEMS. Bimonthly. Edited by F. Columbus. Cambridge Communications Corp., c/o Stevens House, 10 South Prince Street, Lancaster, Pennsylvania 17603.

COMPUTING REVIEW. Monthly. Association for Computing Machinery, 1133 Avenue of the Americas, New York, New York 10036.

CORROSION ABSTRACTS. Bimonthly. National Association of Corrosion Engineers, 2400 W. Loop South, Houston, Texas 77027.

CORROSION SCIENCE. Monthly. University of Manchester, Manchester, England. Pergamon Press, Inc., Maxwell House, Fairview Park, Elmsford, New York 10523.

"An international Journal of the Science and Practice of Corrosion and Protection published under the auspices of the Corrosion Science Society and the Centre Belge d'Etude de la Corrosion."

CRYOGENIC DATA CENTER. Weekly. Commerce Department, National Bureau of Standards, Institute for Basic Standards. Cryogenic Data Center, Boulder, Colorado 80302.

"Current Awareness Service, publications and reports of interest in cryogenics noted, list."

CRYOGENICS AND INDUSTRIAL GASES. 6/yr. Business Communications, Inc., 2800 Euclid Avenue, Cleveland, Ohio 44115.

CRYOGENIC TECHNOLOGY. Bimonthly. Cryogenic Society of America. Technology Publishing Corp., 825 South Barrington Avenue, Los Angeles, California 90049.

CSIR RESEARCH REVIEW. Semiannual. Council for Scientific and Industrial Research, Box 395, Pretoria, South Africa.

CURRENT ABSTRACTS OF CHEMISTRY AND INDEX CHEMICUS/C A C & I C. Weekly. Formerly: INDEX CHEMICUS. Institute for Scientific Information, 325 Chestnut Street, Philadelphia, Pennsylvania 19106.

CURRENT CHEMICAL TRANSLATIONS. Irregular. Transchem, Inc., Chemical Translation Service, Box 669, Knoxville, Tennessee 37901.

CURRENT CONTENTS, CHEMICAL SCIENCES. Weekly. Institute for Scientific Information, 325 Chestnut Street, Philadelphia, Pennsylvania 19106.

CURRENT CONTENTS, ENGINEERING & TECHNOLOGY/C C-E T. Weekly. Institute for Scientific Information, 325 Chestnut Street, Philadelphia, Pennsylvania 19106.

> "Reproduction of the contents pages of approximately 700 international journals."

CURRENT CONTENTS, PHYSICAL & CHEMICAL SCIENCES/C C-P C S. Weekly. Institute for Scientific Information, 325 Chestnut Street, Philadelphia, Pennsylvania 19106.

> "Reproduction of the contents pages of aproximately 800 international journals."

CURRENT INDEX TO CONFERENCE PAPERS. Monthly. CCM Information Corp., 909 Third Avenue, New York, New York 10022.

CURRENT PAPERS ON COMPUTERS & CONTROL/C P C. Monthly. Institute of Electrical and Electronics Engineers, Inc., 345 E. 47th Street, New York, New York 10017.

DATA MANAGEMENT (U.S.). Monthly. Data Processing Management Association, 505 Busse Highway, Park Ridge, Illinois 60068.

DESALINATION ABSTRACTS. Quarterly. Center of Scientific & Technological Information, 84 Hachashmonaim Street, Tel Aviv, Israel.

DESIGN NEWS. Fortnightly. Cahners Publishing Co., Inc., 270 St. Paul Street, Denver, Colorado 80206.

DETERGENTS AND SPECIALTIES. Monthly. Ralph E. Dorland, 4 Second Avenue, Denville, New Jersey 07834.

> "The magazine for the detergent, soap, cosmetic and toiletry, wax, polish and aerosol industries."

DIRECTORY OF PUBLISHED PROCEEDINGS. Series S E M T (Science, Engineering, Medicine and Technology). Monthly. Interdok Corp., 173 Hoastead Avenue, Box 326, Harrison, New York 10528.

EFFLUENT AND WATER TREATMENT JOURNAL. Monthly. Thunderbird Enterprises Ltd., 102 College Road, Harrow, Middlesex, England.

EMPLOYMENT SAFETY AND HEALTH GUIDE. Chicago: Commerce Clearing House, 1974.

"Contains a directory of federal Office of Safety and Health Administration Offices through the U.S."

ENCYCLOPAEDIA CHIMICA INTERNATIONALIS. Annual. Institute for Scientific Information, 325 Chestnut Street, Philadelphia, Pennsylvania 19106.

"An annual cumulation of CURRENT ABSTRACTS OF CHEMISTRY AND INDEX CHEMICUS."

ENGINEERING INDEX. Monthly with annual cumulation. Engineering Index, Inc., 345 East 47th Street, New York, New York 10017.

Abstracts world's engineering and technical literature.

ENGINEERING INDEX CARD SERVICE. Weekly. Engineering Index, Inc., 345 East 47th Street, New York, New York 10017.

"Annotated bibliography of current worldwide engineering literature. 249 divisions of fields of interest. Divisions can be purchased separately. Listing appears on 3 x 5 card and includes title, author, citation and abstract in addition to subject heading. 1200 per week."

ENVIRONMENTAL CONTROL & SAFETY MANAGEMENT. Monthly. Formerly: SAFETY MAINTENANCE. A.M. Best Co., Park Avenue, Morristown, New Jersey 07960.

ENVIRONMENTAL POLLUTION AND CONTROL. Weekly. National Technical Information Service, Springfield, Virginia 22161.

An abstracting service for entire pollution field.

ENVIRONMENTAL SCIENCE AND TECHNOLOGY. Monthly. American Chemical Society, 1155 Sixteenth Street, N.W., Washington, D.C. 20036.

Coverage of industrial environment and pollution control.

ENVIRONMENT INDEX. Annual. Environment Information Center, Inc., 124 East 39th Street, New York, New York 10016.

Indexes all literature pertaining to environment such as journals, books, conference proceedings, and technical reports. Topics covered are pollution, pesticides, work environment safety and related subjects.

FACTORY EQUIPMENT NEWS. Fortnightly. Production Publications (London), Ltd., 10-16 Elm Street, London W.C.1, England.

FIRE RESEARCH ABSTRACTS AND REVIEWS. 3/yr. National Academy of Sciences, Committee on Fire Research, National Research Council, 2101 Constitution Avenue, N.W., Washington, D.C. 20418.

FLUID POWER INTERNATIONAL. Monthly. Morgan-Grampian (Publishers) Ltd., Riverside House, Hough Street, Woolwich, London S.E. 18, England.

FRANKLIN INSTITUTE. Journal. Monthly. Franklin Institute, 20th and Benjamin Franklin Parkway, Philadelphia, Pennsylvania 19103.

FRENCH PATENTS ABSTRACTS. Weekly. Derwent Information Service, Rochdale House, Theobalds Road, London WC1X 8RP, England.

FROM THE STATE CAPITOLS REPORT: SEWAGE AND WASTE DISPOSAL. Monthly. Bethune Jones, 321 Sunset Avenue, Asbury Park, New Jersey 07712.

"Lists developments affecting air and water pollution control matters."

FUTURES. Quarterly. Iliffe-NTP Inc., 300 East 42nd Street, New York, New York 10017.

"The journal of forecasting and planning."

GERMAN PATENTS GAZETTE. Weekly. Derwent Publications Ltd., Rochdale House, 128 Theobalds Road, London WC1X 8RP, England.

GOVERNMENT REPORTS ANNOUNCEMENTS AND GOVERNMENT REPORTS INDEX. Monthly. National Technical Information Service, Springfield, Virginia 22151.

Great Britain. Board of Trade. PATENTS, DESIGNS AND TRADE MARKS. ANNUAL REPORT. H.M.S.O., P.O.B., 569, London S.E.1, England.

GROUND WATER NEWSLETTER. Quarterly. Water Information Center, Inc., Public Relations Department, Water Research Building, Manhasset Isle, Port Washington, Long Island, New York 11050.

Coverage includes all aspects of pollution, underground waste disposal, recharge, management and development.

HUNGARIAN TECHNICAL ABSTRACTS. Quarterly. Hungarian Central Technical Library & Centre for Documentation, Kultura, Box 149, Budapest 62, Hungary.

"Text in English, German and Russian."

HYDATA. Monthly. American Water Resources Association, 206 E. University Avenue, Urbana, Illinois 61801.

Including WATER RESOURCES NEWSLETTER.

ICIREPAT BULLETIN. U.S. Patent Office, Washington, D.C. 20231.

Relates news of international patent scene.

IDEA (U.S.). Bimonthly. Patent, Trademark and Copyright Research Institute, George Washington University, Washington, D.C. 20006.

"Patent, Trademark and Copyright Journal of Research and Education."

INDEX TO REVIEWS, SYMPOSIA VOLUMES AND MONOGRAPHS IN ORGANIC CHEMISTRY FOR THE PERIOD 1940-1960. 3 vols. Elmsford, New York: Pergamon, 1965.

INDEX TRANSLATIONUM 23: INTERNATIONAL BIBLIOGRAPHY OF TRANSLATIONS. New York: Unipub, Inc., 1972.

"Latest issue provides full bibliographic data on 41,322 translated books published in 1970--in every field of learning."

INDUSTRIAL AND ENGINEERING CHEMISTRY. Monthly. American Chemical Society, 1155 16th Street, N.W., Washington, D.C. 20036.

INDUSTRIAL & ENGINEERING CHEMISTRY FUNDAMENTALS. Quarterly. American Chemical Society, 1155 16th Street, N.W., Washington, D.C. 20036.

INDUSTRIAL & ENGINEERING CHEMISTRY PROCESS DESIGN AND DEVELOPMENT. Quarterly. American Chemical Society, 1155 16th Street, N.W., Washington, D.C. 20036.

INDUSTRIAL & ENGINEERING CHEMISTRY PRODUCT RESEARCH AND DEVELOPMENT. Quarterly. American Chemical Society, 1155 16th Street, N.W., Washington, D.C. 20036.

INDUSTRIAL DESIGN. 10/yr. Whitney Publications, Inc., 130 E. 59th Street, New York, New York 10022.

"Designing for industry."

INDUSTRIAL EQUIPMENT, MATERIALS & SERVICES. Monthly. Envoy Journals Ltd., 67 Clerkenwell Road, London E.C.1, England.

INDUSTRIAL EQUIPMENT NEWS (U.S.). Monthly. Thomas Publishing Company, 461 Eighth Avenue, New York, New York 10001.

"What's new in equipment parts, materials and literature and catalogs."

INDUSTRIAL HYGIENE DIGEST. Monthly. Industrial Health Foundation, Inc., Information Services, 5231 Centre Avenue, Pittsburgh, Pennsylvania 15232.

Abstracting service for all aspects of industrial environmental conditions.

INDUSTRIAL MANAGEMENT (U.S.). Monthly. Industrial Management Society, 2217 Tribune Tower, Chicago, Illinois 60611.

INDUSTRIAL SAFETY. Monthly. United Trade Press Ltd., 9 Gough Square, London E.C.4, England.

INDUSTRIAL WATER CONDITIONING. Monthly. Publicom Inc., 17 Sherwood Glace, Greenwich, Connecticut 06830.

INDUSTRIAL WATER ENGINEERING. Monthly. Select Publications, Inc., 900 Northstar Center, Minneapolis, Minnesota 55402.

INDUSTRY WEEK. Weekly. Penton Publishing Co., Penton Building, Cleveland, Ohio 44113.

INFORMATION SCIENCE ABSTRACTS. Bimonthly. Documentation Abstracts, Inc., Box 8510, Philadelphia, Pennsylvania 19101.

Institution of Chemical Engineers. TRANSACTIONS. Quarterly. Institution of Chemical Engineers, 16, Belgrave Square, London SW1X 8PT, England.

INTERNATIONAL AEROSPACE ABSTRACTS. Semimonthly. AIAA Technical Information Service, 750 Third Avenue, New York, New York 10017.

 Lists abstracts of journal literature.

INTERNATIONAL CHEMICAL ENGINEERING. Quarterly. American Institute of Chemical Engineers, 345 E. 47th Street, New York, New York 10017.

 "A quarterly journal of translations from Russia, Eastern Europe, Asia and Japan."

INTERNATIONAL CHEMICAL REGISTER. Monthly. Economic Documentation Office, Graaf Florislaan 30A, Box 505, Hilversum, Netherlands.

 "A monthly buyers' guide."

International Congress of Pure and Applied Chemistry. LECTURES. Biennial. International Union of Pure and Applied Chemistry. Butterworth & Company (Publishers) Ltd., 88 Kingsway, London, WC2B 6AB, England.

INTERNATIONAL JOURNAL OF ENGINEERING SCIENCE. Monthly. Pergamon Press, Inc., Maxwell House, Fairview Park, Elmsford, New York 10523.

 "This Journal contains high quality, original research pertaining to the application of the physical, chemical and mathematical sciences to engineering."

INTERNATIONAL JOURNAL OF ENVIRONMENTAL STUDIES. Quarterly. Gordon and Breach Science Publishers, 12 Bloomsbury Way, London W.C.1, England.

INTERNATIONAL LICENSING. Monthly. International Licensing Ltd., 92 Cannon Lane, Pinner, Middlesex HA5 1HT, England.

 "A monthly bulletin providing an international forum for the negotiation of manufacturing licenses and joint ventures. Publishes details of both new and established products available for manufacturing under license, or joint venture arrangements, from worldwide sources."

International Union of Pure and Applied Chemistry. COMPTES RENDUS OF

IUPAC CONFERENCE. Biennial. International Union of Pure and Applied Chemistry, Bank Court Chembers, 2-3 Pound Way, Cowley Centre, Oxford OX4 3YF, England.

> Text in English or French.

INTERPAS MONTHLY PATENT DATA-BULLETINS. Monthly. Interpas (International Patent Service), Buitenahven 25, Hertogenbosch, Netherlands.

> "Text mostly in English; French and German titles in original language."

ISRAEL STATE RECORDS-PATENTS AND DESIGNS JOURNAL. Monthly. Israel Patent Office, Ministry of Justice, Government Printer, Jerusalem, Israel.

> "Part A: Patents, Trademarks and Copyrights; Part B: Patents and Designs. Text in English and Hebrew."

ISRAEL STATE RECORDS-TRADE MARKS JOURNAL. Monthly. Israel Patent Office, Ministry of Justice, Government Printer, Jerusalem, Israel.

> Text in English and Hebrew.

JAPANESE MONTHLY PATENT DATA. Monthly. International Patent Service, (Interpas) N.V., Buitenhaven 25, Ben Bosch, Netherlands.

> "Contains the English title of all Japanese patents issued during preceding month."

JAPANESE PATENTS ABSTRACTS. Weekly. Derwent Publications Ltd., Rochdale House, 128 Theobalds Road, London W.C.1, England.

> "Chemical coverage only. Text in English."

Japan Information Center of Science and Technology. CURRENT BIBLIOGRAPHY ON SCIENCE AND TECHNOLOGY: COMPLETE CHEMICAL ABSTRACTS OF JAPAN (Japan). Monthly. Japan Information Center of Science and Technology, 2-5-2 Nagata-cho, Chiyoda-ku, Tokyo, Japan.

JAPAN PATENT NEWS. Monthly. Japan Patent Center, Inc., Box 72 Shitaya P.O., Tokyo, Japan. Distributor for the Americas, Europe & Africa: Interpas (International Patent Service) Douglaslaan 5, Rosmalen, Netherlands.

JOB SAFETY & HEALTH REPORT. Weekly. Business Publications, Inc., P.O. Box 1067, Silver Spring, Maryland 20910.

> "Federal and state legislation...standards...regulations...research activities...enforcement cases...are a few of the topics covered bi-weekly in the informative JS&HR."

JOURNAL OF APPLIED CHEMISTRY. Monthly. Society of Chemical Industry, 14 Belgrave Square, London S.W.1, England.

JOURNAL OF CHEMICAL DOCUMENTATION. Quarterly. American Chemical Society, 1155 16th Street, N.W., Washington, D.C. 20036.

JOURNAL OF CHEMICAL ENGINEERING OF JAPAN. Quarterly. Society of Chemical Engineers, Kyotitsu Building 6-19, Kohinata 4-chome, Bunkyo-ku, Tokyo, Japan.

JOURNAL OF ENVIRONMENTAL SCIENCES. Bimonthly. Institute of Environmental Sciences, 940 E. Northwest Highway, Mt. Prospect, Illinois 60056.

JOURNAL OF FIRE AND FLAMMABILITY. Quarterly. Technomic Publishing Co., Inc., 265 W. State Street, Westport, Connecticut 06880.

"Authoritative source of new information on: The Flammability of Materials • Fire Exposures and Characteristics • Fire and Flammability in Various Environments • Text Methods Apparatus, and Results, etc.

JOURNAL OF HUMAN ECOLOGY. Quarterly. Human Ecological Society, Box 146, Elsah, Illinois 62028.

JOURNAL OF LUBRICATION TECHNOLOGY. Quarterly. American Society of Mechanical Engineers, United Engineering Center, 345 E. 47th Street, New York, New York 10017.

"Series F - ASME Transactions."

JOURNAL OF PHYSICAL AND CHEMICAL REFERENCE DATA. Quarterly. American Chemical Society, 1155 16th Street, N.W., Washington, D.C. 20036.

"Designed to continue information formerly presented in the INTERNATIONAL CRITICAL TABLES."

JOURNAL OF SAFETY RESEARCH. Quarterly. National Safety Council, 425 North Michigan Avenue, Chicago, Illinois 60611.

JOURNAL ON HUMAN ECOLOGY. Quarterly. Human Ecological Society, Box 146, Elsah, Illinois 62028.

LEAD ABSTRACTS. Monthly. Lead Development Association, 34 Berkeley Square, London W.1, England.

MALAWI PATENT JOURNAL AND TRADE MARKS JOURNAL. Monthly. Malawi Government Printing Department, Box 37, Zomba, Malawi.

MANAGEMENT ABSTRACTS. Quarterly. British Institute of Management, Management House, Park Street, London WC2B SPT, England.

MANAGEMENT INDEX. Monthly, except July. Keith Business Library, Box 453, Ottawa, Canada.

"The international monthly guide to new American-Canadian-British sources of business and technical information."

MANAGEMENT REVIEW. Monthly. American Management Association, 135 West 50th Street, New York, New York 10020.

MANAGEMENT SCIENCE. Bimonthly. Institute of Management Sciences, P.O. Box 6112, Providence, Rhode Island 02904.

MANAGEMENT TODAY. Monthly. Management Publications Ltd., Gillow House, 5 Winsley Street, London WIN 8AP, England.

Supersedes: MANAGER.

MANUFACTURING CHEMIST AND AEROSOL NEWS. Monthly. Morgan-Grampian Publications Ltd., 25 Essex Street, London WC2, England.

MARKETING INFORMATION GUIDE. Monthly. U.S. Department of Commerce BDSA. Superintendent of Documents, Washington, D.C. 20402.

METRON (ENG.). Monthly. Permagon Press, Inc., Elmsford, New York 10523.

"Measurement, control, automation."

MODERN CHEMICALS. Bimonthly. Modern Chemicals Publishing Co., Box 810, Red Bank, New Jersey 07701.

MONTHLY CATALOG. U.S. Superintendent of Documents, Government Printing Office, Washington, D.C. 20402.

Muszaki Lapszemle Kemia. VEGYIPAR/TECHNICAL ABSTRACTS CHEMICAL INDUSTRY. Monthly. Orszagos Muszaki Konyvtar es Dokumentacios, Lozpont, Reviczky u.6, Budapest 8, Hungary.

NATIONAL SAFETY NEWS. Monthly. National Safety Council, 425 North Michigan Avenue, Chicago, Illinois 60611.

News source for safety engineers and those in safety field.

NETHERLANDS PATENTS REPORT. Weekly List. Derwent Publications Ltd., Rochdale House, 128 Theobalds Road, London WC1X 8RP, England.

NEW EQUIPMENT DIGEST. Monthly. Penton Publishing Co., Penton Building, Cleveland, Ohio 44113.

"Equipment, materials, processes, designs, literature."

NEW EQUIPMENT NEWS (CANADA). Monthly. Canadian Engineering Publications Ltd., 46 St. Clair Avenue, East, Toronto 7, Canada.

NEW EQUIPMENT NEWS (SOUTH AFRICA). Monthly. Viking Publications (Pty.) Ltd., Box 47062, Parklands, Transvalt, South Africa.

NEW ZEALAND PATENT OFFICE JOURNAL. Monthly. Government Printing Office, Private-Bag, Wellington, New Zealand.

NOISE CONTROL REPORT. Biweekly. Business Publishers Inc., P.O. Box 1067, Silver Spring, Maryland 20910.

"Covers significant developments at city, state, and federal levels... legislative, technical, research...standards to control noise on the job and in the overall environment."

NON-DESTRUCTIVE TESTING (ENG.). Bimonthly. IPC Science and Technology Press Ltd., IPC House, 32 High Street, Guildford, Surrey, England. U.S. address: IPC Inc., 300 E. 42nd Street, New York, New York 10017.

NOTES FROM THE PATENT OFFICE. 1970. Monthly. U.S. Trademark Association, 6 East 45th Street, New York, New York 10017.

Looseleaf service containing material excerpted from TRADEMARK REPORTER.

OCCUPATIONAL HAZARDS. Monthly. 614 Superior Avenue West, Cleveland, Ohio 44113.

Covers industrial security, hygiene, safety and related areas.

OCCUPATIONAL SAFETY AND HEALTH. Monthly. OCCUPATIONAL SAFETY AND HEALTH SUPPLEMENT. Royal Society for the Prevention of Accidents, 52 Grosvenor Gardens, London, S.W.1, England.

OCCUPATIONAL SAFETY AND HEALTH ABSTRACTS. Monthly. International Labor Office, 917 15th Street, N.W., Washington, D.C. 20005.

Available in card form.

OCCUPATIONAL SAFETY AND HEALTH REPORTER. Monthly. Bureau of National Affairs, 1231 25th Street, N.W., Washington, D.C. 20037.

OFFICIAL JOURNAL. (Patents) Weekly. Patent Office, St. Mary Cray, Orpington, Kent, BR5 3RD, England.

ORGANIC SYNTHESES. Annual. John Wiley & Sons, 605 Third Avenue, New York, New York 10016.

"An annual publication of satisfactory methods for the preparation of organic chemicals. Part of a continuing series on synthetic method."

PANDEX CURRENT INDEX OF SCIENTIFIC AND TECHNICAL LITERATURE. Fortnightly. Macmillan Information Co., 866 Third Avenue, New York, New York 10022.

Patent and Trademark Institute of Canada. ANNUAL PROCEEDINGS. Patent and Trademark Institute of Canada, P.O. Box 553, Station B, Ottawa 4, Ontario, Canada.

PATENT AND TRADEMARK REVIEW. Monthly. Clark Boardman Co., Ltd., 435 Hudson Street, New York, New York 10014.

"Monthly, the text of new laws and official rules, proposed legislation, and important decisions affecting intellectual property on a world-wide basis. Published 11 times annually."

PATENT-DIGEST INTERNACIONAL. Monthly. Antoni Guilleumas Brosa, Calle Parque 3, apto. 1381, Barcelona-2, Spain.

PATENT EXCHANGE NEWS AND REPORTER. Bimonthly. Patent Exchange Inc., 26 Broadway, New York, New York 10004.

PATENT JOURNAL INCLUDING TRADEMARKS AND MODELS. Monthly. Government Printer, Bosman Street, P-Bag 85, Pretoria, South Africa.

"Text in Afrikaans and English."

PATENT LAW REVIEW. Annual. Sage Hill Publishers, Inc., 116 Washington Avenue, Albany, New York 12210.

PATENT LICENSING GAZETTE. Bimonthly. Techni Research Associates, Professional Center Building, Willow Grove, Pennsylvania 19090.

PATENT OFFICE RECORD (CANADA). Weekly. Commissioner of Patents. Information Canada, Publishing Division, Ottawa, Canada.

"Text and title in English and French."

PATENT OFFICES BULLETIN. Irregular. Committee for International Cooperation in Information Retrieval, Among Examining Patent Offices, 32 Chemin des Colombelles, 1211 Geneva 20, Switzerland.

PATENT OFFICE SOCIETY. JOURNAL. Monthly. 104 Academy Avenue, Federalsburg, Maryland 21632.

PATENTS THROUGHOUT THE WORLD. Supplemented three times annually. Clark Boardman Co., Ltd., 435 Hudson Street, New York, New York 10014.

"The patent laws of the world digested and uniformly arranged. Contains information concerning: law; conventions; patentee; novelty; opposition; taxes; working; compulsory licenses; marking; assignment; documents required."

PHYSICS ABSTRACTS. Section A: Science Abstracts. Biweekly. Installation of Electrical Engineers, Savoy Place, London WC2R 0BL, England.

POLLUTION ABSTRACTS. Bimonthly. Pollution Abstracts, Inc., P.O. Box 2369, La Jolla, California 92037.

POLLUTION ENGINEERING. Monthly. W. Arthur West Publishing Co., 35 Mason Street, Greenwich, Connecticut 06830.

POLLUTION EQUIPMENT NEWS. Bimonthly. 8550 Babcock Boulevard, Pittsburgh, Pennsylvania 15237.

PREDICASTS ABSTRACTING SERVICE. Quarterly. Predicasts, 200 University Circle Research Center, 11001 Cedar Avenue, Cleveland, Ohio 44106.

Economic information on chemicals.

PRINCIPAL CHEMICAL AND CHEMICALLY RELATED TERMS USED IN UNITED STATES PATENTS. Annual. Information for Industry, Inc., 1000 Connecticut Avenue, N.W., Washington, D.C.

Thesaurus of patent terms.

PROCEEDINGS IN PRINT. Bimonthly. Proceedings in Print, Inc., Box 247, Mattapan, Massachusetts 02426.

"Lists and indexes of proceedings of conferences, symposia, etc. in all subject areas, and in all languages."

PROCESS ENGINEERING, PLANT AND CONTROL. Monthly. Morgan-Grampian (Publishers) Ltd., 28 Essex Street, Strand, London W.C.2, England.

PRODUCT DESIGN AND DEVELOPMENT. Monthly. Chilton Co., Chestnut & 56th Streets, Philadelphia, Pennsylvania 19139.

PRODUCT DESIGN ENGINEERING. Monthly. Design Engineering Publications Ltd., Hermes House, 89 Blackfriars Road, London S.E. 1, England.

PROFESSIONAL ENGINEER (U.S.). Monthly. National Society of Professional Engineers, 2029 K Street, N.W., Washington, D.C. 20006.

PROGRESS IN HEAT AND MASS TRANSFER. Continuing series. 6 vols. to date. New York: Pergamon.

Contains papers on the subject from various conferences.

PROGRESS IN SEPARATION AND PURIFICATION. Vol. 4. New York: Wiley-Interscience, 1973.

PROGRESS IN THE CHEMISTRY OF FATS AND OTHER LIQUIDS. Continuing series. New York: Pergamon.

Critical reviews of current developments in the field.

PROTECTION. 11/year. Alan Osborne & Associates, 113 Blackheath Park, London, SE3, England.

Official journal of the Institution of Industrial Safety, Offices of Great Britain.

PUBLIC WORKS. Monthly. Public Works Journal Corp., 200 South Broad Street, Ridgewood, New Jersey 07451.

QUALITY ASSURANCE. Quarterly. Institute of Quality Assurance, 146 Cromwell Road, Kensington, London, SW7 4EF, England.

"The management and engineering magazine for quality control, product reliability and vendor evaluation."

QUARTERLY REVIEWS. Quarterly. Chemical Society, Burlington House, London, W1V 0BN, England.

RECENT ADVANCES IN ENGINEERING SCIENCE. Vol. 5, in 2 parts. Edited by A. Eringen. Proceedings of the Sixth Annual Meeting of the Society of Engineering Science held at Princeton University, Princeton, New Jersey, November 11-13, 1968. New York: Gordon & Breach, 1970.

RECENT ADVANCES IN LIQUID-LIQUID EXTRACTION. New York: Pergamon, 1972.

"This book brings into one volume a comprehensive treatment of many topics associated with liquid extraction that are otherwise only found scattered about the literature in unorganized form."

RECENT DEVELOPMENTS IN SEPARATION SCIENCE. Cleveland, Ohio: CRC Press, 1972.

RECORD OF CHEMICAL PROGRESS. Quarterly. Plenum Press, 226 West 17th Street, New York, New York 10011.

RESEARCH INDEX. Fortnightly. Business Surveys Ltd., The Mead, Wallington, Surrey, England.

REPORTS OF PATENT, DESIGN, TRADE MARK AND OTHER CASES. Irregular. Patent Office (Sale Branch), St. Mary Cray, Orpington, Kent, BR5 BRD, England.

REPORTS ON THE PROGRESS OF APPLIED CHEMISTRY. Annual. Society of Chemical Industry, 14 Belgrave Avenue, London, S.W.1, England.

REUSE/RECYCLE. Monthly. Technomic Publishing Company, 265 West State Street, Westport, Connecticut 06880.

"New Recycling Processes...New Products from Wastes • Recycling Technology and Economics • Municipal, Industrial, and Consumer Waste Recycling, etc."

REVIEWS IN ANALYTICAL CHEMISTRY. Quarterly. Freund Publishing House, Ltd., Box 35010, Tel Aviv, Israel.

REVIEWS OF PURE AND APPLIED CHEMISTRY. Quarterly. Royal Australian Chemical Institute, Headquarters: Clunies Ross House, 191 Royal Parade, Parkville 3052, Australia.

REVIEWS ON COATINGS AND CORROSION. Quarterly. Freund Publishing House, Ltd., P.O.B. 35010, Tel Aviv, Israel.

S.A. CHEMICAL & ENGINEERING AGE. Monthly. Chemical Industry Publications (Pty.), Box 3080, 6th Floor, B.P. Centre, 36 Kerk Street, Johannesburg, South Africa.

S.A. CHEMICAL PROCESSING. Bimonthly. Thomson Publications, South Africa (Pty.) Ltd., Box 80, Cape Town, South Africa.

SAFETY JOURNAL. Monthly. J.R. Young, Box 19, Anderson, South Carolina.

SAFETY NEWS. Monthly. Royal Society for the Prevention of Accidents, 52 Grosvenor Gardens, London, S.W. 1, England.

SAFETY NEWSLETTER - CHEMICAL SECTION. Monthly. National Safety Council, Chicago, Illinois 60611.

> Provides chemical information sheets telling properties, hazards and precautions of various toxic materials.

SAFETY PRODUCTS NEWS. Bimonthly. Industrial Safety Products, Equipment & Systems, Ames Publishing Co., One West Olney Avenue, Philadelphia, Pennsylvania 19120.

SAFETY STANDARDS. Bimonthly. Superintendent of Documents, Government Printing Office, Washington, D.C. 20402.

SANITARY ENGINEERING RESEARCH LABORATORY NEWS QUARTERLY. University of California, College of Engineering, Sanitary Engineering Research Laboratory, Richmond Field Station, 1301 S. 46th Street, Richmond, California 94804.

SCIENCE CITATION INDEX. Quarterly with Annual Cumulation. Philadelphia: Institute for Scientific Information.

> "Subtitle: An international interdisciplinary index to the literature of science, medicine, agriculture and technology. A computer-produced index, which provides access to related articles by indicating sources in which a known article by a given author has been cited."

SCIENTIFIC AND TECHNICAL AEROSPACE REPORTS. Semimonthly. National Aeronautics and Space Administration, Scientific and Technical Information Division, Washington, D.C. 20546.

> Indexes NASA report literature.

SCIENTIFIC MEETINGS. Quarterly. Special Libraries Association, 235 Park Avenue, New York, New York 10003.

> "Lists information about future conferences and conventions, arranged alphabetically by society and university, chronologically, geographically, and by subject."

SCI-TECH NEWS. Science-Technology Division, Special Libraries Association. Jet Propulsion Laboratory Library, 4800 Oak Grove Drive, Pasadena, California 91103.

> Regularly lists bibliographies, many of which are government sponsored research projects.

SCI-TECH QUARTERLY INDEX TO U.S. GOVERNMENT TRANSLATIONS. CCM Information Corp., 909 3rd Avenue, New York, New York 10022.

SEARCH MONTHLY ABSTRACT SERVICE. Compendium Publishers International Corp. 2175 Lemoine Avenue, Fort Lee, New Jersey 07024.

> Has following divisions related to chemical industry:

> SEARCH: INORGANIC CHEMICALS DIVISION.

> SEARCH: METALS DIVISION.

> SEARCH: NON-METALLIC MINERALS DIVISION.

> SEARCH: OILS, FATS & WAXES DIVISION.

> SEARCH: ORGANIC CHEMICALS DIVISION.

> SEARCH: CHEMICAL MATERIALS & PRODUCTS DIVISION.

> SEARCH: COAL, COKE & MINERAL TARS DIVISION.

> SEARCH: DYES, PIGMENTS & COATINGS DIVISION.

> SEARCH: ESSENTIAL OILS, SOAPS & TOILETRIES DIVISION.

SELECTED WATER RESOURCES ABSTRACTS. Semimonthly. Water Resources Scientific Information Center, Office Water Resources Research, Department of the Interior. National Technical Information Service, U.S. Department of Commerce, Springfield, Virginia 22151.

SOAP/COSMETICS/CHEMICAL SPECIALTIES. Monthly. MacNair-Dorland Co., Inc. 101 West 31st Street, New York, New York 10001.

> Covers the production and marketing of soaps, detergents, toiletries and cosmetics and other chemical specialties.

SOCMA PATENT LIST. Annual. Synthetic Organic Chemical Manufacturing Association, 261 Madison Avenue, New York, New York 10016.

SOLID-LIQUID FLOW ABSTRACTS. Irregular. British Hydromechanics Research Association, Cranfield, Bedford, MK43 0AJ, England.

SOLID WASTE REPORT. Biewekly. Business Publishers Inc., P.O. Box 1067, Silver Spring, Maryland 20910.

> "Reports on federal, state and local legislation-grants and contracts-products-conferences...etc."

SOLVENT EXTRACTION REVIEWS. Semiannual. Marcel Dekker, 95 Madison Avenue, New York, New York 10016.

Summarizes latest developments in solvent extraction.

SOVIET INVENTIONS ILLUSTRATED. Weekly. Derwent Publications Ltd., Rochdale House, 128 Theobalds Road, London, WC1X 8RP, England.

STANDARDS: MONTHLY ADDITIONS. Formerly: ISI STANDARDS: MONTHLY ADDITIONS. Indian Standards Institution, Manak Bhavan, 9 Bahadur Shah Zafar Marg, New Delhi 1, India.

STANDARDS ACTION. Fortnightly. American Standards Institute, 1430 Broadway, New York, New York 10018.

STANDARDS AND SPECIFICATIONS INFORMATION BULLETIN. Weekly. National Standards Association, Inc., 1321 Fourteenth Street, N.W., Washington, D.C. 20005.

Lists all new federal, military, Air Force-Navy aeronautical and aerospace standards.

STANDARDS ENGINEERING. Bimonthly. Box 7507, Philadelphia, Pennsylvania 19101.

Standards Engineers Society. PROCEEDINGS OF ANNUAL MEETING. Standards Engineers Society, Box 7507, Philadelphia, Pennsylvania 19101.

THEORETICAL CHEMICAL ENGINEERING ABSTRACTS. Bimonthly. Chemical Engineering Abstracts, Box 146, Liverpool, L69 2BL, England.

TODAY'S SAFETY GUIDES ABSTRACTS. U.S. Department of Labor, Bureau of Labor Standards, Office of Occupational Safety, Washington, D.C. 20210.

Available in card format.

TOXICITY BIBLIOGRAPHY. Quarterly. Superintendent of Documents, U.S. Government Printing Office, Washington, D.C. 20402.

TRADEMARKS JOURNAL (CANADA). Weekly. Information Canada, Publishing Division, Ottawa, K1A 0S9, Canada.

Text in English and French.

TRADEMARK ALERT. Weekly. TCR Service, Inc., 140 Sylvan Avenue, Englewood Cliffs, New Jersey 07632.

THE TRADEMARK REPORTER. Monthly. U.S. Trademark Association, 6 E. 45th Street, New York, New York 10017.

Authoritative reporting service for this field.

TRADEMARKS JOURNAL (ENG.). Weekly. Patent Office, St. Mary Cray, Orpington, Kent, BR5 3RD, England.

TRADEMARKS THROUGHOUT THE WORLD. By White and Ravenscroft. 3/yr. Trade Activities, Inc., 374 Madison Avenue, New York, New York 10017.

TRANSATOM BULLETIN. Monthly. Euratom, CID Agence et Messageries de la Presse S.A., 1 rue de la Petite, ille Brussels, Belgium.

TRANSDEX. Quarterly. Macmillan Information Corp., 866 Third Avenue, New York, New York 10022.

   Guide and index to JPRS translations.

TRANSLATIONS REGISTER-INDEX. Monthly. National Translation Center, John Crerar Library, 35 West 33rd Street, Chicago, Illinois 60616.

   Lists new translations received at the national translation center.

ULRICH'S INTERNATIONAL PERIODICAL DIRECTORY. Annual. R.R. Bowker Co., 1180 Avenue of the Americas, New York, New York 10036.

U.S. CHEMICAL PATENTS. Weekly. U.S. Federal Government. Godfrey Memorial Library, 134 Newfield Street, Middletown, Connecticut 06457.

U.S. Department of Commerce. BUSINESS SERVICE CHECKLIST. Weekly. Government Printing Office, Washington, D.C. 20042.

_____. CHEMICAL INDUSTRY REPORT. Quarterly. Government Printing Office, Washington, D.C. 20042.

_____. CHEMICALS, QUARTERLY INDUSTRY REPORT. Government Printing Office, Washington, D.C. 20042.

_____. INDUSTRIAL GASES. Prepared in Industry Division, U.S. Department of Commerce, Washington, D.C. 20230.

_____. TRADEMARK, RULES OF PRACTICE OF PATENT OFFICE WITH FORMS AND STATUTES, 1973. 8th ed., rev. Government Printing Office, Washington, D.C. 20042.

U.S. Federal Supply Service. INDEX OF FEDERAL SPECIFICATIONS, STANDARDS AND HANDBOOKS. Annual. Government Printing Office, Washington, D.C. 20042.

   "Contains lists of Federal and Interim Federal Specifications and Standards; Military Specifications and Standards; Federal Handbooks; and General Services Administration Specifications."

U.S. Library of Congress. MONTHLY CHECKLIST OF STATE PUBLICATIONS. Monthly. Government Printing Office, Washington, D.C. 20042.

U.S. National Bureau of Standards. COMMERCIAL STANDARDS. Irregular. Government Printing Office, Washington, D.C. 20042.

_____. NATIONAL STANDARDS REFERENCE DATA SYSTEM. Irregular. Government Printing Office, Washington, D.C. 20042.

_____. PRODUCT STANDARDS. Irregular. Government Printing Office, Washington, D.C. 20042.

U.S. Patent Office. CLASSIFICATION BULLETINS. Irregular. Commissioner of Patents, Washington, D.C. 20231.

_____. INDEX OF PATENTS ISSUED FROM THE PATENT OFFICE. Annual. Government Printing Office, Washington, D.C. 20042.

_____. INDEX OF TRADEMARKS ISSUED FROM THE PATENT OFFICE. Annual. Government Printing Office, Washington, D.C. 20042.

_____. OFFICIAL GAZETTE. Weekly. Government Printing Office, Washington, D.C. 20042.

_____. PATENT LAWS. Irregular. Government Printing Office, Washington, D.C. 20042.

_____. TRADEMARK RULES OF PRACTICE OF THE PATENT OFFICE WITH FORMS AND STATUTES. Irregular. Government Printing Office, Washington, D.C. 20042.

UNITED STATES PATENTS QUARTERLY. Weekly. Bureau of National Affairs, Inc., 1231 25th Street, N.W., Washington, D.C. 20037.

U.S. PATENTS REPORT WEEKLY SERVICE. Derwent Publications, Ltd., Rochdale House, 128 Theobalds Road, London, WC1X 8RP, England.

URETHANE ABSTRACTS. Monthly. Prepared by the Franklin Institute. Technomic Publishing Co., Inc., 265 West State Street, N.W., Westport, Connecticut 06880.

> "Each month URETHANE ABSTRACTS reviews current articles,
> papers, patents, government reports, and other literature on
> urethanes--both from the U.S. and around the world."

VERFAHRENSTECHNIK BERICHTE [Chemical and process engineering abstracts]. Bimonthly. Verlag Chemie GMBH, Pappelallee 3, 3 Post Fach, 1260 Weinheim/ Bergstrasse, West Germany.

WATER AND POLLUTION CONTROL. Monthly. Southam Business Publications, Ltd., 1450 Don Mills Road, Ontario, Canada.

WATER AND AIR POLLUTION NEWSLETTER. Monthly. International Executive Newsletters Co., 52 rue du Progress, 1000 Brussels, Belgium and 35 West Elm Street, Littleton, New Hampshire 03561.

WATER AND WASTE DIGEST. Monthly. Scranton Publishing, 35 E. Wacker, Chicago, Illinois 60611.

WATER & WASTES ENGINEERING. Monthly. Reuben H. Donnelley Co., 466 Lexington Avenue, New York, New York 10017.

Contents: Water resources and pollution control.

WATER CONDITIONING. Monthly. Publicom, Inc., 17 Sherwood Place, Greenwich, Connecticut 06830.

WATER POLLUTION ABSTRACTS. Monthly. Department of the Environment, Her Majesty's Stationery Office, Atlantic House, Holborn Viaduct, London, E.C.1, England.

WATER POLLUTION CONTROL. Bimonthly. Institute of Water Pollution Control, Ledson House, 53 London Road, Maidstone, Kent, England.

WATER POLLUTION CONTROL FEDERATION. JOURNAL. Monthly. 3900 Wisconsin Avenue, N.W., Washington, D.C. 20016.

"Includes abstracts on a wide range of topics in water pollution control."

WATER RESEARCH. Monthly. Pergamon Press, Fairview Park, Elmsford, New York 10523.

Journal of the International Association on Water Pollution Research.

WATER RESOURCES ABSTRACTS. Monthly. American Water Resources Association, 206 East University Avenue, Urbana, Illinois 61801.

WATER RESOURCES BULLETIN. Bimonthly. Including WATER RESOURCES NEWSLETTER. American Water Resources Association, 206 East University Avenue, Urbana, Illinois 61801.

WATER RESOURCES NEWSLETTER. Bimonthly. American Water Resources Association, 206 East University Avenue, Urbana, Illinois 61801.

WHAT'S NEW IN CHEMICAL PROCESSING EQUIPMENT. Quarterly. Putnam, 111 East Dearborn Street, Chicago, Illinois 60611.

WORLD INDEX OF SCIENTIFIC TRANSLATIONS AND HISTORY OF TRANS-LATIONS NOTIFIED TO ETC. Monthly. European Translations Centre, Doelenstraat 101 Delft, Netherlands.

"Primarily concerned with translations from Russian and other Slavic languages, Chinese, and Japanese."

WORLD MEETINGS: UNITED STATES AND CANADA. Quarterly. World Meetings Information Center, 824 Boylston Street, Chestnut Hill, Massachusetts 02167.

"Gives information on scientific, engineering, and medical meetings that will take place in the United States and Canada during the two-year period following the date of each issue."

WORLD MEETINGS OUTSIDE U.S.A. AND CANADA. Quarterly. World Meetings Information Center, 824 Boylston Street, Chestnut Hill, Massachusetts 02167.

Provides details on meetings to come of interest to scientists and engineers.

WORLD SURFACE COATINGS ABSTRACTS. Monthly. Chorley & Pickersgill, Ltd., Amberley House, Norfolk Street, Strand, London, W.C. 2, England.

ZINC ABSTRACTS. Monthly. Zinc Development Association, 34 Berkeley Square, London, W. 1, England.

9. Reviewing Services

Includes serial publications which provide coverage of research developments on a continuing basis.

ADVANCES IN CHEMICAL ENGINEERING. New York: Academic Press, 1970.

"Critical reviews bringing standard topics up to date and summarizing new developments for chemical engineers."

ADVANCES IN CORROSION SCIENCE AND TECHNOLOGY. New York: Plenum, 1970.

ADVANCES IN CRYOGENIC ENGINEERING. Vol. 16. New York: Plenum, 1971.

Proceedings of cryogenic engineering conferences.

ADVANCES IN ELECTROCHEMISTRY AND ELECTROCHEMICAL ENGINEERING. Electrochemistry. New York: Interscience, Division of Wiley, 1970.

Reviews latest developments on the subject.

ADVANCES IN HEAT TRANSFER. Vol. 9. New York: Academic Press, 1973.

This is one volume in a continuing series.

ADVANCES IN HIGH PRESSURE RESEARCH. New York: Academic Press, 1970 (c1969).

"Reviews recent work on topics of current interest to research scientists and engineers."

ADVANCES IN HYDROSCIENCE. Academic Press, 111 Fifth Avenue, New York, New York 10003.

Continuing series on hydrology and wave motion in fluids.

ANNUAL REPORTS ON PROGRESS OF CHEMISTRY. Chemical Society, Burlington House, Piccadilly, London, WIV OBN, England.

ANNUAL REVIEW OF FLUID MECHANICS. Vol. 5. 1973. Annual Reviews Inc., Palo Alto, California 94306.

ANNUAL REVIEW OF PHYSICAL CHEMISTRY. Annual Reviews, Inc., Palo Alto, California 94306.

CHEMICAL COMMUNICATIONS. Semimonthly. Chemical Society, Burlington House, London WIV OBN, England.

A journal for urgent preliminary accounts of important chemical research.

CHEMICAL REVIEWS. Bimonthly. American Chemical Society, 1155 16th Street, N.W., Washington, D.C. 20036.

Chemical Society. QUARTERLY REVIEWS. Chemical Society, Burlington House, Piccadilly, London, WIV OBN, England.

_____. "Section A: General, Physical, and Inorganic Chemistry." ANNUAL REPORTS OF THE PROGRESS OF CHEMISTRY. Chemical Society, Burlington House, London, WIV OBN, England.

CHEMICAL TECHNOLOGY REVIEW. Annual. Park Ridge, New Jersey: Noyes Data Corp.

CIS INDEX. Monthly looseleaf service. Congressional Information Service, 500 Montgomery Building, Washington, D.C. 20014.

DOCUMENTS OF INTERNATIONAL ORGANIZATIONS: A BIBLIOGRAPHIC HANDBOOK COVERING THE UNITED NATIONS AND OTHER INTERGOVERN- MENTAL ORGANIZATIONS. Chicago: American Library Association, 1973.

A guide to international organizations, their programs and publications.

ORGANIC SYNTHESIS; AN ANNUAL PUBLICATION OF SATISFACTORY METHODS FOR THE PREPARATION OF ORGANIC CHEMICALS. Vol. 52. New York: Wiley, 1972.

Part of a continuing series on synthetic method.

PROGRESS IN HEAT AND MASS TRANSFER. Continuing series. 6 vols. to date. Elmsford, New York: Pergamon.

Contains papers on the subject from various conferences.

PROGRESS IN SEPARATION AND PURIFICATION. Vol. 4. New York: Wiley-Interscience, 1973.

PROGRESS IN THE CHEMISTRY OF FATS AND OTHER LIQUIDS. Elmsford, New York: Pergamon. Continuing series.

Critical reviews of current developments in the field.

QUARTERLY REVIEWS. Chemical Society, Burlington House, London, W1V OBN, England.

RECENT ADVANCES IN ENGINEERING SCIENCE. Vol. 5. (In 2 parts.) Proceedings of the Sixth Annual Meeting of the Society of Engineering Science held at Princeton University, Princeton, New Jersey, November 11-13, 1968. New York: Gordon & Breach, 1970.

RECENT ADVANCES IN LIQUID-LIQUID EXTRACTION. Elmsford, New York: Pergamon Press, 1971.

"This book brings into one volume a comprehensive treatment of many topics associated with liquid extraction that are otherwise only found scattered about the literature in unorganized form."

RECENT DEVELOPMENTS IN SEPARATION SCIENCE. Cleveland, Ohio: CRC Press, 1972.

RECORD OF CHEMICAL PROGRESS. Quarterly. New York: Plenum.

REPORTS ON THE PROGRESS OF APPLIED CHEMISTRY. New York: Academic, 1972.

REVIEWS IN ANALYTICAL CHEMISTRY. Quarterly. Freund Publishing House Ltd., Box 35010, Tel Aviv, Israel.

REVIEWS OF PURE AND APPLIED CHEMISTRY. Quarterly. Royal Australian Chemical Institute, Headquarters Clunies Ross House, 191 Royal Parade, Parkville 3052, Australia.

REVIEWS ON COATINGS AND CORROSION. Quarterly. Freund Publishing House Ltd., Box 35010, Tel Aviv, Israel.

SCIENTIFIC AND TECHNICAL AEROSPACE REPORTS. A semimonthly abstract journal with indexes. (STAR) National Aeronautics and Space Administration, Scientific and Technical Information Division, Washington, D.C. 20546.

Indexes NASA report literature.

SCI-TECH NEWS. Quarterly. Science-Technology Division, Special Libraries Association, Jet Propulsion Laboratory Library, 4800 Oak Grove Drive, Pasadena, California 91103.

Regularly lists bibliographies, many of which are government sponsored research reports.

Society of Chemical Industry, London. REPORTS ON THE PROGRESS OF APPLIED CHEMISTRY. Annual. Society of Chemical Industry, 14 Belgrave Avenue, London, SW 1, England.

SOLVENT EXTRACTION REVIEWS. Vol 1. New York: Dekker, 1971.

Summarizes latest developments in solvent extraction.

SURVEY OF PROGRESS IN CHEMISTRY. Irregular. New York: Academic.

U.S. Superintendent of Documents. MONTHLY CATALOG. Washington, D.C.: Government Printing Office.

a. GUIDES TO REVIEWS

CHEMICAL ABSTRACTS. Weekly. Chemical Abstracts Service, Ohio State University, Columbus, Ohio 43210.

STANDARD PERIODICAL DIRECTORY (& SUPPLEMENT). 4th ed. New York: Oxbridge Publishing Co., 1973.

ULRICH'S INTERNATIONAL PERIODICAL DIRECTORY. Annual. 2 vols. New York: R.R. Bowker Co.

UNESCO. LISTE DE "MISES AU POINT" NUELLES SUR LES PROGRES DE LA SCIENCE ET DE LA TECHNIQUE [List of annual reviews of progress in science and technology]. 2nd ed. Paris: 1969.

10. Document and Report Literature

This part contains lists of Document and Report literature bibliographies and services. Document and report titles on chemical process industries have been interfiled with items in Part 6 - Monographs, Reports, and Symposia - of this guide.

a. LISTS

E. Biggert. "Federal Government Information Sources." CHEMICAL ENGINEERING, May 15, 1972, pp. 103-14.

_____. "State and City Information Sources." CHEMICAL ENGINEERING, August 21, 1972, pp. 94-103.

MONTHLY CHECKLIST OF STATE PUBLICATIONS. Washington, D.C.: Government Printing Office.

b. CLEARINGHOUSES

Department of Defense. Documentation Center, Alexandria, Virginia 22314. Classified specifications and standards are released only under the usual security restrictions.

National Technical Information Service. Springfield, Virginia 22151.

Superintendent of Documents. Government Printing Office, Washington, D.C. 20042.

11. Sources of Information on Masters and Doctoral Theses

DISSERTATION ABSTRACTS INTERNATIONAL. Section B: Physical Sciences and Technology. Monthly. Xerox University Microfilms, 300 North Zeeb Road, Ann Arbor, Michigan 48106.

Covers nearly all dissertations from U.S. institutions since 1861.

MASTER ABSTRACTS. Quarterly. Xerox University Microfilms, 300 North Zeeb Road, Ann Arbor, Michigan 48106.

These service publications are available for reference use in large public and academic libraries. Assistance in obtaining photocopies or microfilm copies of these masters and doctoral theses and in searching for these materials by subject is available from Xerox University Microfilms, 300 North Zeeb Road, Ann Arbor, Michigan 48106.

12. Sources of Information on Statistics and Costs

a. STATISTICS

i. Guides and Handbooks

AMERICAN STATISTICS INDEX. Annual. Pt. 1 Index. Pt. 2 Abstracts. Congressional Information Service, 600 Montgomery Building, Washington, D.C. 20014.

Comprehensive guide and index to the statistical publications of the U.S. Government. Kept current by supplements.

Bowker, A. ENGINEERING STATISTICS. 2nd ed. Englewood Cliffs, New Jersey: Prentice-Hall, 1972.

A basic work on use of statistics in engineering.

Dovies, O., ed. STATISTICAL METHODS IN RESEARCH AND PRODUCTION. Edinburgh: Oliver and Boyd, 1970.

ii. Journals and Serial Publications

CURRENT INDUSTRIAL REPORTS. Monthly. Industry Division Bureau of Census, Washington, D.C. 20233.

Daniel, J. "1972 and Beyond: the Chemical Industry; Questions and Answers." CHEMICAL ENGINEERING, Vol. 79, January 10, 1972, pp. 88-92.

"Facts and Figures for the Chemical Industry." CHEMICAL AND ENGINEERING NEWS, Vol. 53, June 2, 1975, pp. 29-66.

MONTHLY ENERGY REVIEW. Government Printing Office, Washington, D.C. 20042.

SEARCH: CIP Marketing & Statistics Division. Monthly Abstracting Service of Compendium Publishers, International Corp., 2175 Lemoine Avenue, Fort Lee, New Jersey 07024.

STATISTICAL ABSTRACTS OF UNITED STATES. Annual. NATIONAL DATA BOOK AND GUIDE TO SOURCES. Government Printing Office, Washington, D.C. 20042.

SYNTHETIC ORGANIC CHEMICALS, UNITED STATES PRODUCTION AND SALES. Annual. Government Printing Office, Washington, D.C. 20042.

U.S. Census Bureau. DEPARTMENT OF COMMERCE ANNUAL SURVEY OF MANUFACTURERS. Government Printing Office, Washington, D.C. 20042.

U.S. International Trade Commission. SYNTHETIC ORGANIC CHEMICALS, UNITED STATES PRODUCTION AND SALES. Government Printing Office, Washington, D.C. 20042.

iii. Current Journals
Which Include Related Statistical Information

The following journals and serial publications frequently contain statistics related to this field. See the heading Journals in this section for addresses.

AMERICAN OIL CHEMISTS' SOCIETY. Journal

BRITISH CHEMICAL ENGINEERING

CANADIAN CHEMICAL PROCESSING

CANADIAN OCCUPATIONAL SAFETY

CHEMICAL AGE

CHEMICAL ENGINEER AND TRANSACTIONS OF THE INSTITUTION OF CHEMICAL ENGINEERS

CHEMICAL ENGINEERING

CHEMICAL HORIZONS. Weekly Reports

CHEMICAL INDUSTRY REPORT

CHEMICAL MARKET ABSTRACTS

CHEMICAL PROCESSING AND ENGINEERING

CHEMICAL TECHNOLOGY

CLEAN AIR AND WATER NEWS

DATA MANAGEMENT

DETERGENTS AND SPECIALTIES

ENVIRONMENTAL SCIENCE AND TECHNOLOGY

FIRE RESEARCH ABSTRACTS AND REVIEWS

FRANKLIN INSTITUTE JOURNAL

IDEA

INDUSTRY WEEK

JOURNAL OF SAFETY RESEARCH

MANAGEMENT TODAY

PRODUCT DESIGN ENGINEERING

PROFESSIONAL ENGINEER

QUALITY ASSURANCE

SOUTH AFRICAN CHEMICAL AND ENGINEERING AGE

STANDARDS ENGINEERING

WATER AND AIR POLLUTION NEWSLETTER

WATER RESEARCH

b. COSTS

i. Guides and Handbooks

Aries, R. CHEMICAL ENGINEERING COST ESTIMATION. New York: McGraw-Hill, 1955.

Bauman, Herman C. FUNDAMENTALS OF COST ENGINEERING IN THE CHEMICAL INDUSTRY. Illus. New York: Van Nostrand-Reinhold, 1964.

CHEMICAL ECONOMICS HANDBOOK. Monthly. Chemical Information Services, Stanford Research Institute, Menlo Park, California 94025.

> 16 vol. looseleaf set. Installments with current data update statistical detail in the handbook.

Chemical Engineering. MODERN COST-ENGINEERING TECHNIQUE. New York: McGraw-Hill, 1970.

CHEMICAL MARKETING REPORTER. Weekly. Schnell Publishing Co., 100 Church Street, New York, New York 10007.

CHEMICAL PRICING PATTERNS; COMPARISONS OF ANNUAL HIGH AND LOW PRICES OF 1250 KEY CHEMICALS AND RELATED PROCESS MATERIALS FOR THE YEARS OF 1952-1967. New York: Schnell Publishing Co., 1968.

Freemantle, M. THE CHEMIST IN INDUSTRY. New York: Oxford, 1975.

Grant, E. PRINCIPLES OF ENGINEERING ECONOMY. 5th ed. New York: Ronald Press, 1970.

> "Defining alternatives and predicting their consequences. The need for criteria and analytical procedures. Interest formulas. Equivalent uniform annual cash flow. Calculating an unknown interest rate."

A GUIDE TO THE ECONOMIC EVALUATION OF PROJECTS. By D. Allen. London: Institution of Chemical Engineers, 1972.

Happel, J., et al. CHEMICAL PROCESS ECONOMICS. 2nd ed. New York: M. Dekker, 1975.

Jelen, F. COST AND OPTIMIZATION ENGINEERING. New York: McGraw-Hill, 1970.

> "Presents an introduction to cost and optimization engineering, including engineering economy."

Page, J. ESTIMATOR'S MANUAL OF EQUIPMENT AND INSTALLATION COSTS. Houston, Texas: Gulf Publishing Co., 1963.

Peters, M. PLANT DESIGN AND ECONOMICS FOR CHEMICAL ENGINEERS. 2nd ed. New York: McGraw-Hill, 1968.

Popper, H. MODERN COST ENGINEERING TECHNIQUES. New York: McGraw-Hill, 1970.

> "A collection of cost engineering techniques which have been described in CHEMICAL ENGINEERING magazine during the last few years."

THE PREDICASTS BASEBOOK. 1973. Predicasts, 200 University Circle Research Center, 1101 Cedar Avenue, Cleveland, Ohio 44106.

> "Contains annual data since 1960 and growth rates for 12,000 significant economic and industrial time series. Included are detailed data on production, prices, income, shipments, sales, foreign trade and employment."

SYMPOSIUM ON COST VERSUS PERFORMANCE IN EQUIPMENT USED IN UNIT OPERATIONS, 1966. London: Institute of Chemical Engineering, 1968.

Teknekron, Inc. ASSESSMENT OF THE COSTS AND CAPABILITIES OF WATER POLLUTION CONTROL TECHNOLOGY FOR THE STEAM ELECTRIC POWER INDUSTRY. Rept. no. PB 251 372. Springfield, Virginia: National Technical Information Service, 1976.

U.S. Department of Defense. ENGINEER MANUAL EM (SERIES). ENGINEER-ING AND DESIGN: COST ESTIMATES, PLANNING AND DESIGN STAGES. Washington, D.C.: Government Printing Office, 1972.

Available in Government Document Depository libraries.

_____. ENGINEERING DESIGN HANDBOOK: SYSTEM ANALYSIS AND COST-EFFECTIVENESS. Washington, D.C.: Government Printing Office, April 1971.

U.S. Department of Interior, Office of Saline Water. DESALTING COST CAL-CULATING PROCEDURES. Report no. 555. Washington, D.C.: Government Printing Office, 1970.

ii. Journals and Serial Publications

Bresier, S. "Cost Estimating by Computer." CHEMICAL ENGINEERING, Vol. 79, May 29, 1972, pp. 84-86.

Capello, V. "Simplifying Scaleup Cost Estimation." CHEMICAL ENGINEER-ING, Vol. 79, August 7, 1972, p. 99.

CHEMICAL ENGINEERING. Monthly. McGraw-Hill Publishing Co., 1221 Avenue of the Americas, New York, New York 10020.

Regularly includes cost data.

CHEMICAL PRICING PATTERNS. 3rd ed. New York: Schnell Publishing Co., 1973.

A comparison of annual high and low prices for 1250 chemicals and related processes for years 1952-1970.

COST ENGINEERING. Quarterly. Industrial Research Service, Inc., Masonic Building, Dover, New Hampshire 03820.

Provides continual source of data on changes in costs in engineer-ing related fields.

"Cost Literature, 1970." COST ENGINEERING, Vol. 16, January 1971, pp. 6-39.

Doane, E. "Plants Investment vs. Profitability." CHEMICAL TECHNOLOGY, Vol. 2, November 1972, pp. 652-65.

Epstein, L. "Cost of Standard-sized Reactors and Storage Tanks." CHEMICAL ENGINEERING, Vol. 78, October 18, 1971, pp. 160-61.

Guthrie, K. "Capital and Operating Costs for 54 Chemical Processes." CHEMICAL ENGINEERING, Vol. 77, June 15, 1970, pp. 140-56.

_____. "Pump and Valve Costs." CHEMICAL ENGINEERING, Vol. 78, January 11, 1971, pp. 151-59.

Jenckes, L. "How to Estimate Operating Costs and Depreciation." CHEMICAL ENGINEERING, Vol. 77, December 14, 1970, p. 168.

Katell, S. "Cost Engineer in R and D." CHEMICAL TECHNOLOGY, Vol. 1, November 1971, pp. 648-49.

Kennedy, R. "Annotated Bibliography of Estimation Procedures Useful in Engineering Economy." ENGINEERING ECONOMIST, Spring 1971, pp. 211-17.

"Plant Operations and Loss Prevention; Special Report." CHEMICAL ENGINEERING PROGRESS, Vol. 68, May 1972, pp. 41-69.

Popper, H. "CPI's Cost of Meeting Environmental Standards." CHEMICAL ENGINEERING, Vol. 78, August 23, 1971, pp. 106-8.

Sommerville, R. "New Method Gives Quick Accurate Estimate of Distillation Costs." CHEMICAL ENGINEERING, Vol. 79, May 1, 1972, pp. 71-76.

Thorsen, D. "7-year Surge in the CE Cost Indexes." CHEMICAL ENGINEERING, Vol. 79, November 13, 1972, p. 168.

WORLD-PRODUCT COSTS. Irregular. Predicasts, Inc., 200 University Circle Research Center, 11001 Cedar Avenue, Cleveland, Ohio 44106.

   A subscription service of market forecasts by product in areas of agriculture, wood and paper, food, chemicals, polymers, primary metals, etc.

Zanker, A. "Estimating Cooling Tower Costs from Operating Data." CHEMICAL ENGINEERING, Vol. 79, June 12, 1972, pp. 118-19.

For additional information on costs see the subject headings "Chemical Engineering-Costs" and "Chemicals-Costs" in Abstracting and Indexing Publications listed in this Guide.

13. Sources of Information on Standards, Specifications, and Patents

a. STANDARDS AND SPECIFICATIONS

i. Guides and Handbooks

American National Standards Institute. CATALOG OF STANDARDS. Annual. New York: American National Standards Institute.

American Society for Testing and Materials. BOOK OF A.S.T.M. STANDARDS, WITH RELATED MATERIAL. 32 vols. Philadelphia: American

Society for Testing and Materials.

Revised annually.

Booth, S. STANDARDIZATION ACTIVITIES IN THE UNITED STATES; A DE-SCRIPTIVE DIRECTORY. National Bureau of Standards, Miscellaneous Publication 230. Washington, D.C.: Government Printing Office, 1960.

British Standards Institution. YEARBOOK. Annual. London: British Standards Institution.

"The Yearbook contains lists of British standards in numerical order, with descriptions and a subject index. Kept up to data by B.S.I. News (issued monthly)."

International Organization for Standardization. ISO CATALOGUE. Annual. Geneva: International Standards Organization.

Jessup, W. LAW AND SPECIFICATIONS FOR ENGINEERS AND SCIENTISTS. Englewood Cliffs, New Jersey: Prentice-Hall, 1963.

Struglia, Erasmus J. STANDARDS AND SPECIFICATIONS INFORMATION SOURCES. Detroit: Gale Research Co., 1965.

U.S. Department of Defense. INDEX OF SPECIFICATIONS AND STANDARDS. Monthly with annual cumulations. Washington, D.C.: Government Printing Office.

U.S. General Services Administration. GUIDE TO SPECIFICATIONS AND STANDARDS OF THE FEDERAL GOVERNMENT. Washington, D.C.: Government Printing Office, 1968.

_____. INDEX OF FEDERAL SPECIFICATIONS AND STANDARDS. Monthly with annual cumulations. Washington, D.C.: Government Printing Office.

U.S. National Bureau of Standards. DIRECTORY OF STANDARDS LABORATORIES, 1971. National Conference of Standards Laboratories, National Bureau of Standards, Gaithersburg, Maryland 20760.

_____. DIRECTORY OF UNITED STATES STANDARDIZATION ACTIVITIES. Washington, D.C.: Government Printing Office, 1975.

_____. INDEX OF U.S. VOLUNTARY ENGINEERING STANDARDS, COVERING THOSE STANDARDS, SPECIFICATIONS, TEST METHODS, AND RECOMMENDED PRACTICES ISSUED BY NATIONAL STANDARDIZATION ORGANIZATIONS IN THE UNITED STATES. Edited by William J. Slattery. Gaithersburg, Maryland: December 1972.

_____. INDEX TO REPORTS OF NATIONAL CONFERENCE ON WEIGHTS AND MEASURES FROM 1st-56th, 1905-71. Edited by Frances C. Bell. Gaithersburg, Maryland: February 1973.

_____. LIST OF PUBLICATIONS, DIRECTORY OF UNITED STATES STAN-DARDIZATION ACTIVITIES. Irregular. Washington, D.C.: Government Printing Office.

_____. STANDARD REFERENCE MATERIALS: 1973 SRM PRICE LIST, SUPPLE-MENT. Gaithersburg, Maryland: 1973.

An additional source of standards information is:

U.S. National Bureau of Standards. Office of Engineering, Standards Services, Information Section, Washington, D.C. 20234.

ii. Journals and Serial Publications

ANSI REPORTER/STANDARDS ACTION. Fortnightly. Formerly: STANDARDS INSTITUTE REPORTER. American National Standards Institute, 1430 Broadway, New York, New York 10018.

STANDARDS: MONTHLY ADDITIONS. Monthly. Formerly: ISI STANDARDS: MONTHLY ADDITIONS. Indian Standards Institution, Manak Bhavan, 9 Bahadur Shah Zafar Marg, New Delhi 1, India.

STANDARDS ACTION. Fortnightly. American Standards Institute, 1430 Broad-way, New York, New York 10018, 1970.

STANDARDS AND SPECIFICATIONS INFORMATION BULLETIN. Weekly. National Standards Association, Inc., 1321 Fourteenth Street, N.W., Washing-ton, D.C. 20005.

> Lists all new federal, military, Air Force-Navy aeronautical and national aerospace standards.

STANDARDS ENGINEERING. Bimonthly. Box 7507, Philadelphia, Pennsyl-vania 19101.

Standards Engineers Society. PROCEEDINGS OF ANNUAL MEETING. 1951. Standard Engineers Society, Box 7507, Philadelphia, Pennsylvania 19101.

U.S. Department of Defense. INDEX OF SPECIFICATIONS AND STANDARDS. Monthly. Washington, D.C.: Government Printing Office.

U.S. Federal Supply Service. INDEX OF FEDERAL SPECIFICATIONS, STAN-DARDS AND HANDBOOKS. Annual. Washington, D.C.: Government Printing Office.

> "Contains lists of Federal and Interim Federal Specifications and Standards; Military Specifications and Standards; Federal Hand-books; and General Services Administration Specifications."

U.S. National Bureau of Standards. COMMERCIAL STANDARDS. Irregular. Washington, D.C.: Government Printing Office.

_____. NATIONAL STANDARD REFERENCE DATA SYSTEM. Irregular. Washington, D.C.: Government Printing Office.

_____. PRODUCT STANDARDS. Irregular. Washington, D.C.: Government Printing Office.

Specifications, standards, codes and regulations are available for individual states from the official state document printing offices and from local and municipal government agencies.

b. PATENTS AND TRADEMARKS

i. Guides and Handbooks

Arnold, T. INVENTION PROTECTION FOR PRACTICING ENGINEERS. New York: Barnes and Noble, 1971.

Introduction to U.S. patent system.

Calvert, R., ed. THE ENCYCLOPEDIA OF PATENT PRACTICE AND INVENTION MANAGEMENT. New York: Reinhold, 1964.

"A comprehensive statement of the principles and procedures in solicitation, enforcement and licensing of patents and recognition and utilization of inventions."

GENERAL INFORMATION CONCERNING PATENTS. Washington, D.C.: Government Printing Office, 1969.

Kase, F. FOREIGN PATENTS: A GUIDE TO OFFICIAL PATENT LITERATURE. Dobbs Ferry, New York: Oceana, 1972.

"A guide for those who have to consult foreign patent material in search of knowledge of prior art; it is not a handbook of foreign patent law or procedure."

_____. TRADEMARKS. New York: Oceana, 1974.

Liebesny, R., ed. MAINLY ON PATENTS: THE USE OF INDUSTRIAL PROPERTY AND ITS LITERATURE. Hamden, Connecticut: Shoe String, 1973.

For those interested in foreign patents.

MARTINDALE-HUBBELL LAW DIRECTORY. Annual. Martindale-Hubbell Co., Summit, New Jersey 07901.

Includes patent attorneys.

National Paint and Coatings Association. TRADEMARK DIRECTORY. Triennial. National Paint and Coatings Association, 1500 Rhode Island Avenue N.W., Washington, D.C. 20005.

Newby, F. HOW TO FIND OUT ABOUT PATENTS. New York: Pergamon, 1967.

"Patent Rewards; Special Report." CHEMICAL ENGINEERING PROGRESS, Vol. 67, November 1971, pp. 19-32.

PATENTS FOR CHEMICAL INVENTIONS. Advances in Chemistry Series 46. Washington, D.C.: American Chemical Society, 1964.

QUESTIONS AND ANSWERS ABOUT PATENTS: ANSWERS TO QUESTIONS FREQUENTLY ASKED ABOUT PATENTS. Washington, D.C.: U.S. Patent and Trademark Office, 1975.

STATE TRADEMARK REGISTRATION MANUAL, 1966. U.S. Trademark Association, 6 East 45 Street, New York, New York 10017.

TRADEMARK MANAGEMENT: A GUIDE FOR BUSINESSMEN. 4th ed. New York: U.S. Trademark Association, 1968.

U.S. Patent Office. CONCORDANCE, UNITED STATES PATENT CLASSIFICATION TO INTERNATIONAL PATENT CLASSIFICATION. 2nd ed. Washington, D.C.: Government Printing Office, November 1971.

_____. DIRECTORY OF REGISTERED PATENT ATTORNEYS AND AGENTS ARRANGED BY STATES AND COUNTIES. Annual. Washington, D.C.: Government Printing Office.

_____. GENERAL INFORMATION CONCERNING PATENTS. Annual. Washington, D.C.: Government Printing Office.

_____. INDEX OF PATENTS ISSUED FROM PATENT OFFICE, PART 1, LIST OF PATENTEES. Annual. Washington, D.C.: Government Printing Office.

_____. INDEX OF PATENTS ISSUED FROM PATENT OFFICE, PART 2, INDEX TO SUBJECTS OF INVENTION. Annual. Washington, D.C.: Government Printing Office.

_____. PATENTS AND INVENTIONS. Washington, D.C.: Government Printing Office, 1974.

"Information aid for inventors, step-by-step guide to help inventors decide whether to apply for patent, how to obtain patent protection, and promote his invention."

UNITED STATES PATENT PREVIEWS, 1965-1970. New York: R.R. Bowker, 1972.

UNITERM INDEX OF U.S. CHEMICAL PATENTS INFORMATION FOR INDUSTRY, INC. 1970. 1000 Connecticut Avenue, N.W., Washington, D.C. 20036.

Includes chemical patents for years 1950-1968.

White, W., et al. TRADEMARKS THROUGHOUT THE WORLD. Rev. ed.
New York: Boardman, 1974.

ii. Journals and Serial Publications

ABSTRACTS FROM TECHNICAL AND PATENT PUBLICATIONS. Weekly.
British Internal Combustion Engine Research Institute, 111/112 Buckingham Avenue,
Slough, Bucks, England.

BRITISH PATENTS ABSTRACTS. Weekly. Derwent Publications Ltd., Rochdale
House, 128 Theobalds Road, London, WC1X 8RP, England.

"Central Patents Index" is a service of Derwent Publications Ltd., Rochdale
House, 128 Theobalds Road, London, WC1X 8RP, England.

This service includes European and U.S. Patents in chemical, electrical and mechanical engineering fields. Copies of patents in
microfiche format are available, as well as details from patents on
cards, magnetic tapes and through computer searches.

Committee For International Cooperation in Information Retrieval Among Examining Patent Offices. PROCEEDINGS OF ANNUAL MEETINGS. Committee
for International Cooperation in Information Retrieval Among Examining Patent
Offices, 32 Chemin des Colombelles, 1211 Geneva 10, Switzerland.

FOREIGN PATENT INDEX. 1973. IFI Plenum Data Co., 2001 Jefferson
Davis Highway, Arlington, Virginia 22202.

Indexes patents from CHEMICAL ABSTRACTS and other sources.
Updated by supplements.

FRENCH PATENTS ABSTRACTS. Weekly. Derwent Information Service, Rochdale House, Theobalds Road, London, WC1, England.

GERMAN PATENTS GAZETTE. Weekly. Derwent Publications Ltd., Rochdale
House, 128 Theobalds Road, London, WC1X 8RP, England.

Great Britain. Board of Trade. PATENTS, DESIGNS AND TRADE MARKS.
ANNUAL REPORT H.M.S.O., P.O.B. 569, London, S.E.1, England.

ICIREPAT BULLETIN. Irregular. U.S. Patent Office, Washington, D.C.
20231.

Relates news of international patent scene.

IDEA (U.S.). PATENT, TRADEMARK AND COPYRIGHT JOURNAL OF RE-
SEARCH AND EDUCATION. 5/yr. Patent, Trademark and Copyright Research
Institute, George Washington University, Washington, D.C. 20006.

INDEX OF TRADEMARKS ISSUED FROM THE PATENT OFFICE. 3/yr. Government Printing Office, Washington, D.C. 20042.

INTERNATIONAL LICENSING. Monthly. International Licensing Ltd., 92 Cannon Lane, Pinner, Middlesex, HA5 1HT, England.

"A monthly bulletin providing an international forum for the negotiation of manufacturing licenses and joint ventures. Publishes details of both new and established products available for manufacturing under license, or joint venture arrangements, from worldwide sources."

INTERPAS PATENT DATA-BULLETINS. Monthly. Interpas (International Patent Service), Buitenhaven 25, Hertogenbosch, Netherlands.

Text mostly in English; French and German titles in original language.

ISRAEL STATE RECORDS-PATENTS AND DESIGNS JOURNAL. Monthly. Israel Patent Office, Ministry of Justice, Government Printer, Jerusalem, Israel.

Part A: Patents, Trademarks and Copyrights; Part B: Patents and Designs. Text in English and Hebrew.

ISRAEL STATE RECORDS-TRADE MARKS JOURNAL. Monthly. Israel Patent Office, Ministry of Justice, Government Printer, Jerusalem, Israel.

Text in English and Hebrew.

JAPANESE PATENTS ABSTRACTS. Weekly. Derwent Publications Ltd., Rochdale House, 128 Theobalds Road, London, WC1X 8RP, England.

MALAWI PATENT JOURNAL AND TRADE MARKS JOURNAL. Monthly. Malawi Government Printing Department, Box 37, Zomba, Malawi.

NETHERLAND PATENTS REPORT. Weekly List. Derwent Publications Ltd., Rochdale House, 128 Theobalds Road, London, SC1X 8RP, England.

NEW ZEALAND PATENT OFFICE JOURNAL. Monthly. Government Printing Office, Private Bag, Wellington, New Zealand.

OFFICIAL JOURNAL. (Patents) Weekly. Patent Office, St. Mary Cray, Orpington, Kent, BR5 3RD, England.

PATENT ABSTRACT BULLETIN. Semimonthly. Xerox Corporation, Systems and Services for Libraries, Xerox Square 007, Rochester, New York 14603.

A service for libraries.

Patent and Trademark Institute of Canada. ANNUAL PROCEEDINGS. Patent and Trademark Institute of Canada, P.O. Box 553, Station B, Ottawa 4, Ontario, Canada.

PATENT AND TRADEMARK REVIEW. Published 11 times annually. Clark Boardman Company, Ltd., 435 Hudson Street, New York, New York 10014.

The text of new laws and official rules, proposed legislation, and important decisions affecting intellectual property on a world-wide basis.

PATENT-DIGEST INTERNACIONAL. Monthly. Antoni Guilleumas Brosa, Calle Parque 3, apto. 1381, Barcelona-2, Spain.

PATENT EXCHANGE NEWS AND REPORTER. Bimonthly. Patent Exchange Inc., 26 Broadway, New York, New York 10004.

PATENT JOURNAL INCLUDING TRADEMARKS AND MODELS. Monthly. Government Printer, Bosman Street, Private Bag, 85, Pretoria, South Africa.

Text in Afrikaans and English.

PATENT LAW REVIEW. Annual. Sage Hill Publishers, Inc., 116 Washington Avenue, Albany, New York 12210.

PATENT LICENSING GAZETTE. Bimonthly. Techni Research Associates, Professional Center Building, Willow Grove, Pennsylvania 19090.

PATENT OFFICE RECORD (CANADA). Weekly. Commissioner of Patents. Information Canada, Publishing Division, Ottawa, Canada.

Text and title in English and French.

Patent Offices. BULLETIN. Irregular. Committee for International Cooperation in Information Retrieval Among Examining Patent Offices, 32 Chemin des Colombelles, 1211 Geneva 20, Switzerland.

PATENT OFFICE SOCIETY. JOURNAL. Monthly. 104 Academy Avenue, Federalsburg, Maryland 21632.

PATENTS THROUGHOUT THE WORLD. Supplemented three times annually. Clark Boardman Company, Ltd., 435 Hudson Street, New York, New York 10014.

"The patent laws of the world digested and uniformly arranged. Contains information concerning: law, conventions; patentee; novelty; opposition; taxes; working; compulsory licenses; marking; assignment; documents required."

PRINCIPAL CHEMICAL AND CHEMICALLY RELATED TERMS USED IN UNITED STATES PATENTS. Annual. Information for Industry, Inc., 1000 Connecticut Avenue, N.W., Washington, D.C.

Thesaurus of patent terms.

PRODUCTS LIST CIRCULAR. Monthly. Washington, D.C.: Government Printing Office.

"Presents information on royalty-free nonexclusive Government owned patents as well as available private patents."

SOVIET INVENTIONS ILLUSTRATED. Monthly service in English. Derwent Publications Ltd., Rochdale House, 128 Theobalds Road, London, WC1X 8RP, England.

STATE TRADEMARK STATUTES. Irregular. U.S. Trademark Association, 6 East 45th Street, New York, New York 10017.

> Looseleaf service containing trademark statutes of all 50 states. Supplements bring it up to 1970.

TRADEMARK ALERT. Weekly. TCR Service, Inc., 140 Sylvan Avenue, Englewood Cliffs, New Jersey 07632.

THE TRADEMARK REPORTER. Monthly. U.S. Trademark Association, 6 East 45th Street, New York, New York 10017.

> Authoritative reporting service for this field.

TRADE MARKS JOURNAL (CANADA). Weekly. Information Canada, Publishing Division, Ottawa, K1A OS9, Canada.

> Text in English and French.

TRADE MARKS JOURNAL (ENG.). Weekly. Patent Office, St. Mary Cray, Orpington, Kent, BR5 3RD, England.

TRADEMARKS THROUGHOUT THE WORLD: PRACTICE AND PROCEDURE FOR SECURING TRADEMARK PROTECTION INTERNATIONALLY. Annual. Clark Boardman Co., Ltd., 435 Hudson Street, New York, New York 10014.

> The trademark laws of the world digested and uniformly arranged: who may apply; what can be registered; classification of good; duration; renewal; etc.

U.S. CHEMICAL PATENTS. Weekly. U.S. Federal Government. Godfrey Memorial Library, 134 Newfield Street, Middletown, Connecticut 06457.

U.S. Department of Commerce. TRADEMARK, RULES OF PRACTICE OF PATENT OFFICE WITH FORMS AND STATUTES. Irregular. Washington, D.C.: Government Printing Office.

U.S. Patent Office. CLASSIFICATION BULLETINS. Irregular. Commissioner of Patents, Washington, D.C. 20231.

_____. GENERAL INFORMATION CONCERNING TRADEMARKS. Irregular. Washington, D.C.: Government Printing Office.

_____. INDEX OF PATENTS ISSUED FROM THE PATENT OFFICE. Annual. Washington, D.C.: Government Printing Office.

_____. INDEX OF TRADEMARKS ISSUED FROM THE PATENT OFFICE. Annual. Washington, D.C.: Government Printing Office.

_____.   OFFICIAL GAZETTE, PATENTS.   Weekly; annual index.   Washington, D.C.:   Government Printing Office.

Weekly listing of patent abstracts.

_____.   OFFICIAL GAZETTE, TRADEMARKS.   Weekly.   Washington, D.C.: Government Printing Office.

_____.   PATENT LAWS.   Irregular.   Washington, D.C.:   Government Printing Office.

_____.   RESEARCH AND DEVELOPMENT REPORTS.   Irregular.   National Technical Information Service, Springfield, Virginia   22150.

_____.   TRADEMARK RULES OF PRACTICE OF THE PATENT OFFICE WITH FORMS AND STATUTES.   Irregular.   Washington, D.C.:   Government Printing Office.

UNITED STATES PATENTS QUARTERLY.   Weekly.   Edited by Alan B. Bagley. Washington, D.C.:   Bureau of National Affairs, Inc.

U.S. PATENTS REPORT.   Weekly service.   Derwent Publications, Ltd., Rochdale House, 128 Theobalds Road, London, WC1X 8RP, England.

c. CURRENT JOURNALS
WHICH INCLUDE RELATED PATENT LISTINGS

The following journals and serial publications frequently list patents.   See the heading Journals in this section for addresses.

ABSTRACT REVIEWS

AMERICAN OIL CHEMISTS' SOCIETY.   Journal

APPLIED SCIENCE AND TECHNOLOGY INDEX

AUSTRALIA CHEMICAL PROCESSING

CHEMICAL ABSTRACTS

CHEMICAL AGE

CHEMICAL MARKET ABSTRACTS

CHEMICAL PROCESSING AND ENGINEERING

CHEMICAL TECHNOLOGY

CHEMICAL TECHNOLOGY REVIEW

COLOR ENGINEERING

COMPUTER ABSTRACTS

COMPUTER AND INFORMATION SYSTEMS

DESALINATION ABSTRACTS

DESIGN NEWS

DETERGENTS AND SPECIALTIES

ENGINEERING INDEX

INTERNATIONAL CHEMICAL REGISTER

JOURNAL OF SAFETY RESEARCH

QUALITY ASSURANCE

SCIENCE CITATION INDEX

SOAP/COSMETICS/CHEMICAL SPECIALTIES

U.S. GOVERNMENT REPORT ANNOUNCEMENT

URETHANE ABSTRACTS

WATER AND AIR POLLUTION NEWSLETTER

14. Sources of Information on
Translations, Meetings, and Proceeedings of Symposia

a. TRANSLATIONS

i. Guides

A GUIDE TO SCIENTIFIC AND TECHNICAL JOURNALS IN TRANSLATION.
New York: Special Libraries Association, 1968.

> This directory is useful in learning of foreign journals translated
> cover-to-cover.

National Translations Center. CONSOLIDATED INDEX OF TRANSLATIONS
INTO ENGLISH. New York: Special Libraries Association, 1969.

> "Brings together information on the availability of translations
> which have appeared in a number of different lists issued by dif-
> ferent agencies at different times...Included are translation bibli-
> ographies, selective translation journals, and collections of transla-
> tions."

ii. Journals and Serial Publications

CURRENT CHEMICAL TRANSLATIONS. Irregular. Transchem, Inc., Chemical
Translation Service, Box 669, Knoxville, Tennessee 37901.

Index Translationum. Annual. Unipub, Inc., Box 433, New York, New York
10016.

> Provides full bibliographic data on translated books published in
> every field of learning.

SCI-TECH QUARTERLY INDEX TO U.S. GOVERNMENT TRANSLATIONS. CCM Information Corporation, 909 Third Avenue, New York, New York 10022.

TRANSATOM BULLETIN. Monthly. Euratom, CID Agence et Messageries de la Presse SA, 1 rue de la Petite, Ille, Brussels, Belgium.

Guide to nuclear literature in translation.

TRANSDEX. Monthly. CAM Information Corporation, Subsidiary of Crowell Collier and Macmillan Inc., 909 Third Avenue, New York, New York 10022.

Guide and index to Joint Publications Research Service translations.

TRANSLATIONS REGISTER-INDEX. Semimonthly. National Translation Center, John Crerar Library, 35 West 33rd Street, Chicago, Illinois 60616.

Lists new translations received at the SLA Translation Center.

U.S. Joint Publications Research Service, Arlington, Virginia.

Publishes translations of scientific and technical materials from Eastern Europe, Russia and China. These are listed in the MONTH-LY CATALOG of the Superintendent of Documents.

WORLD INDEX OF SCIENTIFIC TRANSLATIONS. Quarterly with annual cumulations. Delft, European Translations Centre, The Netherlands.

"Primarily concerned with translations from Russian and other Slavic languages, Chinese, and Japanese."

b. MEETINGS AND SYMPOSIA (ANNOUNCEMENTS)

CALLS FOR PAPERS. Weekly. CCM Information Corporation, 909 Third Avenue, New York, New York 10022.

Lists meetings upcoming to alert scientists and engineers.

SCIENTIFIC MEETINGS. Quarterly. Special Libraries Association, 235 Park Avenue, New York, New York 10003.

"Lists information about future conferences and conventions, arranged alphabetically by society and university, chronologically, geographically, and by subject."

WORLD MEETINGS: UNITED STATES AND CANADA. Quarterly. World Meetings Information Center, 825 Boylston Street, Chestnut Hill, Massachusetts 02167.

"Gives information on scientific, engineering, and medical meetings that will take place in the United States and Canada during the two-year period following the date of each issue."

WORLD MEETINGS OUTSIDE USA AND CANADA. Quarterly. Newton Centre, Massachusetts, Technical Meetings Information Service.

Provides details on meetings to come of interest to scientists and engineers.

c. PROCEEDINGS OF SYMPOSIA

CURRENT INDEX TO CONFERENCE PAPERS. Monthly. CCM Information Corporation, 9909 Third Avenue, New York, New York 10022.

DIRECTORY OF PUBLISHED PROCEEDINGS. Series SEMT - Science, Engineering, Medicine and Technology. Monthly. Interdok Corporation, 173 Halstead Avenue, Box 326, Harrison, New York 10528.

MEETING-CONVENTION REPORTING SERVICE: "EUROPE TODAY." Available from Yarsley Research Laboratories, Ltd., Clayton Road, Chessington, Surrey, England.

Provides in-depth reports on important meetings of prominent organizations in chemical and plastics industries.

PROCEEDINGS IN PRINT. Bimonthly. Proceedings in Print, Inc., Box 247, Mattapan, Massachusetts 02126.

"Lists and indexes of proceedings of conferences, symposia, etc. in all subject areas, and in all languages."

15. Selected Literature Sources of
Pollution Control and Safety Information in Chemical Process Industries

a. AIR POLLUTION AND GENERAL MATERIALS ON POLLUTION CONTROL

Contains titles on air pollution control and those whose scope includes both air and water pollution as related to chemical process industries.

i. Guides and Handbooks

AIR AND GAS CLEANUP EQUIPMENT. 2nd ed. Park Ridge, New Jersey: Noyes Data Corporation, 1972.

"This book gives basic technical information and specifications pertaining to commercial equipment currently available."

AIR POLLUTION REFERENCE LIBRARY. American Conference of Governmental Industrial Hygienists, 1014 Broadway, Cincinnati, Ohio 45202.

"A compilation of references on pollution including lists of books, handbooks, journals and periodicals, and other references."

American Society for Engineering Education. Engineering Libraries Division. GUIDE TO LITERATURE OF ENVIRONMENTAL SCIENCES. Washington, D.C.: 1973.

Baturin, V. FUNDAMENTALS OF INDUSTRIAL VENTILATION. 3rd ed. International Series of Monographs in Heating, Ventilation and Refrigeration, vol. 8. New York: Pergamon, 1972.

Bregman, J. HANDBOOK OF WATER RESOURCES AND POLLUTION PROBLEMS. New York: Van Nostrand Reinhold, 1976.

CONFERENCE ON ENVIRONMENTAL ASPECTS OF NONCONVENTIONAL ENERGY SOURCES, PROCEEDINGS. Denver, Colorado, February 1976. Hinsdale, Illinois: American Nuclear Society, 1976.

CONVERSION OF REFUSE TO ENERGY. Proceedings, 1st International Conference and Technical Exhibition, Montreaux, Switzerland, November 1975. Piscataway, New Jersey: Institute of Electrical and Electronic Engineers, 1976.

COST EFFECTIVENESS IN POLLUTION CONTROL. Westport, Connecticut: Technomic Publishing Company, Inc., 1972.

> "The economics of pollution control planning, management, and technology are examined here. A wide range of air, water, and solid waste pollution problems confronting government and industry today are explored. Topics include: emission control costs, equipment evaluation, plant site location, waste water reclamation, monitoring, utilization of municipal refuse, and recycling programs."

Dricoll, J. SAMPLING AND EVALUATING GASSES FROM STACKS. Ann Arbor, Michigan: Humphrey, 1972.

ENVIRONMENT INFORMATION ACCESS. Environment & Information Center, Inc., Government Affairs Department, New York, New York.

GAS SCRUBBERS USED IN POLLUTION CONTROL. By K. Werner, et al. Rept. no. NTIS/PS-76/0050. Springfield, Virginia: National Technical Information Service, 1976.

> Citations from NTIS data base. Covers 1964-76.

HANDBOOK OF ENVIRONMENTAL CONTROL. 4 vols. Cleveland: CRC Press, 1972.

INDUSTRIAL MARKETS FOR AIR POLLUTION EQUIPMENT. Predicasts, Inc., 200 University Circle, 11001 Cedar Avenue, Cleveland, Ohio 44106.

> "Analyzes air pollution equipment expenditures by industry. Also reviews role of test and measuring systems, and data are presented for the years 1958, 1963, 1967 and 1970 with projections for 1975, 1980 and 1985."

Jones, H. FINE DUST AND PARTICULATES REMOVAL. Park Ridge, New Jersey: Noyes Data Corporation, 1972.

> "Deals with the control of large and small particle size dust with special emphasis on airborne particles which measure from 0.01 to

2.0 microns. The book is based on government reports and U.S. patents."

Lawrence, A. NITROGEN OXIDES EMISSION CONTROL 1972. Park Ridge, New Jersey: Noyes Data Corporation.

Based on review of patent literature.

Lund, H. INDUSTRIAL POLLUTION CONTROL HANDBOOK. New York: McGraw-Hill, 1972.

Practical reference tool for engineers.

POLLUTION CONTROL TECHNOLOGY FOR PESTICIDE FORMULATORS AND PACKAGERS. By T.L. Ferguson. Washington, D.C.: Government Printing Office, 1975.

Purkiss, B. BIOTECHNOLOGY OF INDUSTRIAL WATER CONSERVATION. London: Mills and Boon, Ltd., 1972.

Ross, R. AIR POLLUTION AND INDUSTRY. New York: Van Nostrand-Reinhold, 1972.

Guide to air pollution causes and solutions in industrial plants.

Schildhauer, G., comp. ENVIRONMENTAL INFORMATION SOURCES: ENGINEERING AND INDUSTRIAL APPLICATIONS. New York: Special Libraries Association, 1972.

Slack, A. SULFUR DIOXIDE REMOVAL FROM WASTE GASES 1971. Park Ridge, New Jersey: Noyes Data Corporation.

"Reviews the problems of smelter operators, power plants, refineries, sulfur acid plants, and Claus process sulfur plants."

Stecher, P. G. HYDROGEN SULFIDE REMOVAL PROCESSES 1972. Park Ridge, New Jersey: Noyes Data Corporation.

SULFUR DIOXIDE CONTROL. Vol. 1: 1964-72. Rept. no. NTIS/PS 75/456. Vol. 2: 1973-76. Rept. no. NTIS/PS 76/0519. Springfield, Virginia: National Technical Information Service, 1976.

A bibliography with abstracts.

Stauss, W., ed. AIR POLLUTION CONTROL. PART I. New York: Interscience, Division of Wiley, 1971.

"Presents authoritative reviews of specific fields currently of major importance."

Tuwiner, S., ed. ENVIRONMENTAL SCIENCE TECHNOLOGY INFORMATION RESOURCES 1973. Park Ridge, New Jersey: Noyes Data Corporation.

"Section I includes the proceedings and panel discussions of the Symposium on Environmental Science Technology Information Resources sponsored by the Chemical International Information Center, and held at the Chemists' Club, New York, on April 28, 1972.

Section II comprises selected papers presented at the National Environmental Information Symposium, sponsored by the Environmental Protection Agency (EPA), held at Cincinnati, Ohio, September 24–27, 1972.

Section III is a bibliography of basic governmental, institutional, and organizational documents assembled by the United Nations Conference on the Human Environment, held at Stockholm, Sweden, June 5–16, 1972."

U.S. Department of Commerce. ENVIRONMENTAL POLLUTION AND CONTROL. National Technical Information Service, Springfield, Virginia 22151.

U.S. Department of Health, Education and Welfare. ENVIRONMENTAL HEALTH SERIES AP: AIR POLLUTION. Public Health Service, Division of Air Pollution, Washington, D.C. 20201.

_____. HANDBOOK OF AIR POLLUTION. Public Health Service, National Center for Air Pollution Control, Durham, North Carolina 27701.

U.S. Environmental Protection Agency. AIR POLLUTION ASPECTS OF EMISSION SOURCES: BOILERS; BIBLIOGRAPHY WITH ABSTRACTS. May 1972. Air Pollution Technical Information Center, Research Triangle Park, North Carolina 27711.

_____. COST OF CONVENTIONAL AND ADVANCED TREATMENT OF WASTEWATERS. by Robert Smith. Washington, D.C.: July 1968.

_____. CURRENT LAWS, STATUES AND EXECUTIVE ORDERS. Washington, D.C.: Government Printing Office, January 1972.

_____. DIRECTORY OF INFORMATION SOURCES. Washington, D.C.: April 1972.

_____. ENVIRONMENTAL BIBLIOGRAPHY. Compiled by the Editorial Research Section, Office of Public Affairs. Washington, D.C.: Government Printing Office, 1972.

_____. GUIDE TO RESEARCH IN AIR POLLUTION; PROJECTS ACTIVE IN CALENDAR YEAR 1972. 8th ed. Compiled by Center for Air Environment Studies of Pennsylvania State University. Rockville, Maryland: 1973.

U.S. Government Printing Office. PRICE LIST. Vol. 88: Ecology. Water and Air Pollution, Environmental Pollution, Beautification. Washington, D.C.: Government Printing Office.

Contains list of documents available.

U.S. Library of Congress. National Referral Service. A DIRECTORY OF IN-FORMATION RESOURCES IN THE UNITED STATES: WATER. Washington, D.C.: Government Printing Office, 1966.

U.S. National Air Pollution Control Administration. A GUIDE TO AIR POLLU-TION RESEARCH. Washington, D.C.: Government Printing Office, 1969, 1970.

_____. NITROGEN OXIDES: AN ANNOTATED BIBLIOGRAPHY 1970. Washington, D.C.: Government Printing Office.

U.S. National Industrial Pollution Control Council. CHEMICAL INDUSTRY AND POLLUTION CONTROL. Washington, D.C.: Government Printing Office, June 1971.

U.S. Smithsonian Institution. THE GEOLOGICAL AND ECOLOGICAL EFFECTS OF AIR POLLUTION IN TROPICAL WATERS. AN ANNOTATED BIBLIOGRAPHY. Rept. no. PB 248 899. Springfield, Virginia: National Technical Information Service, 1975.

WORLD ENVIRONMENTAL DIRECTORY. Silver Spring, Maryland: Business Publishers, 1975.

ii. Abstracting and Indexing Services

ABSTRACTS OF AIR AND WATER CONSERVATION LITERATURE AND PATENTS. Weekly. American Petroleum Institute, Central Abstracting and Indexing Service, 1271 Avenue of the Americas, New York, New York 10020.

ABSTRACTS ON HEALTH EFFECTS OF ENVIRONMENTAL POLLUTANTS. Month-ly. Biosciences Information Service of Biological Abstracts, 2100 Arch Street, Philadelphia, Pennsylvania 19103.

ACCESS. Bimonthly. Environment Information Service, 124 East 39th Street, New York, New York 10016.

> Computerized information service for environmental area literature searches, special studies on contract basis; hard copy abstract ser-vice (bimonthly) as well as microfiche of governmental reports on subscription.

AIR POLLUTION ABSTRACTS. Monthly. United Kingdom, Department of Trade and Industry, Warren Spring Laboratory, Gunnels Wood Road, Stevenage, Herts, SG1 2BX, England.

AIR POLLUTION ABSTRACTS (U.S.). Monthly. Air Pollution Control Office, Box 12055, Research Triangle Park, North Carolina 27709.

AIR POLLUTION TITLES. 6 issues per year. Center for Air Environment Studies. The Pennsylvania State University, 226 Chemical Engineering II, University Park, Pennsylvania 16802.

AIR QUALITY DATA STATISTICS. Monthly. Government Printing Office, Washington, D.C. 20042.

APCA ABSTRACT. Monthly. Air Pollution Control Association, 4400 Fifth Avenue, Pittsburgh, Pennsylvania 15213.

ENVIRONMENTAL ENGINEERING. New York: McGraw-Hill, 1972.

A special deskbook edition of the May 8, 1972, issue of CHEMICAL ENGINEERING.

ENVIRONMENT INDEX. Annual. Environment Information Center, Inc., 124 East 39th Street, New York, New York 10016.

Indexes all literature pertaining to environment such as journals, books, conference proceedings and technical reports. Topics covered are pollution, pesticides, work environment safety and related subjects.

POLLUTION ABSTRACTS. Bimonthly. Pollution Abstracts, Inc., P.O. Box 2369, La Jolla, California 92307.

PREDI-BRIEFS ABSTRACTING SERVICE. Monthly. POLLUTION CONTROL. Predicasts, 200 University Circle Research Center, 11001 Cedar Avenue, Cleveland, Ohio 44106.

"Topics include water supply and use, water purification, water softening, fresh water, desalted water, water pollution, sewage systems, refuse systems, air pollution control, noise abatement, air and water treatment equipment, and water treatment chemicals."

iii. Journals and Serial Publications

ADVANCES IN ECOLOGICAL RESEARCH. Irregular. Academic Press, 111 Fifth Avenue, New York, New York 10003.

ADVANCES IN ENVIRONMENTAL SCIENCES AND TECHNOLOGY. Irregular. John Wiley & Sons, Inc., 605 Third Avenue, New York, New York 10016.

AIR POLLUTION CONTROL ASSOCIATION. JOURNAL. 4400 Fifth Avenue, Pittsburgh, Pennsylvania 15213.

AIR/WATER POLLUTION REPORT. Weekly. Business Publications, Inc., P.O. Box 1067, Silver Spring, Maryland 20910.

"Covering significant aspects of environmental problems."

ATMOSPHERIC ENVIRONMENT. Quarterly. Pergamon Press, Inc., Maxwell House, Fairview Park, Elmsford, New York 10523.

CHEMOSPHERE. Bimonthly. Pergamon Press, Inc., Maxwell House, Fairview Park, Elmsford, New York 10523.

"CHEMOSPHERE offers rapid dissemination of original articles related to environmental affairs and reports scientific investigations in the fields of chemistry, physics, and biology.

CLEAN AIR AND WATER NEWS. Weekly. Commerce Clearing House Inc., 4025 West Peterson Avenue, Chicago, Illinois 60646.

"News about air and water pollution, treatment, waste and disposal."

ENVIRONMENTAL SCIENCE AND TECHNOLOGY. Monthly. American Chemical Society Publications, 1155 16th Street, N.W., Washington, D.C. 20036.

INTERNATIONAL JOURNAL OF ENVIRONMENTAL STUDIES. Quarterly. Gordon and Breach Publishers, 1 Park Avenue, New York, New York 10016.

JOURNAL OF ENVIRONMENTAL SCIENCES. Bimonthly. Institute of Environmental Sciences, 940 E. Northwest Highway, Mt. Prospect, Illinois 60056.

JOURNAL OF HUMAN ECOLOGY. Quarterly. Human Ecological Society, Box 146, Elsah, Illinois 62028.

JOURNAL ON HUMAN ECOLOGY. Quarterly. Human Ecological Society, Box 146, Elsah, Illinois 62028.

POLLUTION ENGINEERING. Monthly. W. Arthur West Publishing Co., 35 Mason Street, Greenwich, Connecticut 06830.

POLLUTION EQUIPMENT NEWS. Bimonthly. Richard Rimbach, Sr., 8550 Babcock Boulevard, Pittsburgh, Pennsylvania 15237.

REUSE/RECYCLE. Monthly. Technomic Publishing Company, Inc., 265 W. State Street, Westport, Connecticut 06880.

"New recycling processes...new products from wastes · recycling technology and economics · municipal, industrial, and consumer waste recycling, etc."

iv. Products and Manufacturers Information

COMMERCIAL WATER/WASTEWATER CATALOG. Annual. Greenwich, Connecticut: Publicom Inc.

Supplies vendor information.

MACRAE'S BLUE BOOK. Annual. MacRae's Blue Book Company, Western Springs, Illinois.

POLLUTION CONTROL COMPANIES USA 1972. Park Ridge, New Jersey: Noyes Data Corporation.

"Provides a marketing guide to the U.S. pollution control industry."

THOMAS REGISTER OF AMERICAN MANUFACTURERS. Annual. New York: Thomas Publishing Company.

b. WATER POLLUTION CONTROL

i. Guides and Handbooks

Advances in Water Pollution Research, 5th International Conference. San Francisco, California, July 27-31, 1970 and Honolulu, Hawaii, August 2-5, 1970. Sponsored by International Association on Water Pollution Research. PROCEEDINGS. 2 vols. New York: Pergamon, 1971.

Biological Waste Treatment. Symposium. PROCEEDINGS. Biotechnology and Bioengineering Symposium No. 2. Held during the 160th National Meeting of the American Chemical Society, Chicago, Illinois, September 16, 1970. New York: Wiley-Interscience, 1971.

Boby, W. WATER TREATMENT DATA; A HANDBOOK FOR CHEMISTS AND ENGINEERS IN INDUSTRY. London: Hutchinson, 1965.

CHEMICAL ENGINEERING APPLICATIONS IN SOLID WASTE TREATMENT, Papers at Meetings. Sponsored by American Institute of Chemical Engineers. (AICHE Symposium series, vol. 68, no. 122, 1972.) American Institute of Chemical Engineers, 345 East 47th Street, New York, New York 10017.

Ciaccio, L., ed. WATER AND WATER POLLUTION HANDBOOK. New York: Dekker, 1971.

An overview of the subject of water pollution control.

DESIGN HANDBOOK OF WASTEWATER SYSTEMS: DOMESTIC, INDUSTRIAL, COMMERCIAL. Westport, Connecticut: Technomic Publishers, 1971.

DISCHARGE OF INDUSTRIAL EFFLUENTS TO MUNICIPAL SEWERAGE SYSTEMS. Symposium sponsored by Institute of Water Pollution Control, London, England. Institute of Water Pollution Control, Ledson House, 53 London Road, Maidstone, Kent, England, 1971.

ENVIRONMENTAL MERCURY CONTAMINATION. Conference sponsored by University of Michigan, Ann Arbor, Michigan. Ann Arbor, Michigan: Ann Arbor Science Publishers, 1972.

Flinn, J., ed. MEMBRANE SCIENCE AND TECHNOLOGY; INDUSTRIAL, BIOLOGICAL, AND WASTE TREATMENT PROCESSES. Proceedings of a symposium held at the Columbus Laboratories of Battelle Memorial Institute, Columbus, Ohio, October 20-21, 1969. New York: Plenum, 1971.

"The use of membrane diffusion as a tool for separating and characterizing naturally occurring polymers. Anomalous transport of penetrants in polymeric membranes. Application and theory of membrane processes for biological and other macromolecular solutions."

INDUSTRIAL PROCESS DESIGN FOR POLLUTION CONTROL. 4th Workshop, sponsored by American Institute of Chemical Engineers, Charleston, West Virginia. New York: American Institute of Chemical Engineers, 1972.

INDUSTRIAL WASTE. 4th Mid-Atlantic Conference, Baltimore, Maryland; Philadelphia, Pennsylvania; Newark, Delaware; November 18-20, 1970. PROCEEDINGS. Newark, Delaware: University of Delaware, Department of Civil Engineering, 1971.

> Sponsored by the University of Delaware, Johns Hopkins University, University of Maryland, University of Pennsylvania, Drexel Institute of Technology, Delaware Department of Natural Resources & Environmental Control, Delaware Department of Health & Social Services, and American Water Resources Association.

INDUSTRIAL WASTE WATER. A Symposium on Recent Developments sponsored by Royal Australian Chemical Institute, and Institution of Engineers, Australia, Sydney, N.S.W., Australia. Institution of Engineers, Australia, Science House, 157 Gloucester Street, Sydney 2000, Australia, 1972.

James, R. SEWAGE SLUDGE TREATMENT 1972. Park Ridge, New Jersey: Noyes Data Corporation.

> "This book describes 119 patent-based processes relating to the treatment of sewage sludge."

Jones, H. DETERGENTS AND POLLUTION; PROBLEMS AND TECHNOLOGICAL SOLUTIONS. Park Ridge, New Jersey: Noyes Data Corporation, 1972.

> Overview of industry.

_____. ENVIRONMENTAL CONTROL IN THE INORGANIC CHEMICAL INDUSTRY, 1972. Park Ridge, New Jersey: Noyes Data Corporation, 1972.

> Contains detail on waste disposal in chemical plants as well as U.S. patents in this area.

_____. MERCURY POLLUTION CONTROL 1971. Park Ridge, New Jersey: Noyes Data Corporation.

> "Book explains in detail what measures can be taken to prevent further pollution and to remove existing contamination."

McCoy, J. THE CHEMICAL TREATMENT OF COOLING WATER. New York: Chemical Publishing Company, Inc., 1973.

> Provides details on water treating chemicals and processes.

Meyer, J. AQUATIC HERBICIDES AND ALGAECIDES 1971. Park Ridge, New Jersey: Noyes Data Corporation.

> "This book gathers the latest technology for producing or using aquatic herbicides and algaecides, based on U.S. patent literature."

Organization for Economic Cooperation and Development. Directorate for Scientific Affairs. DIRECTORY OF WATER POLLUTION RESEARCH LABORATORIES. Paris: 1965.

RECENT DEVELOPMENTS IN INDUSTRIAL POLLUTION CONTROL. Westport, Connecticut: Technomic Publishing Company, Inc., 1971.

> "Advanced processes and equipment for solving or abating a variety of today's industrial pollution problems are described in detail by leading authorities in this field. Pollution problems related to plastics, coal, petroleum, chemicals, metals, fertilizers, sewage, refuse, and land use are covered."

REUSE AND RECYCLE OF WASTES. Westport, Connecticut: Technomic Publishing Company, Inc., 1971.

> "Methods and equipment for reclaiming and reusing various industrial and municipal wastes are examined from the perspectives of both economics and technology. Topics and materials discussed include aluminum products, composting, compaction, shredding, acid hydrolysis, packaging materials, and plating wastes."

SLUDGE TREATMENT & DISPOSAL. 4th Public Health Engineering Conference, sponsored by Loughborough University of Technology, Loughborough, England. Department of Civil Engineering, Loughborough University of Technology, Loughborough, Leicester, LE1 3TU, England, 1971.

SOLID-LIQUID FLOW IN PIPES & ITS APPLICATION TO THE COLLECTION & REMOVAL OF SOLID WASTES. International Symposium, Philadelphia, Pennsylvania, March 4-6, 1968. Sponsored by the University of Pennsylvania. Issued as: ADVANCES IN SOLID-LIQUID FLOW IN PIPES & ITS APPLICATION. New York: Pergamon, 1971.

SOLID WASTE DISPOSAL EQUIPMENT. Cleveland, Ohio: Predicasts, Inc., August 1972.

> "Analyzes flow of solid waste from material consumption to ultimate disposal. Markets for solid waste disposal equipment are discussed." Also forecast through 1985.

Standard, W. COOLING TOWERS; PRINCIPLES AND PRACTICE. 2nd ed. Birmingham, England: Carter Thermal Engineering, 1970.

> Covers theory and practice of cooling towers.

U.S. Congress. Senate. OCEAN WASTE DISPOSAL. Hearings before the U.S. Senate Committee on Commerce, Washington, D.C., March 2-22, and April 28, 1971.

U.S. Environmental Protection Agency. BIOLOGICAL REMOVAL OF CARBON AND NITROGEN COMPOUNDS FROM COKE PLANT WASTES WITH LIST OF REFERENCES. By John E. Barker and R.J. Thompson. Washington, D.C.: Government Printing Office, April 1973.

U.S. Environmental Protection Agency. Effluent Guidelines Division. DEVEL-OPMENT DOCUMENT FOR BEST TECHNOLOGY AVAILABLE FOR THE LOCA-TION, DESIGN, CONSTRUCTION, AND CAPACITY OF COOLING WATER INTAKE STRUCTURES FOR MINIMIZING ADVERSE ENVIRONMENTAL IMPACT. By D. Barnes. Rept. no. PB 253-573. Springfield, Virginia: National Technical Information Service, 1976.

_____. FEDERAL WATER POLLUTION CONTROL ACT AMENDMENTS OF 1972, HIGHLIGHTS. Washington, D.C.: January 1973.

Concerns environmental law.

_____. PRIMER ON WASTE WATER TREATMENT. Rev. ed. Washington, D.C.: Government Printing Office, March 1971.

Some illustrations are colored.

_____. PROJECTED WASTEWATER TREATMENT COSTS IN ORGANIC CHEMI-CAL INDUSTRY WITH BIBLIOGRAPHY. Washington, D.C.: Government Printing Office, July 1971.

_____. SOLID WASTE MANAGEMENT, LIST OF AVAILABLE LITERATURE. October 1971. Publications Distribution Unit, Solid Waste Management Office, 5555 Ridge Avenue, Cincinnati, Ohio 45213.

_____. SOLID WASTE MANAGEMENT, LIST OF AVAILABLE LITERATURE. Washington, D.C.: Government Printing Office, October 1972.

U.S. Environmental Protection Agency. Technology Transfer. HANDBOOK FOR ANALYTICAL QUALITY CONTROL IN WATER AND WASTEWATER LABO-RATORIES. By Analytical Quality Control Laboratory, National Environmental Research Center, Cincinnati, Ohio. Washington, D.C.: June 1972.

U.S. Environmental Protection Agency. Water Programs Office. GLOSSARY OF WATER RESOURCE TERMS. By Olga Adler Titelbaum. Rockville, Mary-land: April 1970.

U.S. National Industrial Pollution Control Council. CHEMICAL INDUSTRY AND POLLUTION CONTROL. Washington, D.C.: Government Printing Office, June 1971.

Water Pollution Control Act of 1972. ECONOMIC IMPACTS, CHEMICAL AND ALLIED INDUSTRIES. Rept. no. PB 252289. Springfield, Virginia: National Technical Information Service, 1976.

Zajic, J. WATER POLLUTION; DISPOSAL AND REFUSE. 2 vols. New York: M. Dekker, 1971.

ii. Abstracting and Indexing Services

SELECTED WATER RESOURCES ABSTRACTS. Semimonthly. (Water Resources Scientific Information Center, Office Water Resources Research, Department of the Interior) National Technical Information Service, U.S. Department of Commerce, Springfield, Virginia 22151.

U.S. Environmental Protection Agency. Office of Solid Wastes Management Programs. PATENT ABSTRACTS: SOLID WASTE MANAGEMENT. 1945-1969. (INTERNATIONAL). Washington, D.C.: Government Printing Office.

WATER POLLUTION ABSTRACTS. Monthly. Department of the Environment, Her Majesty's Stationery Office, Atlantic House, Holborn Viaduct, London, EC1, England.

WATER RESOURCES ABSTRACTS. Monthly. American Water Resources Association, 206 East University Avenue, Urbana, Illinois 61801.

Also see General Abstracting and Indexing Services on this subject:

APPLIED SCIENCE AND TECHNOLOGY INDEX

CHEMICAL ABSTRACTS

ENGINEERING INDEX

INDEX MEDICUS

iii. Journals and Serial Publications

ADVANCES IN BIOLOGICAL WASTE TREATMENT. Pergamon Press, Inc., 44-01 21st Street, Long Island City, New York 11101.

American Water Resources Association. PROCEEDINGS. Annual. American Water Resources Association, 206 East University Avenue, Urbana, Illinois 61801.

AMERICAN WATER WORKS ASSOCIATION JOURNAL. American Water Works Association, 2 Park Avenue, New York, New York 10012.

CLEAN WATER REPORT. Biweekly. Business Publishers Inc., P.O. Box 1067, Silver Spring, Maryland 20910.

"Market opportunities in water pollution abatement...government controls-water research programs-demonstration results...and reuse."

EFFLUENT AND WATER TREATMENT JOURNAL. Monthly. Thunderbird Enterprises Ltd., 102 College Road, Harrow, Middlesex, England.

FROM THE STATE CAPITOLS REPORT: SEWAGE AND WASTE DISPOSAL. Monthly. Bethune Jones, 321 Sunset Avenue, Asbury Park, New Jersey 07712.

"Lists developments affecting air and water pollution control and related matters."

THE GROUND WATER NEWSLETTER. Semimonthly. Water Information Center, Inc., 44 Sintsink Drive, East Port Washington, New York 10050.

> Coverage includes "all aspects of pollution, undergroundwaste disposal, recharge, management, development."

HYDATA (including WATER RESOURCES NEWSLETTER). Monthly. American Water Resources Association, 206 East University Avenue, Urbana, Illinois 61801.

INDUSTRIAL WATER CONDITIONING. Monthly. Publicom Publishing, Inc., 17 Sherwood Place, Greenwich, Connecticut 06830.

PUBLIC WORKS. Monthly. Public Works Publications, 200 South Broad Street, Ridgewood, New Jersey 07451.

SANITARY ENGINEERING RESEARCH LABORATORY NEWS QUARTERLY. University of California, College of Engineering, Sanitary Engineering Research Laboratory, Richmond Field Station, 1301 South 46th Street, Richmond, California 94804.

SOLID WASTE REPORT. Biweekly. Business Publishers, Inc., P.O. Box 1067, Silver Spring, Maryland 20910.

> "Reports on federal, state and local legislation-grants and contracts-products-conferences...etc."

WATER AND POLLUTION CONTROL. Monthly. Southam Business Publications, Ltd., 1450 Don Mills Road, Ontario, Canada.

WATER AND WASTE DIGEST. Monthly. Scranton Publishing, 35 E. Wacker, Chicago, Illinois 60611.

WATER AND WASTES ENGINEERING. Monthly. Reuben H. Donnelly Company, 466 Lexington Avenue, New York, New York 10017.

> Contents: Water resources and pollution control.

WATER CONDITIONING. Monthly. Publicom Inc., 17 Sherwood Place, Greenwich, Connecticut 06830.

WATER POLLUTION. Pergamon Press, 44-01 21st Street, Long Island City, New York 11101.

WATER POLLUTION CONTROL. Bimonthly. Institute of Water Pollution Control, 49-55 Victoria Street, London, S.W.1, England.

WATER POLLUTION CONTROL ACTIVITIES AND THE INTERSTATE AIR POLLUTION PROGRAM. Report of the Interstate Sanitation Commission, 10 Columbus Circle, New York, New York 10019.

WATER POLLUTION CONTROL FEDERATION. JOURNAL. Monthly. Water Pollution Control Federation, 3900 Wisconsin Avenue, N.W., Washington, D.C. 20016.

WATER RESEARCH. Monthly. International Association on Water Pollution Research. Pergamon Press Inc., Journals Department, Maxwell House, Fairview Park, Elmsford, New York 10523.

WATER RESOURCES BULLETIN. Bimonthly. (Including WATER RESOURCES NEWSLETTER.) American Water Resources Association, 206 East University Avenue, Urbana, Illinois 61801.

WATER RESOURCES NEWSLETTER. Bimonthly. American Water Resources Association, 206 East University Avenue, Urbana, Illinois 61801.

iv. Products and Manufacturers Information

THE ENVIRONMENTAL WASTES CONTROL MANUAL AND CATALOG FILE. Public Works, Publications/City, County and State, 200 South Broad Street, Ridgewood, New Jersey 07451.

INDUSTRIAL WATER ENGINEERING PURCHASING DIRECTORY. Select Publications, Inc., 900 Northstar Center, Minneapolis, Minnesota 55406.

THE SEWAGE MANUAL AND CATALOG FILE. Public Works Journal Corporation, 200 South Broad Street, Ridgewood, New Jersey 07451.

WASTE WATER CLEANUP EQUIPMENT. 2nd ed. Park Ridge, New Jersey: Noyes Data Corporation, 1973.

> Details on latest equipment available based on vendor drawings and catalogs.

WATER AND POLLUTION CONTROL DIRECTORY. Annual. Southam Business Publications, Ltd., 1450 Don Mills Road, Ontario, Canada.

> Complete guides and buyers directory for environmental services in Canada.

WATER AND WATER ENGINEERING BUYERS DIRECTORY. Annual. Reuben H. Donnelly Corporation, 466 Lexington Avenue, New York, New York 10017.

WATER POLLUTION CONTROL EQUIPMENT. Cleveland, Ohio: Predicasts, Inc., 1972.

> "Investigates water use and pollution control spending by economic sectors. Gives historical (since 1960) and projected (to 1985) data for over a dozen types of water pollution control equipment by end-use."

WATER TREATMENT CHEMICALS. Cleveland, Ohio: Predicasts, Inc., 1972.

> "Compare alternative processes for water and wastewater treatment. Gives historical (since 1960) and projected (to 1985) usage and markets for chemicals by type for both neutralization and separation."

c. SAFETY

i. Guides and Handbooks

ACCIDENT PREVENTION MANUAL FOR INDUSTRIAL OPERATIONS. 1969. National Safety Council, 425 Michigan Avenue, Chicago, Illinois 60611.

ACTIVITIES OF FEDERAL AGENCIES CONCERNING SELECTED HIGH VOLUME CHEMICALS. Washington, D.C.: U.S. Environmental Protection Agency, Office of Toxic Substances, 1975.

BEST'S SAFETY DIRECTORY. 1970. A.M. Best Company, Park Avenue, Morristown, New Jersey 07960.

Browning, E. TOXICITY OF INDUSTRIAL METALS. 2nd ed. London: Butterworth, 1969.

> "A survey of the main features of the occurrence, preparation, physical and chemical properties, metabolism and toxicology of the principal metals encountered in modern industry."

British Cryogenics Council. CRYOGENICS SAFETY MANUAL. London: 1970.

> "Produced for the guidance of all who are concerned with the operation and maintenance of plant for producing, storing and handling commercial gases which liquefy at relatively low temperatures."

Burklin, C. SAFETY STANDARDS, CODES AND PRACTICES FOR PLANT DESIGN. 3-part article. 1) "Design Consideration" in CHEMICAL ENGINEERING, October 2, 1972, pp. 56-63. 2) "Testing Standards and Insurance Associations" in CHEMICAL ENGINEERING, October 16, 1972, pp. 113-20. 3) "Professional and Trade Associations and Government Agencies" in CHEMICAL ENGINEERING, November 13, 1972, pp. 143-55.

> Provides a guide to safety information sources complete with names, addresses and brief description of their service programs.

Buschman, R. LOSS PREVENTION AND SAFETY PROMOTION IN THE PROCESS INDUSTRIES. New York: American Elsevier, 1975.

CHEMICAL ENGINEERING PROGRESS. SAFETY IN AIR AND AMMONIA PLANTS. New York: American Institute of Chemical Engineers.

CHEMICAL HAZARDS RESPONSE INFORMATION SYSTEM FOR MULTIMODAL ACCIDENTS (CHRISMA), REEVALUATION OF CHRIS FOR ALL MODES OF TRANSPORTATION. Final Report. Springfield, Virginia: National Technical Information Service, 1975.

FACTORY MUTUAL SYSTEM: HANDBOOK OF INDUSTRIAL LOSS PREVENTION. 2nd ed. New York: McGraw-Hill, 1967.

"Details on how to protect industrial plants and processes against damage by fire, explosions, lightning, wind, and earthquakes."

Faulkner, L. HANDBOOK OF INDUSTRIAL NOISE CONTROL. New York: Industrial Press, 1976.

Fawcett, H. SAFETY AND ACCIDENT PREVENTION IN CHEMICAL OPERA-TIONS. New York: John Wiley & Sons, 1965.

Hammer, W. HANDBOOK OF SYSTEM AND PRODUCT SAFETY. Edgewood Cliffs, New Jersey: Prentice Hall, 1973.

Presents principles and mathematical models for use in improving plant and product safety.

Handley, W., ed. INDUSTRIAL SAFETY HANDBOOK. New York: McGraw-Hill, 1970.

"Communicating the safety message. Chemical handling. Commer-cial vehicles. Entry into confined spaces. Fire control: automatic. Mechanical safety, etc."

HAZARDOUS CHEMICALS HANDLING AND DISPOSAL. Park Ridge, New Jersey: Noyes Data Corp., 1971.

Institution of Fire Engineers. HAZARDOUS LOADS. 2nd ed. Leicester, England: 1972.

"A quick reference guide, giving possible dangers and the action to be taken at incidents involving hazardous loads in transit."

International Loss Prevention Symposium, 1974. LOSS PREVENTION AND SAFETY PROMOTION IN THE PROCESS INDUSTRIES. New York: American Elsevier Co., 1974.

Lyons, J. THE CHEMISTRY AND USES OF FIRE RETARDANTS. New York: Interscience, Division of Wiley, 1970.

Useful information for safety applications in chemical plants.

MANUAL FOR SAFE HANDLING OF INFLAMMABLE AND COMBUSTABLE LIQUIDS AND OTHER HAZARDOUS PRODUCTS. Washington, D.C.: U.S. Coast Guard, Transportation Department, 1976.

Midwest Research Institute. SAMPLING AND ANALYSES OF SELECTED TOXIC SUBSTANCES. Task 1A-Hexachlorobenzene. By T. Spigarelli, et al. Rept. no. PB 253 794. Springfield, Virginia: National Technical Information Service, 1976.

National Safety Council. HANDBOOK OF OCCUPATIONAL SAFETY AND HEALTH. Chicago: 1975.

Peck, T. GUIDE TO OCCUPATIONAL SAFETY AND HEALTH INFORMATION SOURCES. Detroit: Gale Research Company, 1973.

Peterson, D. TECHNIQUES OF SAFETY MANAGEMENT. New York: McGraw-Hill, 1971.

"Stressing safety psychology and safety systems rather than details of safety engineering or safety promotion."

Plunkett, E., ed. HANDBOOK OF INDUSTRIAL TOXICOLOGY. 2nd ed. New York: Chemical Publishing Co., 1976.

PROCEEDINGS OF THE INTERNATIONAL SYMPOSIUM ON THE CARRIAGE OF DANGEROUS GOODS. Rotterdam, The Netherlands, May 27-30, 1968. 1969. Available from Mrs. G.B.C. Meelis, c/o Scheepvaartinspectie (Inspection of Navigation), Afdeling Gevaarlijke Ladingen, Hoge Prins Willemstraat 26, Den Haag.

Collection of papers containing valuable information on safety aspects of chemical industries.

PROGRESS IN HAZARDOUS CHEMICALS HANDLING & DISPOSAL. Vol. 3. Park Ridge, New Jersey: Noyes Data Corp., 1972.

Rodgers, W. INTRODUCTION TO SYSTEM SAFETY ENGINEERING. New York: Wiley, 1971.

Sax, N. DANGEROUS PROPERTIES OF INDUSTRIAL MATERIAL. New York: Van Nostrand-Reinhold Publishing Co., 1975.

Steere, N., ed. CRC HANDBOOK OF LABORATORY SAFETY. 2nd ed. Cleveland: Chemical Rubber Co., 1971.

SYMPOSIUM ON CHEMICAL PROCESS HAZARDS WITH SPECIAL REFERENCE TO PLANT DESIGN. Manchester, England, 1967. London: Institution of Chemical Engineers.

Symposium on Hazardous Chemicals Handling and Disposal, 1st. Lafayette, Indiana, 1970. HAZARDOUS CHEMICALS HANDLING AND DISPOSAL: THE PROCEEDINGS. Robert H.L. Howe, chief editor. Park Ridge, New Jersey: Noyes Data Corp., 1970.

Symposium on Hazardous Chemicals Handling and Disposal, 2nd. Indiana, 1971. HAZARDOUS CHEMICALS HANDLING AND DISPOSAL: THE PROCEEDINGS. Park Ridge, New Jersey: Noyes Data Corp., 1971.

TRW Systems Group, Redondo Beach, California. ASSESSMENT OF INDUSTRIAL HAZARDOUS WASTE PRACTICES, ORGANIC CHEMICALS, PESTICIDES, AND

EXPLOSIVES INDUSTRIES. By G. Gruber. Rept. no. PB 251 307. Spring-field, Virginia: National Technical Information Service, 1976.

U.S. Coast Guard. Transportation Department. CHRIS (CONDENSED GUIDE TO CHEMICAL HAZARDS). Washington, D.C.: 1974.

_____. EVALUATION OF THE HAZARD OF BULK WATER TRANSPORTATION OF INDUSTRIAL CHEMICALS (A TENTATIVE GUIDE). Springfield, Virginia: National Technical Information Service, 1973.

U.S. Congress. House of Representatives. Government Operations Committee. TRANSPORTATION OF HAZARDOUS MATERIALS. Hearings before subcommit-tee, 92nd Congress, 1st and 2nd sessions, November 17, 1971-June 28, 1972. Washington, D.C.: Government Printing Office, 1973.

U.S. Department of Health, Education and Welfare, National Institute for Occupational Safety and Health. TOXIC SUBSTANCES LIST. Annual. Wash-ington, D.C.: Government Printing Office.

U.S. Department of Labor. Occupational Safety and Health. GENERAL IN-DUSTRY STANDARDS AND INTERPRETATIONS. CHANGE. Vol. 1. Wash-ington, D.C.: Government Printing Office, 1975.

U.S. Environmental Protection Agency. PHARMACEUTICAL INDUSTRY: HAZARDOUS WASTE GENERATION, TREATMENT AND DISPOSAL. Washing-ton, D.C.: Government Printing Office, 1976.

Vervalin, C., ed. FIRE PROTECTION MANUAL FOR HYDROCARBON PRO-CESSING PLANTS. 2nd ed. Houston, Texas: Gulf Publishing Co., 1973.

ii. Abstracting and Indexing Services

FIRE RESEARCH ABSTRACTS AND REVIEWS. 3/yr. National Academy of Sciences, Committee on Fire Research Council, 2101 Constitution Avenue, N.W., Washington, D.C. 20418.

INDUSTRIAL HYGIENE DIGEST. Monthly. Industrial Health Foundation, Inc., Information Services, 5231 Centre Avenue, Pittsburgh, Pennsylvania 15232.

Abstracting service for all aspects of industrial environmental con-ditions.

OCCUPATIONAL SAFETY AND HEALTH. Monthly with supplements. Royal Society for the Prevention of Accidents, 6 Buckingham Place, London, SW1E 6HR, England.

OCCUPATIONAL SAFETY AND HEALTH ABSTRACTS. Monthly. International Labor Office, 917 15th Street, N.W., Washington, D.C. 20005.

Available in card format.

TODAY'S SAFETY GUIDES ABSTRACTS. U.S. Department of Labor, Bureau of Labor Standards, Office of Occupational Safety, Washington, D.C. 20210.

Available in card format.

TOXICITY BIBLIOGRAPHY. Quarterly. Government Printing Office, Washington, D.C. 20402.

iii. Journals and Serial Publications

AMERICAN SOCIETY OF SAFETY ENGINEERS. JOURNAL. Monthly. American Society of Safety Engineers, 850 Busse Highway, Park Ridge, Illinois 60068.

CANADIAN OCCUPATIONAL SAFETY. Bimonthly. Parkins Publishing Co., Ltd., 1215 Greene Avenue, Montreal, 1215, Quebec, Canada.

CHEMICAL NEWSLETTER. National Safety Council, 425 North Michigan Avenue, Chicago, Illinois 60611.

EMPLOYMENT SAFETY AND HEALTH GUIDE. 1971. Commerce Clearing House, 420 W. Glenlake Avenue, Chicago, Illinois 60646.

Contains a directory of federal Office of Safety and Health Administration offices through the U.S.

INDUSTRIAL SAFETY. Monthly. United Trade Press, Ltd., 9 Gough Square, London, EC4, England.

JOB SAFETY & HEALTH REPORT. Biweekly. Business Publications, Inc., P.O. Box 1067, Silver Spring, Maryland 20910.

"Federal and state legislation...standards...regulations...research activities...enforcement cases...are a few of the topics covered bi-weekly in the formative JS&HR."

JOURNAL OF FIRE AND FLAMMABILITY. Quarterly. Technomic Publishing Co., Inc., 265 West State Street, Westport, Connecticut 06880.

"Authoritative source of new information on: • The Flammability of Materials • Fire Exposures and Characteristics • Fire and Flammability in Various Environments • Text Methods Apparatus, and Results, etc."

JOURNAL OF SAFETY RESEARCH. Quarterly. National Safety Council, 425 North Michigan Avenue, Chicago, Illinois 60611.

NATIONAL SAFETY NEWS. Monthly. National Safety Council, 425 North Michigan Avenue, Chicago, Illinois 60611.

NOISE CONTROL REPORT. Biweekly. Business Publishers, Inc., P.O. Box 1067, Silver Spring, Maryland 20910.

"Covers significant developments at city, state, and federal levels
...legislative, technical, research...standards to control noise on
the job and in the overall environment."

OCCUPATIONAL HAZARDS. Monthly. Occupational Hazards, 614 Superior
Avenue, West, Cleveland, Ohio 44113.

Covers industrial security, hygiene, safety and related areas.

OCCUPATIONAL SAFETY AND HEALTH. Monthly. Occupational Safety and
Health Supplement. Royal Society for the Prevention of Accidents, 52 Grosvenor Gardens, London, S.W. 1, England.

PROTECTION: THE JOURNAL OF LOSS PREVENTION. 11/yr. Alan Osborne
& Associates, 1/113 Blackheath Park, London, SE3 OHA, England.

Official journal of Institution of Industrial Safety Officers of Great
Britain.

SAFETY JOURNAL. Monthly. J.R. Young, Box 19, Anderson, South Carolina
19622.

SAFETY NEWS. Monthly. Royal Society for the Prevention of Accidents, 52
Grosvenor Gardens, London, SW1, England.

SAFETY NEWSLETTER - CHEMICAL SECTION. Monthly. National Safety
Council, Chicago, Illinois 60611.

Provides chemical information sheets telling properties, hazards and
precautions of various toxic materials.

SAFETY PRODUCTS NEWS. Bimonthly. Industrial Safety Products, Equipment
& Systems, Ames Publishing Co., One West Olney Avenue, Philadelphia, Pennsylvania 19120.

SAFETY STANDARDS. Bimonthly. Government Printing Office, Washington,
D.C. 20402.

iv. Products and Manufacturers Information

MACRAE'S BLUE BOOK. Annual. MacRae's Blue Book Co., Western Springs,
Illinois.

Product and Manufacturer Information is also available from National Safety
Council, 425 N. Michigan Avenue, Chicago, Illinois 60611.

THOMAS REGISTER OF AMERICAN MANUFACTURERS. Annual. New York:
Thomas Publishing Co.

16. Management Information Sources

This part has been added to Section I of this guide to provide a brief overview of management literature and reference sources for all chemical related industries represented in this Guide.

An attempt has been made to cite source materials in areas which seem significant to management of major industries and where possible items specifically related to chemical process industries have been included.

This part like all the others in this guide should be considered selective in content and useful as a bridge to the extensive information materials available on this subject.

a. GENERAL

i. Guides and Handbooks

American Institute of Consulting Engineers. ENGINEERING CONSULTANTS; DIRECTORY. 5th ed. New York: American Institute of Consulting Engineers, 1970.

> Consulting engineers - (members of AICE) and Consulting engineering firms.

Cetron, M. TECHNOLOGICAL FORECASTING; A PRACTICAL APPROACH. New York: Gordon & Breach, 1969.

> "The conceptual framework of technological forecasting, forecasting technology, forecasting methodologies, selection, and presentations of forecasting information. Problems and pitfalls."

Chironis, N. MANAGEMENT GUIDE FOR ENGINEERS AND TECHNICAL ADMINISTRATORS. New York: McGraw-Hill, 1969.

> "Consists of about 100 key articles drawn from numerous publications including PRODUCT ENGINEERING, BUSINESS WEEK, AMERICAN MACHINIST which discuss engineering office practices and management techniques useful to engineers in any field."

Commerce Clearing House, Inc. GOVERNMENT CONTRACTS GUIDE. 1968. 420 Lexington Avenue, New York, New York 10017.

CONSULTANTS AND CONSULTING ORGANIZATIONS DIRECTORY: A REFERENCE GUIDE TO CONCERNS AND INDIVIDUALS ENGAGED IN CONSULTATION FOR BUSINESS, INDUSTRY, AND GOVERNMENT. 3rd ed. Detroit, Michigan: Gale Research Company, 1976.

> "Contains descriptive details on 5,041 firms and individuals. Arranged alphabetically by names of firms, the main section offers descriptive profile statements on each organization, including address, names of chief executive officers, locations of branches, and

specialties. The second part is arranged under 146 subjects so
that users can easily determine which firms are engaged in specific
fields of consulting in their geographic area."

DIRECTORY OF CONSULTING SERVICES. Biennial publication. Association
of Consulting Chemists and Chemical Engineers, Inc., 50 East 41st Street, New
York, New York 10016.

"Directory consists of an alpha-geographic listing of 120 members
of the Association of Consulting Chemists and Chemical Engineers,
Inc."

Dunham, C. CONTRACTS, SPECIFICATIONS AND LAW FOR ENGINEERS.
New York: McGraw-Hill, 1971.

ENGINEERING CONSULTANTS. 1970. American Institute of Consulting Engineers, 345 East 47th Street, New York, New York 10017.

ENGINEERS'S GUIDE TO MANAGEMENT TECHNIQUES. Edited by D. Lock.
New York: Cahners, 1973.

Greene, J., ed. PRODUCTION AND INVENTORY CONTROL HANDBOOK.
New York: McGraw-Hill, 1970.

"Provides a thorough and detailed source of data and information
for operating managers, engineers, as well as other personnel involved in the manufacturing process."

Gross, H. HOW TO DO BUSINESS WITH THE GOVERNMENT. New York:
Pilot Books, 1969.

Karger, D. MANAGING ENGINEERING AND RESEARCH; THE PRINCIPLES
AND PROBLEMS OF MANAGING THE PLANNING, DEVELOPMENT AND
EXECUTION OF ENGINEERING, AND RESEARCH ACTIVITIES. 2nd ed.
New York: Industrial Press, 1969.

Provides practical guide to engineering research management and
its varied aspects.

Laidlaw, R. ENGINEERING LAW. Ontario, Canada: University of Toronto
Press, 1958.

Lobstein, R. GUIDE TO CHEMICAL PLANT PLANNING. Park Ridge, New
Jersey: Noyes Data Corp., 1969.

Martino, R. CRITICAL PATH NETWORKS. New York: McGraw-Hill, 1970.

"A basic presentation of network analysis techniques, explains in
direct language why and how networks are constructed, and how
they make possible the effective management and control of many
business and industrial operations."

_____. INTEGRATED MANUFACTURING SYSTEMS. New York: McGraw-Hill, 1972.

> "Designed for a specific audience and a specific purpose, this book is not only directed to business men and systems analysts, but to anyone interested in an overview of the entire manufacturing process."

Maynard, H. HANDBOOK OF MODERN MANUFACTURING MANAGEMENT. New York: McGraw-Hill, 1970.

> "Prepared for plant operations manager. Deals with the human, technical and financial aspects of manufacturing management."

Midge, A. VALUE ENGINEERING. New York: McGraw-Hill, 1971.

> "Describes the latest methods and technologies developed within the approach in recent years."

Nord, M. LEGAL PROBLEMS IN ENGINEERING. New York: Wiley, 1956.

O'Brien, J. CONTRACTOR'S MANAGEMENT HANDBOOK. New York: McGraw-Hill, 1971.

> A handbook of importance to engineers in management. Comprehensive coverage.

Peters, M. PLANT DESIGN AND ECONOMICS FOR CHEMICAL ENGINEERS. New York: McGraw-Hill, 1968.

Pilorough, L. INSPECTION OF CHEMICAL PLANT. Chemical and Process Engineering Series. London: Leonard Hill, 1971.

> Details on design, construction operation and maintenance of chemical plants.

Rossnagel, W. CHECKLISTS FOR MANAGEMENT, ENGINEERING, MANUFACTURING AND PRODUCT ASSURANCE. Vol. 1. Management Checklists. New York: Spartan, 1971.

> Useful management guide providing system of checklists as technique in research projects.

Simpson, L. LAW FOR ENGINEERS AND ARCHITECTS. St. Paul, Minnesota: West Publishing Co., 1958.

Vaughn, R. LEGAL ASPECTS OF ENGINEERING. Englewood Cliffs, New Jersey: Prentice Hall, 1962.

Weber, E., et al., eds. TECHNOLOGY FORECAST FOR 1960. New York: Van Nostrand Reinhold, 1971.

Wilson, E. FROM IDEA TO WORKING MODEL. New York: Wiley, 1970.

> Guide to system design techniques.

ii. Abstracting and Indexing Services

MANAGEMENT ABSTRACTS. Quarterly. British Institute of Management, Management House, Park Street, London, WC2B SPT, England.

MANAGEMENT INDEX. Monthly except July. Keith Business Library, Box 453, Ottawa, Canada.

"The international monthly guide to new American-Canadian-British sources of business and technical information."

MARKETING INFORMATION GUIDE. Monthly. Superintendent of Documents, Washington, D.C. 20402.

PERSONNEL MANAGEMENT ABSTRACTS. Quarterly. University of Michigan, Bureau of Industrial Relations, Graduate School of Business Administration, Ann Arbor, Michigan 48104.

iii. Selected Related Journal and Serial Titles

CHEMICAL INDUSTRY NOTES. Weekly. Chemical Abstracts Service, Akron, Ohio 43210.

Digest news of interest to chemical marketing and management personnel.

CHEMICAL MARKETING REPORTER. Weekly. Schnell Publishing Co., 100 Church Street, New York, New York 10007.

FUTURES. Quarterly. IPC Science & Technology Press Ltd., 32 High Street, Guildford, Surrey, England and Iliffe-NTP Inc., 300 East 42nd Street, New York, New York 10017.

"The journal of forecasting and planning."

INDUSTRIAL MANAGEMENT (U.S.). Monthly. Industrial Management Society, 2217 Tribune Tower, Chicago, Illinois 60611.

MANAGEMENT REVIEW. Monthly. Management Association, Inc., 135 West 50th Street, New York, New York 10020.

MANAGEMENT SCIENCE. Monthly. Institute of Management Sciences, P.O. Box 6112, Providence, Rhode Island 02904.

MANAGEMENT TODAY. Monthly. Management Publications Ltd., Gillow House, 5 Winsley Street, London, WIN 8AP, England.

VALUE ENGINEERING DIGEST/DEFENSE CONTRACT GUIDE; AND VALUE ENGINEERING GUIDE. Fortnightly. Washington, D.C.: Sci/Tech Digest.

b. COMPUTERS

i. Guides and Handbooks

American Society for Engineering Education. Engineering Libraries Division. GUIDE TO LITERATURE OF COMPUTERS. Washington, D.C.: 1968.

CHEMICAL FORECASTS BY COMPUTER - TOP 200 PRODUCTS. Westport, Connecticut: Technomic Publishing Co., Inc., 1972.

> "Presents authoritative computer forecasts. Includes year-by-year 5-year production forecasts for 211 of the most important chemical products, given in both tabular and graph form. A 15-year statistical history of sales and production is also given for each product. Histories and forecasts are also presented for 12 major groups of chemicals."

DATAMATION 1972. EDP Industry Directory. Annual. Barrington, Illinois: Technical Publishing Co.

> A catalog of EDP products and services.

ENCYCLOPEDIA OF INFORMATION SYSTEMS AND SERVICES. Ann Arbor, Michigan: Edwards Brothers, 1973.

Hodge, B. COMPUTERS FOR ENGINEERS. New York: McGraw-Hill, 1969.

> "For engineers who have had no formal training in computers."

Hovanessian, S. DIGITAL COMPUTER METHODS IN ENGINEERING. New York: McGraw-Hill, 1969.

> "Guide for self-study in digital computer programs, written in FORTRAN and BASIC programming languages."

Silvester, E. COMPUTER PROCESS CONTROL-GUIDEBOOK FOR MANAGEMENT. Westport, Connecticut: Technomic Publishing Co., Inc., 1970.

> "Guide to computer applications in the chemical industry. Includes procedures of feasibility study, design, implementation, continuing management, and a 5-year case history."

Sippl, C. COMPUTER DICTIONARY AND HANDBOOK. 2nd ed. New York: Sams Publishing Co., 1972.

Weik, M. STANDARD DICTIONARY OF COMPUTERS AND INFORMATION PROCESSING. Rochelle Park, New Jersey: Hayden Book Co., 1973.

> Includes 10,000 hardware definitions and mentions 500 computer models.

ii. Abstracting and Indexing Services

COMPUTER ABSTRACTS. Monthly. Edited by B.A. Fancourt. Technical Information Co., Ltd., Box 59, St. Helier, Jersey, British Channel Islands.

COMPUTING REVIEWS. Monthly. Association for Computing Machinery, 1133 Avenue of the Americas, New York, New York 10036.

CURRENT PAPERS ON COMPUTERS & CONTROL/C P C. Institute of Electrical and Electronics Engineers, Inc., 345 East 47th Street, New York, New York 10017.

Also see General Abstracting and Indexing Services on this subject:

APPLIED SCIENCE AND TECHNOLOGY INDEX

CHEMICAL ABSTRACTS

ENGINEERING INDEX

INDEX MEDICUS

iii. Selected Related Journal and Serial Titles

COMPUTER AND INFORMATION SYSTEMS. Bimonthly. Cambridge Communications Corp., c/o Stevens House, 10 South Prince Street, Lancaster, Pennsylvania.

"An abstract journal pertaining to the theory, design, fabrication and application of computer and information systems."

DATA MANAGEMENT (U.S.). Monthly. Data Processing Management Association, 505 Busse Highway, Park Ridge, Illinois 60068.

METRON (ENG.). Monthly. Pergamon Press Inc., Maxwell House, Fairview Park, Elmford, New York 10523.

Measurement, control, automation.

c. MARKET SURVEYS

CHEMICAL MARKETS AFFECTED BY THE ENVIRONMENT. Proceedings of the 5th International Conference of the European Chemical Marketing Research Association, Palais Auersperg, Vienna 20-22d, September 1971. London: European Association for Industrial Marketing Research, 1972.

MARKETING AND MANAGEMENT: A WORLD REGISTER OF ORGANIZATIONS. 1st ed. Edited by I. Anderson. Detroit, Michigan: Gale Research Company, 1969.

THE PAINT INDUSTRY. Cleveland, Ohio: Predicasts, Inc., 1973.

"Analysis of historical and projected (1975 and 1980) demand for paint by vehicle, pigment, solvent, package and end-use. Emphasizes new developments."

PROCESS CONTROL EQUIPMENT. Cleveland, Ohio: Predicasts, Inc., 1973.

"Comprehensive analysis of the industry's producers, economics, products and markets. Historical and projected (1975 and 1980) data given for 14 types of equipment and 11 end-use markets."

d. HANDLING AND PROCESSING TECHNICAL INFORMATION

i. Guides and Handbooks

ABSTRACTING SERVICES IN SCIENCE, TECHNOLOGY, MEDICINE, AGRICULTURE, SOCIAL SCIENCES, HUMANITIES. The Hague, Netherlands: International Federation for Documentation, 1965.

American Chemical Society. Division of Chemical Literature. Washington, D.C.: American Chemical Society, 1961.

"Searching the chemical literature. Revised and enlarged."

American National Standards Institute. DIRECTORY OF ENGINEERING DOCUMENTATION SOURCES. New York: 1973.

Benjamin, R. CONTROL OF THE INFORMATION SYSTEM DEVELOPMENT CYCLE. New York: Interscience, Division of Wiley, 1971.

"Concept of structural standards. Basic principles for managing the systems development process."

Biggert, E. "Federal Government Information Sources." CHEMICAL ENGINEERING, May 15, 1972, pp. 103-14.

_____. "State and City Information Sources." CHEMICAL ENGINEERING, August 21, 1972, pp. 94-103.

Bottle, R. USE OF THE CHEMICAL LITERATURE. London: Butterworth, 1962.

CHEMICAL ENGINEERING THESAURUS. New York: American Institute of Chemical Engineers, 1961.

Clarke, E. A GUIDE TO TECHNICAL LITERATURE PRODUCTION: A CONCISE HANDBOOK OF PRODUCTION METHODS. River Forest, Illinois: TW Publishing, 1961.

"A discussion of the production of technical literature by groups or teams of scientists or writers."

DIRECTORY OF COMPUTERIZED INFORMATION IN SCIENCE AND TECHNOLOGY. New York: Science Associates/International Inc., 1968.

A DIRECTORY OF INFORMATION RESOURCES IN THE UNITED STATES: PHYSICAL SCIENCES, ENGINEERING, 1971. U.S. Library of Congress, National Referral Service. Washington, D.C.: Government Printing Office.

Dreyfuss, H. SYMBOL SOURCEBOOK: AN AUTHORITATIVE GUIDE TO IN-TERNATIONAL GRAPHIC SYMBOLS. New York: McGraw-Hill, 1972.

> "Guide to graphic symbols used internationally in business, indus-try, and the sciences."

Elias, A., ed. TECHNICAL INFORMATION CENTER ADMINISTRATION. 3 vols. Rochelle Park, New Jersey: Hayden Book Company, 1973.

> Presents overall treatment of Technical Information Center opera-tions, routines and management.

Engineers Joint Council. THESAURUS OF ENGINEERING AND SCIENTIFIC TERMS. New York: 1969.

FEDERAL LIBRARY RESOURCES: A USER'S GUIDE TO RESEARCH COLLEC-TIONS. Compiled by M. Benton. New York: Science Associates/Inter-national, Inc., 1973.

GUIDE TO LOCATING U.S. GOVERNMENT TECHNICAL INFORMATION, TECHNOLOGY AND PATENTS, 1972. Chicago: TTA Information Services, 1972.

A GUIDE TO THE WORLD'S ABSTRACTING AND INDEXING SERVICES IN SCIENCE AND TECHNOLOGY. Washington, D.C.: Library of Congress, 1963.

Harwell, G. TECHNICAL COMMUNICATION. New York: Macmillan, 1960.

> "A handbook on the principles of effective treatment of letters, reports, articles, and speeches, primarily for the field of engi-neering but applicable as well to business, medicine, forestry, and the physical sciences."

Houghton, Bernard. TECHNICAL INFORMATION SOURCES. 2nd ed. Hamden, Connecticut: Linnet Books, 1972.

Jourdan, S., et al, eds. HANDBOOK OF TECHNICAL WRITING PRACTICES. New York: Interscience, Division of Wiley, 1971.

> "Covers the wide spectrum of modern technical writing practices in all aspects of technical writing."

Kobe, K. CHEMICAL ENGINEERING REPORTS: HOW TO SEARCH THE LITERATURE AND PREPARE A REPORT. 4th ed. New York: Interscience, 1957.

Magnan, G. USING TECHNICAL ART; AN INDUSTRY GUIDE. New York: Interscience, Division of Wiley, 1970.

A useful guide for engineers who are preparing graphic material.

Maizell, R. ABSTRACTING SCIENTIFIC AND TECHNICAL LITERATURE; AN INTRODUCTORY GUIDE AND TEXT FOR SCIENTISTS, ABSTRACTORS, AND MANAGEMENT. New York: Interscience, Division of Wiley, 1971.

Passman, S. SCIENTIFIC AND TECHNOLOGICAL COMMUNICATION. New York: Pergamon, 1969.

Survey of scientific and technical communication methods.

THE SOURCE DIRECTORY OF PREDICASTS, INC. Annual. Cleveland, Ohio: Predicasts, Inc.

"Contains bibliographic information on 6000 documents and services needed by business information specialists, market researchers and executives."

Special Libraries Association. BIBLIOGRAPHY OF ENGINEERING ABSTRACTING SERVICES. New York: 1955.

TAYLOR'S ENCYCLOPEDIA OF GOVERNMENT OFFICIALS; FEDERAL AND STATE. Dallas: Political Research, Inc., 1973.

Trelease, S. HOW TO WRITE SCIENTIFIC AND TECHNICAL PAPERS. 3rd ed. Baltimore: Williams and Wilkins, 1958.

United Nations Educational, Scientific and Cultural Organization. WORLD GUIDE TO SCIENCE INFORMATION AND DOCUMENTATION SERVICES. Paris: 1965.

Serves as a guide to programs and services.

_____. WORLD GUIDE TO TECHNICAL INFORMATION AND DOCUMENTATION SERVICES. Paris: 1969.

A companion to UNESCO's World guide to science information and documentation services.

U.S. GOVERNMENT MANUAL. Annual. Government Printing Office, Washington, D.C. 20402.

ii. Abstracting and Indexing Services

INFORMATION SCIENCE ABSTRACTS. Bimonthly. Lipetz. Documentation Abstracts, Inc., Box 8510, Philadelphia, Pennsylvania 19101.

Also see General Abstracting and Indexing Services on this subject:

APPLIED SCIENCE AND TECHNOLOGY INDEX

CHEMICAL ABSTRACTS

ENGINEERING INDEX

INDEX MEDICUS

iii. Selected Related Journal and Serial Titles

JOURNAL OF CHEMICAL DOCUMENTATION. Quarterly. American Chemical Society, 1155 16th Street, N.W., Washington, D.C. 20036.

17. Sources of Information on Laws and Legislation

Laws, codes, ordinances, regulations which pertain to chemical industries are contained in various official publications of governmental agencies. The following are suggested sources for locating such information for the chemical industries in general. Specific source materials are listed in the other sections of this Guide.

a. FEDERAL

FEDERAL REGISTER. Daily publication listing new regulations and revisions. Updates Code of Federal Regulations.

Available on a subscription basis from GPO.

Library of Congress. DIGEST OF PUBLIC GENERAL BILLS AND REGULATIONS.

Available on subscription basis from GPO.

UNITED STATES CODE. 1970 ed. and supplements. Government Printing Office, Washington, D.C. 20402.

Code of federal regulations. Continuing series issued throughout the year. Available on a subscription basis from GPO.

U.S. Government Printing Office, Washington, D.C. 20402. PRICE LIST SERIES (contains lists of documents available). No. 10. LAWS, RULES AND REGULATIONS.

b. STATE

Library of Congress. MONTHLY CHECKLIST OF STATE PUBLICATIONS. Available on a subscription basis from Government Printing Office.

c. LOCAL AND MUNICIPAL

Contact local government offices for codes, ordinances and regulations.

All of these information source materials listed here are usually available in the reference departments of large public libraries.

Law Library collections have abstracting and indexing services which would provide additional citations and information on legal matters in the chemical industries.

Section II

AGRICULTURAL ENGINEERING

Section II

AGRICULTURAL ENGINEERING

This section covers those areas of the total agricultural engineering process which particularly pertain to chemical engineering and to chemical process and supply industries.

A. ORGANIZATIONS AND ASSOCIATIONS

1. Governmental Agencies

a. FEDERAL

THE CONGRESS

A. HOUSE COMMITTEE ON AGRICULTURE - Longworth House Office Building, Room 1301, New Jersey and Independence Avenues S.E., Washington, D.C. 20515.

> Areas of interest: Primarily laws dealing with general farm policy, including: the adulteration of seeds; control of insect pests; agricultural and industrial chemistry; agricultural colleges and experiment stations; agricultural education and extension service; agricultural production, marketing, and price stabilization; animal industry and diseases of animals; crop insurance and soil conservation; dairy industry; entomology and plant quarantine; farm credit and security; plant industry; soils and agricultural engineering; etc. The Committee also has jurisdiction over legislation relating to the general operations of the Department of Agriculture, the Commodity Credit Corporation, and the Farm Credit Administration.

> Publications: Transcripts of hearings, legislative calendars, committee prints.

> Information services: Answers inquiries.

B. SENATE COMMITTEE ON AGRICULTURE AND FORESTRY - Old Senate Office Building, Room 324, Constitution Avenue and 1st Street N.E., Washington, D.C. 20510.

> Areas of interest: Agriculture generally; inspection of livestock and meat products; animal industry and diseases of animals; adulteration of seeds; insect pests; agricultural colleges and experiment

stations; agricultural economics and research; agricultural and industrial chemistry; dairy industry; plant industry, soils, and agricultural engineering; agricultural educational extension services; extension of farm credit and farm security; rural electrification; agricultural production and marketing; stabilization of prices of agricultural products; crop insurance; soil conservation, etc.

Holdings: Reference collection.

Publications: Transcripts of hearings.

Information services: Answers inquiries; distributes transcripts of hearings; permits onsite use of reference collection.

DEPARTMENT OF AGRICULTURE

A. CURRENT RESEARCH INFORMATION SYSTEM (CRIS) - Science and Education Staff, South Building, Room 6818, 14th Street and Independence Avenue S.W., Washington, D.C. 20250.

A computer-based information storage and retrieval system, CRIS is designed to improve communications among agricultural research scientists, especially with regard to research work currently under way, and to provide research managers with up-to-date and coordinated information on the total research programs of USDA and the State Agricultural Stations.

Areas of interest: Current agricultural research activities of the U.S. Department of Agriculture (USDA), the State Agricultural Experiment Stations, and other cooperating institutions.

Holdings: About 24,000 Research Work Unit (project) descriptions, spanning the total research effort sponsored by six USDA agencies (Agricultural Research Service, Forest Service, Economic Research Service, Cooperative State Research Service, Farmer Cooperative Service, and Statistical Reporting Service), 53 State Agricultural Experiment Stations, 16 former Negro Land-Grant Institutions, and 9 Forestry Schools. Information stored on magnetic tape includes a brief description of the Research Work Unit including title, objectives, plan of work, current progress, the more important publications, name of principal investigator, performing institution, and cooperators, amount of money allocated and sources of funds, and manpower allocations by types (scientists, professional support, and other personnel).

Publications: Primarily computer printouts for internal use.

Information services: Answers inquiries regarding research projects or refers inquirers to persons performing the research.

B. FARMERS HOME ADMINISTRATION - 14th Street and Independence Avenue S.W., Washington, D.C. 20250.

Areas of interest: Financial and management assistance to farmers, local organizations, and public agencies; community water supply systems; sewage and waste disposal systems; soil conservation measures; land-use shifts to grassland and forestry; recreation areas;

watershed projects; rural area renewal and development.

Publications: Pamphlets.

Information services: Answers inquiries regarding credit and grant services.

C.   OFFICE OF INFORMATION - 14th Street and Independence Avenue S.W., Washington, D.C. 20250.

Areas of interest: Department of Agriculture activities.

Information services: The Office of Information disseminates general information on Department activities and serves as a channel for all official USDA information. The Office also administers the publications program of the Department of Agriculture and produces and distributes motion pictures, exhibits, photographs, slide sets, and other visual aids. Technical inquiries are generally referred to specialists of the Department for answers.

D.   AGRICULTURAL RESEARCH SERVICE -

1.   Eastern Marketing and Nutrition Research Division - Wilson Street, Route 2, Durant, Oklahoma 74701.

Areas of interest: Predicting the effects of agricultural practices on water quality in the United States and maintaining desirable water quality for agricultural and other uses; movement of nutrients, pesticides, agricultural chemicals, and agricultural wastes into farm ponds, waterways, and ground water; effect of soil and water conservation practices on water quality in agricultural watersheds; effects of pollutants on agricultural uses of water; regulation of water quality in ponds, waterways, and ground water; methods for conducting water quality investigations related to agriculture.

Information services; Answers inquiries.

2.   Cotton Research Center - 4207 East Broadway, Phoenix, Arizona 85040.

Areas of interest: Cotton (genetics, physiology, pathology); nematology (cotton, citrus, alfalfa); weed control in cotton.

Publications: Articles in scientific journals.

Information services: Answers inquiries relating to the Center's research work; makes referrals.

3.   Eastern Utilization Research and Development Division - 600 East Mermaid Lane, Philadelphia, Pennsylvania 19118.

Areas of interest: Chemistry; biochemistry; chemical engineering; dairy products; detergents; food technology; fruit and vegetable utilization; hides and leather; industrial wastes; meat; milk; oils and fats; polymers; proteins; spectrum analysis; tobacco.

Publications: Directories; reports, monthly list of new accessions.

Information services: Answers inquiries; provides reference services; permits onsite use of collection by technically and professionally trained personnel.

4. Field Crops and Animal Products Research Branch (Clemson, S.C.) - Market Quality Research Division, Clemson University, P.O. Box 792, Clemson, South Carolina 29631.

Areas of interest: Physical and chemical research necessary to develop methods, techniques, instruments, and apparatus for rapid representative sampling and accurate objective measures of quality factors in cotton and cotton products. Works to develop standards.

Information services: Answers inquiries, provides reference assistance, and makes referrals.

5. Human Nutrition Research Division - Agricultural Research Center, Beltsville, Maryland 20705.

Areas of interest: Human metabolism and requirements for nutrients; functions and metabolism of nutrients; nutrient value of foods; experimental nutrition.

Information services: Answers inquiries (those that can be answered with prepared materials are referred to the Service's Information Division).

6. Information Division - East-West Highway and Belcrest Road, Hyattsville, Maryland 20782.

Areas of interest: Agricultural research.

Information services: Answers inquiries; makes referrals.

7. Market Quality Research Division - Federal Center Building, East-West Highway and Belcrest Road, Hyattsville, Maryland 20782.

Areas of interest: Quality characteristics for grading and inspecting agricultural commodities, including horticultural crops, field crops, and animal products; protection of agricultural commodities from deterioration and damage from insects, diseases, and other hazards.

Information services: Answers inquiries; provides consulting and advisory services. Requests for information may be directed to Division headquarters or to any of the Division's branch offices.

8. Northern Utilization Research and Development Division - 1815 North University Street, Peoria, Illinois 61604.

Areas of interest: Chemical, microbiological, and engineering research and development on food and industrial uses of corn, wheat, grain sorghum, other cereal grains, soybeans, flaxseed, erucic acid-containing oilseed crops, and new crops.

Information services: Answers inquiries; provides reference services; makes interlibrary loans; permits onsite use of collection by technically and professionally trained personnel.

9. Pesticides Regulation Division - U.S. Department of Agriculture, Washington, D.C. 20250.

Areas of interest: Review of safety and efficacy data submitted by applicants for registration of pesticide chemicals.

Information services: Answers letters of inquiry about registered pesticides.

10. Snake River Conservation Research Center - Route 1, Box 186, Kimberly, Idaho 83341.

Areas of interest: Soil and water management problems prevalent on irrigated lands and associated dryland farms and rangelands in the upper Snake River Valley and similar agricultural areas; soil-water-plant chemistry; soil physics; irrigation engineering (automatic control devices for irrigation, water-measuring devices, hydraulics of surface irrigation methods and return flow systems, sprinkler irrigation, irrigation efficiencies); mulches; slick spot soils; agro-meteorology and irrigation scheduling.

Publications: Articles in technical journals and in special publications of the Department of Agriculture and the University of Idaho (a list is available).

11. Soil and Water Conservation Research Division - Plant Industry Station, Beltsville, Maryland 20705.

Areas of interest: Watershed engineering (watershed hydrology, ground-water recharge, sedimentation, hydraulics of water control structures); water management (water erosion control and stream channel stabilization, water conservation, irrigation, drainage salinity, remote sensing, wind erosion control); soil management (soil tillage, structure and fertility, pesticide residues, animal wastes disposal, water pollution, soil-plant relationships, micro-climatology).

Publications: HYDROLOGIC DATA FOR EXPERIMENTAL AGRI-CULTURAL WATERSHEDS IN THE UNITED STATES; books, bibliographies.

Information services: Answers inquiries concerning the Division's research program; makes referrals.

12. Southeastern Utilization Research and Development Division - P.O. Box 5677, Athens, Georgia 30604.

Areas of interest: Research on production, utilization, and marketing of agricultural commodities and products, including fruits, vegetables, tree nuts, forage and other feed products, oilseeds, poultry, and livestock products.

Information services: Answers inquiries; provides consulting and reference services.

13. Southern Utilization Research and Development Division - P.O. Box 19687, New Orleans, Louisiana 70119; Location: 1100 Robert E. Lee Boulevard, New Orleans, Louisiana.

Areas of interest: Utilization research on cotton, cottonseed, peanuts, rice, pine gum, citrus fruits, sweet potatoes, cucumbers, and other Southern-grown vegetables.

Publications: LIST OF PUBLICATIONS AND PATENTS WITH AB-STRACTS (semiannual, available free on request); occasional bibliographies.

Information services: Answers inquiries; provides consulting services; provides reprints of technical papers reporting the Division's research findings; makes interlibrary loans; permits onsite use of collection by technically and professionally trained personnel.

14. Western Utilization Research and Development Division - 800 Buchanan Street, Albany, California 94710.

Areas of interest: Basic and applied research on food, feed, and fibers, including the following commodities: Western fruits (citrus and other subtropical and tropical fruits); tree nuts; vegetables; poultry products; forage crops; wheat; rice; barley; wool and mohair; dry beans and peas; castor, safflower, and western oilseeds. Project interests include product and process development and improvement to promote greater uses of farm crops, enhance rural area development, promote foreign trade, reduce hunger at home and abroad, reduce pollution of air, water, and soil, and improve quality and safety of consumer products.

Publications: LIST OF PUBLICATIONS AND PATENTS WITH AB - STRACTS (semiannual, available free on request); directories, bibliographies, reports.

Information services: Answers inquiries; provides consulting services; makes interlibrary loans; permits onsite use of collection by technically and professionally trained personnel.

E. SOIL CONSERVATION SERVICE

Information Division - 14th Street and Independence Avenue S.W., Washington, D.C. 20250.

Areas of interest: Soil and water conservation; soil classification and geography; soil chemistry; agronomy; plant materials; range and woodland conservation and improvement; watershed protection; flood prevention; erosion control; water yield and use; irrigation and drainage; channel improvement; sediment yield and transport; small watershed hydrology.

Publications: Bulletins, handbooks.

Information services: Answers inquiries; makes referrals; permits onsite use of collection by approved visitors.

DEPARTMENT OF DEFENSE

U.S. ARMY

A. CORPS OF ENGINEERS - Waterways Experiment Station, Soil Mechanics Information Analysis Center, P.O. Box 631, Vicksburg, Mississippi 39180.

Acquires, evaluates and synthesizes literature and information in this subject field on a worldwide basis.

Information services: Cor piles bibliographies, literature searching, state-of-the-art studies, referrals and special analyses for particular inquiries.

B.   DIRECTORATE OF AEROBIOLOGY AND EVALUATION LABORATORIES - Technical Information Division, ATTN: SMXFD-AE-T, Fort Detrick, Maryland 21701.

Areas of interest: Microbiology (aerobiology, bacteriology); biochemistry; aerosols (physics); biophysics; engineering (aerospace, biological, chemical, industrial, mechanical, sanitary); agronomy; plant pathology; plant physiology.

Publications: Technical reports (130 a year); abstracts bulletin (quarterly); accessions list (monthly); annual report; bibliographies, brochures, motion pictures, exhibits.

Information services: Answers inquiries; makes referrals; provides technical editing, consulting, reference, literature-searching, abstracting, translation, and duplication services; makes interlibrary loans. Services are available to government agencies, their contractors, and to a limited extent, the public.

C.   ENGINEER AGENCY FOR RESOURCES INVENTORIES REFERENCE BRANCH - 4701 North Sangamore Road, Washington, D.C. 20016.

Areas of interest: Aerial photo surveys and surveying techniques; agrarian reform; agricultural cooperatives; agricultural crops and livestock; agricultural sciences; area description (general); boundaries; climate; communications systems and facilities; conservation; dams; economic development and planning; drainage; ecology; economic geography and geology; environmental factors; fauna; fertilizers; fish and fisheries; floods and flood control; flora; forestry and forest products; etc.

Publications: SELECTED ACCESSIONS (weekly, available on written request); resource atlases (not generally available for public distribution); special studies, reports, papers and manuals (some available on request); bibliographies.

Information services: Answers inquiries; provides advisory, reference, and limited bibliographic services; makes limited interlibrary loans.

DEPARTMENT OF STATE

AGENCY FOR INTERNATIONAL DEVELOPMENT

Public Services Division - Information Staff, Office of Legislative and Public Affairs, Agency for International Development, Washington, D.C. 20523.

Areas of interest: Science and technology for application in the less developed countries, including agriculture, rural development, engineering, food assistance, food from the sea, public health, housing and urban development, nutrition and child feeding, and population; background information on investment opportunities in less developed countries.

Holdings: Technical reports, surveys, studies, etc. prepared in support of specific technical assistance projects.

Publications: INDEX TO CATALOG OF INVESTMENT INFORMA-TION (a listing of available reports and studies sponsored by the Agency and other public, international, and private institutions on economic and social conditions, legal requirements, current industry and market status, feasibility investigations, and other investment factors); INDUSTRY PROFILES (400 individual publications giving basic information for the establishment in less developed countries of small and medium-sized plants in selected industries.)

Information services: Answers inquiries; makes referrals.

ENVIRONMENTAL PROTECTION AGENCY

PESTICIDES REGISTRATION BRANCH - 200 C Street S.W., Washington, D.C.

Maintains a pesticide registrations branch which is particularly concerned with the environmental effects of pesticides.

Information services: Answers inquiries; fields reference-research questions; makes referrals; collects data.

NATIONAL RESEARCH COUNCIL

A. DIVISION OF BIOLOGY AND AGRICULTURE

1. Agricultural Board - 2101 Constitution Avenue N.W., Washington, D.C. 10418.

Areas of interest: Agricultural problems and research needs; application of research discoveries; agricultural products; animal feeds; animal health and nutrition; pesticides.

Publications: Technical reports.

Information services: Answers inquiries from research investigators.

2. Food and Nutrition Board - Division of Biology and Agriculture, 2101 Constitution Avenue N.W., Washington, D.C. 10418.

Areas of interest: Food and nutritional science, including protein nutrition, fats, child nutrition, and dietary allowances.

Publications: Technical reports.

Information services: Answers inquiries from research investigators.

b. STATE AND MUNICIPAL

See this topic in Section I of this guide.

2. Professional Societies, Trade Associations, and Other Scientific-Technological Related Organizations

AMERICAN AGRICULTURAL ECONOMICS ASSOCIATION - University of Kentucky, Lexington, Kentucky 40503.

Purpose: To promote the development of systematic knowledge of agricultural economics for the purpose of improving agriculture.

Publications: AMERICAN JOURNAL OF AGRICULTURAL ECONOMICS, five issues per year.

Information services: Answers inquiries; collects data; makes referrals.

AMERICAN DAIRY SCIENCE ASSOCIATION - 113 North Neil Street, Champaign, Illinois 61820.

Purpose: Serves the dairy and dairy-related industries by stimulating experimentation and research for application and dissemination to the dairy industry. Acts as clearinghouse for dairy related industry information. Publishes original research, reviews and information.

Publications: JOURNAL OF DAIRY SCIENCE, monthly.

Information services: Answers inquiries; makes referrals; collects data and material on dairy science.

AMERICAN FARM BUREAU RESEARCH FOUNDATION (AFBRF) - 1000 Merchandise Mart, Chicago, Illinois 60654.

Purpose: "To improve profit opportunities in the farming business through advancement of scientific knowledge." Sponsors agricultural research by contracting work with universities, public research agencies and private research firms, in such areas as production and marketing, quality and yield improvement, mechanization, farm safety, conservation of natural resources, improved or better use of fertilizers, insecticides and herbicides and ways of increasing net farm income. Promotes use of basic research on farms and farming sites. Maintains library of research reports. Supported by the American Farm Bureau Federation and by individual donations by farmers and others.

AMERICAN FEED MANUFACTURERS ASSOCIATION, INC. - 53 West Jackson Street, Chicago, Illinois 60604.

Areas of interest: Animal and poultry feeds; feed production and marketing; animal nutrition; livestock management; industry careers.

Publications: Feed production books and manuals, booklets, brochures, reports.

Information services: Answers inquiries free.

AMERICAN GUERNSEY CATTLE CLUB, INC. - Main Street, Peterborough, New Hampshire 03458.

Areas of interest: Registration of Guernsey cattle, maintenance of pedigree files.

Information services: Answers inquiries; provides consulting, reference, translation, and duplication services; makes referrals; permits onsite use of collection. Services are provided free to the general public.

AMERICAN INSTITUTE OF CROP ECOLOGY - 809 Dale Drive, Silver Spring, Maryland 20910.

Areas of interest: Bioclimatology and agroclimatology; comparison between analogous climatological areas in the United States and other countries and between analogous areas outside the United States; physical and plant geography; crop ecology; crop-climate relationships.

Holdings: Meteorological records and biological data from various parts of the world, including phenological records of pure line crop varieties, clones, inbred lines, and hybrids, as well as their principal diseases and pests under similar and diverse environmental conditions; maps, charts, project records, tables, technical reports, books, monographs, journals, abstracts.

Publications: Technical reports and surveys of climatology and crop ecology of various countries. A list of publications and a descriptive brochure are available on request.

Information services: Answers brief inquiries free; provides consulting and translating services; performs research at cost.

AMERICAN RICE GROWERS COOPERATIVE ASSOCIATION - 211 Pioneer Building, Lake Charles, Louisiana 70601.

A federation of 25 local farmers' cooperatives engaged in processing, storage, and marketing of rice.

Publication: WEEKLY RICE MARKET REVIEW.

Information services: Answers inquiries; makes referrals within Association.

AMERICAN SEED RESEARCH FOUNDATION - Executive Building, Suite 964, 1030 15th Street N.W., Washington, D.C. 20005.

An organization of breeders, producers and distributors of seeds.

Purpose: Seeks to advance seed technology by supporting research on seeds.

Publications: SEARCH (quarterly).

Information services: Provides information on seed research in answer to inquiries. Makes referrals.

AMERICAN SOCIETY FOR HORTICULTURAL SCIENCE - P.O. Box 109, 914 Main Street, St. Joseph, Michigan 49085.

Purpose: To further national and international interest in horticulture. Promotes horticultural research through publications and meetings, etc.

Publications: JOURNAL OF THE AMERICAN SOCIETY FOR HORTICULTURAL SCIENCE (bimonthly); HORTSCIENCE (bimonthly); ANNUAL REPORT and MEMBERSHIP DIRECTORY, published as separate supplements of HORTSCIENCE.

Information activities: Answers inquiries; makes referrals; collects information in its field.

AMERICAN SOCIETY OF AGRICULTURAL ENGINEERS - 2950 Niles Road, St. Joseph, Michigan 49085.

Areas of interest: All aspects of agricultural engineering, including farm structures and structural equipment; animal shelters; design of tractors and other machinery; farm power (electrical equipment and wiring practice); soil and water management; hydrology; feed and food drying, processing, storage, and handling; instrumentation; radiation.

Publications: AGRICULTURAL ENGINEERING (monthly); TRANSACTIONS OF THE ASAE (quarterly); AGRICULTURAL ENGINEERS YEARBOOK, which includes ASAE standards, recommendations, and data; bibliographies; survey reports; conference and seminar proceedings.

Information services: Answers technical inquiries by citing ASAE publications or by referring inquirers to ASAE members.

AMERICAN SOCIETY OF AGRONOMY - 677 South Segoe Road, Madison, Wisconsin 53711.

The Society is closely associated with the Crop Science Society of America and the Soil Science Society of America.

Areas of interest: Agronomic education; agronomic extension; land use and management; climatology; soil physics; soil chemistry; soil fertility; plant nutrition; fertilizer technology and use; soil microbiology; soil genesis, morphology, and classification; soil and water management and conservation; forest and range soils; crop breeding, genetics, and cytology; crop physiology and metabolism; crop ecology, production, and management; crop quality and utilization; turfgrass management; seed production.

Publications: AGRONOMY JOURNAL (bimonthly); CROPS AND SOILS (9 issues a year); AGRONOMY ABSTRACTS (annual); agronomy monographs (such as METHODS OF SOIL ANALYSIS, SOIL NITROGEN, WHEAT); special publication series (such as FORAGE ECONOMICS, POTASSIUM, PHYSIOLOGY OF CROP YIELD, and NUTRIENT MOBILITY IN SOILS).

Information services: Answers inquiries; makes referrals; distributes some free publications; makes other publications available by subscription and purchase.

AMERICAN SOCIETY OF SUGAR BEET TECHNOLOGISTS - 156 South College Avenue, Fort Collins, Colorado 80521.

Areas of interest: Production and processing of sugar beets; research related to the sugar beet industry, including agronomy, genetics in terms of variety improvements, entomology, plant pathology, agricultural engineering, and beet sugar chemistry.

Publications: JOURNAL OF THE AMERICAN SOCIETY OF SUGAR BEET TECHNOLOGISTS (quarterly).

Information services: Answers technical inquiries or refers inquiries to member specialists; makes available unpublished data in the Society's files; provides loans and photocopies of documents on request. A nominal charge is made for volume orders.

ASSOCIATION OF AMERICAN FEED CONTROL OFFICIALS - c/o Mr. Reed McDonald, Secretary, P.O. Box 3160, College Station, Texas 77840.

Areas of interest: Animal feed ingredient definitions; rules and regulations covering labeling of animal feeds.

Publications: ASSOCIATION OF AMERICAN FEED CONTROL OFFICIALS (annual).

Information services: Answers inquiries; provides short lists of literature citations in response to specific inquiries; makes referrals.

ASSOCIATION OF OFFICIAL ANALYTICAL CHEMISTS - Box 540, Benjamin Franklin Station, Washington, D.C. 20044. Location: 200 C Street S.W., Washington, D.C.

Areas of interest: Analytical chemistry; chemical, microbiological, and organoleptic methods of analysis for beverages, foods and food additives, vitamins and other nutrients, marine products, oils and fats, drugs, cosmetics, disinfectants, fertilizers, pesticide formulations and residues, radioactivity, and hazardous substances.

Publications: JOURNAL OF THE ASSOCIATION OF OFFICIAL ANALYTICAL CHEMISTS (bimonthly); OFFICIAL METHODS OF

ANALYSIS OF THE AOAC (every 5 years, with annual supplements). A complete list of AOAC publications is available.

Information services: Answers inquiries; makes referrals; provides reference services.

ASSOCIATION OF OFFICIAL SEED ANALYSTS - c/o Dr. William N. Rice, Secretary-Treasurer, Seed Laboratory, West Experiment Station, University of Massachusetts, Amherst, Massachusetts 01002.

Areas of interest: Seed analysis, storage, physiology, longevity, and identification; seed testing for purity, germination, and dormancy; chemical tests for viability and vigor; seed laws; crop seeds.

Publications: ASSOCIATION OF OFFICIAL SEED ANALYSTS NEWSLETTER (quarterly); annual proceedings.

Information services: Answers inquiries; provides consulting and identification services free or for fee; provides short lists of literature citations in response to specific inquiries; makes referrals.

BEET SUGAR DEVELOPMENT FOUNDATION - 156 South College Avenue, Fort Collins, Colorado 80521.

Independent nonprofit research and membership organization supported by beet sugar industry of United States and Canada.

Areas of interest: Sugar beet production, beet sugar processing and byproduct utilization. Collects reports and non-printed material and disseminates information on research data in these fields.

Publication: JOURNAL OF AMERICAN SOCIETY OF SUGAR BEET TECHNOLOGISTS (quarterly).

Information services: Answers inquiries; assists with reference research searching; makes referrals.

BIO-DYNAMIC FARMING AND GARDENING ASSOCIATION, INC. - R.D. 1, Stroudsburg, Pennsylvania 18360.

Biochemical Research Laboratory, Threefold Farm, Spring Valley, New York 10977.

Areas of interest: The Association promotes the biodynamic method of agriculture for the purpose of improving nutrition and health. It is interested in soil conservation, balanced fertility, the principles of ecology, and a stable diversified agriculture through the use of biodynamic treatment of manures, composts, and green manures to raise their fertilizer and humus forming qualities. The Biochemical Research Laboratory conducts research for improvements in composting and humus formation and specializes in their microbiology.

Publications: BIO-DYNAMICS (quarterly).

Information services: Information is made available through literature and conferences. A consulting and testing service on soils, crops, and farm and commercial composting is provided, for a fee, by the Biochemical Research Laboratory.

CALIFORNIA FERTILIZER ASSOCIATION - 222 Watt Avenue, Suite B-7, Sacramento, California 95825.

"The Association acts as legislative advocate for the fertilizer industry and maintains liaison between the industry and scientists in the University of California, the State colleges, junior colleges, and State control officials."

Areas of interest: Fertilizers for California soils and crops; soil fertility; fertilizer transportation, safety, application, and equipment.

Publications: WESTERN FERTILIZER HANDBOOK; reports, proceedings, directory, booklets, brochures.

Information services: Answers inquiries; makes referrals.

COUNCIL FOR AGRICULTURAL AND CHEMURGIC RESEARCH - 350 Fifth Avenue, New York, New York 10001

Membership is made up of farmers, scientists, and industrialists interested in research on the industrial, chemical uses of farm products. Fosters the discovery of: "New, non-food uses for farm crops, their residues and by-products; new and profitable uses for previously unused plant materials; new crops that farmers may grow profitably; more valuable uses for presently used crops, through chemurgic upgrading."

Publication: CHEMURGIC DIGEST, 8/year.

Information services: Answers inquiries; makes referrals; assists with reference/research programs in its area of technical specialty.

CROP QUALITY COUNCIL - 828 Midland Bank Building, Minneapolis, Minnesota 55401.

An organization of individuals and industrial firms interested in agricultural crop improvement and protection. Sponsors research in control of pests affecting food, feed, oil, and forage crops. Control of the cereal rusts and the improvement of crop quality is a special area of interest.

Information services: Answers inquiries; makes referrals.

CROP SCIENCE SOCIETY OF AMERICA - 677 South Segoe Road, Madison, Wisconsin 53711.

Membership consists of plant breeders, physiologists, ecologists, crop production specialists, seed technologists, turf grass specialists, and others interested in improvement, management, and use of field crops.

Purpose: The advancement of research, teaching of all basic and applied phases of the crop sciences. To promote communication and cooperation among organizations and societies similarly interested in the improvement, production, management and utilization of field crops.

Publication: JOURNAL OF ENVIRONMENTAL QUALITY (quarterly).

Information services: Answers inquiries; makes referrals.

DAIRY AND FOOD INDUSTRIES SUPPLY ASSOCIATION - 5530 Wisconsin Avenue, Bethesda, Maryland 20015.

Areas of interest: Sanitary standards for food and dairy processing equipment and materials; technology of processing dairy products; market research statistics.

Publications: The Association participates in the writing of 3-A Sanitary Standards for food and dairy processing equipment and materials, in collaboration with the Dairy Industry Committee, the U.S. Public Health Service, and the International Association of Milk, Food, and Environmental Sanitarians. These standards are originally published officially in the JOURNAL OF MILK AND FOOD TECHNOLOGY.

Information Services: Technical inquiries are answered by members of the Association or referred to specialists. Pertinent reprints of published 3-A Sanitary Standards are furnished in reply to inquiries.

DAIRY COUNCIL OF THE TWIN CITIES - Hillsborough Office Building, Suite 220, 2353 North Rice Street, St. Paul, Minnesota 55113.

Areas of interest: Food and nutrition.

Publications: NUTRITION NEWS (4 issues a year); DAIRY COUNCIL DIGEST (6 issues a year); booklets, leaflets, posters, motion pictures, filmstrips, transparencies.

Information services: Answers inquiries; provides consulting and reference services; conducts nutrition education workshops and seminars. Services are primarily for professional health personnel and fees may be charged to persons living outside the Twin Cities area.

DAIRY HERD IMPROVEMENT ASSOCIATION - Building 263, Agricultural Research Center, Beltsville, Maryland 20705.

A research group in the Agricultural Research Service, U.S. Department of Agriculture. Collects and analyzes production testing data on individual cows to improve the management and breeding program of participating dairymen. Conducts research in population genetics of dairy cattle; compiles statistics on milk production.

Publications: DHIA SIRE SUMMARY LIST (3/year); DHIA COW PERFORMANCE INDEX (3/year); LACTATION AVERAGES (annual); DAIRY HERD IMPROVEMENT LETTER (irregular).

Information services: Answers inquiries; makes referrals.

FARM AND INDUSTRIAL EQUIPMENT INSTITUTE - 410 North Michigan Avenue, Chicago, Illinois 60611.

Areas of interest: Farm and industrial machinery and equipment, including tractors, combines, mowers, discs, harvesting machinery, balers, power sprayers and dusters, hand sprayers and dusters,

industrial loaders, backhoes and other materials handling or land-
scape maintenance equipment, milking machines, poultry equipment, livestock equipment, barn equipment such as cleaners,
feeders and silo unloaders, and crop dryers; farm safety; soil and
water conservation; foreign trade.

Publications: THE FIEI LETTER (quarterly); INDUSTRIAL EQUIP-
MENT SPECIFICATION DEFINITIONS and THE MODERN WAY
TO EFFICIENT MILKING (booklets); membership directory (annual);
retail sales reports (monthly).

Information services: Answers inquiries; provides reference and
consulting services.

FERTILIZER INDUSTRY ROUND TABLE - c/o Mr. Housden L. Marshall,
Secretary-Treasurer, 112 Dumbarton Road, Baltimore, Maryland 21212.

Areas of interest: Fertilizer production, technology and research,
including aspects of granulation, ammoniation, sampling, compati-
bility, physical properties, chemical properties and reactions, and
machinery and instruments used in the industry; chemical analyses
of fertilizers.

Publications: PROCEEDINGS (annual).

Information services: Answers inquiries; provides consulting services; makes referrals; prepares analyses. A fee may be charged
for extensive services.

THE FERTILIZER INSTITUTE - 1015 18th Street N.W., Washington, D.C.
20036.

Areas of interest: Soil fertility and plant nutrition; chemical
control of fertilizers; fertilizer application; economics of fertilizer
use; water use with fertilizers in agriculture; industry production
and financial statistics.

Publications: FERTILIZER PROGRESS (bimonthly).

Information services: Answers technical inquiries; makes referrals
to other specialists or organizations. Information is provided for
use in the United States only. Inquiries should be directed to
the Director, Communications.

HAWAIIAN SUGAR PLANTERS' ASSOCIATION EXPERIMENT STATION -
1527 Keeaumaku Street, Honolulu, Hawaii 96822.

"Integral unit of Hawaiian Sugar Planters' Association, an independent unin-
corporated nonprofit cooperative organization. Supported by the sugar industry."

Areas of interest: Agronomy, chemistry, entomology, agricultural
engineering, sugarcane breeding, plant pathology, physiology,
biochemistry, sugar technology and weed control.

Activities: Serves as research center for Hawaiian sugar industry
and conducts basic and applied research in areas affecting sugarcane production, harvesting, improvement and processing.

Publication: HAWAIIAN PLANTERS' RECORD (irregular). Maintains a sugar research library of 75,000 volumes.

Information services: Answers inquiries; makes referrals.

NATIONAL AGRICULTURAL CHEMICALS ASSOCIATION - 1155 15th Street N.W., Washington, D.C. 20005.

Areas of interest: Agricultural chemistry; pesticides; toxicology.

Publications: NAC NEWS AND PESTICIDE REVIEW (bimonthly, with an annual issue carrying FDA tolerances).

Information services: Answers inquiries; makes referrals; provides consulting services.

NATIONAL AGRICULTURAL INSTITUTE (NAI) - 1632 K Street, N.W., Suite 302, Washington, D.C. 20006.

Membership includes farm commodity and supply organizations.

Purpose: "To promote improved understanding of agriculture's problems and contributions in an increasingly urban-oriented society." Conducts political, economic and social studies related to agriculture.

Publication: AGRICULTURE USA (monthly).

Information services: Answers inquiries; makes referrals.

NATIONAL ASSOCIATION COUNTY AGRICULTURAL AGENTS (NACAA) - Court House, Valley City, North Dakota 58072.

Membership includes county agricultural agents and extension workers. Serves and promotes interests of county agents through the following committees: Extension Programs; 4-H Young Men and Women; Professional Training; Public Information; Recognitions and Awards; State Relations.

Publications: COUNTY AGENT (quarterly); COUNTY AGENTS DIRECTORY (annual).

Information services: Answers inquiries; makes referrals.

NATIONAL RENDERERS ASSOCIATION, INC. - 3150 Des Plaines Avenue, Des Plaines, Illinois 60018.

Areas of interest: Rendering industry; processing of animal and poultry byproducts; domestic and foreign marketing; uses for products; research on new uses of industry products; animal and poultry nutrition; control of sanitation in plants; air and water pollution problems.

Publications: NRA NEWSLETTER (monthly); NRA YEARBOOK (annual); publications guide, pamphlets and other trading rules and reports relating to the rendering industry and its products, directories, product specifications.

Information services: Answers inquiries free; provides consulting services free to Association members and at cost to others. Three educational motion pictures on the industry are available free to anyone.

NATIONAL SOYBEAN CROP IMPROVEMENT COUNCIL - 211 South Race Street, Urbana Illinois 61801.

Areas of interest: Soybean production practices in the United States.

Holdings: Reference collection of journals, motion pictures, photos, maps, charts, graphs, pamphlets, news releases, and clippings.

Publications: Pamphlets.

Information services: Answers inquiries; makes referrals.

POTASH INSTITUTE OF NORTH AMERICA, INC. - 1649 Tullie Circle N.E., Atlanta, Georgia 30329.

Areas of interest: Use of potash in fertilizers; factors, especially nutrition, affecting efficient production of crops; soil fertility; fertilization of pastures, row-crops, orchards, and forests; plant food utilization.

Publications: BETTER CROPS WITH PLANT FOOD (magazine, 4 issues a year); POTASH NEWSLETTER (4 a year, regionalized).

Information services: Answers inquiries; provides consulting and limited bibliographic services; provides speakers for meetings; makes field demonstrations; permits onsite use of collection.

POTATO ASSOCIATION OF AMERICA - New Jersey Agricultural Experiment Station, New Brunswick, New Jersey 08903.

Areas of interest: Research related to Irish potatoes, their production and handling, entomology and plant pathology, soils and fertilizers, food technology, seed potato certification, utilization, blight and virus diseases, nutrition, genetics, and cytology.

Publications: AMERICAN POTATO JOURNAL (monthly).

Information services: Questions are answered or referred to specialists. Photocopies and reprints may be provided in answer to inquiries; usually, no fees are charged for these documents. Extensive literature searches are generally not undertaken.

THE POULTRY SCIENCE ASSOCIATION, INC. - Texas A & M University, College Station, Texas 77843.

Areas of interest: All aspects of poultry production, management.

Publications: POULTRY SCIENCE (bimonthly).

Information activities: Answers inquiries; makes referrals.

PURE BRED DAIRY CATTLE ASSOCIATION - Box 126, Peterborough, New Hampshire 03458.

Areas of interest: Pure bred dairy cattle; artificial insemination.

Information services: Answers inquiries; provides reference services; makes referrals. The Association is the source of supply for certain forms and literature used by all member breed organizations.

SOCIETY OF COMMERCIAL SEED TECHNOLOGISTS - c/o Colborn Seed Testing Service, 2600 Woods Boulevard, Lincoln, Nebraska 68502.

A society of analysts of grains; field, garden, flower and grass seeds; and fine turf grasses for the seed industry.

Activities: Accomplished through the following committees: Herbarium, Research; Rules for Seed Testing.

Publications: SEED TECHNOLOGIST NEWS (quarterly); CONFERENCE PROCEEDINGS (annual).

Information services: Provides assistance with inquiries; makes referrals.

SOIL CONSERVATION SOCIETY OF AMERICA - 7515 Northeast Ankeny Road, Ankeny, Iowa 50021.

Areas of interest: Soil conservation and related areas, such as renewable natural resources, water conservation and management, and land use.

Publications: JOURNAL OF SOIL AND WATER CONSERVATION (bimonthly).

Information services: Makes referrals.

SOIL SCIENCE FOUNDATION - 1305 East Main Street, Lakeland, Florida 33801.

Areas of interest: Soil analysis and fertilization; fertilization of citrus groves; Florida agriculture; herbicides.

Publications: Periodicals, pamphlets, reports, proceedings, reviews.

Information services: Answers inquiries; provides consulting services. All services are provided at cost.

SOIL SCIENCE SOCIETY OF AMERICA - 677 South Segoe Road, Madison, Wisconsin 53711.

The Society is closely affiliated with the American Society of Agronomy and the Crop Science Society of America. It also has liaison representatives to the American Forage and Grassland Council, National Academy of Sciences-- National Research Council, American Institute of Planners, American Society of Planning Officials, American Society of Agricultural Engineers, American

Society of Civil Engineers, American Society for Testing and Materials, Clay Minerals Society, International Association for Quaternary Research, and the International Society of Soil Science.

> Areas of interest: Soil physics; soil chemistry; soil genesis, morphology, and classification; soil and water management; fertilizer technology and use. In addition, committees of the Society cover such topics as training of soil scientists, terminology, information retrieval, microform, translations, particle size and distribution, soil testing and plant analysis, water resources, and geomorphology.

> Publications: SOIL SCIENCE SOCIETY OF AMERICA PROCEEDINGS (bimonthly); the following special publications and books: CHANGING PATTERNS IN FERTILIZER USE; SELECTED PAPERS IN SOIL FORMATION AND CLASSIFICATION; SOIL TESTING & PLANT ANALYSIS; GLOSSARY OF SOIL SCIENCE TERMS. Publications are available by purchase or subscription.

> Information services: Answers inquiries; makes referrals.

UNITED STATES BEET SUGAR ASSOCIATION - 1156 15th Street N.W., Suite 1019, Washington, D.C. 20005.

> Areas of interest: Sugar; beet sugar; sugar beets.

> Information services: In response to inquiries, the Association provides general, nontechnical, nonscientific information about the sugar beet industry. For scientific and technical information, inquiries should be directed to the American Society of Sugar Beet Technologists, described elsewhere in this directory.

UNITED STATES COMMITTEE ON IRRIGATION, DRAINAGE, AND FLOOD CONTROL - P.O. Box 15326, Denver, Colorado 80215.

> Areas of interest: Irrigation practices and systems; design of irrigation works; irrigation economics and reclamation; water quality; water requirements; sedimentation; land classification; saline and alkaline soils; drainage requirements and economics; flood hydrology and prevention; river training; water law.

> Publications: ANNUAL REPORT; ANNUAL BULLETIN; TRIENNIAL TRANSACTIONS OF INTERNATIONAL CONGRESSES ON IRRIGATION AND DRAINAGE; MULTI-LINGUAL DICTIONARY ON IRRIGATION AND DRAINAGE; PROCEEDINGS OF THE FIRST INTER-SOCIETY CONFERENCE ON IRRIGATION AND DRAINAGE; IRRIGATION IN THE WORLD -- A GLOBAL REVIEW; reports.

> Information services: Answers inquiries on a limited basis; provides minor reference services.

WESTERN AGRICULTURAL CHEMICALS ASSOCIATION - 111 Capitol Mall, Suite 102, Sacramento, California 95814.

An organization of producers, formulators, retailers and applicators of basic pesticide chemicals, and suppliers of solvents diluents, emulsifiers and containers.

Publication: NEWSLETTER (45/year).

Information services: Answers inquiries; makes referrals.

3. Research Centers

AGRICULTURAL BOARD (Research) - 2101 Constitution Avenue N.W., Washington, D.C. 20418.

Membership made up of industrial firms, universities, scientific societies and institutions, federal agencies, state experiment stations.

> Purpose: To bring together the scientific talent of government, industry and universities to study the diverse problems of agriculture; to gather facts and evaluate present knowledge in relation to existing agricultural policies and practices; to ascertain trends in agricultural research and suggest areas in which research needs to be done to disseminate knowledge and applications of research discoveries to technological practice.

> Publication: NEWSLETTER (bimonthly). Also publishes research reports, nutritional series for domestic animals, other agricultural studies.

> Information services: Accepts inquiries for information; provides referral service.

AGRICULTURAL RESEARCH CENTER, INC. - 1305 East Main Street, Lakeland, Florida 33801.

A partial successor to the Soil Science Foundation, the Agricultural Research Center, Inc. is a nonprofit organization engaged in practical and limited basic research. Commercial services on a fee basis are provided by the other successor, Applied Agricultural Research, Inc., which also operates a private laboratory.

> Areas of interest: Soil analysis and fertilization, including fertilization of citrus groves; Florida agriculture; pesticides; herbicides.

> Publications: Pamphlets, reports, proceedings, reviews.

> Information services: Answers inquiries and provides limited consulting services.

AGRICULTURAL RESEARCH INSTITUTE - Joseph Henry Building, Room 835, 2100 Pennsylvania Avenue N.W., Washington, D.C. 20037.

> Purposes: To provide for liaison among federal, state, industrial and other groups engaged in agricultural research; to promote the objectives of the Agricultural Board of the National Research Council, which are to advance and interpret scientific knowledge

pertaining to agriculture, and to initiate recommendations relative to agricultural research; to assist the National Academy of Sciences -- National Research Council and governmental agencies by providing, on request, information relating to agriculture; to define and publicize specific problem areas of agriculture; to disseminate agricultural facts and conclusions in the public interest.

Publications: ARI NEWSLETTER (quarterly); proceedings of annual meetings and numerous specialized publications on agricultural matters.

CROP PROTECTION INSTITUTE - P.O. Drawer "S", Durham, New Hampshire 03824.

Conducts biological research and development in insecticides, fungicides, nematicides, bioregulants, and fabric protectants. Bioassay methods for quality control. Laboratory and greenhouse facilities. Performs laboratory, greenhouse and field plot evaluations. Consultants to agricultural chemical industry.

4. International Organizations

ASSOCIATION FOR THE OILSEEDS, ANIMAL AND VEGETABLE OILS AND FATS AND DERIVATES TRADE IN THE EEC - c/o Nederlandse Vereniging voor de groothandel in olien vetten en aanverwante artikelen, Oppert 34, Rotterdam-3001, Rotterdam, Netherlands.

COLLABORATIVE INTERNATIONAL PESTICIDES ANALYTIC COUNCIL LTD. (CIPAC) - Ministry of Agriculture, Fisheries and Food Plant Pathology Laboratory, Hatching Green, Harpenden, Herts, United Kingdom.

Purpose: To organize collaborative work for preparing methods of analysis for determining the chemical composition and physical properties of materials and their formulations used in crop protection, insecticides, fungicides, herbicides, rodenticides, attractants and repellents; promote international agreement on methods for correlating biological efficacy with chemical and physical tests; methods for residue analysis are expressly excluded.

Publications: CIPAC HANDBOOK, VOL. 1. ANALYSIS OF TECHNICAL AND FORMULATED PESTICIDES (1970).

COMMONWEALTH AGRICULTURAL BUREAUX - Farnham House, Farnham Royal, Slough, SL2 3BN, United Kingdom, T. Farnham Common 2281, C Comag Slough.

Purpose: Provides a comprehensive information service for research workers in the agricultural sciences including animal health and forestry. Consists of 13 technical institutes and bureaux under the general supervision of each of the member governments.

Institute of Entomology, R.G. Fennah, 56 Queen's Gate, London, SW7 5JR. T. 584-0067.

Mycological Institute, A. Johnstone, Ferry Lane, Kew, Surrey, T. 940-4086/7.

Institute of Biological Control, F.J. Simmonds, Gordon Street, Curepe, Trinidad, West Indies. T. 94111. C Biocontrol Trinidad.

Institute of Helminthology, Miss S.M. Willmott, The White House, 103 St. Peter's Street, St. Albans, Herts. T. 52126.

Bureau of Agricultural Economics, J.O. Jones, Dartington House, Little Clarendon Street, Oxford, OX1 2HH. T. 59829.

Bureau of Animal Breeding and Genetics, J.D. Turton, The King's Buildings, West Mains Road, Edinburgh, EH9 3JX. T. NEWington 6901.

Bureau of Animal Health, M.R. Dhanda, Central Veterinary Laboratory, New Haw, Weybridge, Surrey. T.BYFleet 42826.

Bureau of Animal Nutrition, Miss D.L. Duncan, Rowett Research Institute, Bucksburn, Aberdeen, AB29SB. T. BUCksburn 2162.

Bureau of Dairy Science and Technology, E.J. Mann, National Institute of Research in Dairying, Shinfield, Reading, RG29AT. T.883895.

Forestry Bureau, P.G. Beak, Commonwealth Forestry Institute, South Parks Road, Oxford, OX1 3RD. T. 57185.

Bureau of Horticulture and Plantation Crops, G.E. Tidbury, East Malling Research Station, Nr. Maidstone, Kent. T. West Malling 3033.

Bureau of Pasture and Field Crops, P.J. Boyle, Hurley, Nr. Maidenhead, Berks, T. Hurley 363-366.

DAIRY SOCIETY INTERNATIONAL - 30 F Street N.W., Washington, D.C. 20001.

Purpose: The extension of dairy and dairy industrial enterprises internationally through an interchange and dissemination of scientific, technological, economic, dietary and other relevant information.

Publications: BULLETIN of recent world-wide technical, bibliographical and miscellaneous dairy information; CONFIDENTIAL REPORT to members, in English; PROCEEDINGS OF FIRST WORLD CONGRESS FOR MILK UTILIZATION, Basic Handbooks on Dairy Techniques (English and Spanish); GLOSSARY OF DAIRY TERMINOLOGY (French-English; Spanish-English); MARKET FRONTIER NEWS; THE DAIRY SITUATION REVIEW; TONED MILK -- THE WHY AND HOW OF IT; UNITED STATES BUTTER -- A QUALITY PRODUCT; INSTANT NONFAT DRY MILK RECIPES; annual reports.

EUROPEAN ASSOCIATION FOR POTATO RESEARCH - P.O. Box 20, Wageningen, Netherlands.

Purpose: Improve research concerned with subjects related to the potato.

Publications: POTATO RESEARCH (quarterly). Conference proceedings.

EUROPEAN COMMITTEE OF ASSOCIATIONS OF MANUFACTURERS OF AGRICULTURAL MACHINERY - 19 rue Jacques Bingen, 75-Paris 17e, France.

Purpose: To study technical and economic problems of the industry; concerned with the most suitable methods for the development and protection of members' common and international interests; exchange and disseminate information of interest to members.

EUROPEAN FEDERATION OF COMPOUND ANIMAL FEEDING STUFFS MANUFACTURERS - SGAP Namur, 223 rue de la Loi, 1000 Brussels, T.34. 39.70. Belgium.

INTER-AMERICAN COMMITTEE FOR AGRICULTURAL DEVELOPMENT - 1725 Eye Street N.W., Room 414, Washington, D.C. 20006.

Purpose: To fill gaps in information and identify problem areas that obstruct agricultural development, giving special attention to such factors as agricultural education, research and extension, farm credit, marketing, input items; advise when requested by the Latin American governments in the formulation of their agricultural development and agrarian reform policies.

Publications: REPORT AND OFFICIAL DOCUMENTS OF THE MEETING OF HIGH LEVEL EXPERTS IN AGRICULTURE (1961) in English and Spanish; DOCUMENTS AND PROCEEDINGS OF THE MEETING FOR THE STUDY OF AGRICULTURAL EDUCATION, RESEARCH AND EXTENSION IN LATIN AMERICA (1964) in English and Spanish; Regional Report, Selected Bibliography and individual Country Reports (English and national language) of the Inventory of Information Basic to the Planning of Agricultural Development in Latin America. Research papers issued by the CIDA Program for the Study of Land Tenure and Land Reform nos. 1-17.

INTER-AMERICAN INSTITUTE OF AGRICULTURAL SCIENCES - Apartado 10281, San Jose, Costa Rica.

Purpose: As specialized agricultural agency of the Organization of American States (OAS), endeavors to promote economic and social development through graduate teaching, research, technical assistance, consultation services and communications in the field of agriculture and rural life; concerned with problems of plant and animal production and socio-economic development to improve the welfare of the rural population through technical assistance to national institutions; directs project 206 -- Inter-American Program for Rural Development and Agrarian Reform of the Secretariat of the OAS, sponsored by the Inter-American Economic and Social Council (IAECOSOC).

Publications: TURRIALBA (quarterly); DASONOMIA INTERAMERI-CANA (quarterly); CACAO (quarterly) (in Spanish and English); DESARROLLO RURAL EN LOS AMERICAS (3 a year); BOLETIN BIBLIOGRAFICO AGRIOLA (quarterly); series of bibliographies, technical bulletins, teaching materials, manuals and texts.

INTERNATIONAL ASSOCIATION OF AGRICULTURAL ECONOMISTS - 600 South Michigan Avenue, Chicago, Illinois 60605.

An organization of agricultural economists engaged in research, teaching, or administrative work in 78 countries. Promotes development of the science of agricultural economics in different countries, and fosters application of the results of economic investigation of agricultural processes to improve economic and social conditions. Sponsors triennial conferences to discuss subjects bearing on individual and public farm problems and international problems relating to agriculture and research.

Publications: Proceedings of Conferences (triennial); INTERNA-TIONAL JOURNAL OF AGRARIAN AFFAIRS (irregular).

INTERNATIONAL ASSOCIATION FOR CEREAL CHEMISTRY - Schmidgasse 3-7, 2320 Schwechat, Austria.

Purpose: Research on the problems of technical research in cereal and flour chemistry and adjacent fields; unification of test methods and their application for the purpose of elaboration of standardized procedures for this research work.

INTERNATIONAL CENTRE FOR SCIENTIFIC AND TECHNICAL INFORMATION IN AGRICULTURE AND FORESTRY - Slezska, 7, Prague, Czechoslovakia.

Purpose: To coordinate the activities of central agricultural information institutes and libraries among the COMECON countries for the exchange of information on the latest scientific and technological developments.

Publications: BULLETIN DER LANDWIRTSCHAFTLICHEN INFORMATIONSZENTREN DER MITGLIEDERSTAATEN DES RGW (quarterly).

INTERNATIONAL COMMISSION OF AGRICULTURAL ENGINEERING - 10-12 rue du Capitaine Menard, 75-Paris 15e, France.

Purpose: To link agricultural engineering specialists in different countries; promote, encourage and coordinate scientific and technical research; improve training methods and act as documentation center.

Publications: Reports of section meetings; congress reports.

INTERNATIONAL CONFEDERATION OF EUROPEAN SUGAR-BEET GROWERS - 29 rue du General Foy, 75-Paris 8e, France.

Purpose: Foster interests of sugar beet growers in the international sphere; centralize information on all economic and technical questions concerning sugar beet growing and its bearing on the sugar industry.

Publications: BETTERAVIERS EUROPEENS (every 2 years).

INTERNATIONAL CONFEDERATION OF TECHNICAL AGRICULTURE ENGINEERS - 24 Beethovenstrasse, 8002 Zurich, Switzerland T. 36.68.48.

Purpose: Unite technical agriculturists in all countries in order to protect their professional interests on an international level.

Publications: LA TECHNIQUE AGRICOLE; Bulletin du CIA; conference reports.

INTERNATIONAL FEDERATION OF AGRICULTURAL PRODUCERS - 910 17th Street N.W., Room 401, Washington, D.C. 20006.

Federation of 54 farm organizations in 40 countries. To promote the well-being of all who obtain their livelihood from the land and to assure to them adequate and stable remuneration; to promote exchange of information and ideas; to encourage efficiency of production, processing, and marketing of agricultural commodities, etc.

Publications: IFAP NEWS (monthly); WORLD AGRICULTURE (quarterly); INFORMATION AND LIAISON BULLETIN FOR FARM ORGANIZATIONS IN AFRICA AND ASIA (quarterly); conference proceeding.

INTERNATIONAL INSTITUTE FOR SUGAR BEET RESEARCH - Beauduinstraat 150, 3300 Tienen (Tirlemont), Belgium T. 016/828 75.

Purpose: Promote research on sugar beets in general; organize meetings of specialists, scientists and other persons interested in the improvement of sugar beets; disseminate research details on innovations, procedures and methods applicable to sugar beet research.

Publications: IIRB (quarterly journal); congress reports (1958-64); sugar beet glossary.

INTERNATIONAL SUGAR RESEARCH FOUNDATION - 7316 Wisconsin Avenue, Bethesda, Maryland 20014.

Separately incorporated research organization, serving as sugar utilization research arm of world sugar industry; membership organization of associations, producers, processors and refiners of beet and cane sugar, comprising 35 companies and associations on four continents.

Areas of interest: Uses of sugar, other sweetening agents and sugar byproducts in foods and beverages, in nonfoods and as ingredients or raw materials in chemical and manufacturing industries, also studies of place and value of sugar in nutrition and public health.

Activities: Disseminates information resulting from its research with respect to use, purpose, utility and effects of sugar and other sweetening agents.

Publication: ISRF BULLETIN.

NITROGEN STUDY CENTRE - Bleicherweg 33, 8002 Zurich, Switzerland.

Purpose: Scientific and practical study of methods to ensure a rational and increasing use of nitrogen fertilizers throughout the world, especially with a view to intensifying agricultural production in the developing countries.

5. Other Information Sources and Programs

a. SPECIALIZED INFORMATION SERVICES

AG/PACK - Science Associates/International, Inc., 23 East 26th Street, New York, New York 10010.

"AG/PACK" (AGricultural Personal Alerting Card Kits) is a new weekly service of 3 x 5 card current awareness developed by Scientific Documentation Centre, Dunfermline, U.K. It offers individual "packs" that cover virtually every area of interest to agricultural researchers and librarians; the Centre operates one of the world's most comprehensive literature searching efforts and offers weekly card services in many fields. U.S. address above.

INFORMATION COMPANY OF AMERICA - 1011 Lewis Tower, 225 South 15th Street, Philadelphia, Pennsylvania 19102.

Areas of interest: Production and marketing of information retrieval programs. Research and development.

Publication: UPDATE/AGRICULTURAL CHEMICALS (biweekly).

B. LITERATURE

1. Guides and Handbooks

a. GENERAL

AGRICULTURAL CHEMICAL MANUAL. Annual. North Carolina State University, Division of Continuing Education, Box 5125, Raleigh, North Carolina 27607.

AGRICULTURAL ENGINEERS YEARBOOK. Annual. American Society of Agricultural Engineers, 2950 Niles Road, St. Joseph, Michigan 49085.

AGRICULTURE HANDBOOKS. Annual. U.S. Department of Agriculture. Washington, D.C.: Government Printing Office.

Association of American Pesticide Control Officials. PESTICIDE CHEMICALS OFFICIAL COMPENDIUM. Irregular. Association of American Pesticide Control Officials, Inc., Kansas State Board of Agriculture, State Office Building, Topeka, Kansas 66612.

Association of Official Agricultural Chemists. OFFICIAL METHODS OF ANALYSIS. Quinquennial, 1970, 11th ed. Association of Official Analytical Chemists, Box 540, Benjamin Franklin Station, Washington, D.C. 20044.

Blanck, F. HANDBOOK OF FOOD AND AGRICULTURE. New York, Reinhold, 1955.

FARM CHEMICALS (periodical). Annual. FARM CHEMICALS HANDBOOK. Meister Publishing Co., 37841 Euclid Avenue, Willoughby, Ohio 44094.

FERTILIZER HANDBOOK. 2nd ed. Washington, D.C.: Fertilizer Institute, 1976.

HERBICIDE HANDBOOK. 3rd ed. Champaign, Illinois: Weed Science Society of America, 1974.

Imperial Chemical Industries, Agriculture Division. CATALYST HANDBOOK. London: Wolfe Scientific, 1970.

> "Contents are 'Catalytic activity of solids'...'The structural engineering of catalysts'...'Catalyst testing'...'Desulphurisation'... 'Hydrocarbon-reforming catalysts'...'Removal of carbon monoxide' ...etc."

Jurgenson, E. HANDBOOK OF LIVESTOCK EQUIPMENT. Danville, Illinois: Interstate Publishers Co., 1971.

Kezdi, A. HANDBOOK OF SOIL MECHANICS. N.Y.: Elsevier, 1974.

MIDWEST FARM PLANNING MANUAL. 3rd ed. Edited by Sidney C. James. Ames, Iowa: The Iowa State University Press, 1973.

Mortenson, W. THE FARM MANAGEMENT HANDBOOK. 5th ed. Danville, Illinois: Interstate Publishers Co., 1972.

Page, B. THE INSECTICIDE, HERBICIDE, FUNGICIDE, QUICK GUIDE AND DATE BOOK. Annual. Thompson Publications, Box 5601, Fresno, California 93755.

Richey, C., et. al. AGRICULTURAL ENGINEER'S HANDBOOK. New York: McGraw-Hill, 1961.

> "Gives under one cover the theory and practice for the various areas of agricultural engineering, and the application of engineering to the problems of agriculture."

Smith, H. FARM MACHINERY AND EQUIPMENT. 6th ed. New York: McGraw-Hill, 1976.

United Nations. Food and Agricultural Organization. ANIMAL HEALTH
YEARBOOK, 1971. Unipub, Inc., Box 433, New York, New York 10018.

> "Detailed information on occurrence, distribution, control of more
> than 200 diseases."

U.S. Department of Agriculture. GUIDE FOR INTERPRETING ENGINEER
USES OF SOILS (with list of literature cited). Washington, D.C.: Government Printing Office, November, 1971.

_____. 423. HANDBOOK OF AGRICULTURAL CHARTS, 1971 (November).
Economic Research Statistical Reporting, Agricultural Research, Foreign Agricultural Service. Washington, D.C.: Government Printing Office.

_____. SCS NATIONAL ENGINEERING HANDBOOK: SEC. 16, DRAINAGE
OF AGRICULTURAL LAND (with lists of references), 1971. In looseleaf
form.

_____. USDA SUMMARY OF REGISTERED AGRICULTURAL PESTICIDE CHEM-
ICAL USES. Irregular. Pesticides Regulation Division, Department of Agriculture, Washington, D.C. 20250.

U.S. Department of Agriculture. Economic Research Service. ENERGY IN
U.S. AGRICULTURE: A COMPENDIUM OF ENERGY RESEARCH PROJECTS.
By J. Rathivell, et al. Rept. no. PB 247 642. Springfield, Virginia: National
Technical Information Service, 1976.

U.S. Environmental Protection Agency. EPA COMPENDIUM OF REGISTERED
PESTICIDES. Irregular. Washington, D.C.: Government Printing Office.

> Basic manual with periodic updates.

U.S. SOIL CONSERVATION SERVICE NATIONAL ENGINEERING HAND-
BOOK. Irregular. U.S. Soil Conservation Service, c/o Department of
Agriculture, Washington, D.C. 20250.

Utah State University. USE OF GEOTHERMAL WATER FOR AGRICULTURE.
By A. Bishop. Rept. no. ANCR - 1221. Springfield, Virginia: National
Technical Information Service, 1975.

WORLD ATLAS OF AGRICULTURE. International Association of Agricultural
Economists. Monographs edited by the Committee for the World Atlas of
Agriculture, Novara, Italy, Instituto Geografico de Agostini, 1969.

> Provides extensive maps & tables showing agricultural products,
> population, geographical information, etc.

2. Dictionaries

a. TECHNICAL

U.S. Department of Agriculture. DICTIONARY OF INTERNATIONAL AGRI-

CULTURAL TRADE. Compiled by Harry W. Henderson. Foreign Agricultural Service. Washington, D.C.: Government Printing Office, April 1971.

Winburne, J., et al, eds. DICTIONARY OF AGRICULTURAL & ALLIED TERMINOLOGY. East Lansing, Michigan: Michigan State University Press, 1962.

b. FOREIGN LANGUAGE

DICTIONARY OF SOIL MECHANICS IN FOUR LANGUAGES: ENGLISH/ AMERICAN, FRENCH, DUTCH, AND GERMAN. Compiled and arranged by A.D. Visser. Amsterdam: Elsevier, 1965.

Haensch, G. DICTIONARY OF AGRICULTURE. German - English - French - Spanish. 3rd ed. Amsterdam: Elsevier, 1966.

Kratochvil, V. AGRICULTURAL DICTIONARY IN EIGHT LANGUAGES: RUSSIAN-BULGARIAN-CZECH-POLISH-HUNGARIAN-ROUMANIAN-GERMAN-ENGLISH. 2 vols. New York.: International Publications Service, 1972.

Usovskii, B.N., et al. COMPREHENSIVE RUSSIAN-ENGLISH AGRICULTURAL DICTIONARY. 2nd ed., revised and enlarged. Oxford: Pergamon, 1967.

3. Directories

a. GENERAL

Mann, S. EUROPEAN AGRICULTURAL CHEMICAL SURVEY. Park Ridge, New Jersey: Noyes Data Corp., 1969.

National Research Council. Committee on Animal Nutrition. Subcommittee on Feed Composition. ATLAS OF NUTRITIONAL DATA ON UNITED STATES AND CANADIAN FEEDS. Subcommittee on Feed Composition, Committee on Animal Nutrition, Agricultural Board, National Research Council, U.S., and Committee on Feed Composition, Research Branch, Department of Agriculture, Canada. Washington, D.C.: National Academy of Science, 1971 (1972).

U.S. Department of Agriculture. National Agricultural Library. DIRECTORY OF INFORMATION RESOURCES IN AGRICULTURE AND BIOLOGY, 1971. Washington, D.C.: Government Printing Office.

_____. Foreign Agricultural Service. FOOD AND AGRICULTURAL EXPORT DIRECTORY/YEAR. Annual (FASM--201). Available from Export Trade Services Division, Foreign Agricultural Service, U.S. Department of Agriculture, Washington, D.C. 20250.

_____. LIST OF CHEMICAL COMPOUNDS AUTHORIZED FOR USE UNDER USDA POULTRY, RABBIT AND EGG PRODUCTS INSPECTION PROGRAMS.

Rev. ed. Consumer Protection Programs, Technical Service Division. A 88.6:
C42/97. Washington, D.C.: Government Printing Office, October 1970.

_____. THE LOOK OF OUR LAND, AIRPHOTO ATLAS OF RURAL UNITED
STATES, EAST AND SOUTH (with list of selected references). Compiled by
Simon Baker and Henry W. Dill, Jr. Washington, D.C.: Government Print-
ing Office, April 1971.

_____. YEARBOOK OF AGRICULTURE. Annual since 1895. Washington,
D.C.: Government Printing Office.

Useful information source.

U.S. National Agricultural Library. AGRICULTURAL SCIENCES INFORMA-
TION NETWORK. National Agricultural Library. (A17.2:Sci2). Beltsville,
Maryland: Department of Agriculture, May 1970.

b. BIOGRAPHY

AGRICULTURAL ENGINEERS YEARBOOK. Annual. American Society of
Agricultural Engineers, 420 South Main Street, St. Joseph, Michigan 49085.

"Contains a roster of membership of the American Society of
Agricultural Engineers and a list of agricultural trade associations,
consulting agricultural engineers, and a directory of suppliers and
products for agricultural machinery."

BULLETIN OF THE ENTOMOLOGICAL SOCIETY OF AMERICA. Biannual.
Entomological Society of America, 4603 Calvert Road, College Park, Maryland
20740.

"Contains an alphabetical list of approximately 6,000 individual
and company members of the Entomological Society of America
--pest control specialists, scientists interested in insects, consultants, airplane sprayers, official entomologists and plant patholo-
gists, insecticide and fungicide manufacturers."

COUNTY AGENTS DIRECTORY. Annual. C.L. Mast, Jr., and Associates,
2041 Vardon Lane, Flossmoor, Illinois.

"Directory presents a compilation of approximately 15,000 county
agents, home demonstration agents, 4-H Club agents, state extension specialists, land-grant colleges and experiment stations."

FARM EQUIPMENT DIRECTORY. Annual. Farm Journals Ltd., London, EC4,
England.

Institution of Agricultural Engineers. Annual. YEARBOOK & MEMBERSHIP
DIRECTORY. The Institution of Agricultural Engineers, Hertfordshire, England.

WHO'S WHO IN THE AGRICULTURAL INSTITUTE OF CANADA. Annual.
Agricultural Institute of Canada, 151 Slater Street, Ottawa 4, Ontario, Canada.

c. PRODUCTS AND MANUFACTURERS

i. United States

AGRICULTURAL CHEMICALS. Rev. ed. Indianapolis: Thomson Publications, 1975.

FARM AND POWER EQUIPMENT. Directory of Products, Trade Names and Manufacturers Index. Annual. St. Louis: National Farm and Power Services.

IMPLEMENT TRACTOR PRODUCT FILE (Formerly IMPLEMENT AND TRACTOR PRODUCT FILE). Annual. Intertec Publishing Corp., 1014 Wyandotte Street, Kansas City, Missouri 64105.

MACRAE'S BLUEBOOK. Annual. MacRae's Bluebook, Western Springs, Illinois 60558.

THOMAS REGISTER OF MANUFACTURERS. Annual. New York: Thomas Publishing Co.

ii. International

COMMERCIAL GROWERS' DIRECTORY & BUYERS GUIDE. New York: British Book Centre, 1972.

FARM EQUIPMENT BUYERS' GUIDE. Annual. Agricultural Press Ltd., 161-166 Fleet Street, London, EC4, England (distributed in the United States by Iliffe - N.T.P. Inc., Sales Division, 205 East 42nd Street, New York, New York, 10017.)

FARM EQUIPMENT DIRECTORY. Annual. Text in English and French. Southam Business Publications, Ltd., 1450 Don Mills Road, Don Mills, Ontario, Canada.

THE GREEN BOOK. Annual. London: Thomas Reed Industrial Press.

Annual directory of farm and forestry equipment and suppliers.

United Nations Food and Agriculture Organization. TRADE YEARBOOK. Annual. U.N., F.A.O., Rome, Italy.

WHERE TO BUY. Annual. AGRICULTURAL SUPPLIES. John Adams House, London, WC2, England.

WORLD DIRECTORY OF FERTILIZER MANUFACTURERS. 2nd ed. International Publications Service, 1972.

See also Journal Literature in Part 8 of this section for additional products and manufacturers information.

d. LABORATORIES AND RESEARCH CENTERS

AGRICULTURAL RESEARCH INDEX. 5th ed. Guernsey, England: Francis Hodgson, 1970.

Contains a directory of research institutions.

THE AGRICULTURAL RESEARCH SERVICE. London: Agricultural Research Council, 1969.

A directory of research agencies in England.

DIRECTORY·OF INFORMATION RESOURCES IN AGRICULTURE AND BIOLOGY. National Agricultural Library. U.S. Department of Agriculture. Beltsville, Maryland: 1971.

EUROPEAN RESEARCH INDEX. 2nd ed. Guernsey, England: Francis Hodgson, 1969.

INDUSTRIAL RESEARCH LABORATORIES OF THE UNITED STATES. 3rd ed. New York: R.R. Bowker, 1970.

RESEARCH CENTERS DIRECTORY. 5th ed. Detroit: Gale Research Co., 1975.

Updated by subscription service, NEW RESEARCH CENTERS.

SURVEY OF THE WORLD AGRICULTURAL DOCUMENTATION SERVICES. Edited by H. Buntrock. Rome: Food and Agriculture Organization, 1970.

4. Encyclopedias

No entry.

5. Bibliographies

a. SUBJECT RELATED BIBLIOGRAPHIES

AMERICAN BIBLIOGRAPHY OF AGRICULTURAL ECONOMICS. Biweekly. American Agricultural Economics Documentation Center, Room 1505 S Building, Washington, D.C. 20250.

American Society for Engineering Education. GUIDE TO LITERATURE OF AGRICULTURAL ENGINEERING. 1969. Engineering Libraries Division, The Society, Washington, D.C. 20036.

BIBLIOGRAPHY OF SOIL SCIENCE, FERTILIZERS AND GENERAL AGRONOMY, 1959-1962. Farnham Royal, England: Commonwealth Agricultural Bureaux, 1964.

Blanchard, J. "Agriculture." LIBRARY TRENDS, Vol. 15, (April 1967), pp. 880-95.

"This paper is one of a collection on the theme 'Bibliography: current state and future trends.' It describes the various services and developments in the documentation and bibliography of agricultural literature in the widest sense. There is a bibliography of 111 references."

Blanchard, J., et al. LITERATURE OF AGRICULTURAL RESEARCH. 2nd ed. Berkeley: University of California Press, in press.

Bottle, R. "The Literature of Applied Biology (Pharmacy, Food and Agriculture)." THE USE OF BIOLOGICAL LITERATURE. London: Butterworth, 1966.

Bush, E. AGRICULTURE: A BIBLIOGRAPHIC GUIDE. 2 vols. London: Macdonald & Co., 1974.

Commonwealth Forestry Bureau. CARD TITLE SERVICE. Weekly. Commonwealth Forestry Bureau, South Parks R, Oxford, England.

Gittinger, J. THE LITERATURE OF AGRICULTURAL PLANNING. Washington, D.C.: National Planning Association, 1966.

Harvey, A., comp. "Agricultural Research Literature: A Bibliography of Selected Guides and Periodicals." In AGRICULTURAL RESEARCH INDEX, 5th ed., vol. 2, pp. 957-1043. Guernsey, England: Francis Hodgson, 1970.

International Federation for Documentation. ABSTRACTING SERVICES: Vol. 1: SCIENCE, TECHNOLOGY, MEDICINE, AGRICULTURE. 2nd ed. The Hague: 1969.

United Nations Food and Agriculture Organization. FOOD AND AGRICULTURAL INDUSTRIES, ANNOTATED BIBLIOGRAPHY. New York: Unipub Co., 1970.

U.S. Department of Agriculture. LIST OF AVAILABLE PUBLICATIONS OF DEPARTMENT OF AGRICULTURE. Compiled by Mattie W. Johnson. A 21.9/8:11/8. Washington, D.C.: Government Printing Office, 1971.

_____. POTENTIAL FOR GENETIC SUPPRESSION OF INSECT POPULATIONS BY THEIR ADAPTATIONS TO CLIMATE (with list of literature cited). By W. Klassen, J.F. Creech, and R.A. Bell. October, 1970.) Agricultural Research Service. Prepared in cooperation with North Dakota State University. Government Printing Office.

_____. SEDIMENTATION. ANNOTATED BIBLIOGRAPHY OF FOREIGN LITERATURE SURVEY. Prepared for the U.S. Department of Agriculture and the National Science Foundation by Israel Program for Scientific Translations, Irregular. Springfield, Virginia: National Technical Information Service.

_____. AMS (series). AVAILABLE PUBLICATIONS OF USDA's AGRICULTURAL MARKETING SERVICE. January, 1973. U.S. Department of Agriculture, Washington, D.C. 20250.

U.S. National Agricultural Library. CARD CATALOGS OF NATIONAL AGRICULTURAL LIBRARY, HOW TO USE THEM. Washington, D.C.: January, 1971.

_____. THE NATIONAL AGRICULTURAL LIBRARY CATALOG, 1966-70. 10 vols. Totowa, New Jersey: Rowman and Littlefield, 1972.

Vara, A. FOOD AND BEVERAGE INDUSTRIES: A BIBLIOGRAPHY AND GUIDEBOOK. Detroit: Gale Research Co., 1970.

> Section 2 is on agriculture and covers the following topics: land, water supply, marketing, and agribusiness.

b. RELATED BIBLIOGRAPHIES IN CURRENT JOURNALS

The following journals frequently contain book reviews and/or bibliographies. See the heading Journals and Serial Publications in this section for additional details.

AGRICULTURAL ENGINEERING

AGRICULTURE (CANADA)

BIBLIOGRAPHY OF AGRICULTURE

DAIRY SCIENCE ABSTRACTS

FARM CHEMICALS

FARM TECHNOLOGY

FIELD CROP ABSTRACTS

FOOD RESEARCH INSTITUTE STUDIES IN AGRICULTURAL ECONOMICS, TRADE, AND DEVELOPMENT

HORTICULTURAL ABSTRACTS

INTERNATIONAL SOCIETY OF SOIL SCIENCE BULLETIN

JOURNAL OF AGRICULTURAL ENGINEERING RESEARCH

JOURNAL OF SOIL SCIENCE

JOURNAL OF STORED PRODUCTS RESEARCH

JOURNAL OF THE SCIENCE OF FOOD AND AGRICULTURE

PESTICIDE SCIENCE

REVIEW OF APPLIED ENTOMOLOGY

SOIL SCIENCE

SOIL SCIENCE OF AMERICA. PROCEEDINGS

See also this subject in the Section I of this Guide.

6. Monographs, Reports, and Symposia Proceedings

American Society of Agronomy. AGRICULTURAL ANHYDRONS, AMMONIA. Madison, Wisconsin: 1968.

American Society of Agricultural Engineers, Annual Meeting. PAPERS. Sponsor: American Society of Agricultural Engineers, 2950 Niles Road, St. Joseph, Michigan 49085.

Baver, Leonard D., et al. SOIL PHYSICS. 4th ed. New York: Wiley, 1972.

> Complete discussion of soil, structure and properties, erosion, and management necessary to preserve soil.

Casper, M. LIQUID FERTILIZERS 1973. Noyes Data Corp., Park Ridge, New Jersey 07656.

> An overview of this phase of industry listing trends and patents.

Cooke, G. FERTILIZING FOR MAXIMUM YIELD. New York: Hafner, 1972.

> Intended for British audiences but contains useful detail for others in agricultural engineering.

DOSIMETRY IN AGRICULTURE, INDUSTRY, BIOLOGY & MEDICINE. Symposium, Vienna, Asutria. Sponsor: International Atomic Energy Agency, and World Health Organization. New York: Unipub, Inc., 1973.

FARM, CONSTRUCTION & INDUSTRIAL MACHINERY & POWERPLANT. National Combined meetings, Milwaukee, Wisconsin. Sponsor: Society of Automotive Engineers, 2 Pennsylvania Plaza, New York, New York 10001.'

FATE OF ORGANIC PESTICIDES IN THE AQUATIC ENVIRONMENT. Symposium of the Division of Pesticide Chemistry at 161st meeting, Los Angeles, California. Sponsor: American Chemical Society, 1155 16th Street, N.W. Washington, D.C. 20036. (ADVANCES IN CHEMISTRY SERIES, no. 111.) Edited by Samuel Denton Faust, 1972.

FERTILIZER TECHNOLOGY AND USE. Editing committee, R.A. Olson, editor-in-chief. 2nd ed. Madison, Wisconsin: Soil Science Society of America, 1971.

Fiftieth Congress, International Association of Seed Crushers, Killarney, Ireland, 1973. International Association of Seed Crushers, 1 Watergate, London, EC4, England.

Foley, R. DAIRY CATTLE; PRINCIPLES, PRACTICES, PROBLEMS, PROFITS. Philadelphia: Lea & Febiger, 1972.

> An up to date reference guide to dairy farming.

Goring, C., ed. ORGANIC CHEMICALS IN THE SOIL ENVIRONMENT. 2 vols. New York: Dekker, 1972.

> Covers chemicals already in soil as well as those added by pesticides, fertilizers, fumigants, etc.

International Atomic Energy Agency. NUCLEAR TECHNIQUES FOR STUDYING PESTICIDE RESIDUE PROBLEMS. Proceedings of a panel on isotopic tracer and radioactivation techniques for studying residue problems with particular reference to food entering international trade, organized by the joint FAO/IAEA division of Atomic Energy in Food and Agriculture, and held in Vienna, 16-20 December 1968. New York: Unipub, 1971.

International Commission on Irrigation and Drainage, 6th Congress. Nyaya Marg, Chanakyapuri, New Delhi 110021, India, 1966. 4 vols.

International IUPAC. HERBICIDES, FUNGICIDES, FORMULATION CHEMISTRY. PROCEEDINGS. Edited by A.S. Tahori. Congress of Pesticide Chemistry, 2nd, Tel Aviv, 1971. New York: Gordon & Breach, 1972.

International Symposium on Soil-Water Physics and Technology, Rehovot, 1971. PROCEEDINGS; OPTIMIZING THE SOIL PHYSICAL ENVIRONMENT TOWARD GREATER CROP YIELDS. Edited by D. Hillel. New York: Academic, 1972.

Kepner, R. PRINCIPLES OF FARM MACHINERY. 2nd ed. Westport, Connecticut: Avi, 1972.

Kipps, M. THE PRODUCTION OF FIELD CROPS. 6th ed. New York: McGraw-Hill, 1970.

> Crop production, grain and cash crops, and forage crops are discussed, and new developments, concepts, and practices are given ample coverage.

Kohnke, H. SOIL PHYSICS (AGRICULTURAL SCIENCES SERIES). New York: McGraw-Hill, 1968.

> Covers both fundamentals and the most recent advances and techniques in the physics of water and its relation to soil.

MAPLE PRODUCTS. 8th Conference, Boyne Falls, Michigan. Sponsor: U.S. Agricultural Research Service, Washington, D.C. 20250. ARS Publication no. 73-73; EMN (Eastern Marketing & Nutrition Research Division) Publication 3618, 1972.

Meltzer, Y. HORMONAL AND ATTRACTANT PESTICIDE TECHNOLOGY 1971. Park Ridge, New Jersey: Noyes Data Corp.

> Based on review of technical literature and recent patents.

Mohsenin, N. PHYSICAL PROPERTIES OF PLANT AND ANIMAL MATERIALS. Vol. I. New York: Gordon & Breach, 1970.

"Analyzes the physical properties of a variety of bio-materials. These properties would have application in the design of machines, structures, processes and quality evaluation and control. The materials covered include raw plants and animal materials such as grain and seeds, forage and silage, fruits and vegetables, meats, eggs and dairy products, as well as organic fluids and some processed food materials."

National Research Council, Agricultural Research Institute. PROCEEDINGS. Washington, D.C.: 1966.

NEW FERTILIZER MATERIALS 1968. Park Ridge, New Jersey: Noyes Data Corp.

"Ureaform, Crotonylidene & Isobutylidene Diurea, Triple Superphosphate, Ammonium Phosphates, Nitrophosphates, Nitrate of Potash, Potassium Phosphates and Metaphosphates, Magnesium, Sulfur Fertilizers, Oxamide, Urea Nitrate and Phosphate, Hydrides of Phosphorus, Red Phosphorus. Applications."

OPTIMIZING THE SOIL PHYSICAL ENVIRONMENT TOWARD GREATER CROP YIELDS. Proceedings of an invitational panel convened during the International Symposium on Soil-Water Physics & Technology held at Hebrew University, August 29-September 5, 1971. Edited by Daniel Hillel. New York: Academic, 1972.

PESTICIDAL FORMULATIONS RESEARCH. American Chemical Society. ADVANCES IN CHEMISTRY SERIES 86. Washington, D.C.: The Society, 1969.

PESTICIDE RESIDUES IN FOOD. Joint FAO/WHO Meeting. Sponsor: Food & Agriculture Organization, and World Health Organization, Geneva, Switzerland. World Health Organization, Palais des Nations, CH-1211 Geneva 27, Switzerland (WHO Technical report series, no. 502, FAO Agricultural studies no. 88).

Pesticide Terminal Residues, International Symposium. Sponsor: International Union of Pure & Applied Chemistry, Tel Aviv, Israel. Publisher: Butterworths & Co., Ltd., London. U.S. Distributor: Crane, Russak & Co., 52 Vanderbilt Avenue, New York, New York 10017. Published as: PURE & APPLIED CHEMISTRY, SUPPLEMENT. Edited by A.S. Tahori. 1972.

PIG PRODUCTION. 18th Easter School in Agricultural Science. Sponsor: University of Nottingham, Nottingham, England. Edited by D.J.A. Cole. University Park, Pennsylvania: Pennsylvania State University Press, 1972.

PLANT BREEDING FOR RESISTANCE TO ANIMAL PESTS. International EPPO/EUCARPIA conference. Sponsor: European & Mediterranean Plant Protection Organization, and European Association for Research on Plant Breeding, Rostanga-Svalov, Sweden. Organisation Europeenne et Mediterraneenne pour la Protection des Plantes, 1 rue le Notre, 75016 Paris, France. EPPO Publications, Series A no. 54. REPORT, 1970.

Powell, R. CONTROLLED RELEASE FERTILIZERS 1968. Park Ridge, New Jersey: Noyes Data Corp.

"This book offers complete technical data on numerous processes and products in this field."

RADIOTRACER STUDIES OF CHEMICAL RESIDUES IN FOOD & AGRICULTURE. Combined Panel and Research Coordination Meeting. Sponsor: Joint FAO/IAEA DIVISION of Atomic Energy in Food & Agriculture. Publisher: International Atomic Energy Agency, Vienna. U.S. Distributor: Unipub, Inc., P.O. Box 433, New York, New York 10016. PANEL PROCEEDINGS SERIES, STI/PUB/332, 1972.

THE ROLE OF FERTILIZATION IN THE INTENSIFICATION OF AGRICULTURAL PRODUCTION. Symposium. Sponsor: International Potash Institute, Antibes, France. International Potash Institute, Zieglerstrasse 30, CH-3000 Berne 14, Switzerland. Edited by M.G. Drouineau. 1970.

Sittig, M. AGRICULTURAL CHEMICALS MANUFACTURE 1971. Park Ridge, New Jersey: Noyes Data Corp.

"172 manufacturing process and product descriptions are given."

Slack, A. FERTILIZER DEVELOPMENTS AND TRENDS. Park Ridge, New Jersey: Noyes Data Corp., 1968.

"R & D Trends, Ammonia, Ammonium Nitrate, Sulfate, Urea, Slow Release Nitrogen, Other Nitrogen Fertilizers, Phosphoric Acid, Ammonium Phosphate, Nitric & Superphosphates, Thermal and Other Phosphate Processes, etc."

_____, ed. AMMONIA, Pt 1. FERTILIZER SCIENCE & TECHNOLOGY SERIES. Vol. 2. New York: Dekker, 1972.

SOIL MOISTURE & IRRIGATION STUDIES II. 2nd panel on the use of nuclear techniques in soil physics & irrigation studies. Sponsor: Joint FAO/IAEA Division of Atomic Energy in Food & Agriculture, Vienna, Austria. Publisher: International Atomic Energy Agency, Vienna. U.S. Distributor: Unipub, Inc., P.O. Box 433, New York, New York 10016. Panel proceedings series, STI/PUB/327. 1973.

Sowers, G. INTRODUCTION TO SOIL MECHANICS AND FOUNDATIONS. 3rd ed. New York: Macmillan, 1970.

An overall view of soil engineering. Useful to practitioners as well as those new to field.

SULPHUR IN AGRICULTURE (LE SOUFRE EN AGRICULTURE), International Symposium on. Sponsor: Institut National de la Recherche Agronomique, Association Francaise pour l'Etude du Sol, and Sulphur Institute, Versailles, France. Institut National de la Recherche Agronomique, Route de Saint Cry, F78-Versailles,

France. ANNALES AGRONOMIQUES, Numero hors series, 1972; Publication 72-1.

TRACE ELEMENTS IN SOILS & CROPS. Conference. Sponsor: National Agricultural Advisory Service, London, England. Publisher: Her Majesty's Stationery Office, London. U.S. Distributor: Pendragon House, 899 Broadway Avenue, Redwood City, California 94063. Minister of Agriculture, Fisheries & Food, TECHNICAL BULLETIN NO. 21. 1971.

Treshow, M. ENVIRONMENT AND PLANT RESPONSE. New York: McGraw-Hill, 1970.

"Considers all factors of the physical environment which affect plant development. Discusses such topics as air pollution and pesticides, and describes, on both morphological and physiological bases, the role of temperature, water, light, soil, and atmosphere on plant environment."

U.S. Atomic Energy Commission. Agricultural Research Laboratory operated by University of Tennessee Agricultural Experiment Station for Atomic Energy Commission, ANNUAL PROGRESS REPORT, JANUARY 1 - DECEMBER 31, 1971. July 1972.

U.S. Congress House Agriculture Committee. FOOD COSTS, FARM PRICES, COMPILATION OF INFORMATION RELATING TO AGRICULTURE. 92nd Congress, 1st Session; July 1, 1971. Committee print, 92nd Congress, 1st session. Y 4.Ag8/1: F 73/22/971.

U.S. Department of Agriculture. COOPERATIVES IN AGRIBUSINESS. Rev. ed. A89.4/2:33/2. Farmer Cooperative Service. Washington, D.C.: Government Printing Office, October 1972.

_____. DIRECTORY OF STATE-USDA RURAL DEVELOPMENT COMMITTEES, 1973. Prepared by Community Resource Development Staff. April 1973. A 43.2:R 88/7/973. Extension Service, Washington, D.C. 20250.

_____. DRAINAGE OF AGRICULTURAL LAND. Port Washington, New York: Water Information Center, 1973.

"A practical handbook for the planning, design, construction and maintenance of agricultural drainage systems."

_____. FAO: ITS ORGANIZATION AND WORK AND UNITED STATES PARTICIPATION. FASM93. By R.W. Phillips. Revised 1969. Foreign Agricultural Service, U.S. Department of Agriculture, Washington, D.C. 20520.

_____. KNOW YOUR AGRICULTURAL RESEARCH SERVICE. April, 1972. Agricultural Research Service, U.S. Department of Agriculture, Washington, D.C. 20250.

_____. PEANUT STOCKS AND PROCESSING SEASONAL REPORT, AUGUST 1970. July 1971. Revised September 1971. U.S. Department of Agriculture, Washington, D.C. 20250.

_____. RESPONSE OF SEVERAL CROPS TO 6 HERBICIDES IN IRRIGATION WATER, WITH LIST OF LITERATURE CITED. By V.E. Bruns, J.M. Hodgson, and H.F. Arle. Agricultural Research Service. Prepared in cooperation with Washington, Montana, and Arizona Agricultural Experiment Stations. Washington, D.C. Government Printing Office, December 1972.

_____. SCIENCE STUDY AID. CHEMICAL CONTROL OF PLANT GROWTH, WITH BIBLIOGRAPHY. Information Division, Agricultural Research Service. Washington, D.C.: Government Printing Office, December 1972.

_____. SYNTHETICS AND SUBSTITUTES FOR AGRICULTURAL PRODUCTS, PROJECTIONS FOR 1980. By W.W. Gallimore. March, 1972. Economic Research Service, U.S. Department of Agriculture, Washington, D.C. 20250.

_____. TOWARD THE NEW, REPORT ON BETTER FOODS AND NUTRITION FROM AGRICULTURAL RESEARCH. Prepared by Agricultural Research Service. Washington, D.C.: Government Printing Office, April 1970.

U.S. Environmental Protection Agency. PREDICTION MODELING FOR SALINITY CONTROL IN IRRIGATION RETURN FLOWS, STATE OF THE ART REVIEW WITH BIBLIOGRAPHIES. By A.G. Hornsby. Washington, D.C.: Government Printing Office, 1973.

U.S. National Agricultural Library. NATIONAL AGRICULTURAL LIBRARY. GUIDE TO SERVICE. Rev. ed. Washington, D.C.: October 1972.

U.S. Tennessee Valley Authority. FERTILIZER MARKETING IN CHANGING AGRICULTURE. Proceedings, Fertilizer Production & Marketing Conference, October 1-3, 1969, Memphis, Tennessee. Washington, D.C. 20444.

WASTE MANAGEMENT RESEARCH. 4th Agricultural Waste Management Conference. Sponsor: Cornell University, New York College of Agriculture & Life Sciences; and U.S. Environmental Protection Agency, Ithaca, New York. Graphics Management Corporation, 1101 16th Street, N.W., Washington, D.C. 20036, 1972.

7. Abstracting and Indexing Services

a. GENERAL

APPLIED SCIENCE AND TECHNOLOGY INDEX

CHEMICAL ABSTRACTS

ENGINEERING INDEX

See this subject in Section I of this guide for details on these publications.

b. SPECIFIC TO AGRICULTURAL ENGINEERING

ABSTRACTS FOR THE ADVANCEMENT OF INDUSTRIAL UTILIZATION OF

CEREAL GRAINS. Bimonthly. Washington State University, College of Engineering Research Division, Pullman, Washington 99163.

AGRICULTURAL RESEARCH INDEX; INCLUDING FORESTRY, FISHERIES AND FOOD. 5th ed. 2 vols. Guernsey, England: Francis Hodgson, 1970.

AGRONOMY ABSTRACTS. Annual. American Society of Agronomy, 677 South Segoe Road, Madison, Wisconsin 53711.

ANIMAL BREEDING ABSTRACTS. Quarterly. Commonwealth Bureau of Animal Breeding and Genetics. Commonwealth Agricultural Bureaux, Farnham Royal, Bucks, England.

> An abstract of world literature.

BIBLIOGRAPHY OF AGRICULTURE. U.S. Department of Agriculture, National Agricultural Library. CCM Information Corp., 909 Third Avenue, New York, New York 10022.

BIOLOGICAL AND AGRICULTURAL INDEX. Monthly except August. H.W. Wilson Co., 950 University Avenue, Bronx, New York 10452.

> A subject index to English language periodicals in fields related to biology and agriculture.

BULLETIN BIBLIOGRAPHIQUE INTERNATIONAL DE MACHINISME AGRICOLE/ INTERNATIONAL FARM MACHINERY ABSTRACTS. Quarterly. Commission Internationale de Genie Rural. Text in English and French. Centre International de Documentation du Machinisme Agricole, Parc de Tourvoie, 92 Antony, Hauts-De-Seine, France.

BULLETIN SIGNALETIQUE. Paris, CNRS. Section 380: Sciences agricoles, zootechnie, phytriatrie et phytopharmacie, aliments et industries alimentaires. Classified arrangement. Cumulative indexes.

CURRENT CONTENTS, AGRICULTURAL, FOOD & VETERINARY SCIENCES/ CC-AFV. Weekly. Institute for Scientific Information, 325 Chestnut Street, Philadelphia, Pennsylvania 19106. Author Index.

> "Reproduction of the contents pages of approximately 700 international journals."

DAIRY SCIENCE ABSTRACTS. Monthly. Commonwealth Bureau of Dairy Science and Technology. Commonwealth Agricultural Bureaux, Farnham Royal, Bucks, England.

FAO DOCUMENTATION - CURRENT INDEX. Monthly. Rome: FAO Documentation Center.

> Subject index to FAO publications; semiannual cumulations.

FERTILIZER ABSTRACTS. Monthly. Muscle Shoals, Alabama, Tennessee Valley Authority, National Fertilizer Development Center, Technical Library.

"Abstracts of articles from technical journals and patents prepared by TVA staff members, or selected from Chemical Abstracts."

FIELD CROP ABSTRACTS. Quarterly. Commonwealth Bureau of Pastures and Field Crops, Commonwealth Agricultural Bureaux, Farnham Royal, Bucks, England.

Compiled from world literature.

FOOD AND AGRICULTURE ORGANIZATION OF THE UNITED NATIONS. FAO agricultural development papers. Irregular. Text in English, French, and Spanish. Food and Agriculture Organization of the United Nations, via delle Terme di Caracalla, Rome 00100, Italy (distributed in the United States by Unipub, Inc., 650 First Avenue, P.O. Box 433, New York, New York 10016).

GEO ABSTRACTS. Section B. Biogeography & Climatology, GEO Abstracts, University of East Anglia, Norwich, NOR 88C, England. Six issues a year.

Contains references of interest to agricultural engineers.

HEALTH ASPECTS OF PESTICIDES ABSTRACT BULLETIN. Monthly. EP 5.9. Washington, D.C.: Government Printing Office.

HERBAGE ABSTRACTS. Quarterly. Commonwealth Bureau of Pastures and Field Crops. Commonwealth Agricultural Bureaux, Farnham Royal, Bucks, England.

Abstract journal on grassland husbandry and fodder crop production.

HORTICULTURAL ABSTRACTS. Quarterly. Commonwealth Bureau of Horticulture and Plantation Crops, East Malling, Kent. Commonwealth Agricultural Bureaux, Farnham Royal, Bucks, England.

Compiled from world literature on temperate and tropical fruits, vegetables, ornamentals, plantation crops.

PESTICIDES ABSTRACTS. Monthly. Washington, D.C.: Government Printing Office.

PLANT BREEDING ABSTRACTS. Quarterly. Commonwealth Bureau of Plant Breeding and Genetics. Commonwealth Agricultural Bureaus, Farnham Royal, Bucks, England.

Predi-Briefs. ABSTRACTING SERVICE. Monthly. AGRICULTURAL CHEMICALS. Predicasts, 200 University Circle Research Center, 11001 Cedar Aveue, Cleveland, Ohio 44106.

"This packet contains information on fertilizers, nitrogen materials (ammonia, urea, etc.) potash, phosphates, phosphate rock, sulfur, chemical and fertilizer minerals and pesticides."

REVIEW OF APPLIED ENTOMOLOGY. Monthly. Commonwealth Institute of Entomology, 56 Queen's Gate, London, SW 7, England.

REVIEW OF PLANT PATHOLOGY. Monthly. Commonwealth Mycological Institute, Ferry Lane, Kew, Surrey, England.

SCIENCE AND TECHNOLOGY RESEARCH ABSTRACTS: AGRICULTURE QUARTERLY. G.K. Hall & Co., 70 Lincoln Street, Boston, Massachusetts 02111.

SEARCH: FERTILIZERS DIVISION. Monthly. Compendium Publishers International Corporation, 2175 Lemoine Avenue, Fort Lee, New Jersey 07024.

SEARCH: PESTICIDES DIVISION. Monthly. Compendium Publishers International Corporation, 2175 Lemoine Avenue, Fort Lee, New Jersey 07024.

SOILS AND FERTILIZERS. Bimonthly. Commonwealth Bureau of Soils. Commonwealth Agricultural Bureaux, Farnham Royal, Bucks, England.

Abstracts of world literature.

WEED ABSTRACTS. Bimonthly. Agricultural Research Council, Weed Research Organization. Commonwealth Agricultural Bureaux, Farnham Royal, Slough, SL2 3BN, Bucks, England.

Compiled from world literature.

WEEKLY GOVERNMENT ABSTRACTS. AGRICULTURE AND FOOD. Washington, D.C.: Government Printing Office.

WORLD AGRICULTURAL ECONOMICS AND RURAL SOCIOLOGY ABSTRACTS. Quarterly. Commonwealth Agricultural Bureaux, Farnham Royal, Bucks, England.

Abstracts of world literature.

WORLD FISHERIES ABSTRACTS. Quarterly. Food and Agriculture Organization of the United Nations, Via delle Terme di Caracalla, Rome 00100, Italy.

c. RELATED ABSTRACTS IN CURRENT JOURNALS

The following journals frequently provide abstracts. See the heading Journals and Serial Publications in this section for additional details.

AGRICULTURE (CANADA)

DAIRY AND ICE CREAM FIELD

JOURNAL OF THE SCIENCE OF FOOD AND AGRICULTURE

8. Journals and Serial Publications

a. DIRECTORIES AND LISTS

AYER DIRECTORY OF PUBLICATIONS. Annual. Ayer Press, West Washington Square, Philadelphia, Pennsylvania 19106.

Boalch, D., ed. CURRENT AGRICULTURAL SERIALS; A WORLD LIST OF SERIALS IN AGRICULTURE AND RELATED SUBJECTS (EXCLUDING FORESTRY AND FISHERIES) CURRENT IN 1964. Oxford, England: International Association of Agricultural Librarians and Documentalists, 1967. 2 vols. Vol. 1, 1965; vol. 2, 1967.

Brown, D. "Agricultural periodicals." LIBRARY TRENDS, Vol. 10, (January 1962), pp. 405-13.

"Part of a special issue on Current Trends in U.S. Periodical Publishing."

STANDARD PERIODICALS DIRECTORY (and supplement). New York: Oxbridge Publishing Company.

ULRICH'S INTERNATIONAL PERIODICALS DIRECTORY. Annual. 2 vols. New York: R.R. Bowker Company.

b. JOURNAL AND SERIAL TITLES

ABSTRACTS FOR THE ADVANCEMENT OF INDUSTRIAL UTILIZATION OF CEREAL GRAINS. Bimonthly. Washington State University, College of Engineering Research Division, Pullman, Washington 99163.

ADC NEWSLETTER. 5-6/year. Edited by Arthur T. Mosher. Agricultural Development Council, Inc., 630 Fifth Avenue, New York, New York 10020.

ADVANCES IN AGRONOMY. Annual. Prepaid under the auspices of the American Society of Agronomy. New York: Academic Press.

AG-CEM MONTHLY. Cedar Grove, New Jersey: George F. Farrell.

AGRICHEMICAL AGE. Monthly. Beeler Publishing Co., 3030 Bridgeway, Sausalito, California 94965.

Coverage includes production and use of agrichemicals.

AGRICULTURAL ENGINEERING. Monthly. American Society of Agricultural Engineers, P.O. Box 229, St. Joseph, Michigan 49085.

AGRICULTURAL MACHINERY JOURNAL. Monthly. Agricultural Press Ltd., 161-166 Fleet Street, London, EC4, England.

AGRICULTURAL MARKETING. Monthly. Consumer and Marketing Service and Food and Nutrition Service, Agriculture Department. A 88.26/3. Washington, D.C.: Government Printing Office.

AGRICULTURAL OUTLOOK DIGEST. Monthly. Economic Research Service, U.S. Department of Agriculture, Washington, D.C. 20250.

Agricultural pesticide society. Annual. PROCEEDINGS. Annual Meeting. Agricultural Pesticide Society, Research Institute, Canada Agriculture University, Sub Post Office, London 72, Ontario, Canada.

AGRICULTURAL PRICES. Monthly. Crop Reporting Board, Statistical Reporting Service, Agriculture Department, A 92.16. Washington, D.C.: Government Printing Office.

AGRICULTURAL RESEARCH. Monthly. Agricultural Research Service, Agriculture Department, A 77.12. Washington, D.C.: Government Printing Office.

AGRICULTURAL RESEARCH INDEX. Irregular. 5th ed. 2 vols. Guernsey, England: Francis Hodgson, 1950.

   Including forestry, fisheries, and food..

AGRICULTURAL SCIENCE REVIEW. Quarterly. Edited by Ward W. Konkle. Cooperative State Research Service, Agriculture Department. A 94.11. Washington, D.C.: Government Printing Office.

AGRICULTURAL STATISTICS, ENGLAND AND WALES. Annual. Great Britain, Ministry of Agriculture, Fisheries, and Food. H.M.S.O., P.O. Box 569, London, SE 1, England.

AGRICULTURAL STATISTICS, UNITED KINGDOM. Annual. Great Britain. Ministry of Agriculture, Fisheries and Food; Department of Agriculture and Fisheries for Scotland; Ministry of Agriculture, Northern Ireland. H.M.S.O., P.O. Box 569, London, SE 1, England.

AGRICULTURE (CANADA). Text in English and French. Quarterly. Order of Agrologists of Quebec, 262 Henri-Bourassa Boulevard, W., Montreal 357, Quebec, Canada.

AGRICULTURE INFORMATION BULLETINS. Irregular. U.S. Department of Agriculture. Washington, D.C.: Government Printing Office.

AGRONOMY ABSTRACTS. Annual. American Society of Agronomy, 677 South Segoe Road, Madison, Wisconsin 53711.

ANIMAL BREEDING ABSTRACTS. Quarterly. Commonwealth Bureau of Animal Breeding and Genetics, Commonwealth Agricultural Bureaux, Farnham Royal, Bucks, England.

   An abstract of world literature.

ANIMAL NUTRITION AND HEALTH. Monthly. Beeler Publishing Corporation, 3030 Bridgeway, Sausalito, California 94965.

ANNUAL REVIEW OF ENTOMOLOGY. Vol. 18 (1973). Annual Reviews Inc., 4139 El Camino Way, Palo Alto, California 94306.

ANNUAL REVIEW OF PHYTOPATHOLOGY. Annual Reviews Inc., 4139 El Camino Way, Palo Alto, California 94306.

ANNUAL REVIEW OF PLANT PHYSIOLOGY. Vol. 24 (1973). Annual Reviews Inc., 4139 El Camino Way Palo Alto, California 94306.

ASAE TRANSACTIONS. Quarterly. American Society of Agricultural Engineers, 2950 Niles Road, Box 229, St. Joseph, Michigan 49085.

Association of American Pesticide Control Officials. OFFICAL PUBLICATION. Annual. Association of American Pesticide Control Officials, Inc., Kansas State Board of Agriculture, State Office Building, Topeka, Kansas 66612.

BIBLIOGRAPHY OF AGRICULTURE. Monthly. U.S. Department of Agriculture, National Agricultural Library. CCM Information Corporation, 909 Third Avenue, New York, New York 10022.

BIOLOGICAL AND AGRICULTURAL INDEX. Monthly except August. H.W. Wilson Company, 950 University Avenue, Bronx, New York 10452.

 A subject index to English language periodicals in fields related to biology and agriculture.

BULLETIN BIBLIOGRAPHIQUE INTERNATIONAL DE MACHINISME AGRICOLE/ INTERNATIONAL FARM MACHINERY ABSTRACTS. Quarterly. Commission Internationale de Gente Rural. Text in English and French. Centre International de Documentation du Machinisme Agricole, Parc de Tourvoie, 92 Antony, Hauts-De-Seine, France.

BULLETIN SIGNALETIQUE. Monthly. Section 380: Agronomie - Zootechnie - Phytopathologie - Industries - Alimentaires. 330 F. Centre National de La Recherche Scientifique 26 Rue Boyer 75 971, Paris, 20. France.

 Classified arrangement. Cumulative indexes.

CANADA AGRICULTURE (CANADA). Quarterly. Text in English and French. Canada Department of Agriculture, Ottawa, Ontario, Canada.

Canada. Bureau of Statistics. INDEX OF FARM PRODUCTION. Annual. Bureau of Statistics, Ottawa, Canada.

Canada. Department of Agriculture. Analytical Chemistry Research Service. RESEARCH REPORT. Irregular, 1959-66. Department of Agriculture, Analytical Chemistry Research Service, Ottawa, Canada.

CANADIAN FARM EQUIPMENT DEALER. Monthly. Southam Business Publications Ltd., 1450 Don Mills Road, Ontario, Canada.

CHEMURGIC DIGEST. Quarterly. Council for Agricultural and Chemurgic Research, 350 Fifth Avenue, New York, New York 10001.

COMMERCIAL REVIEW. Weekly. Commercial Review, Inc., 1812 N.W. Kearney Street, Portland, Oregon 97209.

CURRENT CONTENTS, AGRICULTURAL, FOOD & VETERINARY SCIENCES/ CC-AFV. Weekly. Institute for Scientific Information, 325 Chestnut Street, Philadelphia, Pennsylvania 19106.

"Reproduction of the contents pages of approximately 700 international journals."

CURRENT INDUSTRIAL REPORTS. Monthly. Commerce Department, Social and Economic Statistics Administration, Census Bureau. Washington, D.C.: Government Printing Office.

Includes reports of interest to agricultural industry.

DAIRY AND ICE CREAM FIELD. Monthly. Magazines for Industry, 777 Third Avenue, New York, New York 10017.

DAIRY SCIENCE ABSTRACTS. Monthly. Commonwealth Bureau of Dairy Science and Technology. Commonwealth Agricultural Bureaux, Farnham Royal, Bucks, England.

ERDA/ENGINEERING RESEARCH AND DEVELOPEMENT IN AGRICULTURE. Quarterly. Supplements. Canada Department of Agriculture, Engineering Research Service, Research Branch, Ottowa 3, Ontario, Canada.

FAO DOCUMENTATION - CURRENT INDEX. Monthly. Rome: FAO Documentation Center.

Subject index to FAO publications; semiannual cumulations.

FARM AND POWER EQUIPMENT. Monthly. NRFEA Publications, Inc., 2340 Hampton Avenue, St. Louis, Missouri 63139.

Trade publication for the industry.

FARM CHEMICALS. Monthly. Farm Chemicals, 37841 Euclid Avenue, Willoughby, Ohio 44094.

Includes research uses and marketing of production farm chemicals.

FARM ENGINEERING INDUSTRY. Monthly. Farm Engineering Industry Publications, Ltd., 64a Lansdowne Road, South Woodford, London, E. 18, England.

FARM EQUIPMENT. Monthly. Johnson Hill's Press, Inc., 1233 Janesville Avenue, Fort Atkinson, Wisconsin 53538.

FARM INDUSTRY NEWS. 8 issues/year. Webb Publishing Company, 1999 Shepard Road, St. Paul, Minnesota 55116.

FARM TECHNOLOGY. 5 issues/year. Meister Publishing Company, 37841 Euclid Avenue, Wiloughby, Ohio 44094.

FEED/GRAIN EQUIPMENT TIMES. 10 issues/year. Johnson Hill's Press, Inc., 1233 Janesville Avenue, Fort Atkinson, Wisconsin 53538.

FEEDLOT MANAGEMENT. Monthly. Miller Publishing Company, 2501 Wayzata Boulevard, Minneapolis, Minnesota 55440.

FERTILIZER ABSTRACTS. Monthly. Muscle Shoals, Alabama, Tennessee Valley Authority, National Fertilizer Development Center, Technical Library.

> "Abstracts of articles from technical journals and patents prepared by TVA staff members, or selected from Chemical Abstracts."

FERTILIZER PROGRESS. Bimonthly. Fertilizer Institute, 1015 18th St. N.W., Washington, D.C. 20036.

FERTILIZER SOLUTIONS. Bimonthly. National Fertilizer Solutions Association, 910 Lehmann Building, Peoria, Illinois 61602.

FIELD CROP ABSTRACTS. Quarterly. Commonwealth Bureau of Pastures and Field Crops. Commonwealth Agricultural Bureaux, Farnham Royal, Bucks, England.

> Compiled from world literature.

Food and Agriculture Organization of the United Nations. CONFERENCE REPORT. Biennial. 15th session. Food and Agriculture Organization of the United Nations, Via delle Terme di Caracalla, Rome 00100, Italy. Distributed in the United States by UNIPUB, Inc., 650 First Avenue, P.O. Box 433, New York, New York 10016.

_____. FAO AGRICULTURAL DEVELOPMENT PAPERS. Irregular. Text in English, French and Spanish. 1969, no. 91. Food and Agriculture Organization of the United Nations, Via delle Terme di Caracalla, Rome 00100, Italy. Distributed in the United States by UNIPUB, Inc., 650 First Avenue, P.O. Box 433, New York, New York 10016.

_____. FAO AGRICULTURAL STUDIES. Irregular. Text in English, French, and Spanish. 1969, no. 81. Food and Agriculture Organization of the United Nations, Via delle Terme di Caracalla, Rome 00100, Italy. Distributed in the United States by UNIPUB, Inc., 650 First Avenue, P.O. Box 433, New York, New York 10016.

FOOD RESEARCH INSTITUTE STUDIES IN AGRICULTURAL ECONOMICS, TRADE, AND DEVELOPMENT. 3 vols./year. Food Research Institute, Stanford University, Stanford, California 94305.

FOREIGN AGRICULTURE. Weekly. Department of Agriculture, Foreign Agricultural Service, Government Printing Office.

From the state capitols reports:

AGRICULTURE AND FOOD PRODUCTS. Weekly. Bethune Jones, 321 Sunset Avenue, Asbury Park, New Jersey 07712.

> "Concerned with regulation of production processing and marketing of food products."

GEO ABSTRACTS. Section B - Biography and Climatology. 6 issues/year.
GEO abstracts, University of East Anglia, Norwich, NOR 88C, England.

   Contains references of interest to agricultural engineers.

HARDWARE AND FARM EQUIPMENT. Monthly. Western Retail Implement
and Hardware Association, 638 - 40th West 39th Street, Kansas City, Missouri
64111.

HEALTH ASPECTS OF PESTICIDES ABSTRACT BULLETIN. Monthly. Washington, D.C.: Government Printing Office.

HERBAGE ABSTRACTS. Quarterly. Commonwealth Bureau of Pastures and
Field Crops, Commonwealth Agricultural Bureaux, Farnham Royal, Bucks,
England.

   An abstract journal on grassland husbandry and fodder crop
   production.

HORTICULTURAL ABSTRACTS. Quarterly. Commonwealth Bureau of Horticulture and Planatation Crops, East Malling, Kent. Commonwealth Agricultural Bureaux, Farnham Royal, Bucks, England.

   Compiled from world literature on temperate and tropical fruits,
   vegetables, ornamentals, plantation crops.

INTERNATIONAL SOCIETY OF SOIL SCIENCE BULLETIN. Semiannual.
Text in English, French and German. International Society of Soil Science,
63 Mauritskade, Amsterdam, Netherlands.

IRRIGATION AND POWER ABSTRACTS. Bimonthly. Central Board of Irrigation and Power, Kasturba Gandhi Marg, New Delhi, India.

JOURNAL OF AGRICULTURAL AND FOOD CHEMISTRY. Bimonthly. American Chemical Society, 1155 - 16th Street, N.W., Washington, D.C. 20036.

JOURNAL OF AGRICULTURAL ENGINEERING RESEARCH. Quarterly.
British Society for Research in Agricultural Engineering. Academic Press
Inc., 111 Fifth Avenue, New York, New York 10003.

JOURNAL OF AGRICULTURAL SCIENCE. Bimonthly. Cambridge University
Press, 32 E. 57th Street, New York, New York 10022.

JOURNAL OF SOIL SCIENCE (British Society of Soil Science). Quarterly.
Edinburgh School of Agriculture, West Mains Road, Edinburgh, EH9 3JG,
Scotland.

JOURNAL OF STORED PRODUCTS RESEARCH. Vol. 6, 1970. Quarterly.
Pergamon Press Inc., Journals Department, Maxwell House, Fairview Park,
Elmsford, New York 10523.

JOURNAL OF THE SCIENCE OF FOOD AND AGRICULTURE. Monthly.

Including abstracts. Society of Chemical Industry, 14 Belgrave Square, London, SW 1, England.

PESTICIDE BIOCHEMISTRY AND PHYSIOLOGY. Quarterly. Academic Press, Inc., 111 Fifth Avenue, New York, New York 10003.

An international journal.

PESTICIDE SCIENCE. Bimonthly. Society of Chemical Industry, 14 Belgrave Square, London SW 1, England.

A journal of international research and technology on crop protection and pest control.

PLANT BREEDING ABSTRACTS. Quarterly. Commonwealth Bureau of Plant Breeding and Genetics. Commonwealth Agricultural Bureaux, Farnham Royal, Bucks, England.

Predi-briefs. ABSTRACTING SERVICE. Monthly. AGRICULTURAL CHEMICALS. Predicasts, 200 University Circle Research Center, 11001 Cedar Avenue, Cleveland, Ohio 44106.

"This packet contains information on fertilizers, nitrogen materials (ammonia, urea, .etc.) potash, phosphates, phosphate rock, sulfur, chemical and fertilizer minerals and pesticides."

RESIDUE REVIEWS. Irregular. Springer-Verlag, 175 Fifth Avenue, New York, New York 10010.

"Residues of pesticides and other foreign chemicals in foods and feeds."

REVIEW OF APPLIED ENTOMOLOGY. Monthly. Series A: Agricultural. Commonwealth Institute of Entomology, 56 Queen's Gate, London, SW 7 England.

REVIEW OF PLANT PATHOLOGY. Monthly. Commonwealth Mycological Institute, Ferry Lane, Kew, Surrey, England.

SEARCH: FERTILIZERS DIVISION. Monthly. Compendium Publishers International Corporation, 2175 Lemoine Avenue, Fort Lee, New Jersey 07024.

SEARCH: PESTICIDES DIVISION. Monthly. Compendium Publishers International Corporation, 2175 Lemoine Avenue, Fort Lee, New Jersey 07024.

SOIL SCIENCE. Monthly. Williams & Wilkins Company, 428 East Preston Street, Baltimore, Maryland 21202.

SOIL SCIENCE AND PLANT NUTRITION. Quarterly. Japan Publications Trading Company Ltd., Box 5030, Tokyo International, Tokyo, Japan.

SOILS AND FERTILIZERS. Bimonthly. Commonwealth Bureau of Soils. Commonwealth Agricultural Bureaux, Farnham Royal, Bucks, England.

Abstracts of world literature.

Soil Science Society of America. PROCEEDINGS. Bimonthly. Soil Science Society of America, 677 South Segoe Road, Madison, Wisconsin 53711.

U.S. AGRICULTURAL RESEARCH SERVICE ANNUAL REPORT. Agricultural Research Service, Publications Branch, Room 5149, South Building, Washington, D.C. 20250.

U.S. Department of Agriculture. AGRICULTURAL STATISTICS. Annual. Washington, D.C.: Government Printing Office.

_____. REPORT OF THE SECRETARY OF AGRICULTURE. Annual. Office of Management Services, Division of Information, Department of Agriculture, Room 1447, South Building, Washington, D.C. 20250.

_____. TECHNICAL BULLETINS. Irregular. Washington, D.C.: Government Printing Office.

U.S. Soil Conservation Service. SOIL SURVEY INVESTIGATION REPORTS. Irregular. Washington, D.C.: Government Printing Office.

_____. TECHINCAL PUBLICATIONS. Irregular. U.S. Soil Conservation Service, c/o Department of Agriculture, Washington, D.C. 20250.

U.S. Tennessee Valley Authority. CHEMICAL ENGINEERING BULLETINS. Irregular. Tennessee Valley Authority, Public Information Office, Knoxville, Tennessee 37902.

UPDATE/AGRICULTURAL CHEMICALS. Biweekly. Information Company of America, 1011 Lewis Tower, 225 South 15th Street, Philadelphia, Pennsylvania 19102.

WEED ABSTRACTS. Bimonthly. Agricultural Research Council, Weed Research Organization. Commonwealth Agricultural Bureaux, Farnham Royal, Slough, SL2 3BN, Bucks, England.

Compiled from world literature.

WORLD AGRICULTURAL ECONOMICS AND RURAL SOCIOLOGY ABTRACTS. Quarterly. Commonwealth Agricultural Bureaux, Farnham Royal, Bucks, England.

Abstracts of world literature.

WORLD FISHERIES ABSTRACTS. Quarterly. Food and Agriculture Organization of the United Nations, Via delle Terme di Caracalla, Rome 00100, Italy.

9. Reviewing Services

ADVANCES IN AGRONOMY. Annual. New York: Academic Press.

Prepared under the auspices of the American Society of Agronomy.

ANNUAL REVIEW OF ENTOMOLOGY. Vol. 18 (1973). Annual Reviews Inc., 4139 El Camino Way, Palo Alto, California 94306.

ANNUAL REVIEW OF PHYTOPATHOLOGY. Annual Reviews Inc., 4139 El Camino Way, Palo Alto, California 94306.

ANNUAL REVIEW OF PLANT PHYSIOLOGY. Annual Reviews Inc., 4139 El Camino Way, Palo Alto, California 94306.

RESIDUE REVIEWS. Irregular. New York: Springer-Verlag.

U.S. Department of Agriculture. AGRICULTURAL INFORMATION BULLETIN. Prepared by Elizabeth D. Bollt, et al. A1.75:355. Washington, D.C.: Government Printing Office, June 1972.

> Consumer products by design, reports on new foods, fabrics, and materials from agricultural research.

10. Document and Report Literature

a. LISTS

MONTHLY CATALOG. U.S. Superintendent of Documents. Government Printing Office.

U.S. GOVERNMENT REPORTS ANNOUNCEMENTS AND U.S. GOVERNMENT REPORTS INDEX. Monthly publication of the National Technical Information Service, Springfield, Virginia 22151.

b. CLEARINGHOUSES

U.S. Superintendent of Documents. Government Printing Office, Washington, D.C. 20402.

National Agricultural Library, Beltsville, Maryland 20705.

National Technical Information Service, Springfield, Virginia 22151.

11. Sources of Information on Masters and Doctoral Theses

DISSERTATION ABSTRACTS INTERNATIONAL. Section B. Physical Sciences and Technology, University Microfilms, Xerox Company, 300 North Zeeb Road, Ann Arbor, Michigan 48106.

MASTERS THESES IN THE PURE AND APPLIED SCIENCES. Accepted by colleges and universities of the United States. Lafayette, Indiana, School of Mechanical Engineering, Purdue University. Annual. (Thermophysical Properties Research Center.)

12. Sources of Information on Statistics and Costs

a. STATISTICS

i. Guides and Handbooks

U.S. Department of Agriculture. MAJOR STATISTICAL SERIES OF THE U.S. DEPARTMENT OF AGRICULTURE: HOW THEY ARE CONSTRUCTED AND USED. AGRICULTURE HANDBOOK NO. 365. 1970-72. A 1.76:365/v. 1-11. Volume 1. AGRICULTURE PRICES AND PARITY; Volume 2. AGRICULTURAL PRODUCTION AND EFFICIENCY; Volume 3. GROSS AND NET FARM INCOME; Volume 4. AGRICULTURAL MARKETING COSTS AND CHARGES; Volume 5. CONSUMPTION AND UTILIZATION OF AGRICULTURAL PRODUCTS; Volume 6. LAND VALUES AND FARM FINANCE; Volume 7. FARM POPULATION AND EMPLOYMENT; Volume 8. CROP AND LIVESTOCK ESTIMATES; Volume 9. FARMER COOPERATIVES; Volume 10. MARKET NEWS; Volume 11. FOREIGN TRADE, PRODUCTION, AND CONSUMPTION OF AGRICULTURAL PRODUCTS. Washington, D.C.: Government Printing Office.

AGRICULTURAL STATISTICS, ENGLAND AND WALES. Annual. Great Britain, Ministry of Agriculture, Fisheries, and Food. H.M.S.O., P.O.B. 569, London, SE 1, England.

AGRICULTURAL STATISTICS, UNITED KINGDOM. Annual. Great Britain, Ministry of Agriculture, Fisheries, and Food; Department of Agriculture and Fisheries for Scotland; Ministry of Agriculture Northern Ireland. H.M.S.O., P.O.B. 569, London, SE 1, England.

Canada. Bureau of Statistics. INDEX OF FARM PRODUCTION. Annual. Bureau of Statistics, Ottawa, Canada.

U.S. Department of Agriculture. AGRICULTURAL RESEARCH. Monthly. Washington, D.C.: Government Printing Office.

_____. AGRICULTURAL STATISTICS. Annual. Washington, D.C: Government Printing Office.

U.S. Department of Commerce, Census Bureau. FARM MACHINES AND EQUIPMENT, 1971. Prepared in Industry Division. Washington, D.C.: Government Printing Office, September 1972.

WORLD AGRICULTURAL PRODUCTION AND TRADE STATISTICAL REPORT. Monthly. U.S. Department of Agriculture, Foreign Agriculture Service, Information Service Branch, Washington, D.C. 20250.

ii. Current Journals
Which Include Related Statistical Information

The following journals frequently contain statistics related to this field. See the heading Journals and Serial Publications in this section for further details.

ADC NEWSLETTER

AGRICULTURE (CANADA)

CANADA AGRICULTURE

CANADIAN FARM EQUIPMENT DEALER

DAIRY SCIENCE ABSTRACTS

FARM AND POWER EQUIPMENT

FERTILIZER SOLUTIONS

FOOD RESEARCH INSTITUTE STUDIES IN AGRICULTURAL ECONOMICS, TRADE AND DEVELOPMENT

HARDWARE AND FARM EQUIPMENT

JOURNAL OF STORED PRODUCTS RESEARCH

SEARCH: FERTILIZERS DIVISION

SEARCH: PESTICIDES DIVISION

SOIL SCIENCE

WEED ABSTRACTS

b. COSTS

BIBLIOGRAPHY OF AGRICULTURE. Monthly. National Agricultural Library. The Oryx Press, 3930 East Camelback Road, Phoenix, Arizona 85018.

WORLD-PRODUCT COSTS. Predicasts, Inc., 200 University Circle Research Center, 11001 Cedar Avenue, Cleveland, Ohio 44106.

> A subscription service of market forecasts by product in areas of agriculture, wood and paper, food, chemicals, polymers, primary metals, etc.

See also this subject in Section I of this Guide.

13. Sources of Information on Standards, Specifications, and Patents

a. STANDARDS AND SPECIFICATIONS

i. Guides and Handbooks

Struglia Erasmus J. STANDARDS AND SPECIFICATIONS INFORMATION

SOURCES. Detroit: Gale Research Co., 1965.

Coverage is both general and specific to agricultural engineering.

U.S. Department of Commerce. National Bureau of Standards. DIRECTORY OF U.S. STANDARDIZATION ACTIVITIES. Washington, D.C.: Government Printing Office, 1975.

Coverage is both general and specific to agricultural engineering.

Specifications, standards, codes, and regulations are available for individual states from the official state document printing offices and from local and municipal government agencies.

See also section Information Sources on Standards and Patents in Section I of this Guide.

b. PATENTS

i. Guides and Handbooks

Hendrie, R. GRANULATED FERTILIZERS. Park Ridge, New Jersey: Noyes Data Corp., 1976.

Contains patent information.

U.S. Department of Agriculture. LIST OF PUBLICATIONS AND PATENTS WITH ABSTRACTS. Semiannual. Western Marketing and Nutrition Research Division, Berkeley, California 94710.

_____. PUBLICATIONS AND PATENTS, JANUARY-JUNE 1972. Eastern Marketing and Nutrition Research Division, publication no. 3613-3708. July 1972. A 77.17/4:972. Agricultural Research Service, Washington, D.C. 20250.

UNITED STATES FERTILIZER PATENTS. Annual. Park Ridge, New Jersey: Noyes Development Corp.

U.S. Government Printing Office. PRICE LIST SERIES No. 62. Commerce. BUSINESS, PATENTS, TRADEMARKS, AND FOREIGN TRADE. Washington, D.C.: 1975.

Contains list of documents for sale.

ii. Current Journals Which Include Related Patent Listings

The following journals frequently list patents. See the heading Journals and Serial Publications in this section for additional details.

DAIRY SCIENCE ABSTRACTS

FERTILIZER ABSTRACTS

14. Sources of Information on
Translations, Meetings, and Proceedings of Symposia

a. TRANSLATIONS

National Institute of Agricultural Engineering. TRANSLATIONS. Irregular. National Institute of Agricultural Engineering, Wrest Park, Silsoe, Bedford, England.

U.S. Joint Publications Research Service. Arlington, Virginia. Publishes translations of scientific and technical materials from Eastern Europe, Russia, and China. These are listed in the MONTHLY CATALOG of the Superintendent of Documents.

See also this subject in the Section I of this Guide.

15. Selected Literature Sources on
Pollution Control and Safety

a. POLLUTION CONTROL

Agricultural waste management conference. PROCEEDINGS. Annual. Cornell University College of Agriculture, Waste Management Program, Riley Robb Hall, Cornell University, Ithaca, New York 14850.

Agriculture and pollution seminar, 4th. Sponsor: University of Arizona, Tucson, Arizona. Engineering Experiment Station, College of Engineering, University of Arizona, Tucson, Arizona 85721. Report no. 35, 1972.

Agriculture and processing wastes in western region, eastern experiment station collaborators' conference. Philadelphia, Pennsylvania, December 1-3, 1970. Sponsored by the U.S. Department of Agriculture, Agricultural Research Service, Eastern Marketing and Nutrition Research Division. Proceedings issued as Pub 3457 (1971), U.S. Department of Agriculture Eastern Marketing and Nutrition Research Division, Philadelphia, Pennsylvania, 1971.

BIOLOGICAL IMPACT OF PESTICIDES IN THE ENVIRONMENT. Symposium, Corvallis, Oregon, August 18-20, 1969. Sponsored by Oregon State University, Environmental Health Sciences Center. Available from: Corvallis: Oregon State University, Environmental Health Sciences Center, 1970.

CHEMISTRY AND TOXICOLOGY OF AGRICULTURAL CHEMICALS. Food Protection and Toxicology Center, University of California, Davis, California, 1969.

A report of a survey of chemicals in plants and animals.

Ciaccio, L., ed. WATER AND RAW POLLUTION HANDBOOK. 4 vols.
New York: Dekker, 1972.

Of interest to agricultural engineers. Concerns itself with irrigation, natural water pollution, etc.

CONVERSION OF CATTLE FEEDLOT WASTES TO AMMONIA SYNTHESIS
GAS. By J.E. Halligan, et al. Washington, D.C.: Government Printing
Office, 1974.

Development Planning and Research Associates. WATER POLLUTION CONTROL
ACT OF 1972, TECHNOLOGIES AND ECONOMIC IMPACTS. The Feedlot
Industry Rept. no. PB251 214. Springfield, Virginia: National Technical
Information Service, 1975.

ENERGY, AGRICULTURE AND WASTE MANAGEMENT. 7th Agricultural
Waste Management Conference. Proceedings. Ithaca, New York, April 1975.
Ann Arbor, Michigan: Ann Arbor Publishers, 1975.

Fort Valley State College. Division of Agriculture. WATER RESOURCES:
UTILIZATION AND CONSERVATION IN THE ENVIRONMENT. By M.
Blount. Rept. no. PB 247 612. Springfield, Virginia: National Technical
Information Service, 1975.

Graves, R., ed. PROCEEDINGS OF FARM ANIMAL WASTE CONFERENCE;
FARMERS EXPERIENCES, CODES, GUIDELINES, RESEARCH, PROGRESS.
Madison, Wisconsin: University of Wisconsin, 1972.

International IUPAC. Congress of pesticide chemistry, 2nd, Tel-Aviv, 1971.
METHODS IN RESIDUE ANALYSIS; PROCEEDINGS. Edited by A.S. Tahori.
New York: Gordon & Breach, 1971.

Iowa State University. Center for Agricultural and Rural Development. ECO-
NOMIC ASPECTS OF WATER POLLUTION CONTROL ACT OF 1972: IR-
RIGATED AND NON-IRRIGATED AGRICULTURE. Rept. no. PB248 807.
Springfield, Virginia: National Technical Information Service, 1975.

Micronutrients in agriculture symposium. Muscle Shoals, Alabama, April 20-
22, 1971. Sponsored by TVA, and Soil Science Society of America. PRO-
CEEDINGS available from: Soil Science Society of America, Madison,
Wisconsin.

Naegele, J. AIR POLLUTION IN AGRICULTURE. New ed. Advances in
Chemistry Series, Washington, D.C.: American Chemical Society, 1973.

Tahori, A. PESTICIDE CHEMISTRY. 6 vols. New York: Gordon & Breach,
1972.

"Report comprises the Proceedings of the Second International Congress of Pesticide Chemistry sponsored by the International Union of Pure and Applied Chemistry under the auspices of Israel's National Council for Research and Development."

U.S. Congress, Senate appropriations committee. AGRICULTURE-ENVIRON-MENTAL AND CONSUMER PROTECTION APPROPRIATIONS FOR FISCAL YEAR 1973, HEARINGS BEFORE SUBCOMMITTEE, 92d Congress, 2d session. Pt. 1. Secretary of Agriculture, agencies of the Department, Environmental Protection Agency, Farm Credit Administration, Food and Drug Administration, Food and Nutrition Service. 1972. Washington, D.C. 20510.

U.S. Department of Agriculture. AGRICULTURE HANDBOOK. 380. By R. Lillie. Washington, D.C.: Government Printing Office, August 1970.

Air pollutants affecting performance of domestic animals, literature review with lists of literature cited.

_____. 965. POULTRY PROCESSING INDUSTRY, STUDY OF IMPACT OF WATER POLLUTION CONTROL COSTS WITH LIST OF REFERENCES: By J. Vertrees. June 1972. Economic Research Service, Washington, D.C. 20510.

U.S. Department of Health, Education, and Welfare. PESTICIDE ANALYTICAL MANUAL: V.1, METHODS WHICH DETECT MULTIPLE RESIDUES. December 31, 1971 compiled reprint. Issued in looseleaf form. HE 20.4008: P 43/v.1/971.--V.2, METHODS FOR INDIVIDUAL PESTICIDE RESULTS, 1971. Complete reprint: compilation of all revisions and additions issued as of December 31, 1971. Pages, dates, and regulations section notations are unchanged from original issuances. Issued in looseleaf form. Washington, D.C.: Government Printing Office.

U.S. Environmental Protection Agency. 659. AGRICULTURAL BENEFITS AND ENVIRONMENTAL CHANGES RESULTING FROM USE OF DIGESTED SEWAGE SLUDGE ON FIELD CROPS. By T.D. Hinesly. Environmental Protection Agency, 5555 Ridge Avenue, Cincinnati, Ohio 45213.

Interim report on solid waste demonstration project with list of references.

_____. 72-101. BIBLIOGRAPHY OF LIVESTOCK WASTE MANAGEMENT WITH LIST OF REFERENCES. By J.R. Miner. Washington, D.C.: Government Printing Office, December 1973.

_____. 72-061. CHARACTERISTICS OF RAINFALL RUNOFF FROM BEEF CATTLE FEEDLOT WITH LIST OF REFERENCES. By R.D. Kreis. Water Research Center, Ada, Oklahoma. National Environmental Research Center. Washington, D.C.: Government Printing Office, September 1972.

_____. 13040 DEM 01/71. CHARACTERISTICS OF WASTES FROM SOUTH-WESTERN CATTLE FEEDLOTS WITH LIST OF LITERATURE CITED. By Texas

Tech University Water Resources Center, Lubbock, Texas. Washington, D.C: Government Printing Office.

_____. 73-095. DEVELOPMENT AND DEMONSTRATION OF NUTRIENT REMOVAL FROM ANIMAL WASTES WITH LIST OF REFERENCES. By R.C. Loehr, et al. Washington, D.C: Government Printing Office, 1973.

_____. 73-096. DISPOSAL OF CATTLE FEEDLOT WASTES BY PYROLYSIS WITH LIST OF REFERENCES. By W. Garner. Washington, D.C.: Government Printing Office, 1973.

_____. EPA COMPENDIUM OF REGISTERED PESTICIDES. Irregular. Issued in looseleaf form including basic manual and supplementary material for an indefinite period. Available on subscription basis. Pesticides Program Office, Chamblee, Georgia 30341.

_____. 5418. FEEDLOT WASTE MANAGEMENT, WHY AND HOW. Prepared by Missouri River Basin Animal Waste Management Pilot (Steering) Task Group, Kansas City, Missouri, June 1971.

_____. 16080 DMP 03/71. INTERACTION OF HERBICIDES AND SOIL MICRO-ORGANISMS WITH BIBLIOGRAPHIES. By D.C. Torgeson. Washington, D.C.: Government Printing Office, 1971.

_____. 13020 DPB 08/71. MANAGEMENT OF NUTRIENTS ON AGRICULTURAL LAND FOR IMPROVED WATER QUALITY WITH LIST OF REFERENCES. By Department of Agronomy, College of Agriculture and Sciences, Cornell University, Ithaca, New York. Washington, D.C.: Government Printing Office, August 1971.

_____. MANAGING IRRIGATED AGRICULTURE TO IMPROVE WATER QUALITY WITH LISTS OF REFERENCES. Proceedings of National Conference on Managing Irrigated Agriculture to Improve Water Quality, sponsored by Environmental Protection Agency and Colorado State University, May 16-18, 1972. Washington, D.C.: Government Printing Office, 1973.

_____. MERCURIAL PESTICIDES, MAN, AND ENVIRONMENT. 1971. Special Pesticide Review Group, Office of Pesticide Programs, Washington, D.C. 20460.

_____. 73-157. NATIONAL ANIMAL FEEDLOT WASTES RESEARCH PROGRAM WITH LIST OF REFERENCES. National Environmental Research Center, Project 13040 GJU, Program element 1B2039, Washington, D.C.: Government Office, February 1973.

_____. 16060 DOE 04/72. NITRATE IN UNSATURATED ZONE UNDER AGRICULTURAL LANDS WITH LIST OF REFERENCES. By P.F. Pratt. Washington, D.C.: Government Printing Office, April 1972.

_____. 73-171. PROJECTS OF AGRICULTURAL AND MARINE POLLUTION CONTROL SECTION. Washington, D.C.: Government Printing Office, March 1973.

_____. 13020 DGX 08/71. ROLE OF ANIMAL WASTES IN AGRICULTURAL LAND RUNOFF WITH LIST OF REFERENCES. By J.W.D. Robbins. Washington, D.C.: Government Printing Office, August 1971.

_____. WATER POLLUTION CONTROL RESEARCH SERIES. BIO-ENGINEERING ASPECTS OF AGRICULTURAL DRAINAGE, SAN JOAQUIN VALLEY, CALIFORNIA: DESALINATION OF AGRICULTURAL TILE DRAINAGE WITH BIBLIOGRAPHIES. By B.R. Sword. EP 1.16: (nos.) Washington, D.C.: Government Printing Office, May 1971.

U.S. Environmental Protection Agency. Office of Water and Hazardous Materials. Effluent Guidelines Division. BASIC FERTILIZER CHEMICALS SEGMENT OF FERTILIZER MANUFACTURING POINT SOURCE CATEGOR. Washington, D.C.: Government Printing Office, 1974.

U.S. National Industrial Pollution Control Council. FERTILIZERS AND AGRICULTURAL CHEMICALS, SUB-COUNCIL REPORT. Y 3.In 2/8:2F41. Washington, D.C.: Government Printing Office, October 1971.

White-Stevens, R., ed. PESTICIDES IN THE ENVIRONMENT. Vol. I (in two parts), Part II. New York: Dekker, 1971.

> "The analysis of technical pesticides and their formulations, assay procedures for pesticide residues. Tocicology of pesticides to establish proof of safety. Pest resistance to pesticides."

b. SAFETY

LEGAL GUIDE TO OCCUPATION SAFETY AND HEALTH MONTHLY. Clark Boardman Ltd., 435 Hudson Street, New York, New York 10014.

> Comment, interpretation and advice on new legal developments in occupation safety and health.

Peck, Theodore. OCCUPATIONAL SAFETY AND HEALTH: A GUIDE TO INFORMATION SOURCES. Detroit: Gale Research Company, 1974.

> Has information on agriculture.

U.S. Department of Agriculture. SAFE USE AND STORAGE OF FLAMMABLE LIQUIDS AND GASES ON THE FARM. Rev. ed. Washington, D.C.: Government Printing Office, 1968.

> Gives precautions to observe in using and storing gasoline, kerosene, and other flammable liquids and gases.

U.S. HEW Public Health Service. TOXICITY BIBLIOGRAPHY. Quarterly. Washington, D.C.: Government Printing Office.

U.S. Department of Labor. SAFETY AND HEALTH STANDARDS FOR AGRI-CULTURE, 1971. Information for publication was extracted from vol. 36, no. 105, of Federal Register, May 29, 1971. Occupational Safety and Health Administration, Labor Department, Washington, D.C. 20210.

i. Journals and Serial Publications

FARM SAFETY REVIEW MAGAZINE. Bimonthly. National Safety Council, 425 N. Michigan Avenue, Chicago, Illinois 60611.

Provides information on farm safety.

PESTICIDE ABSTRACTS. Monthly. Washington, D.C.: Government Printing Office.

PESTICIDES MONITORING JOURNAL. Quarterly. Washington, D.C.: Government Printing Office.

See also this subject in Section I of this Guide.

16. Management Information Sources

See this topic in Section I of this Guide.

17. Sources of Information on Laws and Legislation in This Field

a. FEDERAL

FEDERAL REGISTER. Daily publication listing new regulations and revisions. Updates Code of Federal Regulations. Available on a subscription basis from Government Printing Office.

Library of Congress. DIGEST OF PUBLIC GENERAL BILLS AND REGULATIONS. Available on subscription basis from Government Printing Office.

UNITED STATES CODE. 1970 edition and supplements. Washington, D.C.: Government Printing Office.

Code of Federal Regulations. Continuing series issued throughout the year. Available on a subscription basis from Government Printing Office.

U.S. Congress. House of Representatives. LAWS RELATING TO AGRICUL-TURE, 1884-1971. Y1.2:AG8/5/971. Washington, D.C.: Government Printing Office, 1971.

U.S. Government Printing Office. PRICE LIST SERIES No. 10. LAWS, RULES AND REGULATIONS. Washington, D.C. 20402.

Contains list of documents for sale.

b. STATE

Library of Congress. MONTHLY CHECKLIST OF STATE PUBLICATIONS. Available on a subscription basis from Government Printing Office.

c. LOCAL AND MUNICIPAL

Contact local government offices for codes, ordinances, and regulations.

All of these information source materials listed here are usually available in the reference department of large public libraries and in law library collections.

Section III

BIOENGINEERING

Section III

BIOENGINEERING

A. ORGANIZATIONS AND ASSOCIATIONS

1. Governmental Agencies

a. FEDERAL

DEPARTMENT OF DEFENSE

A. U.S. AIR FORCE

Aerospace Medical Division - 6570th Aerospace Medical Research Laboratories, Scientific and Technical Information Office; Wright-Patterson Air Force Base, Ohio 45433.

> Areas of Interest: Bioastronautics; biological acoustics; neuro-physiology; biodynamics; effects of vibration and impact; environmental physiology; bioinstrumentation; effects of acceleration; high-altitude physiology; altitude protective devices; aerospace waste management; aerospace toxic hazards; bionics; engineering; psychology; design of control systems; human engineering; vision; visual perception; man-machine interaction; physical anthropology; motor skills; biomechanics; anthropometrics; environmental stress; simulation; training techniques; programmed learning.

> Publications: Technical reports, bibliographies.

> Information Services: Makes referrals to the appropriate scientists or engineers; provides consulting or advisory services; prepares analyses or evaluations for those with a legitimate need to know; permits onsite use of collection; lends films and photographs.

B. U.S. ARMY

Human Engineering Laboratories - Human Engineering Information Analysis Services, Aberdeen Research and Development Center, Aberdeen Proving Ground, Maryland 21005.

> Areas of interest: Human factors; human performance; human

factors engineering; experimental psychology; physiology; biomechanics; biophysics; biochemistry; biology.

Publications: HUMAN FACTORS ENGINEERING DESIGN GUIDES
and HUMAN FACTORS ENGINEERING DESIGN STANDARDS to
support the development of combat vehicles, wheeled vehicles,
communication and electronic systems, and missile systems; comprehensive bibliographies covering the technical literature for the
period 1940 to 1966, published in a series of four volumes; special
bibliographies on short term memory, vigilance, and reaction time;
state-of-the-art reviews.

Information services: Provides critical review of data and performs
technical analysis of special problems as requested by DoD and
various materiel development agencies and their contractors; answers inquiries; provides consulting services and reference assistance
in response to specific requests; provides literature-searching services; permits onsite use of files. Services available to all components of DoD, other federal agencies, and contractors with
certified needs for information in this subject area.

DEPARTMENT OF HEALTH EDUCATION AND WELFARE

Applied Physiology and Bioengineering Study Section - Division of Research
Grants, National Institutes of Health, Bethesda, Maryland 20014.

This organization has the responsibility for reviews of applications
for grants-in-aid relating to research on the application of physiological and bioengineering principles to the body and its systems.

NATIONAL AERONAUTICS AND SPACE ADMINISTRATION

Scientific and Technical Information Division - 300 Seventh Street S.W.,
Washington, D.C. 20546.

Areas of interest: Aerospace science and technology; worldwide
aerospace information (acquisition, processing, reporting, dissemination); high-speed information retrieval; aerospace medicine; effects
of space environment on man and other primates; instrumentation to
measure the physiological and psychological behavior of human
beings and other primates under imposed forces or in engineered
environments; closed ecological systems; behavioral effects of
rotation and acceleration on living organisms; biological effects
of radiation; cosmobiology; exobiology; decontamination, etc.

Publications: The Division produces six technical publication
series: TECHNICAL REPORTS; TECHNICAL NOTES; TECHNICAL
MEMORANDUMS: CONTRACTOR REPORTS: TECHNICAL TRANS-
LATIONS; and SPECIAL PUBLICATIONS (conference proceedings,
handbooks, data compilations, continuing bibliographies on aerospace subjects of current interest, etc.). The Division is also
responsible for four announcement journals, as follows. SCIENTIF-

IC AND TECHNICAL AEROSPACE REPORTS (STAR) is a semimonthly abstracting and indexing journal covering worldwide aerospace report literature. By arrangement between NASA and the
American Institute of Aeronautics and Astronautics, the AIAA
publication INTERNATIONAL AEROSPACE ABSTRACTS is issued
in coordination with STAR, covering books and scientific and trade
journals worldwide to complement STAR'S report coverage. The
RELIABILITY ABSTRACTS AND TECHNICAL REVIEWS cover the
report and published literature on the theory and practice of
reliability as applied to aerospace. COMPUTER PROGRAM AB-
STRACTS lists documented computer programs developed by NASA,
DoD, and AEC. A current-awareness service, SELECTED CURRENT
AEROSPACE NOTICES (SCAN), provides notifications of aerospace
documents under approximately 200 topics.

Information services: NASA-sponsored reports are provided on
initial distribution without charge to NASA offices, centers, contractors, subcontractors, grantees, and consultants; to selected
libraries; to U.S. government agencies and their contractors; to
other organizations participating with NASA in aerospace programs;
and to foreign organizations and governments that exchange publications with NASA. Others may buy paper or microfiche copies
of NASA-sponsored documents and non-copyrighted foreign documents from such sales agencies as the Clearinghouse for Federal
Scientific and Technical Information, Springfield, Virginia. 22151,
and the Superintendent of Documents, U.S. Government Printing
Office, Washington, D.C. 20402.

NATIONAL RESEARCH COUNCIL

A. COMMITTEE ON SPECIFICATIONS AND CRITERIA FOR BIOCHEMICAL
COMPOUNDS - Division of Chemistry and Chemical Technology, 2101 Constitution Avenue, N.W., Washington, D.C. 20418.

Areas of interest: Chemicals for biochemical research (standardization and improvement of quality).

Publications: SPECIFICATIONS AND CRITERIA FOR BIOCHEMICAL
COMPOUNDS.

Information services: The Division of Chemistry and Chemical
Technology refers inquiries to members of the Committee.

B. NATIONAL COMMITTEE FOR THE INTERNATIONAL BIOLOGICAL
PROGRAM - Division of Biology and Agriculture, 2101 Constitution Avenue
N.W., Washington, D.C. 20418.

Areas of interest: Biological basis of productivity and human
welfare.

Publications: Reports of symposia, meetings, and conferences.

Information services: Answers inquiries from research investigators.

C. NATIONAL COMMITTEE FOR THE INTERNATIONAL UNION OF BIOCHEMISTRY - 2101 Constitution Avenue N.W., Washington, D.C. 20418.

Areas of interest: Activities of the International Union of Biochemistry.

Information services: Answers inquiries.

2. Professional Societies, Trade Associations, and Other Scientific-Technological Related Organizations

AMERICAN INSTITUTE OF BIOLOGICAL SCIENCES (AIBS) - 3900 Washington Avenue N.W., Washington, D.C. 20016.

An organization made up of a federation of professional biological associations and individuals with an interest in the life sciences.

Purpose: To promote cohesiveness of effort among persons engaged in biological research, application of biological data, and to promote the relationships of biological sciences with other sciences and with the lay world. This is done through sponsorship of lectureships at academic institutions and acts in advisory capacity to government agencies.

Committees: Behavioral Biology; Bioinstrumentation; Education; Environmental Biology; Exobiology; Hydrobiology; Microbiology; Oceanic Biology; Physiology; Planetary Quarantine; Public Responsibilities; Publications; Radiation; Pasteurization of Foods; Scientific Manpower Commission; Theoretical Biology.

Publications: BIOSCIENCE (monthly); DIRECTORY OF BIOSCIENCE DEPARTMENTS IN THE U.S. AND CANADA (every 2 years); also, symposia proceedings, AEC-AIBS monographs, and select single titles.

Information services: Acts as clearinghouse for biological research knowledge and data. Answers inquiries; makes referrals.

AMERICAN SOCIETY FOR ARTIFICIAL INTERNAL ORGANS - 3600 Spruce-633 Maloney, Philadelphia, Pennsylvania 19104.

Purpose: To promote the increase of knowledge about artificial internal organs and of their utilization.

Publications: TRANSACTIONS OF THE AMERICAN SOCIETY FOR ARTIFICIAL INTERNAL ORGANS (annual).

Information services: Answers inquiries; assists with research questions; makes referrals.

AMERICAN SOCIETY OF BIOLOGICAL CHEMISTS (ASBC) - 9650 Rockville Pike, Bethesda, Maryland 20014.

A professional society of biochemists who have conducted and published original investigations in biological chemistry.

Publication: JOURNAL OF BIOLOGICAL CHEMISTRY (semi-monthly).

Information services: Answers inquiries; makes referrals.

ASSOCIATION FOR THE ADVANCEMENT OF MEDICAL INSTRUMENTATION - 9650 Rockville Pike, Bethesda, Maryland 20014.

Purpose: To advance scientific investigation, research, and technical advancement in the field of medical instrumentation, and to disseminate information in the field of medical instrumentation to the health sciences field. To establish and encourage the establishment of standards in the field of medical instrumentation and to act as a standardizing organization.

Professional activities: Educational: Workshops on the safe use of medical equipment (Electrical Safety in the Hospital).

Publications: JOURNAL OF THE ASSOCIATION FOR THE ADVANCEMENT OF MEDICAL INSTRUMENTATION (six issues per year) AAMI NEWS (six issues per year).

Information services: Answers inquiries; makes referrals.

BIO-INSTRUMENTATION ADVISORY COUNCIL - 3900 Wisconsin Avenue N.W., Washington, D.C. 20016.

Areas of interest: Instrumentation for research and technological applications in the life sciences.

Publications: Section in BIOSCIENCE, the monthly journal of AIBS; "Information Modules" on theory, techniques, technology, and instrumentation (published as needed). A bulletin listing free publications is available.

Information services: Answers inquiries; makes referrals; provides consulting services.

THE BIOMETRIC SOCIETY (Eastern North American Region) - Foster B. Cady, Secretary-Treasurer, 337 Warren Hall, Cornell University, Ithaca, New York 14850.

Purpose: To advance biological science through the development of quantitative theories and the application, development and dissemination of effective mathematical and statistical techniques.

Publications: BIOMETRICS (quarterly).

Information services: Answers inquiries; makes referrals.

BIOPHYSICAL SOCIETY - c/o National Biomedical Research Foundation, 11200 Lockwood Drive, Silver Spring, Maryland 20901.

Purpose: To further development and dissemination of knowledge of biophysics among health sciences profession.

Publications: BIOPHYSICAL JOURNAL (monthly).

Information services: Answers inquiries; makes referrals.

HEALTH PHYSICS SOCIETY - P.O. Box 156, East Weymouth, Massachusetts 01289.

Purpose: To develop scientific knowledge of radiation effects for benefit of mankind, and to learn measures to effectively protect against harmful aspects of radiation.

Publications: HEALTH PHYSICS JOURNAL (monthly).

Information services: Answers inquiries; makes referrals.

HUMAN FACTORS SOCIETY, INC. - Box 1369, Santa Monica, California 90406.

Purpose: The promotion and advancement of knowledge of the human factors involved in design, manufacture and use of machines, systems and devices of all kinds.

Publications: HUMAN FACTORS (bimonthly); HUMAN FACTORS SOCIETY BULLETIN (monthly); DIRECTORY AND YEARBOOK (annual).

Information services: Answers inquiries; makes referrals; assists with reference/research questions.

NATIONAL ASSOCIATION OF BIOENGINEERS (NAB) - 5500 Wabash Avenue, Terre Haute, Indiana 47803.

An organization of professionals actively interested or working in bioengineering or related areas.

Purpose: To provide a unified approach to bioengineering, including electrical, mechanical, material, and environmental engineering where related to the health sciences, and to develop professionalism and improve the education of bioengineers.

Information services: Answers inquiries; makes referrals.

NATIONAL INSTITUTE OF REHABILITATION ENGINEERING - 59 Hamburg Turnpike, Pompton Lakes, New Jersey 07442.

Areas of interest: Invention, design, construction, fitting, and servicing of special custom-made devices for individuals with any specific physical disabilities.

Holdings: Technical data on over 3,600 different technical aides for the physically disabled; collection of books, periodicals, and reports.

Publications: Reports, specifications, descriptive literature, reprints, newspaper articles.

Information services: Answers inquiries; provides consulting, reference, and duplication services; distributes reprints at cost; makes referrals; lends materials in literature collection; investigates new devices for the physically disabled; builds and dispenses all types of equipment to handicapped individuals. Services are available to any physically disabled person without regard to ability to pay as long as funds are available.

3. Research Centers

AMERICAN FOUNDATION FOR BIOLOGICAL RESEARCH - Route 5, Box 137, Madison, Wisconsin 53704.

Areas of interest: Low temperature biology (cryobiology), including cryomedical subjects, cryoengineering, cryosurgery, and cryoophthalmology.

Publications: Research reports.

Information services: Answers inquiries; provides consulting services; makes referrals; disseminates reprints of published papers on request. Services are primarily for scientists interested in cryobiology.

NATIONAL BIOMEDICAL RESEARCH FOUNDATION - 11200 Lockwood Drive, Silver Spring, Maryland 20901.

Areas of interest: Application of computers to biological and medical research; computer-based retrieval systems; medical electronics; electrooptical devices; cumulative protein sequence file; cumulative polynucleotide sequence file.

Publications: COMPUTERS IN BIOLOGY AND MEDICINE and PATTERN RECOGNITION (both quarterly; published by Pergamon Press); Atlas of PROTEIN SEQUENCES (annual). Research results published in technical reports, journal articles, and books.

Information services: Provides reprints of articles on request; provides data card decks on protein sequences.

4. International Organizations

FEDERATION OF EUROPEAN BIOCHEMICAL SOCIETIES - Department of Biochemistry, King's College, Strand, London, WC2, England.

Publications: INFORMATION BULLETIN (3 a year); EUROPEAN JOURNAL OF BIOCHEMISTRY (monthly); FEBS LETTERS (monthly).

INTERNATIONAL ERGONOMICS ASSOCIATION - c/o Dr. F.H. Bonjer, Leyweg 291, The Hague, Netherlands.

Individuals and organizations interested in the scientific study of human work and its environment.

Activities: Established international contacts among specialists; cooperates with employers' associations and trade unions to encourage the practical application of ergonomic sciences in industries.

Publication: ERGONOMICS (bimonthly).

Information services: Answers inquiries; makes referrals.

INTERNATIONAL FEDERATION FOR MEDICAL AND BIOLOGICAL ENGINEERING - Institute of Medical Physics TNO, 45 Da Costakade, Utrecht, Netherlands.

>Purpose: Encourage research and further international cooperation in the field by the dissemination of technical knowledge.

>Publication: MEDICAL AND BIOLOGICAL ENGINEERING (6 a year).

INTERNATIONAL UNION OF BIOLOGICAL SCIENCES (IUBS) - University Botanical Museum, P.O. Box 12, 5014 Bergen, Norway.

>Purpose: To promote development of pure and applied biology by organizing international discussion and publications of research work, facilitating establishment and development of research institutes.

>Publications: Reports of congresses and symposia (37 vols.).

5. Other Information Sources and Programs

a. SPECIALIZED INFORMATION SERVICES

U.S. ARMY

Human Engineering Information and Analysis Service - Aberdeen Research and Development Center, Aberdeen Proving Ground, Aberdeen, Maryland 21005.

Collects, classifies, and disseminates human engineering information through reviews, state-of-the-art reports, and special bibliographies. The Service operates a library of documents and selected abstracts, and provides a data base for human performance contributions to total systems performance.

>Services include: Reference service; state-of-the-art compilations; and bibliographic research.

>Services are available to U.S. Department of Defense, its contractors, and other government agencies with need to know.

B. LITERATURE

1. Guides and Handbooks

a. GENERAL

Altman, P., et al., comp. BIOLOGY DATA BOOK. 2nd ed. 3 vols. Washington, D.C.: Federation of American Societies for Experimental Biology, 1972.

_____. ENVIRONMENTAL BIOLOGY. Washington, D.C.: Federation of American Societies for Experimental Biology, 1966.

_____. GROWTH INCLUDING REPRODUCTION AND MORPHOLOGICAL DEVELOPMENT. Washington, D.C.: Federation of American Societies for Experimental Biology, 1962.

_____. METABOLISM. Washington, D.C.: Federation of American Societies for Experimental Biology, 1968.

BIOMEDICAL ENGINEERING. Edited by J. Brown. Philadelphia: Davis Publishing Co., 1970.

Cooney, D., et al. BIOMEDICAL AND BIOLOGICAL TRANSPORT PROCESSES. New York: Dekker, 1976.

Fleming, D. HANDBOOK OF ENGINEERING IN MEDICINE AND BIOLOGY. 2 vols. Cleveland: Chemical Rubber Co. Press, 1976.

HANDBOOK OF BIOCHEMISTRY. Edited by H. Sober. London: Blackwell Scientific Publications, 1968.

HANDBOOK OF BIOCHEMISTRY AND BIOPHYSICS. Edited by H. Damm. New York: World Publishing Co., 1966.

International Seminar on Biomechanics, 5th. Jyvaskyla, Finland, 1975. PROCEEDINGS. Baltimore: University Park Press, 1976.

Kenedi, R., ed. PERSPECTIVES IN BIOMEDICAL ENGINEERING. Baltimore: University Park Press, 1973.

Research Triangle Institute. APPLICATIONS OF AEROSPACE TECHNOLOGY IN BIOLOGY AND MEDICINE. Rept. no. N 76-13736. Springfield, Virginia: National Technical Information Service, 1975.

U.S. Department of Defense. HUMAN ENGINEERING GUIDE TO EQUIPMENT DESIGN. Rev. ed. Edited by H.P. Van Cott. Washington, D.C.: Government Printing Office, 1972.

U.S. Public Health Service. REPORT OF THE PRESIDENT'S BIOMEDICAL RESEARCH PANEL: SUBMITTED TO THE PRESIDENT AND THE CONGRESS. Washington, D.C.: Government Printing Office, 1976.

2. Dictionaries

a. TECHNICAL

American Chemical Society. CHEMICAL ABSTRACTS SERVICE. CHEMICAL-BIOLOGICAL ACTIVITIES WORD GUIDE. Washington, D.C.: 1966.

Gray, P. THE DICTIONARY OF THE BIOLOGICAL SCIENCES. New York: Reinhold, 1967.

Useful reference tool for bioscience.

Koch, M. BIOMEDICAL THESAURUS AND GUIDE TO CLASSIFICATION. New York: CCM Information Corp, 1972.

> Useful for subject terms for literature searches.

Stenesh, J. DICTIONARY OF BIOCHEMISTRY. New York: Wiley, 1975.

3. Directories

a. BIOGRAPHY

HUMAN FACTORS SOCIETY DIRECTORY. Human Factors Society, Box 1369, Santa Monica, California 90406.

> "Members of the Human Factors Society giving name, mailing address, human factors interests, job title, and company or institutional affiliation."

INTERNATIONAL WHO'S WHO IN MEDICAL ENGINEERING. Annual. International Institute for Medical Electronics and Biological Engineering, Paris, 1967.

b. PRODUCTS AND MANUFACTURERS

DIRECTORY OF BEHAVIOR AND ENVIRONMENTAL DESIGN. Research and Design Institute, Box 307, Providence, Rhode Island 02901. 1972.

> Deals with relationship of human behavior and physical design.

MacRAE'S BLUEBOOK. Annual. MacRae's Bluebook, Western Springs, Illinois 60558.

MARKETING GUIDE AND COMPANY DIRECTORY FOR PATIENT CARE SYSTEMS & LABORATORY EQUIPMENT. Westport, Connecticut: Technomic Publishing Co., 1972.

THOMAS' REGISTER OF MANUFACTURERS. Annual. New York: Thomas Publishing Co.

c. LABORATORIES AND RESEARCH CENTERS

AMERICAN INSTITUTE OF BIOLOGICAL SCIENCES. AIBS Directory of Bioscience Departments and Faculties in the United States and Canada. 2nd ed. Strandsburg, Pennsylvania: Dowden and Ross, Hutchinson, 1975.

EUROPEAN RESEARCH INDEX. 2nd ed. Guernsey, England: Francis Hodgson, 1969.

INDUSTRIAL RESEARCH LABORATORIES OF THE UNITED STATES. 13th ed. New York: R.R. Bowker, 1970.

RESEARCH CENTERS DIRECTORY. 5th ed. Detroit: Gale Research Co., 1975. Updated by subscription service, NEW RESEARCH CENTERS.

4. Encyclopedias

THE ENCYCLOPEDIA OF BIOCHEMISTRY. Edited by R. Williams. New York: Van Nostrand-Reinhold, 1967.

5. Bibliographies

a. GENERAL

Kerker, A. BIOLOGICAL AND BIOMEDICAL RESOURCE LITERATURE. Lafayette, Indiana: Purdue University, 1968.

Smith, R. GUIDE TO THE LITERATURE OF THE LIFE SCIENCES. 8th ed. Minneapolis: Burgess, 1972.

b. BIBLIOGRAPHIES SPECIFIC TO SUBJECT

ARTIFICIAL KIDNEYS. By P.W. Crockett. Report for 1970–1976. Rept. no. NTIS/PS-76/0427/5GA. Springfield, Virginia: National Technical Information Service, 1976.

Citations from the Engineering Index Data Base.

BIOCOMPATIBLE MATERIALS. By E. Harrison. Vol. 1: 1964–1973. Rept. no. NTIS/PS-75/487. Vol. 2: 1974–1976. Rept. no. NTIS/PS-76/0537. Springfield, Virginia: National Technical Information Service, 1976.

Bibliographies with abstracts.

CRYOBIOLOGY. By E. Harrison. Report for 1964–1976. Rept. no. NTIS/PS-76/0557. Springfield, Virginia: National Technical Information Service, 1976.

A bibliography with abstracts.

c. RELATED ABSTRACTS IN CURRENT JOURNALS

The following journals frequently contain book reviews and/or bibliographies on this subject. See the heading Journals – Serial Publications in this section for addresses.

ARCHIVES OF BIOCHEMISTRY AND BIOPHYSICS

BIOCHEMICAL AND BIOPHYSICAL RESEARCH COMMUNICATIONS

See also this subject in the section I of this Guide.

6. Monographs, Reports, and Symposia Proceedings

ADVANCES IN BIOENGINEERING. Symposium at Houston, Texas, 1975. New York: American Society of Mechanical Engineers, 1975.

American Institute of Chemical Engineers, 70th. Atlantic City, 1971. BIO-MATERIALS FOR SKELETAL & CARDIOVASCULAR APPLICATIONS: PROCEEDINGS. Edited by C. Homsy and C.D. Armediades. New York: Wiley, 1972.

ARTIFICIAL KIDNEY PROGRAM. 4th annual contractor's conference. January 20-22, 1971, Washington, D.C. Sponsored by the National Institute of Arthritis and Metabolic Diseases. Washington, D.C.: Government Printing Office, 1971.

BIOMATERIALS: BIOENGINEERING APPLIED TO MATERIALS FOR HARD & SOFT TISSUE REPLACEMENT. Seattle: University of Washington Press, 1971.

BIOMECHANICS: ITS FOUNDATIONS & OBJECTIVES. Symposium. Sponsored by Office of Naval Research, and University of California at San Diego, San Diego, California. Edited by Yuan Chen Fung, Nicholas Perrone, and Max Anliker. Englewood Cliffs, New Jersey: Prentice-Hall, 1972.

BIOMECHANICS SYMPOSIUM. (Properties of biological fluids and solids; Mechanics of tissues and organs). Symposium at Joint Applied Mechanics and Fluids Engineering Conference. Sponsored by American Society of Mechnical Engineers and Georgia Institute of Technology, Atlanta, Georgia. Edited by Y.C. Fung & J.A. Erighton. 1973. American Society of Mechanical Engineers, 345 East 47th Street, New York, New York 10017.

BIOMEDICAL PHYSICS & BIOMATERIALS SCIENCE. Special Summer Program. Sponsored by Massachusetts Institute of Technology and Harvard University, Program in Health Sciences and Technology, Cambridge, Massachusetts. Edited by H. Eugene Stanley. Cambridge: MIT Press, 1972.

Biomedical Symposium, 14th, San Diego, California, 1975. SAN DIEGO BIO-MEDICAL SYMPOSIUM. 1975. P.O. Box 965, San Diego, California 92112.

BIOTECHNOLOGY. Conference, Blacksburg, Virginia, August 14-18, 1967. Sponsored by NASA Langley Research Center, and Virginia Polytechnic Institute (NASA SP 205.) Washington, D.C.: Government Printing Office, 1971.

Bootzin, D., ed. BIOMECHANICS. Proceedings of the first Rock Island Arsenal Biomechanics Symposium, April, 1967. New York: Plenum, 1969.

Clynes, M. BIOMEDICAL ENGINEERING SYSTEMS. New York: McGraw-Hill, 1970.

    Concerned with instrumentation in support of medical program.

Engineering in Medicine and Biology. 24th annual conference, Las Vegas, Nevada, October 31-November 4, 1971. PROCEEDINGS. Washington, D.C.: Alliance for Engineering in Medicine & Biology, 1971.

Enzyme Engineering, Conference at 64th annual meeting. Sponsored by American Institute of Chemical Engineers, San Franciso, California. Edited by: Lemuel E. Wingard, Jr. 1972. Wiley-Interscience Division, 605 Third Avenue, New York, New York 10016. Published as: BIOTECHNOLOGY & BIOENGINEERING, SYMPOSIUM 3.

Fung, Yuan-Cheng, et al. BIOMECHANICS: ITS FOUNDATIONS AND OBJECTIVES. Englewood Cliffs, New Jersey: Prentice-Hall, 1972.

> Proceedings of the symposium held at LaJolla, California, July, 1970. Emphasis on biomedical systems.

Leonard, E., ed. CHEMICAL ENGINEERING IN MEDICINE AND BIOLOGY. New York: Plenum Press, 1967.

McCormick, E. HUMAN FACTORS ENGINEERING. New York: McGraw-Hill, 1970.

Meister, D. HUMAN FACTORS: THEORY AND PRACTICE. New York: Interscience-Wiley, 1971.

> A thorough review of human engineering as a profession.

METHODS AND REFERENCES IN BIOCHEMISTRY AND BIOPHYSICS. Edited by H. Damm. New York: World Publishing Co., 1966.

Muffey, H. BIOMECHANICS. New York: Plenum Publishing Co., 1959.

Reneau, D. CHEMICAL ENGINEERING IN MEDICINE. New ed. Edited by R.F. Gould. ADVANCES IN CHEMISTRY SERIES, no. 118. Washington, D.C.: American Chemical Society, 1973.

Rusten, D. ENGINEERING AND LIVING SYSTEMS: INTERFACES AND OPPORTUNITIES. Cambridge: M.I.T. Press, 1970.

Talbot, S. SYSTEMS PHYSIOLOGY. New York: Wiley & Sons Publishing Co., 1973.

U.S. NASA 7th Annual Conference on Manual Control. PROCEEDINGS. Washington, D.C.: Government Printing Office, 1972.

7. Abstracting and Indexing Services

a. GENERAL

APPLIED SCIENCE AND TECHNOLOGY INDEX

BIOLOGICAL ABSTRACTS

CHEMICAL ABSTRACTS

INDEX MEDICUS

See this part in Section I of this Guide for further details on these publications.

b. ABSTRACTING AND INDEXING SERVICES SPECIFIC TO TOPIC

ART INDEX. Monthly. H.W. Wilson Co., 950 University Avenue, Bronx, New York 10452.

ARTIFICIAL KIDNEY BIBLIOGRAPHY. Quarterly. National Institute of Arthritis and Metabolic Diseases.

BIOENGINEERING ABSTRACTS. Monthly. Marketing Division, Engineering Index, 345 East 47th Street, New York 10017.

BIORESEARCH TODAY: BIOENGINEERING & INSTRUMENTATION MONTHLY. Bio Sciences Information Service of Biological Abstracts, 2100 Arch Street, Philadelphia, Pennsylvania 19103.

BULLETIN SIGNALETIQUE. Centre de Documentation Sciences Humaines, 54 Bd. Raspail, 75260 Paris, France.

CHEMICAL ABSTRACTS BIOCHEMISTRY SECTIONS. Fortnightly. American Chemical Society, Chemical Abstracts Service, 1155 6th Street N.W., Washington, D.C. 20036.

ENGINEERING INDEX. Monthly. Engineering Index Service, 345 East 47th Street, New York, New York 10017.

WEEKLY GOVERNMENT ABSTRACTS. Biomedical Technology and Engineering. Government Printing Office, Washington, D.C., 20042

c. RELATED ABSTRACTS IN CURRENT JOURNALS

The following journals frequently contain abstracts of related literature. See the heading Journals and Serial Publications in this section for additional details.

APPLIED ERGONOMICS

8. Journals and Serial Publications

a. DIRECTORIES AND LISTS

AYER DIRECTORY OF PUBLICATIONS. Annual. Philadelphia: Ayer Press.

STANDARD PERIODICAL DIRECTORY. Biannual. New York: Oxbridge Publishing Co.

ULRICH'S INTERNATIONAL PERIODICAL DIRECTORY. 2 vols. Annual. New York: R.R. Bowker Co.

b. JOURNAL AND SERIAL TITLES

ADVANCES IN BIOCHEMICAL ENGINEERING. Irregular. Springer-Verlag, 175 Fifth Avenue, New York, New York 10010.

ADVANCES IN BIOMEDICAL ENGINEERING. Vol. 3. New York: Academic Press, 1973.

> "A continuing series on engineering applications in medicine."

ADVANCES IN BIOMEDICAL ENGINEERING AND MEDICAL PHYSICS. Irregular. John Wiley & Son Publishing Co., 605 Third Avenue, New York, New York 10016.

> Engineering theory and systems applied to human anatomy and medicine.

ANNUAL REVIEW QF BIOCHEMISTRY. Annual Reviews, Inc., 4139 El Camino Way, Palo Alto, California 94306.

ANNUAL REVIEW OF BIOPHYSICS AND BIOENGINEERING. Vol. 2 (1973). Annual Reviews Inc., 4139 El Camino Way, Palo Alto, California 94306.

APPLIED ERGONOMICS. Quarterly. IPC Science and Technology Press Ltd., IPC House, 32 High Street, Guildford, Surrey, England.

ARCHIVES OF BIOCHEMISTRY AND BIOPHYSICS. Monthly. Academic Press Inc., 111 Fifth Avenue, New York, New York 10003.

ARTIFICIAL KIDNEY BIBLIOGRAPHY. Quarterly. National Institute of Arthritis and Metabolic Diseases, Superintendent of Documents, Government Printing Office, Washington, D.C. 20402.

BIOCHEMICA ET BIOPHYSICA ACTA. Weekly. Elsevier Publishing Co., Box 211, Amsterdam, Netherlands.

> International weekly of biochemistry and biophysics. Text in English, French, and German.

BIOCHEMICAL AND BIOPHYSICAL RESEARCH COMMUNICATIONS. Semi-monthly. Academic Press Inc., 111 Fifth Avenue, New York, New York 10003.

BIOCHEMICAL JOURNAL. Semimonthly. Biochemical Society, 7 Warwick Court, London, WC1R 5DP, England.

BIOCHEMISTRY. Fortnightly. American Chemical Society, 1155 Sixteenth Street N.W., Washington, D.C. 20036.

BIOINORGANIC CHEMISTRY, AN INTERDISCIPLINARY JOURNAL. Quarterly. American Elsevier Publishing Co., Inc., 52 Vanderbilt Avenue, New York, New York 10017.

BIOMECHANICS. Bimonthly. Pergamon Press, Maxwell House, Fairview Park, Elmsford, New York 10523.

BIOMEDICAL ENGINEERING. Monthly. United Trade Press Ltd., 9 Gough Square, London, EC4, England.

BIOMEDICAL INSIGHT. Fortnightly. American Biomedical Information Service, Box 707, Arcadia, California 91006.

BIOPHYSICAL JOURNAL. Monthly. Rockefeller University Press, 66th Street and York Avenue, New York, New York 10021.

BIORESEARCH INDEX. Monthly. BioSciences Information Service of Biological Abstracts. 2100 Arch Street, Philadelphia, Pennsylvania 19103.

> Provides an abstracting service for more specialized areas of the biosciences.

BIOTECHNOLOGY AND BIOENGINEERING. Bimonthly. Interscience Publishers Division, John Wiley & Sons, 605 Third Avenue, New York, New York 10016.

BULLETIN SIGNALETIQUE. Part 320: Biochimie - biophysique - genie biomedical. Monthly. Centre de Documentation du C.N.R.S., 15 Quai Anatole-France, Paris (7e), France.

CANADIAN BIOCHEMICAL SOCIETY BULLETIN. Quarterly. Canadian Biochemical Society, Department of Biochemistry, Faculty of Medicine, Laval University, Quebec, Canada.

> Text in English and French.

CHEMICAL ABSTRACTS BIOCHEMISTRY SECTIONS. Fortnightly. American Chemical Society, Chemical Abstracts Service, 1155 16th Street N.W., Washington, D.C. 20036.

DESIGN QUARTERLY. Walker Art Center, 807 Hennepin Avenue, Minneapolis, Minnesota 55403.

ENGINEERING INDEX. Monthly. Engineering Index Inc., 345 East 47th Street, New York, New York 10017.

THE EUROPEAN JOURNAL OF BIOCHEMISTRY. 11 issues per year. Springer-Verlag, 175 Fifth Avenue, New York, New York 10010.

    Covers field of biochemistry and molecular biology.

HUMAN FACTORS. Bimonthly. Human Factors Society, P.O. Box 1369, Santa Monica, California 90406.

INDEX MEDICUS. Monthly. Washington, D.C.: Government Printing Office.

JOURNAL OF BIOMECHANICS. Bimonthly. Pergamon Press Inc., Maxwell House, Fairview Park, Elmsford, New York 10523.

MEDICAL AND BIOLOGICAL ENGINEERING. Bimonthly. Elmsford, New York: Pergamon Press.

9. Reviewing Services

ADVANCES IN BIOCHEMICAL ENGINEERING. Irregular. Springer-Verlag, 175 Fifth Avenue, New York, New York 10010.

ADVANCES IN BIOLOGICAL AND MEDICAL ENGINEERING. Irregular. Academic Press, Inc., 111 Fifth Avenue, New York, New York 10003.

ADVANCES IN BIOMEDICAL ENGINEERING. Vol. 3. New York: Academic Press, 1973.

    Continuing series on engineering applications in medicine.

ADVANCES IN BIOMEDICAL ENGINEERING AND MEDICAL PHYSICS. Irregular. John Wiley & Son Publishing Co., 605 Third Avenue, New York, New York 10016.

    Engineering theory and systems applied to human anatomy and medicine.

ANNUAL REVIEW OF BIOCHEMISTRY. Annual Reviews Inc., 4139 El Camino Way, Palo Alto, California 94306.

ANNUAL REVIEW OF BIOPHYSICS AND BIOENGINEERING. Vol. 2 (1973). Annual Reviews Inc., 4139 El Camino Way, Palo Alto, California 94306.

BIOCHEMICAL AND BIOPHYSICAL RESEARCH COMMUNICATIONS. Semimonthly. Academic Press Inc., 111 Fifth Avenue, New York, New York 10003.

10. Document and Report Literature

a. LISTS

MONTHLY CATALOG. Superintendent of Documents, Government Printing Office, Washington, D.C. 20402.

U.S. GOVERNMENT REPORTS ANNOUNCEMENTS; and U.S. GOVERNMENT REPORTS INDEX. Monthly. National Technical Information Service, Springfield, Virginia 22151.

b. CLEARINGHOUSES

DEFENSE DOCUMENTATION CENTER. U.S. Department of Defense, Cameron Station, Virginia 22314.

NATIONAL TECHNICAL INFORMATION SERVICE. Springfield, Virginia 22151.

SUPERINTENDENT OF DOCUMENTS. Government Printing Office, Washington, D.C. 20402.

11. Sources of Information on Masters and Doctoral Theses

See this listing in Section I of this Guide.

12. Sources of Information on Statistics and Costs

a. GUIDES AND HANDBOOKS

AMERICAN STATISTICS INDEX. Annual. Part 1. Index; Part 2. Abstracts. Congressional Information Service, 600 Montgomery Building, Washington, D.C. 20014.

"Comprehensive guide and index to the statistical publications of the U.S. Government."

b. CURRENT JOURNALS
WHICH INCLUDE RELATED INFORMATION ON STATISTICS AND COSTS

The following journals frequently contain statistics related to this field. See the heading Journals and Serial Publications in this section for additional details.

BIOMEDICAL INSIGHT. Fortnightly. American Biomedical Information Service, Box 707, Arcadia, California 91006.

CHEMICAL ABSTRACTS. Biochemistry Sections.

See this listing also in Section I of this Guide.

13. Sources of Information on Standards, Specifications, and Patents

a. STANDARDS AND SPECIFICATIONS

National Research Council. Committee on Biological Chemistry. SPECIFICATIONS AND CRITERIA FOR BIOCHEMICAL COMPOUNDS. 2nd ed. Washington, D.C.: National Research Council, 1967.

U.S. National Bureau of Standards. Directory of United States. STANDARDIZATION ACTIVITIES. Washington, D.C.: Government Printing Office, 1975.

Specifications, standards, codes, and regulations are available for individual states from the official state document printing offices and from local and municipal government agencies.

See also this topic in Section I of this Guide.

b. PATENTS

i. Guides and Handbooks

U.S. Government Printing Office. PRICE LIST SERIES. No. 62. COMMERCE. BUSINESS, PATENTS, TRADEMARKS, AND FOREIGN TRADE. Washington, D.C.: 1973.

Contains list of documents for sale.

ii. Current Journals Which Include Related Patent Listings

The following journals frequently list patents. See the heading Journals and Serial Publications in this section for additional details.

CHEMICAL ABSTRACTS. Biochemistry Sections.

See this listing also in Section I of this Guide.

14. Sources of Information on Translations, Meetings, and Proceedings of Symposia

U.S. JOINT PUBLICATIONS RESEARCH SERVICE. Arlington, Virginia.

Publishes translations of scientific and technical materials from Eastern Europe, Russia, and China. These are listed in the Monthly Catalog of the Superintendent of Documents.

See this listing in Section I of this Guide.

15. Selected Literature Sources on Pollution Control and Safety

a. POLLUTION CONTROL

See this topic in Section I of this Guide.

b. .SAFETY

Peck, Theodore. OCCUPATIONAL SAFETY AND HEALTH: A GUIDE TO INFORMATION SOURCES. Detroit: Gale Research Co., 1974.

See these topics also in Section I of this Guide.

16. Management Information Sources and Sources of Information on Laws and Legislation in This Field

See these topics also in Section I of this Guide.

a. FEDERAL

CODE OF FEDERAL REGULATIONS. Government Printing Office.

Continuing series issued throughout the year. Available on a subscription basis from Government Printing Office.

FEDERAL REGISTER. Daily publication listing new regulations revisions. Updates Code of Federal Regulations. Available on a subscription basis from Government Printing Office.

Library of Congress. DIGEST OF PUBLIC GENERAL BILLS AND REGULATIONS. Monthly. Available on a subscription basis from Government Printing Office, Washington, D.C. 20402.

UNITED STATES CODE. 1970 edition and supplements. Irregular. Washington, D.C.: Government Printing Office.

b. STATE

Library of Congress. MONTHLY CHECKLIST OF STATE PUBLICATIONS. Available on a subscription basis from Government Printing Office.

c. LOCAL AND MUNICIPAL

Contact local government offices for codes, ordinances and regulations.

All of these information source materials listed here are usually available in the reference department of large public libraries.

Section IV

FOOD ENGINEERING

Section IV

FOOD ENGINEERING

A. ORGANIZATIONS AND ASSOCIATIONS

1. Governmental Agencies

a. FEDERAL

DEPARTMENT OF HEALTH, EDUCATION AND WELFARE

A. EDITORIAL OFFICE - Bureau of Foods, BF-107, Food and Drug Administration, 200 C Street S.W., Washington, D.C. 20204.

Areas of interest: Technical editing and writing services for all FDA scientific personnel; editing, clearance, and revision of manuscripts by FDA scientists to appear in scientific literature. All scientific disciplines and areas involved in FDA programs are covered.

Publications: FDA BY-LINES (bimonthly, about 50 pages per issue); SELECTED TECHNICAL PUBLICATIONS (reprint compilation published twice yearly, about 500 pages per issue); bibliographies (irregular). A publications list is available.

Information services: Answers inquiries; makes referrals. The publications listed above are available on request to libraries and scientific laboratories.

B. TECHNICAL OPERATIONS STAFF - Bureau of Foods, BF-30, Food and Drug Administration, 200 C Street S.W., Washington, D.C. 20204.

Areas of interest: Chemical and biological data pertaining to food additive substances for petition reviews, cosmetic products, and GRAS (Generally Recognized As Safe) substances.

Holdings: Data on 4,000 chemical compounds pertaining to food, 25,000 formulations pertaining to cosmetic chemicals, and 1,000 chemical compounds pertaining to GRAS substances used in food processing. The file contains chemical names and Chemical Abstracts registry numbers; it provides chemical and biological information for reviewing new petitions. Peripheral files contain data on biological activity and toxicology. Output is in the form of abstracts, citations, Chemical Abstracts registry numbers, and toxicological data with an estimated turn-around time of four days. The files are available to federal agencies on a limited basis for nonproprietary information.

NATIONAL RESEARCH COUNCIL

FOOD AND NUTRITION BOARD - 2101 Constitution Avenue, N.W., Washington, D.C. 20418.

Evaluates nutritional science as it applies to food processing and public health.

Committees include: Amino Acids; Dietary Allowances; Fats in Human Nutrition; Milk; Food Standards; Food Protection (Subcommittees: Carcinogenesis; Food Technology; Pesticides; Toxicology); Protein Malnutrition. Publishes monographs on nutrition, health, food protection, fat, maternal nutrition and child health, artificial sweeteners, fluoride, packaging and other related subjects. Division of: National Academy of Science--National Research Council.

b. STATE AND MUNICIPAL

See this topic in Section I of this Guide.

2. Professional Societies, Trade Associations, and Other Scientific-Technological Related Organizations

AMERICAN ASSOCIATION OF CEREAL CHEMISTS - 3340 Pilot Knob Road, St. Paul, Minnesota 55121.

A professional society of scientists in the cereal processing industry (milling, baking, convenience foods, feeds, etc.).

Purpose: To encourage research on cereal grains, oil seeds, pulses, and related materials, and to study their processing, utilization, and products. To develop and standardize analytical methods used in cereal and seed chemistry and to disseminate scientific and technical information through conferences, workshops, and publications.

Publications: CEREAL SCIENCE TODAY (monthly); CEREAL CHEMISTRY (bimonthly); DIRECTORY OF MEMBERS (annual); also published monographs.

Information services: Answers inquiries; makes referrals.

AMERICAN FROZEN FOOD INSTITUTE - 919 18th Street N.W., Washington, D.C. 20006.

Areas of interest: Production, handling, and marketing of frozen foods.

Publications: Technical service bulletins, newsletters, reports.

Information services: Answers inquiries; provides consulting and reference services. Services are primarily for members of the Institute and are provided free to some users and at cost to others.

AMERICAN INSTITUTE OF BAKING - 400 East Ontario Street, Chicago, Illinois 60611.

Areas of interest: Baking technology; food technology; nutrition and nutrition education; food plant sanitation; cereal chemistry.

Publications: AMERICAN INSTITUTE OF BAKING NEWS (bimonthly); QUARTERLY INDEX OF PERIODICALS; nutrition education teaching materials; technical reports, standards and specifications, reprints, pamphlets, news releases.

Information services: Answers inquiries; provides consulting services; makes referrals; furnishes laboratory tests of bakery ingredients and products; permits onsite use of collection.

AMERICAN INSTITUTE OF NUTRITION - 9650 Rockville Pike, Bethesda, Maryland 20014.

Purpose: To extend the knowledge of nutrition; facilitate personal contact between investigators in nutrition and closely allied fields of interest and to further research in this field.

Publications: JOURNAL OF NUTRITION (monthly); AMERICAN JOURNAL OF CLINICAL NUTRITION (monthly); AIN NUTRITION NOTES (quarterly).

Information services: Answers inquiries; makes referrals.

AMERICAN MEAT INSTITUTE FOUNDATION - 59 East Van Buren Street, Chicago, Illinois 60605.

Areas of interest: Meat packing industry; meat processing; methods of analyzing meat and meat products; nutrition; fats and oils, hides and skins, animal feedstuffs, and other meat byproducts.

Publications: Technical reports, annual report, conference proceedings.

Information services: Answers inquiries; provides consulting services; makes referrals; permits onsite use of literature collection.

AMERICAN OIL CHEMISTS' SOCIETY - 35 East Wacker Drive, Chicago, Illinois 60601.

Organization made up of chemists, biochemists, chemical engineers, research directors, plant personnel, and others in laboratories and chemical process industries concerned with animal, marine and vegetable oils and fats and their extraction, refining, safety, packaging, quality control and use in consumer and industrial products (foods, drugs, paints, waxes, lubricants, soaps, cosmetics, etc.). Publishes laboratory manual on methods of analysis, sponsors short courses, certifies referee chemists, distributes cooperative check samples and sells official reagents.

Publications: JOURNAL OF THE AOCS (monthly); LIPIDS (monthly); DIRECTORY OF MEMBERSHIP (annual); also published significant symposia and short-course lectures.

Information services: Answers inquiries; makes referrals.

AMERICAN SOCIETY OF BAKERY ENGINEERS (ASBE) - Riverside Plaza Building, Room 1921, 2 North Riverside Plaza, Chicago, Illinois 60606.

A professional organization of persons engaged in bakery production; chemists, production supervisors, engineers, technicians, and others from allied fields.

Purpose: To exchange knowledge in bakery production; maintains information service and library of 10,000 references to baking and related subjects.

Publications: ASBE LETTER (quarterly); Proceedings (annual); Bulletins (irregular).

Information services: Answers inquiries; makes referrals.

AMERICAN SOCIETY OF BREWING CHEMISTS - 1201 Waukegan Road, Glenview, Illinois 60025.

A professional organization of chemists in brewing and malting industries.

Program: Develops standard methods of analysis for raw materials, supplies, and products of brewing, malting and related industries.

Publications: BREWING CHEMISTS NEWS LETTER (quarterly); Annual Proceedings; Methods of Analysis of the ASBC (irregular).

Information services: Makes referrals.

AMERICAN SOCIETY OF SUGAR BEET TECHNOLOGISTS - P.O. Box 538, 156 South College Avenue, Fort Collins, Colorado 80521.

A professional society of persons interested in research on sugar beet production and processing.

Publication: JOURNAL OF ASSBT (quarterly).

Information services: Answers inquiries; makes referrals.

COUNCIL FOR AGRICULTURAL AND CHEMURGIC RESEARCH - 350 Fifth Avenue, New York, New York 10001.

Farmers, scientists and industrialists interested in chemurgic research (research on the industrial, especially chemical, uses of farm products). Promotes the discovery of: "New, non-food uses for farm crops, their residues and by-products; new and profitable uses for previously unused plant materials; new crops that farmers may grow profitably; more valuable uses for presently used crops, through chemurgic upgrading." Seeks increased public acceptance of the importance of chemurgy.

Publication: CHEMURGIC DIGEST (8/year).

Formerly: (1955) National Farm Chemurgic Council.

Also known as: Chemurgic Council.

DAIRY AND FOOD INDUSTRIES SUPPLY ASSOCIATION - 5530 Wisconsin Avenue, N.W., Washington, D.C. 20015.

An organization whose membership is made up of manufacturers and distributors of dairy and food processing equipment, machinery and supplies. Cooperates in development of sanitary standards for dairy and egg processing equipment. Compiles market statistics, holds seminars and annual meetings.

Publications: Dairy Processor Market Guide, DFSIA CONCERN (Newsletter), DFSIA REPORTER (quarterly), YOUR MARKET (monthly), DFSIA DIGEST (quarterly) and TECHNICAL BULLETIN (monthly), reports on technical research and advancement in chemical and engineering fields.

DISTILLED SPIRITS INSTITUTE - 1132 Pennsylvania Building, Washington, D.C. 20004.

"Domestic distillers of beverage alcohol for whiskey, vodka, gin and rum; and producers of distillers' dried feeds."

Information services: Answers inquiries, makes referrals and acts as clearinghouse in matters concerning this industry.

Publications: SUMMARY OF STATE LAWS AND REGULATIONS TO DISTILLED SPIRITS (biennial); ANNUAL REPORT: also publishes ANNUAL STATISTICAL REVIEW OF THE DISTILLING INDUSTRY; PUBLIC REVENUES FROM ALCOHOLIC BEVERAGES; RETAIL OUTLETS FOR THE SALE OF DISTILLED SPIRITS.

EVAPORATED MILK ASSOCIATION - 910 17th Street N.W., Washington, D.C. 20006.

Areas of interest: Production technology of evaporated milk; human nutrition and evaporated milk.

Publications: INFORMATION BULLETIN (monthly); STATISTICAL BULLETIN (quarterly).

Information services: Answers brief inquiries; makes referrals.

FOOD FACILITIES CONSULTANTS SOCIETY - P.O. Box 1238, Harrisburg, Pennsylvania 17108; Location: 1517 North 2d Street, Harrisburg, Pennsylvania, Tel: (717) 234-7069.

Areas of Interest: Public and private food facilities engineering, including planning, specifications, cost, construction, sanitary requirements, and maintenance.

Publications: News bulletin.

Information services: Answers inquiries; makes referrals. Consulting services are provided by individual members on fee basis.

FOOD PROCESSING MACHINERY AND SUPPLIES ASSOCIATION - 7758 Wisconsin Avenue N.W., Washington, D.C. 20014.

Areas of interest: Food processing machinery, supplies, and services necessary for industry production; management of national exposition for food processors.

Publications: MEMBERSHIP DIRECTORY with products designation; CONVENTION/EXPOSITION DIRECTORY; periodic newsletters to the industry.

Information services: Answers inquiries; makes referrals; provides consulting marketing information.

INSTITUTE OF FOOD TECHNOLOGISTS - 221 North LaSalle Street, Suite 2120, Chicago, Illinois 60601.

Areas of interest: Food technology; application of science and engineering to the research, production, processing, packaging, distribution, preparation, and utilization of foods; chemistry; biochemistry.

Publications: FOOD TECHNOLOGY (monthly journal); JOURNAL OF FOOD SCIENCE (bimonthly); FOOD SCIENCE AND TECH-NOLOGY ABSTRACTS (monthly); membership directory (annual); reference materials.

Information services: The Institute's Subcommittee on Technical Assistance answers inquiries, especially from developing countries, in connection with food science and technology problems. In-quiries should be directed to: Harold J. Rafson, Chairman, Sub-committee on Technical Assistance, Institute of Food Technologists, 1852 Dale Avenue, Highland Park, Illinois 60035.

INSTITUTE OF SANITATION MANAGEMENT - 1710 Drew Street, Clearwater, Florida 33515.

Areas of interest: Industrial and institutional environmental sani-tation maintenance management, including housekeeping, building maintenance, insect and rodent control, maintenance of personnel facilities, and food industry sanitation; methods, training, and advancements in industrial sanitation.

Publications: PROFESSIONAL SANITATION MANAGEMENT (quarterly); ISM DIGEST; TECHNICAL AND INFORMATION REPORTS (24 a year); TRADE PERIODICALS FOR SANITATION TRAINING AND EDUCATION (annual); brochure, glossary of terminology, membership directory, posters.

Information services: Answers inquiries; makes referrals; provides consulting, reference, and literature-searching services; lends materials; conducts educational seminars, programs, and conferences; operates a speakers bureau.

INTERNATIONAL ASSOCIATION OF ICE CREAM MANUFACTURERS - 1105 Barr Building, Washington, D.C. 20006.

Areas of interest: Ice cream production; merchandising and accounting; management; economics; statistics; nutrition; history of ice cream production.

Publications: PRODUCTION INDEX (annual); SUMMARY FOR ICE CREAM EXECUTIVES; HISTORY OF ICE CREAM; YOUR ANSWERS ABOUT THE ICE CREAM INDUSTRY.

Information services: Answers inquiries; makes referrals.

INTERNATIONAL FOODSERVICE MANUFACTURERS ASSOCIATION - United of America Building, One East Wacker Drive, Chicago, Illinois 60601.

Food manufacturing companies supplying products used by restaurants, hotels, clubs, hospitals, schools, institutions -- all segments of the away-from-home eating market. Activities are aimed at marketing, merchandising, sales training, and market research.

Publications: BUSINESS BAROMETER AND DIGEST (monthly); NEWS FROM IFMA (bimonthly); IFMA LEGAL REPORTER (quarterly); NUTRITION DIGEST (quarterly); IFMA VIEWPOINT (quarterly); MEMBERSHIP DIRECTORY (annual).

Information services: Makes referrals.

INTERNATIONAL SOCIETY OF SUGAR CANE TECHNOLOGISTS - P.O. Box 2450, Honolulu, Hawaii 96804.

An organization of individuals engaged in the growing and production of sugar cane.

Purpose: To promote the discussion of technical problems in the production of sugar cane.

Publication: PROCEEDINGS (triennial).

Information services: Answers inquiries; makes referrals.

MALTING BARLEY IMPROVEMENT ASSOCIATION - 828 North Broadway, Milwaukee, Wisconsin 53202.

Areas of interest: Malting barley breeding and quality evaluation; malting and brewing.

Publications: Annual report, other reports, maps, news releases.

Information services: Answers inquiries; disseminates publications; conducts barley variety identification schools. Services are primarily for barley growers and barley grain handlers.

MILK INDUSTRY FOUNDATION - 910 17th Street, N.W., Washington, D.C. 20006.

Processors and distributors of fluid milk and manufacturers of fluid milk by-products.

Information services: Answers inquiries, compiles statistics, training aids, manuals and is active in providing information about the industry to the public.

Publications: NEWSLETTER (monthly); MILK FACTS (annual); ALERT (irregular); SUMMARY FOR MILK EXECUTIVES (irregular).

NATIONAL ASSOCIATION OF FOOD AND DAIRY EQUIPMENT MANUFACTURERS - 5530 Wisconsin Avenue, Washington, D.C. 20015.

Manufacturers of food and dairy plant equipment and machinery.

Activities: Conducts studies and technical research including testing of supplies, materials and products for the industry.

Information services: Answers inquiries; makes referrals.

NATIONAL ASSOCIATION OF FOOD EQUIPMENT MANUFACTURERS - 111 East Wacker Drive, Chicago, Illinois 60601.

Membership made up of manufacturers of commercial food-service equipment for restaurant, hotel, and institutional use.

Publication: NAFEM NEWSLETTER (monthly).

Information services: Answers inquiries; makes referrals.

NATIONAL CHEESE INSTITUTE - 110 North Franklin Street, Chicago, Illinois 60606.

Manufacturers, processors, assemblers and distributors of cheese of cheese products.

Programs: Research and dissemination of information within industry and to public as activities permit.

Information services: Answers inquiries, makes referrals.

Publication: BULLETIN (semimonthly).

NATIONAL CONFECTIONERS ASSOCIATION - 36 South Wabash Avenue, Chicago, Illinois 60603.

Areas of interest: Candy manufacturing technology; scientific research; plant sanitation; education; packaging and labeling regulations; trade names; import competition; legislation; marketing; statistics; data processing; financial operations; public relations.

Publications: CONFECTIONEWS (monthly newsletter); "The Confectionery Industry Speaks" (annual survey of candy, chocolate, and chewing gum manufacturers); annual financial operations survey of candy manufacturing firms; bulletins on food and drug regulations; import and export statistics; trade mark registrations; credit reports on candy manufacturers' customers.

Information Services: Answers inquiries (technical inquiries should be addressed to Director of Research and Education); makes referrals; provides consulting services; sponsors research programs; annual three-week course in candy technology, marketing and financial operations seminars, and annual convention and exposition of candy-making equipment and services.

NATIONAL DAIRY COUNCIL - 111 North Canal Street, Chicago, Illinois 60606.

Milk producers, milk dealers, manufacturers of butter, cheese, ice cream, dairy equipment and supplies.

Activities: Programs on nutrition research, health education and public relations in the use of milk and its products. Has library of 3100 volumes on nutrition research and education.

Committees: Communications; Nutrition Education; Nutrition Research.

Publications: DAIRY COUNCIL DIGEST (semimonthly); NUTRITION NEWS (semimonthly); DAIRY COUNCILOR (quarterly).

Information services: Answers inquiries; makes referrals.

NATIONAL FRUIT AND SYRUP MANUFACTURERS ASSOCIATION - Suite 2806, 122 East 42nd Street, New York, New York 10017.

An organization of manufacturers of fruits and syrups and home sundae toppings for the ice cream and fountain trades.

Purpose: To maintain communications within the trade, promote advancement and research and represent industry to government and public.

Information activities: In addition to above, answers inquiries, makes referrals.

Publications: BULLETIN (monthly); YEARBOOK and REPORT OF CONVENTION PROCEEDINGS.

NATIONAL NUTRITIONAL FOODS ASSOCIATION (NNFA) - 770 South Brea Boulevard, Suite 226, Brea, California 92621.

Retailers, distributors and manufacturers of specialty and dietary foods.

Publications: NEWS LETTER (monthly); HEALTH FOODS RETAILING (monthly); DIETETIC FOODS INDUSTRY (bimonthly).

Information services: Answers inquiries, makes referrals.

NATIONAL SOFT DRINK ASSOCIATION - 1128 16th Street N.W., Washington, D.C. 20006

Areas of interest: Carbonated soft drinks; nutrition; dental caries; water purity; sugar; synthetic sweeteners; sanitation; equipment; plant layout.

Publications: Brochures.

Information services: Answers inquiries.

NEW YORK SUGAR TRADE LABORATORY - 37 Warren Street, New York, New York 10007.

> Areas of interest: Sugar chemistry; methods of sugar analysis; chemistry of cane molasses, particularly the nature of color bodies formed.

> Publications: Technical reports.

> Information services: Answers inquiries; makes referrals; permits onsite use of collection.

NUTRITION FOUNDATION - 99 Park Avenue, New York, New York 10016.

An organization sponsored by companies within the food and closely related industries, with the purpose of supporting basic research and education in the science of nutrition through provision of grants-in-aid to universities and similar institutions.

> Publication: NUTRITION REVIEWS (monthly).

> Information services: Answers inquiries; makes referrals.

QUALITY BAKERS OF AMERICA COOPERATIVE - 120 West 42d Street, New York, New York 10036.

> Areas of interest: All phases of the baking industry specifically (including processing of flour) and the food industry in general.

> Information services: Answers inquiries; makes referrals; provides analytical, identification, and testing services on baking ingredients.

> Facilities include a headquarters laboratory, mobile laboratory, and access to 143 bakeries for use as pilot plants.

SOCIETY OF SOFT DRINK TECHNOLOGISTS (SSDT) - 1101 16th Street N.W., Washington, D.C. 20036.

An organization of persons engaged in the scientific or technical phases of production, research, or quality control in the soft drink industry.

> Information services: Answers inquiries; makes referrals.

UNITED STATES BREWERS ASSOCIATION, INC. - 535 Fifth Avenue, New York, New York 10017.

> Areas of interest: Beer and ale (brewing, packaging, marketing legal aspects).

> Publications: Nontechnical motion pictures, pamphlet.

> Information services: Makes referrals; provides literature-searching and bibliographic services; permits onsite use of collection.

3. Research Centers

For information on research in Food Engineering, contact organizations listed in parts 1 and 2 of this Section.

4. International Organizations

ASSOCIATION OF NATIONAL BAKERY AND CONFECTIONERY FEDERATIONS OF THE EEC - 83 bd L Mettewie, 1080 Brussels, Belgium.

ASSOCIATION OF POWDERED MILK MANUFACTURERS OF THE EEC - 140 Boulevard Haussmann, 75-Paris 8e, France.

COCOA PRODUCERS' ALLIANCE - 8-10 Yakubu Gowon, P.O. Box 1718, Lagos, Nigeria.

> Purpose: "Examine problems of mutual concern and promote economic and social relations between producing countries; seek by strict control to limit supplies reaching the market; promote consumption; exchange technical and scientific information."

DAIRY SOCIETY INTERNATIONAL - 30 F Street N.W., Washington, D.C. 20001.

An institution made up of scientists, farmers, equipment suppliers, students, processors, and dairy organizations with an interest in increasing world nutrition through use of milk and milk products; membership represents 52 countries.

> Purpose: To raise the standards of world health and world economy through the better utilization and distribution of dairy products.

> Activities: Cooperates with Departments of Agriculture and Commerce to survey overseas markets, introduce education programs, conduct sampling programs and demonstrations of recombined milk and ice cream. Cooperates with Food and Agriculture Organization (FAO) and UNICEF in the United Nations, Food for Peace, AID, other governments, and non-governmental organizations.

> Publications: MARKET FRONTIER NEWS (bimonthly); REPORT TO MEMBERS (irregular); DSI BULLETIN (irregular).

> Information services: Answers inquiries; makes referrals.

EUROPEAN ORGANIZATION OF PRESERVES AND TINNED FRUIT INDUSTRIES SG Emile Toebosch, Av de Cortenbergh 172, 1040 Brussels.T., Belgium.

INTERNATIONAL ASSOCIATION OF MILK, FOOD, AND ENVIRONMENTAL SANITARIANS - P.O. Box 437, Blue Ridge Road, Shelbyville, Indiana 46176.

A professional society of persons working in the field of milk, dairy products, food technology and general sanitation, including: Food and Drug officials; milk and food industry fieldmen and technicians; laboratory workers; sanitary engineers, research, teaching, agriculture and military personnel.

> Purpose: To develop uniform and proper methods of supervision and inspection of dairy farms, milk and milk product plants, food handling establishments, and transportation equipment; examination

of milk, milk products and other foods; improve sanitary methods of production of milk and related food products; develop equipment and supplies to improve the handling of dairy and food products.

Publication: JOURNAL OF MILK AND FOOD TECHNOLOGY (monthly); also publishes sanitary standards, and procedures for investigation of food-borne disease outbreaks.

INTERNATIONAL COMMITTEE ON FOOD SERVICE AND TECHNOLOGY - Department of Food Science and Technology, University of California, Davis, California 95616.

INTERNATIONAL FOOD INFORMATION SERVICE - Commonwealth Agricultural Bureaux, Great Britain.

Abstracting service based on entire contents of FOOD SCIENCE AND TECH-NOLOGY ABSTRACTS is available on computer tapes from two centers in the United States:

COMPUTER SEARCH CENTER - Illinois Institute of Technology Research Institute, 10 West 25th Street, Chicago, Illinois 60616. Contact person: Mr. Peter B. Schimpa, Manager.

NORTH CAROLINA SCIENCE AND TECHNOLOGY RESEARCH CENTER - P.O. Box 12235, Research Triangle Park, North Carolina 27709. Contact person: Mr. Peter Chenery, Director.

Information services: Printouts of abstracts; Selective Dissemination of Information Service; retrospective searching of files; copy and translation services; workshops on services.

INTERNATIONAL FOOD RESEARCH AND EDUCATIONAL CENTER - North Easton, Massachusetts 02356.

An organization which provides educational programs for institutional food service on a worldwide basis. Conducts research into food preparation for large institutions and organizations.

Publications: Publications list includes books and manuals on foods and cookery, purchasing, training, and operation in food service, management of food preparation and service operations, and design and decor in such places.

INTERNATIONAL MILLING ASSOCIATION - 66 rue de la Boetie, 75-Paris 8e, France.

Purpose: Maintain relations, facilitate contacts and strengthen links between all concerned with the milling industry and promote its efficiency.

Publication: BULLETIN (irregular), for members only.

INTERNATIONAL ORGANIZATION OF THE FLAVOR INDUSTRY - 49 sq Marie-Louise, 1040 Brussels, Belgium.

Purpose: Study scientific data relative to aromatic substances, particularly as they affect human health; assemble and study control measures; put this information at the disposition of members and of international organizations.

INTERNATIONAL SUGAR CONFECTIONERY MANUFACTURERS ASSOCIATION 194 rue de Rivoli, 75-Paris ler, France.

Purpose: To collect and supply its members with information concerning general economic and social conditions, customs tariffs, trade operations, relations with international organizations, application of social laws; establish internationally a voluntary code of commercial practice; set up and help any body that may be in the interest of the trade.

Activities: Technical Committee on standardization of methods of analysis.

Publications: BULLETIN, in English, French, German. Booklet on METHODS OF ANALYSIS.

5. Other Information Sources and Programs

a. SPECIALIZED INFORMATION SERVICES

SWIFT AND COMPANY NUTRITION DATA BANK

A computerized nutrition Data Bank has been established which provides comprehensive analytical information for more than 2500 foods. Values included are energy, moisture, protein, fat, carbohydrate, nine vitamins (A, C, B_6, B_{12}, thiamine, riboflavin, niacin, pantothenic acid, and folacin) and 8 minerals (calcium, phosphorus, iron, zinc, sodium, potassium, magnesium, copper). Also fatty acids and amino acid data are compiled for numerous items.

For additional details, contact: Chief Nutritionist, Nutrition Data Bank Service, Research and Development Center, 1919 Swift Drive, Oak Brook, Illinois 60521.

B. LITERATURE

1. Guides and Handbooks

Blanck, F. HANDBOOK OF FOOD AND AGRICULTURE. New York: Van Nostrand-Reinhold, 1955.

CHEMISTRY OF FOOD PACKAGING. Washington, D.C.: American Chemical Society, 1974.

Church, C. FOOD VALUES OF PORTIONS COMMONLY USED. 12th ed. Philadelphia: Lippincott, 1975.

FOOD INDUSTRY YEARBOOK. Profit Press Inc., 170 Fifth Avenue, New York, New York 10010.

> A comprehensive reference on food marketing operations at all levels of the industry.

Furia, T., ed. HANDBOOK OF FOOD ADDITIVES. Cleveland: Chemical Rubber Co., 1972.

Furia, T., et al., ed. FERRAROLI'S HANDBOOK OF FLAVOR INGREDIENTS. 2nd ed. 2 vols. Cleveland: CRC Press, 1975.

Hugot, E. HANDBOOK OF CANE SUGAR ENGINEERING. Translated from the French by G.H. Jenkins. 2nd ed. New York: Elsevier, 1972.

Komarik, S., et al. FOOD PRODUCTS FORMULARY. Vol. 1. Westport, Connecticut: Avi, 1974.

Kutsky, R. HANDBOOK OF VITAMINS AND HORMONES. New York: Van Nostrand, 1973.

Lees, R. LABORATORY HANDBOOK OF METHODS OF FOOD ANALYSIS. 2nd ed. London: Leonard Hill, 1971.

> A useful handbook for laboratory personnel.

Shuey, W. FARINOGRAPH HANDBOOK. St. Paul, Minnesota: American Association of Cereal Chemists, 1972.

> "Handbook on one of the most widely used and researched rheological instruments was prepared under the auspices of the AACC committee on physical testing methods."

Woollen, A. FOOD INDUSTRIES MANUAL. New York Chemical Publishing Co., 1970.

a. TABLES

FOOD ADDITIVE TABLES. Amsterdam: Elsevier Scientific Publications Co., 1975.

2. Dictionaries

Bender, A. DICTIONARY OF NUTRITION AND FOOD TECHNOLOGY. London: Butterworths, 1975.

3. DIRECTORIES

a. BIOGRAPHY

American Meat Science Association. MEMBERSHIP DIRECTORY. American Meat Science Association, National Livestock & Meat Board, 36 South Wabash Avenue, Chicago, Illinois 60603.

"Giving names and addresses of members of the American Meat Science Association. Revised and published annually."

AMERICAN POTATO YEARBOOK. American Potato Yearbook, 834 South Avenue, W., P.O. Box 398, Westfield, New Jersey 07090.

"Contains current information on 500 potato research workers and potato organizations. Revised and published annually."

AMERICAN TOMATO YEARBOOK. American Tomato Yearbook, 834 South Avenue, W., P.O. Box 398, Westfield, New Jersey 07090.

"Names and addresses of persons engaged in tomato research in the United States, with a list of state agricultural colleges, experiment stations, and associations. Revised and published annually."

b. PRODUCTS AND MANUFACTURERS

i. United States

BAKING INDUSTRY BUYING DIRECTORY. Annual. Clissold Publishing Co., 105 West Adams Street, Chicago, Illinois 60603.

"2,500 listings of manufacturers and suppliers of bakery products. Revised and published annually."

BREWER'S DIGEST. Siebel Publishing Co., 4049 West Peterson Avenue, Chicago, Illinois 60646. Annual. January issue. "Buyers Guide and Directory for Brewing Industry."

CANADIAN FOOD AND PACKAGING DIRECTORY. Annual. Publications of Canada, 2813A Eglinton Avenue, Scarborough 701, Ontario, Canada.

DAIRY INDUSTRIES SUPPLIES AND SERVICES BUYING DIRECTORY. Annual. Miller Publishing Co., P.O. Box 67, 2501 Wayzata Boulevard, Minneapolis, Minnesota 55440.

"Guide to dairy industry supply and equipment manufacturers; dairy product manufacturers' associations, agricultural colleges giving instruction in dairy-product processing; federal, state, dominion and province officials; in the United States and Canada charged with enforcement of dairy and food laws; statistics dealing with dairy production. Revised and published annually."

THE FABRICATED FOODS MARKET. New York: Frost & Sullivan "Market Survey," 1976.

FOOD AND BEVERAGE INDUSTRY MARKETING GUIDE, 1971. Noyes Data Corp., Mill Road and Grand Avenue, Park Ridge, New Jersey 07656.

FOOD AND BEVERAGE PROCESSING INDUSTRIES 1971. Noyes Data Corp., Noyes Building, Park Ridge, New Jersey 07656.

> Book is divided into 2 sections. The first section is an alphabetical listing of approximately 3,500 U.S. food firms. The second section is arranged numerically according to zip code with the companies once again listed alphabetically within their proper zip code numbers."

FOOD ENGINEERING issues. Annual: "Buyers Guide to Process Equipment, and Ingredients Buyers Guide." Chilton Co., Radnor, Pennsylvania 19089.

"Guide to the Allied Trades of the Cereal Processing Industries." Annual directory in NORTHWESTERN MILLER, September issue. Miller Publishing Co., P.O. Box 67, 2501 Wayzata Boulevard, Minneapolis, Minnesota 55440.

Institute of Food Technologists. IFT WORLD DIRECTORY AND GUIDE. Annual. Institute of Food Technologists, 221 North La Salle Street, Chicago, Illinois 60601.

MACRAE'S BLUEBOOK. Annual. MacRae's Bluebook, Western Springs, Illinois 60558.

MANUFACTURING CONFECTIONER. "Buying Guide Issue." Annual. Manufacturing Confectioner Publishing Co. 1031 South Boulevard W., Oak Park, Illinois 60302.

THOMAS' REGISTER OF MANUFACTURERS. Annual. New York: Thomas Publishing Co.

U.S. Department of Agriculture. LIST OF CHEMICAL COMPOUNDS AUTHORIZED FOR USE UNDER USDA POULTRY, MEAT, RABBIT AND EGG PRODUCTS INSPECTION PROGRAMS. Revised April 1972. Animal and Plant Health Inspection Service, Government Printing Office, Washington, D.C. 20250.

ii. International

BAKER AND BAKERY MANAGEMENT HANDBOOK AND BUYERS GUIDE. Annual. Trade Publications Ltd., John Adam House, 17-19 John Adam Street, Adelphi, Strand, London, WC2, England.

BINSTED'S DIRECTORY OF FOOD TRADE MARKS AND BRAND NAMES. Bi-annual. Food Trade Review Ltd., 7 Garrick Street, London WC2, England.

EUROPEAN FOOD MARKET RESEARCH SOURCES. Park Ridge, New Jersey: Noyes Data Corp., 1970.

> "References are classified by government statistics and reports, other statistics and reports, trade associations, food trade journals; newspapers and periodicals, directories, advertising statistics, bank reviews, etc."

FOOD GUIDE TO EUROPE. 2nd ed. Park Ridge, New Jersey: Noyes Data Corp., 1972.

> "Describes 1,600 of the leading food processing companies located in 18 countries of Western Europe."

FOOD PROCESSING AND PACKAGING DIRECTORY. Annual. Consumer Industries Press, Ltd., 40 Bowling Green Lane, London E.C.I., England.

FOOD TRADES DIRECTORY AND FOOD BUYER'S YEARBOOK. Annual. International Publications, Inc., 381 Park Avenue, S., New York, New York 10016.

FROZEN FOODS YEARBOOK. Annual. Refrigeration Press, Croydon, Surrey, England.

INTERNATIONAL FOOD REGISTER: MONTHLY BUYER'S GUIDE. Economic Documentation Office, Graaf Florislaan 30a, Box 505, Hilversum, Netherlands.

United Nations. Food and Agriculture Organization. INTERNATIONAL DIRECTORY OF AGRICULTURAL ENGINEERING INSTITUTIONS. Rome: FAO, 1973.

c. LABORATORIES AND RESEARCH CENTERS

EUROPEAN RESEARCH INDEX. 2nd ed. Guernsey, England: Francis Hodgson, 1969.

INDUSTRIAL RESEARCH LABORATORIES OF THE UNITED STATES. 13th ed. New York: R.R. Bowker, 1970.

RESEARCH CENTERS DIRECTORY. 5th ed. Detroit: Gale Research Co., 1975.

> Updated by subscription service, NEW RESEARCH CENTERS.

4. Encyclopedias

Hall, C. ENCYCLOPEDIA OF FOOD ENGINEERING. Westport, Connecticut: Avi, 1971.

THE IFMA ENCYCLOPEDIA OF THE FOODSERVICE INDUSTRY. 2nd ed. Chicago: International Foodservice Manufacturers Association, 1975.

Johnson, A., et al. ENCYCLOPEDIA OF FOOD TECHNOLOGY. Westport, Conn: Avi, 1974.

Wollen, A. FOOD INDUSTRIES MANUAL. Revised and enlarged ed. New York: Chemical Publishing Co., Inc., 1970.

Authoritative encyclopedia of food information.

5. Bibliographies

Baker, E. BIBLOGRAPHY OF FOOD; A SELECT INTERNATIONAL BIBLIOG-RAPHY OF NUTRITION, FOOD AND BEVERAGE TECHNOLOGY AND DISTRIBUTION, 1936-56. New York: Academic Press, 1958.

FOOD PACKAGING AND STORAGE. By E. Harrison. Report for 1970-1976. Rept. no. NTIS/PS-76/0406. Springfield, Virginia: National Technical Information Service, 1976.

A bibliography with abstracts.

Mann, E. EVALUATION OF THE WORLD FOOD LITERATURE: RESULTS OF AN INTERNATIONAL SURVEY. Farnham Royal, England: Commonwealth Agricultural Bureaux, 1967.

Reading University. Department of Food Science. A SELECT BIBLIOGRAPHY AND LIBRARY GUIDE TO THE LITERATURE OF FOOD SCIENCE. Compiled by E. Apling. Reading, England: 1969.

SYNTHETIC FOODS. By E.J. Lehmann. Report for 1964-1976. Rept. no. NTIS/PS-76/0534. Springfield, Virginia: National Technical Information Service, 1976.

A bibliography with abstracts.

U.S. Department of Agriculture. 74-59. BIBLIOGRAPHY AND ABSTRACTS ON POULTRY AND EGG RESEARCH IN WESTERN MARKETING AND NUTRITION RESEARCH DIVISION, 1950-70. Washington, D.C.: Government Printing Office, October 1971.

U.S. Department of Commerce. 15564. GLOBAL MARKET SURVEY: FOOD PROCESSING AND PACKAGING MACHINERY EQUIPMENT WITH BIBLIOGRAPHY. Washington, D.C.: Government Printing Office, August 1971.

U.S. Department of Defense. Army Department. Natick Laboratories, Natick, Massachusetts: Technical report (Army Materiel Command). ATMOSPHERIC STEAM SUPERHEATER FOR STEAM DISTILLATION OF FOOD ANTI-OXIDANTS WITH LIST OF LITERATURE CITED. By John Swift. Dominic Laboratory Series FL-109. AD720212. Springfield, Virginia: National Technical Information Service, 1972.

U.S. Department of Health, Education and Welfare. FDA INFORMATIONAL MATERIALS FOR FOOD AND COSMETIC INDUSTRIES (MOVIES, SLIDES, POSTERS, BOOKLETS AND FACT SHEETS). Washington, D.C.: Food and Drug Administration, 1972.

U.S. Government Printing Office. PRICE LIST SERIES. No. 11. HOME ECONOMICS, FOODS, COOKING. Washington, D.C.: 1972.

    Contains list of documents for sale.

Vara, A. FOOD AND BEVERAGE INDUSTRIES: A BIBLIOGRAPHY AND GUIDEBOOK. Detroit: Gale Research Co., 1970.

a. RELATED BIBLIOGRAPHIES IN CURRENT JOURNALS

The following journals frequently contain book reviews and/or bibliographies. See the heading Journals and Serial Publications in this section for addresses.

The American Oil Chemists Society. JOURNAL

Baking Research Association. ABSTRACTS

BRITISH FOOD JOURNAL

CEREAL CHEMISTRY

FAO DOCUMENTATION - CURRENT INDEX

FOOD AND COSMETICS TOXICOLOGY

FOOD AND LIQUOR DIGEST

FOOD ENGINEERING

FOODPACK

FOOD PROCESSING

FOOD PROCESSING INDUSTRY

FOOD PRODUCT DEVELOPMENT

FOOD TECHNOLOGY

FOOD TRADE REVIEW

FOOD WORLD

INTERNATIONAL SUGAR JOURNAL

MILCHWISSENSCHAFT

See also this subject in the section Chemical Engineering – General.

6. Monographs, Reports, and Symposia Proceedings (Selected)

Arbuckle, W. ICE CREAM. Westport, Connecticut: Avi Publishing, 1972.

"Definitive work on ice cream, ice cream manufacture, and the frozen dessert industry."

Bednarcyk, N. FATTY ACIDS SYNTHESIS AND APPLICATIONS. Park Ridge, New Jersey: Noyes Data Corp., 1973.

Bieber, H., ed. ENGINEERING OF UNCONVENTIONAL PROTEIN PRODUCTION. AICHE Chemical Engineering Progress Symposium, Series, No. 93. New York: American Institute of Chemical Engineers, 1969.

Brennan, J. FOOD ENGINEERING OPERATIONS. New York: Elsevier Publishing Co., 1969.

Cook, L. CHOCOLATE FOR PRODUCTION AND USE. New York: Magazines for Industry, 1972.

"Every phase of chocolate production and use is dealt with, from the growing of the bean to the end product of coating, ice cream novelty, candy, or baked item."

Earle, R. UNIT OPERATIONS IN FOOD PROCESSING. New York: Pergamon, 1966.

8TH MEETING OF THE JOINT FAO/WHO EXPERT COMMITTEE ON NUTRITION. Food and Agriculture Organization and World Health Organization. Publisher: Food and Agriculture Organization, Rome. U.S. Distributor: Unipub, Inc., P.O. Box 433, New York, New York 10016. FAO NUTRITION MEETINGS. REPORT SERIES no. 49; WHO TECH. REPORT SERIES no. 477, 1971.

European Brewery Convention. PROCEEDINGS. The proceedings of the 13th congress, 1971. New York: Elsevier, 1972.

Farrall, A. ENGINEERING FOR DAIRY AND FOOD PRODUCTS. New York: Wiley, 1963.

Findlay, W., ed. MODERN BREWING TECHNOLOGY. New York: Macmillan, 1971.

Comprehensive coverage of brewing field.

FLAVOR CHEMISTRY. ADVANCES IN CHEMISTRY SERIES. 56. Washington, D.C.: American Chemical Society, 1966.

Food and Beverage Instrumentation Symposium, 1st, Montreal, 1972. INSTRU-
MENTATION IN THE FOOD AND BEVERAGE INDUSTRY. Proceedings of the
first international ISA Food Instrumentation Division symposium. Programmed
by ISA's Food Instrumentation Division. Edited by E. Nobrega. Pittsburgh:
Instrument Society of America, 1972.

Food and Bioengineering - Fundamental and Industrial Aspects, Symposia. New
Orleans, Louisiana, March, 1969; and Los Angeles, California, December 1969.
Sponsored by AICHE. Proceedings issued as: AICHE CHEMICAL ENGINEERING
PROGRESS SYMPOSIUM SERIES, vol. 67, no. 108, 1971.

FOOD FOR MANKIND. 2 vols. Westport, Connecticut: Technomic Publishing Co., 1973.

>"Information on food sources and requirements presented in compact,
>easy-to-use form. The presentation is comprehensive in scope with
>numerous data tables and a bibliography."

FOOD SCIENCE AND TECHNOLOGY. 4th International Congress of The International Union of Food Science and Technology, Madrid, Spain, September
1974. Proceedings. 6 vols. 1976. Available from Jaime Roig, 11 Valencia
10, Spain.

Gillies, M. SEAFOOD PROCESSING. Park Ridge, New Jersey: Noyes
Data Corp., 1971.

>Overview of this type of food industry.

Gutcho, M. TEXTURED FOODS AND ALLIED PRODUCTS. Park Ridge, New
Jersey: Noyes Data Corp., 1973.

>"Deals primarily with the preparation of -meat-like foods from
>readily available protein, such as soy protein, peanut protein,
>and wheat gluten." Based on review of patent literature.

Gutterson, M. BAKED GOODS PRODUCTION PROCESSES. Park Ridge,
New Jersey: Noyes Data Corp., 1969.

>"This book describes 201 recent processes for the production of
>baked goods. Based on the patent literature."

_____. FOOD CANNING TECHNIQUES. Park Ridge, New Jersey: Noyes
Data Corp., 1972.

>Based on review of patent literature.

Findlay, W., ed. FRUIT JUICE TECHNOLOGY 1970. Park Ridge, New
Jersey: Noyes Data Corp., 1970.

>"This publication deals with the technology of the non-carbonated
>fruit juice industry from 1960 through 1970 as covered in the U.S.
>patent literature."

_____. FRUIT PROCESSING 1971. Park Ridge, New Jersey, Noyes Data Corp., 1971.

> 140 processes based on review of recent patent literature since 1960.

_____. VEGETABLE PROCESSING 1971. Park Ridge, New Jersey: Noyes Data Corp., 1971.

> Based on patent reviews.

Hart, F. MODERN FOOD ANALYSIS. New York: Springer-Verlag, 1971.

> Wide coverage of topics on food analysis. Useful as a reference guide.

Health & Food, Industry-University Cooperation Symposium. Sponsored by University of Reading, National College of Food Technology, Reading, England. Edited by Gordon G. Birch, Leslie Frank Green, and L.G. Plaskett, 1972. Halsted Press, Division of John Wiley & Sons, 605 Third Avenue, New York, New York 10016.

Hough, J. MALTING AND BREWING SCIENCE. London: Chapman & Hall, 1971.

> "A wealth of material on the whole subject, including much on such topical themes as continuous processes as applied to malting and fermentation."

International Atomic Energy Agency. NUCLEAR TECHNIQUES FOR STUDYING PESTICIDE RESIDUE PROBLEMS. Proceedings of a panel on isotopic tracer and radioactivation techniques for studying residue problems with particular reference to food entering international trade, organized by the joint FAO/IAEA division of Atomic Energy in Food and Agriculture, and held in Vienna, 16-20 December, 1968. New York: Unipub, 1971.

International Commission for Uniform Methods of Sugar Analysis. FIFTEENTH SESSION; PROCEEDINGS. London: 1971.

International Conference on The Amino Acid Fortification of Protein Foods. REPORT. Edited by N.S. Scrimshaw and A.M. Altschul. Cambridge: M.I.T. Press, 1971.

Karmas, E. FRESH MEAT PROCESSING. Park Ridge, New Jersey: Noyes Data Corp., 1970.

> Based on review of patent literature since 1960.

_____. MEAT PRODUCT MANUFACTURE 1970. Park Ridge, New Jersey: Noyes Data Corp., 1970.

> "This Review concerns latest technology in preparing packaged meats in ready-to-cook and ready-to-eat forms."

_____. SAUSAGE PROCESSING. Park Ridge, New Jersey: Noyes Data Corp.; New York: Academic Press, 1971.

Provides a state of the art by giving details on patents since 1960.

McKenzie, H., ed. MILK PROTEINS. New York: Academic Press, 1971.

Covers various aspects of milk proteins.

Matz, S. BAKERY TECHNOLOGY AND ENGINEERING. 2nd ed. Westport, Connecticut: Avi Publishing, 1972.

"Comprehensive guide to the science and technology of baking. Covers the principles of modern, large-scale bakery processing techniques."

National Research Council. Committee on Food Protection. FOOD CHEMI-CALS CODEX. 2nd ed. Washington, D.C.: National Academy of Sciences, 1972.

"Contains 639 monographs. Monographs are provided for chemicals added directly to food to perform some desired function as well as substances not added directly to food but which come into contact with food, such as food processing aids."

Nickerson, J. MICROBIOLOGY OF FOODS AND FOOD PROCESSING. New York: Elsevier, 1972.

"Methods used for the microbiological examination of foods, growth of microorganisms in foods, and preservation of foods by various means."

Noyes, R. PROTEIN FOOD SUPPLEMENTS 1969. Park Ridge, New Jersey: Noyes Data Corp., 1969.

"The 126 processes in this book are organized in eight chapters by raw material source including the important newer processes for producing protein by fermentation of hydrocarbons."

NUTRITION, NATIONAL DEVELOPMENT, & PLANNING. International Conference. Edited by E. Berg, et al. Cambridge: MIT Press, 1973.

Paturau, J. BY-PRODUCTS OF THE CANE SUGAR INDUSTRY; AN INTRODUC-TION TO THEIR INDUSTRIAL UTILIZATION. New York: American Elsevier, 1969.

"Brings together much widely dispersed detail from the literature on cane sugar by-product use."

Pearson, D. CHEMICAL ANALYSIS OF FOODS. 6th ed. London: Churchill, 1970.

A standard reference tool for practitioners and specialists in food chemistry.

Pintauro, N. AGGLOMERATION PROCESSES IN FOOD MANUFACTURE. Park Ridge, New Jersey: Noyes Data Corp., 1972.

Contents include review of agglomeration process as well as list of U.S. patents area since 1960.

_____. FLAVOR TECHNOLOGY. Park Ridge, New Jersey: Noyes Data Corp., 1971.

Based on patent reviews.

Pomeranz, P. FOOD ANALYSIS; THEORY AND PRACTICE. Westport, Connecticut: Avi Publishing, 1971.

Comprehensive treatment of the field. Includes section searching literature.

Pomeranz, Y., ed. WHEAT; CHEMISTRY AND TECHNOLOGY. 2nd ed. St. Paul, Minnesota: American Association of Cereal Chemists, 1971.

"Describes in detail the present status of our knowledge in wheat chemistry and technology and discusses the significance and relevance of recent biochemical findings or new technological processes."

Price, J., ed. THE SCIENCE OF MEAT AND MEAT PRODUCTS. 2nd ed. San Francisco: W.H. Freeman, 1972.

Provides overview of meat processing techniques.

Recent Advances in the Chemistry of Food Irradiation. 2nd Symposium on Food Preservation by Ionizing Radiation, at 161st meeting. Published as RADIATION RESEARCH REVIEWS. Vol. 3, no. 4. Amsterdam: Elsevier, 1972.

RHEOLOGY AND TEXTURE OF FOOD STUFFS. London: Society of the Chemical Industry, 1968.

Rice, Science & Man, symposium. International Rice Research Institute, c/o Manila Hotel, Manila, Philippines. 1972.

Rose, A., ed. THE YEASTS. 3 vols. New York: Academic Press, 1971.

A review of known properties of yeast growth.

SEED PROTEINS -- SYNTHESIS, PROPERTIES, AND PROCESSES. Westport, Connecticut: Avi Publishing, 1972.

Proceedings of an American Chemical Society symposia.

Sfat, M., ed. BIOENGINEERING AND FOOD PROCESSING. AICHE Chemical Progress Symposium Engineering Series, no. 69. New York: American Institute of Chemical Engineers, 1966.

Slade, F. FOOD PROCESSING PLANT. 2 vols. London: L. Hill, 1971.

"Unit operations or processes employed in food manufacture."

Smith, A., ed. SOYBEANS: CHEMISTRY AND TECHNOLOGY. Westport, Connecticut: Avi Publishing, 1972.

Society of the Chemical Industry, London. PRODUCTION AND APPLICATION OF ENZYME PREPARATIONS IN FOOD INDUSTRY. New York: Macmillan, 1971.

_____. TEXTURE IN FOODS. Comprising papers read at a symposium organized by the Food Group and held in London October 13-14, 1958. SCI Mono 7. New York: Macmillan, 1960.

SURFACE ACTIVE LIPIDS ON FOODS. Comprising papers read at a joint symposium organized by the Food Group and the Oils and Fats Group of the Society of the Chemical Industry, March 21-22, 1968, the School of Pharmacy, University of London. London: Society of the Chemical Industry, 1968.

Symposium on Quality Control of Food, London, 1957. THE QUALITY CONTROL OF FOOD. Comprising papers read at a symposium organized by the Food Group and the Association of Public Analysts. SCI mono 8. New York: Macmillan, 1960.

Toxicological Evaluation of Some Enzymes, Modified Starches & Certain Other Substances. Meeting of the Joint FAO/WHO Expert Committee on Food Additives. Sponsored by Food & Agriculture Organization & World Health Organization, Rome, Italy. World Health Organization, Palais des Nations, CH-1211 Geneva 27, Switzerland. (WHO FOOD ADDITIVES SERIES, No. 1; FAO Nutrition meetings report series, No. 50A). 1972.

U.S. Atomic Energy Commission. COO-1592-36. IRRADIATION OF FRUITS AND VEGETABLES, ANNUAL REPORT. April 1968 - July 15, 1970. Prepared by P. Narkakis. Springfield, Virginia: National Technical Information Service, 1971.

_____. UCD-34-P-80-9. RADIATION TECHNOLOGY IN CONJUNCTION WITH POSTHARVEST PROCEDURES AS MEANS OF EXTENDING SHELF LIFE OF FRUITS AND VEGETABLES, ANNUAL REPORT. April 15, 1970 - January 1, 1971. Springfield, Virginia: National Technical Information Service, 1971.

U.S. Department of Agriculture. QUALITY AND ITS PRESERVATION IN PROCEESSED FOODS, PROCEEDINGS. Eastern Experiment Station Collaborators' Conference, October 26-27, 1971. (EMN publication 3718.) Washington, D.C.: Government Printing Office, 1972.

U.S. Department of Agriculture. Marketing research reports. 921. STUDY
OF REFRIGERATION SYSTEMS FOR URBAN FOOD DISTRIBUTION CENTERS.
By R.L. Stahlman. Washington, D.C.: Government Printing Office, 1972.

U.S. Department of Interior. DESALINATION PLANT LABORATORY PRO-
CEDURES MANUAL WITH LIST OF REFERENCES AND BIBLIOGRAPHY. By
R.M. Burd, et al. Washington, D.C.: Government Printing Office.

_____. DESALINATION PLANT OPERATORS CHEMICAL CONTROL GUIDE.
By W.R. Greenway and K.L. Atwood. Water. Washington, D.C.: Government Printing Office, 1973.

_____. 1973-74 SALINE WATER CONVERSION SUMMARY REPORT. Wash-
ington, D.C.: Government Printing Office, 1974.

> Office of Saline Water issues periodically report and manuals.
> For additional details consult the Monthly Catalog of The Super-
> intendent of Documents or contact the Department of Interior.

Webb, B., ed. BYPRODUCTS FROM MILK. 2nd ed. Westport, Connecticut:
Avi Publishing, 1971.

> "This volume is a comprehensive well-documented reference which
> any processor of frozen milk products should find indispensible.
> An extensive list of source material is also included at the end
> of every chapter."

Weiss, G. POULTRY PROCESSING. Park Ridge, New Jersey: Noyes Data
Corp., 1971.

> Reviews 55 processes in 58 U.S. patents issued since 1960.

Whitaker, J. PRINCIPLES OF ENZYMOLOGY FOR THE FOOD SCIENCES.
New York: Dekker, 1972.

Wieland, H. ENZYMES IN FOOD PROCESSING AND PRODUCTS. Park
Ridge, New Jersey: Noyes Data Corp., 1972.

> Based on patent reviews.

Wilcox, G. EGGS, CHEESE AND YOGURT PROCESSING. Park Ridge,
New Jersey: Noyes Data Corp., 1971.

> Provides review of current processes based on recent patents.

_____. MILK, CREAM AND BUTTER TECHNOLOGY. Park Ridge, New
Jersey: Noyes Data Corp., 1971.

> Reviews U.S. Patent literature 1960-70 to provide details on
> present techniques and methods.

Wingard, L., comp. ENZYME ENGINEERING. New York: Interscience
Publishers, 1972.

"Papers in part from the Engineering Foundation conference on 'enzyme engineering,' August 9-13, 1971, Henniker, New Hampshire, and the 64th annual meeting of the American Institute of Chemical Engineers, December 1, 1971, San Francisco, California."

7. Abstracting and Indexing Services

a. GENERAL

APPLIED SCIENCE AND TECHNOLOGY INDEX

BIOLOGICAL ABSTRACTS

CHEMICAL ABSTRACTS

ENGINEERING INDEX

INDEX MEDICUS

See this topic in the Section I of this Guide for further details on these services.

b. SPECIFIC TO TOPIC

ABSTRACTS FOR THE ADVANCEMENT OF INDUSTRIAL UTILIZATION OF CEREAL GRAINS. Bimonthly. Washington State University, College of Engineering Research Division, Pullman, Washington, 99163.

ABSTRACTS FROM CURRENT SCIENTIFIC AND TECHNICAL LITERATURE. Monthly. British Food Manufacturing Industries Research Association, Randalls Road, Leatherhead, Surrey, England.

BAKING RESEARCH ASSOCIATION, ABSTRACTS. Bimonthly. Flour Milling and Baking Research Association, Chorleywood, Rickmansworth, Herts WD3 5SH, England.

Commonwealth Bureau of Dairy Science and Technology. FOOD SCIENCE AND TECHNOLOGY ABSTRACTS. Monthly. Farnham Royal, Slough SL2 3BN England.

CURRENT CONTENTS: AGRICULTURAL FOOD AND VETERINARY SCIENCES. Weekly. Philadelphia, Institute for Scientific Information.

Reproduces tables of contents of key journals.

DAIRY SCIENCE ABSTRACTS. Monthly. Farnham Royal, England: Commonwealth Agricultural Bureaux.

FAO DOCUMENTATION -- CURRENT INDEX. Monthly. Food and Agriculture Organization of the United Nations, Documentation Centre, Via delle Terme di Caracalla, Rome, Italy.

> Text in English, French, and Spanish; summaries in English.

FOOD PROCESSING ABSTRACTS. Weekly. Evanston, Illinois: Lowry-Cocroft Abstracts.

> "Card service, about 1,600 abstracts per annum."

FOOD SCIENCE AND TECHNOLOGY ABSTRACTS. Monthly. International Food Information Service, Commonwealth Agricultural Bureaux, Farnham Royal, Bucks, England.

INTERNATIONAL SUGAR JOURNAL. Monthly. London: International Sugar Journal.

> Approximately 1,000 abstracts.

MILCHWISSENSCHAFT: MILK SCIENCE INTERNATIONAL. Monthly. Verlag Hans Carl, K.G., Breite Gasse 58-60, Nuernberg, West Germany.

> Text in English, French, German, Italian, and Spanish.

Muszaki lapszemle. ELELMISZERIPAR/TECHNICAL ABSTRACTS FOOD INDUSTRY. Monthly. Orszagos Muszaki Konyvtar es Dokumentacios Kozpont, Reviezky u.6., Budapest 8, Hungary.

NUTRITION ABSTRACTS AND REVIEWS. Quarterly. Issued under the direction of the Commonwealth Agricultural Bureaux Council, the Medical Research Council, and the Reid Library. Aberdeen, Scotland: Commonwelath Bureau of Animal Nutrition.

> Annual combined table of contents and author and subject indexes.

Predi-briefs. ABSTRACTING SERVICE. Monthly. FOOD AND BEVERAGES. Predicasts, 200 University Circle Research Center, 11001 Cedar Avenue, Cleveland, Ohio 44106.

> Covers food and beverage market data and the food industry's use of chemical products, food additives, and supplements as well as fats and oils. Also FDA regulatory action.

SEARCH: FOODSTUFFS DIVISION. Monthly. Compendium Publishers International Corp., 2175 Lemoine Avenue, Fort Lee, New Jersey 07024.

SUGAR INDUSTRY ABSTRACTS. Monthly. Keston, Kent, Tate & Lyle, Ltd., for the Sugar Refiners' Association and the British Sugar Corporation. London.

U.S. Food and Drug Administration. CHEMICAL EXPERIENCE ABSTRACTS. Monthly. Washington, D.C.: Food and Drug Administration.

c. RELATED ABSTRACTS IN CURRENT JOURNALS

The following journals frequently provide abstracts. See the heading Journals and Serial Publications in this section for addresses.

AMERICAN OIL CHEMISTS SOCIETY JOURNAL

BRITISH FOOD JOURNAL

CEREAL CHEMISTRY

FOOD AND COSMETICS TOXICOLOGY

FOOD ENGINEERING

FOOD MANUFACTURE

FOOD TRADE REVIEW

INTERNATIONAL SUGAR JOURNAL

JOURNAL OF SCIENCE OF FOOD AND AGRICULTURE

JOURNAL OF THE INSTITUTE FOR BREWING

Southern Marketing and Nutrition Research Division. PUBLICATIONS AND PATENTS

8. Journals and Serial Publications

a. DIRECTORIES AND LISTS

AYERS DIRECTORY OF PUBLICATIONS. Annual. Philadelphia, Pennsylvania: Ayer Press.

STANDARD PERIODICAL DIRECTORY (and supplement). Irregular. New York: Oxbridge Publishing Co.

ULRICH'S INTERNATIONAL PERIODICAL DIRECTORY. 2 vols. Annual. New York: R.R. Bowker Co.

b. JOURNAL AND SERIAL TITLES

ABSTRACTS FOR THE ADVANCEMENT OF INDUSTRIAL UTILIZATION OF CEREAL GRAINS. Bimonthly. Washington State University, College of Engineering Research Division, Pullman, Washington 99163.

ABSTRACTS FROM CURRENT SCIENTIFIC AND TECHNICAL LITERATURE. Monthly. British Food Manufacturing Industries Research Association, Randalls Road, Leatherhead, Surrey, England.

ADVANCES IN CARBOHYDRATE CHEMISTRY AND BIOCHEMISTRY. Edited by R.S. Tipson and D. Horton. New York: Academic Press, 1972.

Reviews latest research results on the subject.

ADVANCES IN FOOD RESEARCH. New York: Academic Press, 1971.

Surveys advances in food research.

ADVANCES IN FOOD RESEARCH. SUPPLEMENT. New York: Academic Press, 1972.

ADVANCES IN PROTEIN CHEMISTRY. New York: Academic Press, 1972.

Critical reviews of progress in field.

The American Oil Chemists' Society. JOURNAL. Monthly. 35 East Wacker Drive, Chicago, Illinois 60601.

American Society of Bakery Engineers. PROCEEDINGS OF THE ANNUAL MEETING. Annual. American Society of Bakery Engineers, Rm. 1921, 2 North Riverside Plaza, Chicago, Illinois 60606.

Baking Research Association. ABSTRACTS. Bimonthly. Flour Milling and Baking Research Association, Chorleywood, Rickmansworth, Herts WD3 5SH, England.

BRITISH FOOD JOURNAL. Monthly. Peterson Publishing Co., Ltd., Peterson House, Livery Street, Birmingham 3, England.

CEREAL CHEMISTRY. Bimonthly. American Association of Cereal Chemists, Inc., 3340 Pilot Knob Road, St. Paul, Minnesota 55121.

Official journal of the association which publishes research news of cereal, oilseeds, and related industries.

CEREAL SCIENCE TODAY. Monthly. American Association of Cereal Chemists, Inc., 3340 Pilot Knob Road, St. Paul, Minnesota 55121.

Supplies current information on applied cereal chemistry and technology.

Commonwealth Bureau of Dairy Science and Technology. FOOD SCIENCE AND TECHNOLOGY ABSTRACTS. Monthly. Commonwealth Agricultural Bureaux, Farnham Royal, Slough, SL2 3BN, England.

CURRENT CONTENTS: AGRICULTURAL FOOD AND VETERINARY SCIENCES. Weekly. Philadelphia: Institute for Scientific Information.

Reproduces tables of contents of key journals.

U.S. Department of Commerce. CURRENT INDUSTRIAL REPORTS. Commerce Department, Social and Economic Statistics Administration, Census Bureau. Monthly. Washington, D.C.: Government Printing Office.

Contains reports of food industry.

DAIRY SCIENCE ABSTRACTS. Monthly with annual compilation. Commonwealth Agricultural Bureaux, Farnham Royal, Slough, SL2 3BN, England.

Compiled from world literature.

FAO DOCUMENTATION - CURRENT INDEX. Monthly. Food and Agriculture Organization of the United Nations, Documentation Centre, Via delle Terme di Caracalla, 00100 Rome, Italy.

Text in English, French, and Spanish; summaries in English.

FINE FOODS AND BEVERAGES INTERNATIONAL. Bimonthly. Gusto Communications, Inc., 360 Lexington Avenue, New York, New York 10017.

FOOD AND AGRICULTURE LEGISLATION. Semiannual. Food and Agriculture Organization of the United Nations, Distribution and Sales Section, Via delle Terme di Caracalla, 00100 Rome, Italy.

Editions in English, French, and Spanish.

FOOD AND EQUIPMENT PRODUCT NEWS. 9 issues per year. Service Industry Communications Inc., 347 Madison Avenue, New York, New York 10017.

FOOD AND LIQUOR DIGEST. Monthly. F.S. Publications, Box 2458, Palos Verdes Peninsula, California 90274.

FOOD AND NUTRITION. Bimonthly. Agriculture Department, Food and Nutrition Service. Washington, D.C.: Government Printing Office.

FOOD AND NUTRITION NEWS. 9 times per year. National Livestock and Meat Board, 36 South Wabash Avenue, Chicago, Illinois 60603.

FOOD CHEMICAL NEWS. Weekly. Food Chemical News, Inc., 601 Warner Building, Washington, D.C. 20004.

FOOD ENGINEERING. Monthly. Chilton Co., Radnor, Pennsylvania 19089.

Covers food processing field aspects of technology, equipment, research, as well as communications from within the industry.

FOOD IN CANADA. Monthly. Maclean-Hunter Ltd., 481 University Avenue, Toronto 2, Ontario, Canada.

FOOD INDUSTRIES REVIEW. Annual. Gower Press Ltd., P.O. Box 5, Epping, Essex, England.

FOOD INGREDIENTS AND EQUIPMENT. Quarterly. Putman Publishing Co., 111 East Delaware Place, Chicago, Illinois 60611.

FOOD MANUFACTURE. Monthly. 28 Essex Street, Strand, London, WC2, England.

FOODPACK. Quarterly. B P S Exhibitions, 6 London Street, London W2, England.

Foodpack news and digest and produce packaging.

FOOD PLANT IDEAS. 10 times per year. Lakewood Publications, Inc., 1645 Hennepin Avenue, Minneapolis, Minnesota 55403.

News of equipment, ingredients, packaging, and supplies.

FOOD PROCESSING. Monthly. Putman Publishing Co., 111 East Delaware Place, Chicago, Illinois 60611.

FOOD PROCESSING ABSTRACTS. Weekly. Evanston, Illinois: Lowry-Cocroft Abstracts.

"Card service, about 1,600 abstracts per annum."

FOOD PROCESSING INDUSTRY. Monthly. IPC Consumer Industries Press Ltd., 40 Bowling Green Lane, London EC1, England.

FOOD PROCESSING REVIEW. Irregular. Park Ridge, New Jersey: Noyes Data Corp.

FOOD PRODUCT DEVELOPMENT. 8 issues per year. Arlington Publishing Co., 2 North Riverside Plaza, Chicago, Illinois 60606.

FOOD SCIENCE. Irregular. Marcel Dekker, Inc., 95 Madison Avenue, New York, New York 10016.

FOOD SCIENCE AND TECHNOLOGY ABSTRACTS. Monthly. International Food Information Service, Commonwealth Agricultural Bureaux, Farnham Royal, Bucks, England.

FOOD TECHNOLOGY. Monthly. Institute of Food Technologists, 221 North LaSalle Street, Chicago, Illinois 60601.

FOOD TRADE REVIEW. Monthly. Food Trade Review Ltd., 7 Garrick Street, London WC2E 9AT, England.

FOOD WORLD. Monthly. 65-66 Turnmill Street, London EC1, England.

FROZEN FOOD AGE. Monthly. Frozen Food Age Publishing Corp., 230 Park Avenue, New York, New York 10017.

"The industry magazine of marketing and merchandising."

FROZEN FOOD MARKETING. Bimonthly. Western Frozen Food Institute, 1467 Echo Park Avenue, Los Angeles, California 90026.

FROZEN FOODS. Monthly. Retain Journals Ltd., 2 Queensway, Redhill, WC2, Surrey, England.

International Association for Cereal Chemistry. MINUTES OF MEETINGS. Irregular. International Association for Cereal Chemistry, Schmidgasse 3-7, A-2320, Schwechat, Austria.

International Commission of Sugar Technology. PROCEEDINGS OF THE GENERAL ASSEMBLY. Every 3-4 years. Commission International Technique de Sucrerie, General Secretary, 1 Aandorenstratt, Tienen, Belgium.

    Text in English, French, German.

_____. REPORTS OF MEETINGS. Irregular. International Commission of Sugar Technology, 1 Aandorenstraat, 3300-Tienen, Belgium.

    Text in English, French, German.

International Congress of Food Science and Technology. Irregular. Institute of Food Technologists, 221 North LaSalle Street, Chicago, Illinois 60607.

INTERNATIONAL SUGAR JOURNAL. Monthly. 23a Easton Street, High Wycombe, Bucks, England.

    A technical and commercial periodical devoted entirely to the sugar industry.

JACOBSEN'S FATS & OILS BULLETIN. Daily. Jacobsen Publishing Co., 300 West Adams Street, Chicago, Illinois 60606.

JOURNAL OF MILK AND FOOD TECHNOLOGY. Monthly. International Association of Milk, Food and Environmental Sanitarians, Inc., P.O. Box 437, Shelbyville, Indiana 46176.

JOURNAL OF THE INSTITUTE OF BREWING. Bimonthly. London: The Institute.

    "Has a section: Abstracts, covering 50 technical journals."

JOURNAL OF THE SCIENCE OF FOOD AND AGRICULTURE. Monthly. London: Society of Chemical Industry.

    "The Journal has included 'Food and Agriculture Abstracts' since 1954."

MANUFACTURING CONFECTIONER. Monthly. Manufacturing Confectioner Publishing Co., 1031 South Blvd., West Oak Park, Illinois 60302.

MILCHWISSENSCHAFT: MILK SCIENCE INTERNATIONAL. Monthly. Verlag Hans Carl, K.G., Breite Gasse 58-60, Nuernberg, West Germany.

    Text in English, French, German, Italian, and Spanish.

MODERN BREWERY AGE. Weekly. Modern Brewery Age Publishing Corp., 80 Lincoln Avenue, Stanford, Connecticut 06904.

Muzaki lapszemle. TECHNICAL ABSTRACTS FOOD INDUSTRY. Monthly. Orszagos Muszaki Konyvtar es Dokumentacios Kozpont, Reviezky u.6., Budapest 8, Hungary.

NUTRITION ABSTRACTS AND REVIEWS. Quarterly. Commonwealth Agricultural Bureaux Council, the Medical Research Council, and the Reid Library. Aberdeen, Scotland: Commonwealth Bureau of Animal Nutrition.

> With annual combined table of contents and author and subject indexes.

Predi-briefs. Abstracting service. Monthly. FOOD AND BEVERAGES. Predicasts, 200 University Circle Research Center, 11001 Cedar Avenue, Cleveland, Ohio 44106.

> "Covers food and beverage market data and the food industry's use of chemical products, food additives and supplements as well as fats and oils. Also FDA regulatory action."

SEARCH: FOODSTUFFS DIVISION. Monthly. Compendium Publishers International Corp., 2175 Lemoine Avenue, Fort Lee, New Jersey 07024.

SUGAR INDUSTRY ABSTRACTS. Monthly. Keston, Kent, Tate & Lyle, Ltd., for the Sugar Refiners' Association and the British Sugar Corporation, London.

SUGAR TECHNOLOGY REVIEWS. Irregular. Elsevier Publishing Co., P.O. Box 211, Amsterdam, Netherlands.

9. Reviewing Services

ADVANCES IN CARBOHYDRATE CHEMISTRY AND BIOCHEMISTRY. Continuing series. Edited by R.S. Tipson and D. Horton. New York: Academic Press, 1972.

> Reviews latest research results on the subject.

ADVANCES IN FOOD RESEARCH. Vol. 19. New York: Academic Press, 1971.

> Surveys advances in food research.

ADVANCES IN FOOD RESEARCH. Supplement. New York: Academic Press, 1972.

ADVANCES IN PROTEIN CHEMISTRY. Irregular. New York: Academic Press.

> Critical reviews of progress in field.

SUGAR TECHNOLOGY REVIEWS. Irregular. Elsevier Publishing Co., P.O. Box 211, Amsterdam, Netherlands.

10. Document and Report Literature

a. LISTS

MONTHLY CATALOG. U.S. Superintendent of Documents, Government Printing Office, Washington, D.C. 20402.

U.S. GOVERNMENT RESEARCH REPORTS ANNOUNCEMENTS, and U.S. GOVERNMENT RESEARCH REPORTS INDEX. Monthly. Springfield, Virginia: National Technical Information Service.

b. CLEARINGHOUSES

National Technical Information Service. Springfield, Virginia 22151.

Superintendent of Documents. U.S. Government Printing Office, Washington, D.C. 20402.

11. Sources of Information on Masters and Doctoral Theses

See this subject in the Section I of this Guide.

12. Sources of Information on Statistics and Costs

a. STATISTICS

i. Guides and Handbooks

AMERICAN STATISTICS INDEX. Annual publication. Pt. 1: Index; Pt. 2: Abstracts. Congressional Information Service, 600 Montgomery Building, Washington, D.C. 20014.

"Comprehensive guide and index to the statistical publications of the U.S. Government."

United Nations. Food and Agriculture Organization. FAO COMMODITY REVIEW AND OUTLOOK. Annual. Unipub, Inc., Box 433, New York, New York 10016.

"Detailed summary of situation and outlook in world agricultural commodity trade -- food and feed products, beverages and tobacco, agricultural raw materials, forestry products."

_____. STATE OF FOOD AND AGRICULTURE. Annual. Unipub, Inc., Box 433, New York, New York 10016.

"Annual report on world food and agriculture situation: production, stocks, demand, trade, food supplies, farm and consumer prices, development plans, education and research."

_____. TRADE YEARBOOK. Annual. Unipub, Inc., Box 433, New York, New York 10016.

"Companion to FAO's Production Yearbook, with latest data on quantities and values of imports and exports for all major agricultural products and requisites."

U.S. Department of Agriculture. Economic Research Service. FATS AND OILS STATISTICS, 1950-71. Statistical Bulletin 489. Washington, D.C.: Government Printing Office, 1972.

U.S. Domestic and International Business Administration. Department of Commerce. FOOD INDUSTRIES DATA SOURCES. Washington, D.C. 20234.

Vara, Albert C. FOOD AND BEVERAGE INDUSTRIES: A BIBLIOGRAPHY AND GUIDEBOOK. Detroit, Michigan: Gale Research Company, 1970.

An extensive guide to source material and information outlets.

ii. Current Journals
Which Include Related Statistical Information

The following journals frequently contain statistics related to this field. See the heading Journals and Serial Publications in this section for further details.

THE AMERICAN OIL CHEMISTS SOCIETY. JOURNAL

FOOD AND AGRICULTURAL LEGISLATION

FOOD PROCESSING INDUSTRY

FOOD PRODUCT DEVELOPMENT

FOOD TECHNOLOGY

FROZEN FOOD AGE

FROZEN FOODS

INTERNATIONAL SUGAR JOURNAL

JACOBSEN'S FATS AND OILS BULLETIN

SEARCH: FOODSTUFFS DIVISION

b. COSTS

Rasmussen, C. "Freezing foods; the cost of quality." ASHRAE JOURNAL, Vol. 14, November 1972, pp. 29-32.

WORLD-PRODUCT COSTS. Predicasts, Inc., 200 University Circle Research Center, 11001 Cedar Avenue, Cleveland, Ohio 44106.

> A subscription service of market forecasts by product in areas of agriculture, wood and paper, food, chemicals, polymers, primary metals, etc.

13. Sources of Information on Standards, Specifications, and Patents

a. STANDARDS AND SPECIFICATIONS

i. Guides and Handbooks

AMERICAN NATIONAL STANDARDS INSTITUTE CATALOG. Annual. American National Standards Institute, 1430 Broadway, New York, New York 10018.

> Lists standards in this subject area. Catalog and standards are available from the Institute.

Struglia, Erasmus J. STANDARDS AND SPECIFICATIONS INFORMATION SOURCES. Detroit: Gale Research Company, 1965.

> Coverage is both general and specific to food engineering.

U.S. Department of Agriculture. USDA GRADE STANDARDS FOR FOOD, HOW THEY ARE DEVELOPED AND USED. January 1973. Information Division, Agricultural Marketing Service, Washington, D.C. 20250.

U.S. Department of Commerce. National Bureau of Standards. DIRECTORY OF U.S. STANDARDIZATION ACTIVITIES. Washington, D.C.: Government Printing Office, 1975.

> Coverage is both general and specific to food engineering.

Specifications, standards, codes, and regulations are available for individual states from the official state document printing offices and from local and municipal government agencies.

b. PATENTS

i. Guides and Handbooks

Gutcho, M. FOOD ACID MANUFACTURE: RECENT DEVELOPMENTS. Park Ridge, New Jersey: Noyes Data Corp., 1975.

Pintauro, N. NUTRITION TECHNOLOGY OF PROCESSED FOODS. Park Ridge, New Jersey: Noyes Data Corp., 1975.

Southern Marketing and Nutrition Research Division. PUBLICATIONS AND PATENTS. Semiannual. U.S. Department of Agriculture, Southern Marketing and Research Division, 1100 Robert E. Lee Blvd., Box 19687, New Orleans, Louisiana 70124.

ii. Current Journals Which Include
Related Patent Listings

The following journals frequently list patents. See the heading Journals and Serial Publications in this section for addresses.

ABSTRACTS FROM CURRENT SCIENTIFIC AND TECHNICAL LITERATURE

THE AMERICAN OIL CHEMISTS SOCIETY. JOURNAL

BAKING RESEARCH ASSOCIATION ABSTRACTS

FOOD ENGINEERING

FOOD IN CANADA

FOOD MANUFACTURE

FOOD TECHNOLOGY

INTERNATIONAL FOOD REGISTER

INTERNATIONAL SUGAR JOURNAL

MILCHWISSENSCHAFT

SUGAR INDUSTRY ABSTRACTS

See also: 1) patent listings in CHEMICAL ABSTRACTS and other abstracting services;   2) the subject of patents in the section I of this Guide.

14. Sources of Information on
Translations, Meetings, and Proceedings of Symposia

a. TRANSLATIONS

U.S. Joint Publications Research Service. Arlington, Virginia. Publishes translations of scientific and technical materials from Eastern Europe, Russia, and China. These are listed in the MONTHLY CATALOG of the Superinten- dent of Documents.

See also this subject in Section I of this Guide.

15. Selected Literature Sources on
Pollution Control and Safety

a. POLLUTION CONTROL

i. Guides and Handbooks

Development Planning and Research Associates, Inc. ECONOMIC ANALYSIS

OF EFFLUENT GUIDELINES - BEET SUGAR INDUSTRY. By M. David, et al. Rept. no. PB 248-844. Springfield, Virginia: National Technical Information Service, 1975.

_____. WATER POLLUTION CONTROL ACT OF 1972. ECONOMIC IMPACTS. DAIRY PRODUCTS. Rept. no. PB251 217. Springfield, Virginia: National Technical Information Service, 1975.

_____. WATER POLLUTION CONTROL ACT OF 1972, ECONOMIC IMPACTS, FRUITS AND VEGETABLES. Rept. no. PB251 216. Springfield, Virginia: National Technical Information Service, 1975.

_____. WATER POLLUTION CONTROL ACT OF 1972, ECONOMIC IMPACTS, GRAIN MILLING. Rept. no. PB251 223. Springfield, Virginia: National Technical Information Service, 1975.

_____. WATER POLLUTION CONTROL ACT OF 1972, ECONOMIC IMPACTS, MEAT PRODUCTS. Rept. no. PB251 215. Springfield, Virginia: National Technical Information Service, 1975.

Food processing waste management, 5th Agricultural waste management conference at Syracuse, New York. PROCEEDINGS. Cornell University, 207 Riley-Robb Hall, Ithaca, New York 14850. 1973.

Food processing wastes, 2nd National Symposium. PROCEEDINGS. Environmental Protection Agency, Rte. 8, Box 116, Highway 70 West, Raleigh, North Carolina 27607. 1971.

FOOD PROCESSING WASTE TREATMENT. By R.J. Brown. Rept. no. NTIS/PS-76/0497. Springfield, Virginia: National Technical Information Service, 1976.

A bibliography with abstracts. Covers 1964-1976.

Jones, H. WASTE DISPOSAL CONTROL IN FRUIT AND VEGETABLE INDUSTRY. Park Ridge, New Jersey: Noyes Development Corp., 1973.

National Canners Association, Berkeley. SOLID WASTE MANAGEMENT IN THE FOOD PROCESSING INDUSTRY (PB-219019). Springfield, Virginia: National Technical Information Service, 1973.

"Report outlines processing procedures for 28 major commodities, and describes quantities of residuals, in-plant handling methods, on-site storage facilities, disposal methods, and by-products. Environmental problems associated with solid waste management and costs incurred in handling, treatment, and disposal are enumerated."

U.S. Environmental Protection Agency. DAIRY FOOD PLANT WASTES AND WASTE TREATMENT PRACTICE. By W.J. Harper, et al. Government Printing Office, Washington, D.C. 20402. 1972.

With bibliographies.

_____. HAWAII SUGAR INDUSTRY WASTE STUDY. Government Printing Office, Washington, D.C. 20402. 1971.

With selected bibliography.

_____. IMPROVEMENT OF TREATMENT OF FOOD INDUSTRY WASTE. By S.B. Tutwiner. Washington, D.C.: Government Printing Office, 1974.

_____. NATIONAL MEAT PACKING WASTE MANAGEMENT RESEARCH AND DEVELOPMENT PROGRAM. By J.L. Witherow, et al. Government Printing Office, Washington, D.C. 20402. 1972.

With list of references.

_____. PILOT PLANT INSTALLATION FOR FUNGAL TREATMENT OF VEG-ETABLE CANNING WASTES. By B.D. Church, et al. Government Printing Office, Washington, D.C. 20402. 1972.

With list of references.

_____. PROCEEDINGS. 3rd national symposium on food processing wastes. Government Printing Office, Washington, D.C. 20402. 1972.

With bibliographies.

_____. PROCEEDINGS. 4th national symposium on food processing wastes. Government Printing Office, Washington, D.C. 20402. 1973.

_____. STATE-OF-ART, SUGARBEET PROCESSING WASTE TREATMENT. By J.A. Fischer, et al. Government Printing Office, Washington, D.C. 20402. 1972.

With lists of references.

_____. TRICKLING FILTER TREATMENT OF FRUIT PROCESSING WASTE WATER. Government Printing Office, Washington, D.C. 20402. 1972.

_____. WHEY EFFLUENT PACKED TOWER TRICKLING FILTRATION. By T.P. Quirk, et al. Government Printing Office, Washington, D.C. 20402. 1972.

With list of references.

U.S. National Industrial Pollution Control Council. POLLUTION CONTROL COSTS AND RESEARCH PRIORITIES IN THE ANIMAL SLAUGHTERING AND PROCESSING INDUSTRIES. Washington, D.C.: Government Printing Office, 1973.

See this subject also in section Chemical Engineering - General.

b. SAFETY

i. Guides and Handbooks

Safety Standards are listed in: CATALOG, AMERICAN NATIONAL STANDARDS INSTITUTE. Annual. New York: The Institute.

Peck, Theodore. OCCUPATIONAL SAFETY AND HEALTH: A GUIDE TO INFORMATION SOURCES. Detroit: Gale Research Company, 1974.

ii. Journals and Serial Publications

FOOD AND COSMETICS TOXICOLOGY. Bimonthly. Pergamon Press Inc., Maxwell House, Fairview Park, Elmsford, New York, 10523.

> "An International Journal published for the British Industrial Biological Research Association. Includes original papers and authoritative reviews on the latest world-wide developments and legislative changes in the toxicology and regulatory control of: food additives and contaminants, cosmetics, toiletries, and household products, food packaging materials, and chemical substances of environmental and occupational importance."

LEGAL GUIDE TO OCCUPATION SAFETY AND HEALTH. Monthly. Clark Boardman Ltd., 435 Hudson Street, New York, New York 10014.

> Comment, interpretation, and advice on new legal developments in occupation Safety and Health.

U.S. Occupational Safety and Health Administration. SAFETY AND HEALTH GUIDELINES FOR GENERAL INDUSTRY. Rept. no. PB239 310 SDS. Springfield, Virginia: National Technical Information Service, 1975.

16. Management Information Sources

See this topic in Section I of this Guide.

17. Sources of Information on Laws and Legislation in This Field

a. FEDERAL

CODE OF FEDERAL REGULATIONS. Government Printing Office.

> Continuing series issued throughout the year. Available on a subscription basis from Government Printing Office.

FEDERAL REGISTER. Daily publication listing new regulations and revisions. Updates CODE OF FEDERAL REGULATIONS. Available on a subscription basis from Government Printing Office.

Library of Congress. DIGEST OF PUBLIC GENERAL BILLS AND REGULATIONS. Monthly. Available on a subscription basis from Government Printing Office, Washington, D.C. 20402.

UNITED STATES CODE, 1970 edition and supplements. Irregular. Government Printing Office, Washington, D.C. 20402.

b. STATE

Library of Congress. MONTHLY CHECKLIST OF STATE PUBLICATIONS. Available on a subscription basis from Government Printing Office.

c. LOCAL AND MUNICIPAL

Contact local government offices for codes, ordinances and regulations.

All of these information source materials listed here are usually available in the reference department of large public libraries.

Section V

MATERIALS, INCLUDING PLASTICS, METALS, CERAMICS,

RUBBERS, AND TEXTILES

Section V

MATERIALS, INCLUDING PLASTICS, METALS, CERAMICS, RUBBERS, AND TEXTILES

A. ORGANIZATIONS AND ASSOCIATIONS

1. Governmental Agencies

a. FEDERAL

DEPARTMENT OF COMMERCE

A. NATIONAL BUREAU OF STANDARDS - Washington, D.C. 20234.

Purpose: The Bureau's overall goal is to strengthen and advance the nation's science and technology and facilitate their effective application for public benefit. To this end, the Bureau conducts research and provides: 1) a basis for the nation's physical measurement system, 2) scientific and technological services for industry and government, 3) a technical basis for equity in trade, 4) technical services to promote public safety, and 5) technical information services.

Applied Technology: The Institute for Applied Technology provides technical services to promote the use of available technology and to facilitate technological innovation in industry and government; cooperates with public and private organizations in the development of technological standards and test methodologies; and provides advisory and research services for federal, state, and local government agencies.

MAJOR FIELD ORGANIZATIONAL UNITS AND STATIONS

Units and Stations	Address
Organizational units at Boulder, Colorado	
Deputy Director, Institute for Basic Standards/Boulder.....................	National Bureau of
Cryogenics Division	Standards, Boulder, Colorado 80302.

Quantum Electronics Division. . . . . (see page 333)
Electromagnetics Division. . . . . (see page 333)
Time and Frequency Division . . . . (see page 333)

Laboratory Astrophysics Division . . . . JILA, University of
Colorado, 80302.

Field stations reporting to Boulder, Colorado
Standard Frequency Stations WWV,
WWVB, and WWVL. . . . . Box 83-E, Route 2,
Fort Collins,
Colorado 80521.

Standard Frequency Station WWVH . . . . Box 417, Kekaha,
Kauai, Hawaii 96752.

Field stations reporting to Washington, D.C.
Master Railway Track Scale Depot. . . . . 5800 W. 69th St.,
Chicago (Clearing),
Illinois 60638.

Visual Landing Aids Field Laboratory. . . . . . Box 193, Arcata,
California 95521.

1. Alloy Data Center - Gaithersburg, Maryland.

Areas of interest: Properties of metals, alloys (primarily binary),
semimetallic materials, and intermetallic compounds. Properties
of interest are structure-insensitive properties, including electronic transport properties, magnetic properties, selected mechanical
properties, resonance properties, quantum description parameters,
electromagnetic radiation, superconductivity, and thermodynamic
properties. The Center also collects and evaluates data in those
areas where special competence exists in the Alloy Physics Section of the Metallurgy Division, e.g., soft X-ray spectra and
Knight shift data.

Holdings: Data.

Publications: ANNOTATED BIBLIOGRAPHY ON SOFT X-RAY
SPECTROSCOPY (NBS Monograph 52, 1952, kept up to date by
the Center); NBS ALLOY DATA CENTER: PERMUTED MATERIALS
INDEX (NBS Special Publication 324, 1970); NBS ALLOY DATA
CENTER: AUTHOR INDEX (available from the Clearinghouse for
Federal Scientific and Technical Information, 1970). A compila-
tion of reference books within the Center's scope, published in
NBS Technical Note 464, is kept up to date by the Center.

Information services: Answers inquiries; makes referrals; provides
reference and literature-searching services.

2. Diffusion in Metals and Alloys Data Center - Gaithersburg, Maryland.

Areas of interest: Diffusion properties of metals and alloys.

Holdings: Data.

Publications: Evaluated compilations of diffusion data will be published.

Information services: Answers inquiries; critically evaluates data. Services are available to U.S. government agencies and their contractors, research and educational institutions, and industry.

3. Inorganic Materials Division - Gaithersburg, Maryland.

Areas of interest: Inorganic chemistry; glass; high-temperature chemistry; solid state chemistry; physical properties; crystallography; solid state physics.

Holdings: Phase diagrams of nonmetallic, inorganic anhydrous substances, compiled from journals, reports, and patents; scientific papers in several European languages.

Publications: PHASE DIAGRAMS FOR CERAMISTS (American Ceramic Society).

Information services: Answers inquiries; provides consulting services to government agencies on a contract basis or without charge.

4. Metallurgy Division - Gaithersburg, Maryland.

Areas of interest: Engineering metallurgy; alloy physics; lattice defects and microstructures; corrosion; metal physics; electrolysis and metal deposition; crystallization of metals.

Information services: Answers inquiries; provides consulting services without charge or under contract to government agencies, the scientific community, and industry.

5. Office of Standard Reference Data - Gaithersburg, Maryland.

Areas of interest: Physical and chemical properties of well characterized substances; thermodynamics and transport data; chemical kinetics; nuclear data; solid state data; colloid and surface chemistry; atomic and molecular data; computer processing of data.

Holdings: A collection of the world's output of evaluated reference data in the physical sciences.

Publications: Tables of evaluated data and critical reviews issued in the NATIONAL STANDARD REFERENCE DATA SERIES and other forms.

Information services: Answers inquiries; makes referrals; permits onsite use of collection. The Office administers the National Standard Reference Data System, the government-wide complex of activities concerned with the compilation and critical evaluation of quantitative information on the physical and chemical properties of substances.

6. Office of Standard Reference Materials - Gaithersburg, Maryland.

Areas of interest: Standard reference materials, including all types of well-characterized materials that can be used to calibrate a measurement system or to produce scientific reference data; inorganic standards; metal standards; organic standards.

Publications: STANDARD MATERIALS ISSUED BY THE NATIONAL BUREAU OF STANDARDS.

Information services: Answers inquiries; provides consulting services. The Office serves as a clearinghouse for standard reference materials activities.

7. Polymers Division - National Bureau of Standards, Washington, D.C. 20234.

Areas of interest: Natural and synthetic polymers (constants, properties, constitution, and structure; correlation between the chemical and physical structure and the properties); performance characteristics of polymeric produc⁣ts; test procedures for polymeric materials; measurement of physical and chemical properties of dental materials; structure of human tooth enamel and dentin; rubber; polymer solutions.

Information services: Conducts physical and chemical testing on polymeric materials for federal and state agencies; cooperates with other government agencies in evaluating the performance characteristics of polymeric products: cooperates with industry, technical societies, government agencies, and national and international standardization bodies in measurement and standardization activities.

DEPARTMENT OF DEFENSE

A. U.S. AIR FORCE

1. Aerospace Research Laboratories - Building 450, Wright-Patterson Air Force Base, Ohio 45433.

Areas of interest: Research in the physical and engineering sciences, particularly in areas of hypersonic aerodynamics; fluid dynamic facilities; energy conversion processes; thermomechanics, including heat transfer and electric arc phenomena; metallurgy and ceramics, including fracture mechanics and electrical properties; chemistry, including organosilicon and metal chelate chemistry, combustion, and surface chemistry; plasma physics; solid state physics of II-VI compounds; general physics, including theoretical, atomic, nuclear structure, and optical physics; applied mathematics, including analysis, statistics, and numerical processes.

Holdings: A library of about 19,000 volumes and 400 journals, including some complete runs.

Publications: An average of 150 journal articles per year and a similar number of ARL reports, both internal and contractor prepared; biannual reports on research; annual lists of publications;

summary of coupling activities; other staff reports.

Information services: Articles and reports are available through the journals or DDC, as appropriate. Requests for specific information should be directed to the ARL Information Office at the above address.

2. School of Engineering - Air Force Institute of Technology (AFIT/EN), Wright-Patterson Air Force Base, Ohio 45433.

Areas of interest: Aerospace engineering; electrical engineering; mathematics; mechanical engineering; mechanics; physics; nuclear engineering; systems management and analysis.

Publications: Reports and theses.

Information services: Answers inquiries; provides advisory services. Services are primarily for the Air Force and other Department of Defense agencies.

B. U.S. ARMY

1. Plastics Technical Evaluation Center (PLASTEC) - Picatinny Arsenal, Dover, New Jersey 07801.

Areas of interest: Plastic materials and composites of interest to the Department of Defense, with emphasis on plastics in structural applications (excluding building construction), electrical and electronic applications, packaging, foams, and mechanical goods applications; test methods and specifications; environmental effects; microbiological effects; processing.

Holdings: Books, journals, periodicals, abstracts, indexes, research reports, specifications and standards, technical bulletins, data sheets, proceedings.

Publications: State-of-the-art reports, indexes, directories, annotated bibliographies.

Information services: Distributes technical data and data evaluations to Department of Defense activities, their designees, or other organizations with suitable arrangements; renders technical advice and assistance on plastics, adhesives, and composites to Department of Defense activities and other organizations where appropriate; performs computer searches on demand.

2. Redstone Scientific Information Center - U.S. Army Missile Command, Redstone Arsenal, Alabama 35809.

Areas of interest: Aeronautics; astronomy and astrophysics; atmospheric sciences; chemistry; electronics and electrical engineering; energy conversion (nonpropulsive); materials; mathematical sciences; mechanical and industrial engineering; methods and equipment; military sciences; missile technology; navigation; communication; detection and countermeasures; nuclear science and technology; ordnance; physics; propulsion and fuels; space technology.

Holdings: The Center contains the central collection of scientific information for the Army Missile Command and for NASA's Marshall Space Flight Center. Its holdings consist of approximately 100,000 books and bound volumes of periodicals, 850,000 technical reports, including half a million reels of microfilm, microfiche, and other microforms, and about 3,000 slides, and motion pictures.

Publications: Technical reports, state-of-the-art reviews, bibliographies, announcement bulletins.

Information services: Answers inquiries; provides reference, literature-searching, translation, and duplication services; provides information on methods for handling and processing data and information; makes interlibrary loans. Services are available to personnel of the Marshall Space Flight Center and the Army Missile Command and their contractors in the Huntsville area. The library collections are accessible for onsite use by others by special arrangement.

3.. Scientific and Technical Information Branch - Picatinny Arsenal, ATTN: SMUPA-RT-S, Dover, New Jersey 07801.

Areas of interest: Explosives; ammunition; packaging; nuclear weapons; propellants; polymers and plastics; pyrotechnics.

Holdings: 50,000 books; 1,200 periodical titles; 150,000 reports; 70,000 microfiche.

Publications: TECHNICAL INFORMATION BULLETIN; reports, bibliographies.

Information services: Answers inquiries; makes referrals; provides reference, literature-searching, translation, and duplication services; makes inter-library loans. Services are available to Department of Defense, other government agencies, and to a limited extent, other groups on a selected basis.

4. Scientific Synthesis Office - U.S. Army Research Office--Durham, Box CM, Duke Station, Durham, North Carolina.

Areas of interest: Ceramics for structural use; lasers; high pressure and temperature processes used to produce new materials; mechanical behavior of materials under dynamic loads; upper atmosphere atomic and plasma physics; corrosion; subsonic aircraft and hovering-type vehicles; aerodynamics of missiles; sensing and identification of objects; hazardous devices and substances; night vision; behavior of composites; brittle fracture; mathematical analysis of nonlinear systems; information sciences.

Holdings: Over 5,000 abstracts.

Publications: MILITARY THEME RESEARCH REVIEW BULLETIN (monthly).

Information services: Provides reference, literature-searching, and abstracting and indexing services; makes referrals. Services are available to Army agencies, contractors, and grantees.

DEPARTMENT OF THE INTERIOR

A. BUREAU OF MINES

1. Albany Metallurgy Research Center - 1450 West Queen Avenue, Albany, Oregon.

Areas of interest: Melting, casting, and forming of high-purity refractory metals by high-energy input; melting processes (arc, induction, electron-beam, and conventional methods); thermodynamic properties of inorganic materials; organometallic reactions; pyrometallurgy; hydrometallurgy; physical metallurgy of high temperature alloys; X-ray analyses of metals at high temperatures; gas-solid reactions; nuclear fuel materials; recovery of metal from industrial wastes and scrap metal products; brittleness in refractory metals and ceramics; ceramic fabrication techniques; superconductivity phenomena; kinetic studies of transition metals and their halides; etc.

Publications: Reports of studies made by the Center are included in the Bureau's series of publications.

Information services: Provides technical assistance and consulting services; provides petrographic, metallographic, and mechanical and physical evaluations as a service to other laboratories.

2. Boulder City Metallurgy Research Laboratory - U.S. Bureau of Mines, 500 Date Street, Boulder City, Nevada 89005.

Areas of interest: Fused-sale electrolysis of titanium, beryllium, zirconium, and vanadium; reprocessing of scrap metals by physical, pyrometallurgical, and electrometallurgical treatment.

Publications: Reports of studies made by the Laboratory are included in the Bureau's series of publications.

Information services: Provides advisory and consulting services in regard to technical questions on metallurgical problems.

3. College Park Metallurgy Research Center - U.S. Bureau of Mines, College Park, Maryland 20740.

Areas of interest: Hydrometallurgy; pyrometallurgy; bacterial metallurgy; electrometallurgy; recovery of alumina from clays and dawsonite associated with oil shale; technical evaluation and cost estimation of metallurgical processes; recovery and recycling processes for secondary metals; electrostatic and magnetic separation of minerals; X-ray and electron probe analyses; recovery of useful materials from incinerator residues; mathematical simulation of metallurgical processes.

Publications: Reports of studies made by the Center are included in the Bureau's series of publications.

Information services: Answers inquiries; provides consulting services; lends reports for short periods. The Center houses an Industrial Water Laboratory which analyzes feed and boiler water for U.S. Government heating plants and makes recommendations for chemical treatment.

4. Office of Mineral Information - U.S. Bureau of Mines, 18th and C Streets N.W., Washington, D.C. 20240.

Areas of interest: Mining; minerals; coal; petroleum; environmental problems linked to extraction, processing, and use of minerals and fuels; mine safety; natural gas; helium; metallurgy; statistics and economics of mineral resources and fuels.

Publications: MINERALS YEARBOOK (annual, in four volumes); BULLETINS; INFORMATION CIRCULARS; REPORTS OF INVESTIGATIONS; MINERAL INDUSTRY SURVEYS (weekly, monthly, quarterly, and annual commodity reports); NEW PUBLICATIONS-- BUREAU OF MINES (monthly list, with annual and five-year cumulations; listings of journal articles by Bureau specialists are included).

Information services: The Office's Division of Public Information answers inquiries and handles distribution of information to public news media through press releases, copies of speeches, and special publications. The Office arranges for the publication of reports produced by all the Bureau's divisions. The Division of Production and Distribution, 4800 Forbes Avenue, Pittsburgh, Pennsylvania 15213, (tel: (412) 621-4500) distributes single copies of the Bureau's free publications and also circulates on a free-loan basis motion pictures produced by the Bureau in cooperation with industry sponsors. A catalog of films is available from the Washington office. Bureau publications for sale may be purchased from the Superintendent of Documents, Washington, D.C. 20402.

5. Reno Metallurgy Research Center - U.S. Bureau of Mines, 1605 Evans Avenue, Reno, Nevada 89505.

Areas of interest: Recovery of gold and other heavy metals from low-grade ores; extraction and separation of rare-earths; fused-salt metallurgy (of vanadium, manganese, titanium, chromium, molybdenum, uranium, thorium, tungsten, hafnium, columbium, beryllium, yttrium, and rare-earth metals); combined pyro- and electro-metallurgical techniques to effect separations; high-temperature techniques in preparation of high-purity common metals (such as nickel and cobalt); electrorefining of metals to obtain ultra-high purity; physical evaluation of properties of metals and their alloys that are a product of research; chelation of copper oxide minerals; processing of nonferrous sulfide to recover elemental sulfur.

Publications: Reports of studies made by the Center are included in the Bureau's series of publications.

Information services: Provides advisory and consulting services in regard to technical questions; provides mineral identification services.

6. Rolla Metallurgy Research Center - 1300 Bishop Street, Rolla, Montana.

Areas of interest: Recovery and utilization of materials from industrial, mining, and municipal wastes; recycling of secondary materials; physical metallurgy of ferrous and nonferrous metals; extractive metallurgy and reconstitution of minerals; development of mineral detection devices.

Publications: Reports of studies made by the Center are included in the Bureau's series of publications listed under Office of Mineral Information.

Information services: Answers inquiries; makes referrals; provides consulting services; provides mineral identification services.

7. Salt Lake City Metallurgy Research Center - U.S. Bureau of Mines, 1600 East 1st South Street, Salt Lake City, Utah 84112.

Areas of interest: Improved recoveries of copper, molybdenum, magnesium, uranium, vanadium, gold, and silver from natural resources of significance via new and improved extractive metallurgy technology; recovery of useful products from solid industrial wastes, scrap automobiles, and electronic scrap; stabilization of fine mineral wastes; recovery of sulphur from copper and lead smelter gas; process development and cost evaluation studies.

Publications: Reports of studies made by the Center are included in the Bureau's series of publications listed under Office of Mineral Information.

Information services: Answers inquiries; makes referrals; provides consulting services in regard to specific problems.

8. Tuscaloosa Metallurgy Research Laboratory - U.S. Bureau of Mines, Box L, University, Alabama 35486.

Areas of interest: Recovery of such industrial minerals in pure form as potash, kyanite, mica, feldspar, phosphate rock, high quality clays, iron ore, and glass sand; microgrinding mineral products to extreme fine sizes; separation by heavy liquid, flotation, and other procedures; product development and uses of waste glass and incinerator residue.

Information services: Provides technical consultations on industrial mineral problems; tests clays.

9. Twin Cities Metallurgy Research Center Library - U.S. Bureau of Mines, Box 1660, Twin Cities, Minnesota 55111.

Areas of interest: Iron ore and copper; mineral resources of Minnesota, Iowa, Indiana, Illinois, Wisconsin, and Michigan; mining and metallurgical technology of raw materials and scrap.

Holdings: 1,000 periodical titles; 5,000 reports; 1,000 books.

Information services: Answers inquiries; provides reference services; lends materials; permits onsite use of collection.

NATIONAL COMMISSION ON MATERIALS POLICY - 1016 16th Street N.W., Washington, D.C. 20036.

The National Commission on Materials Policy was established by Public Law 91-512, approved October 26, 1970. This commission has published a report on materials policy in two parts:

1. Towards a National Materials Policy, basic data and issues, interim report. April 1972. Y3.M41:2n21 (This deals with domestic aspects.)

2. Toward a National Materials Policy, world perspective. 2nd interim report. January 1973. Y3.M41:2n21/973 (Provides a more global viewpoint.)

NATIONAL RESEARCH COUNCIL

NATIONAL MATERIALS ADVISORY BOARD - Division of Engineering, 2101 Constitution Avenue N.W., Washington, D.C. 20418.

Areas of interest: Materials science and technology; metallurgy; organic and inorganic nonmetallic materials; materials processing and fabrication; materials phenomena; education in the materials field.

Publications: Technical reports; annual list of reports.

Information services: On request, provides advice and assistance to government agencies and appropriate organizations in the private sector on matters of materials science and technology affecting the national interest; disseminates information on Board activities and related national concerns; makes referrals.

2. Professional Societies, Trade Associations, and Other Scientific-Technological Related Organizations

a. GENERAL

The following listed organizations have a range of interests which touch upon general aspects of materials technology.

ABRASIVE GRAIN ASSOCIATION - 2130 Keith Building, Cleveland, Ohio 44115.

Areas of interest: Abrasive grain; aluminum oxide; silicon carbide; grinding materials.

Holdings: Reports, standards and specifications, press releases.

Information services: Answers brief inquiries; makes referrals.

AMERICAN SOCIETY FOR NONDESTRUCTIVE TESTING - 914 Chicago Avenue, Evanston, Illinois 60202.

Areas of interest: Techniques for the nondestructive testing of metals, ceramics, wood, plastics, and components, utilizing penetrating radiation, ultrasonics, magnetic particles, eddy currents, fluorescent penetrants, gaging, metrology, and thermographic methods; improved product quality, predictable reliability, and effective process control of nondestructive testing techniques.

Holdings: Books, journals, films, photos.

Publications: MATERIALS EVALUATION (monthly; all volumes since 1942 of this publication and its predecessor, the bimonthly NONDESTRUCTIVE TESTING, are also for sale on microfilm); proceedings, standards and specifications, data handbook, state-of-the-art reviews. A price list is available.

Information services: Answers inquiries; provides reference services; provides duplication services for fee; lends books and films; permits onsite use of collection.

AMERICAN SOCIETY FOR TESTING AND MATERIALS - 1916 Race Street, Philadelphia, Pennsylvania 19103.

Areas of interest: Standardization of specifications and methods of testing materials, including ferrous and nonferrous metals, cement, lime, gypsum, mortar, concrete, mineral aggregates, bituminous materials, soils, asbestos-cement products, masonry units, pipe and drain tile, refractories, ceramic whitewares, porcelain enamel, glass, building stone, thermal insulation, acoustical materials, paper and packaging, adhesives, wood, cellulose, casein, leather, petroleum products and lubricants, paint, naval stores, coal and coke, gaseous fuels, industrial aromatic hydrocarbons, industrial chemicals, plastics, carbon black, textiles, soap, sorptive mineral materials, organic solvents, industrial water, wax polishes, rubber, and electrical insulation; identification of phases from diffraction data, either X-ray or electron, that have been obtained from a powder; spectral absorption data and chemical structure, mass spectral data, and gas chromatographic data.

Publications: MATERIALS RESEARCH AND STANDARDS (monthly); BOOK OF ASTM STANDARDS (annual), consisting of 33 parts; ASTM INDEX TO STANDARDS (annual); ASTM PROCEEDINGS (official annual record); fifty-year index and five-year indexes to ASTM technical papers and reports. A publications price list is available. ASTM is also responsible for the distribution of the POWDER DIFFRACTION DATA FILE compiled by the Joint Committee on Powder Diffraction Standards under the joint auspices of ASTM, the American Crystallographic Association, the British Institute of Physics, and the National Association of

Corrosion Engineers. The POWER DIFFRACTION DATA FILE
(issued on 3 x 5 index cards and in book form for revised sets)
is used for the identification of phases from diffraction data,
either X-ray or electron, that have been obtained from a powdered crystalline material. Powder diffraction data is also
available on magnetic tape, with Fortran programs for computer
searching. Comprehensive indexes are available.

Information services: Answers inquiries. A standards and specifications service provides reference services and referral assistance
without charge.

INSTITUTE FOR COMPOSITE MATERIALS, INC. - P.O. Box 1536, Stow,
Ohio 44224; LOCATION: 877 East Steels Corners Road, Stow, Ohio.

Areas of interest: Applied and fundamental aspects of plastic,
ceramic, elastomeric, and metallic composite materials, including
the technical, economic, environmental, and conservation of
materials aspects; physical, mechanical, and chemical properties
of composite materials; production, uses, and sources of composite
materials.

Holdings: Journals, periodicals, industrial literature, technical
and scientific data, project or administrative data, government
reports, books, abstract journals, technical correspondence.

Publications: MONTHLY NEWSLETTER; THE COMPOSITE
MATERIALS REVIEW (semiannual); manuals, standards and specifications, reports, surveys, directories, bibliographies, handbooks, proceedings of symposiums and workshops.

Information services: Answers technical inquiries; provides consulting, reference, literature-searching, abstracting, and duplication services; makes referrals; provides information on current
research, development, test, and evaluation projects; furnishes
data and data compilations; prepares specifications and standards;
conducts symposiums, workshops, seminars, and conferences; permits onsite use of collection. Services are available only to
Institute members, except by special action of the board of
directors, and to selected government agencies.

INSTITUTE OF ENVIRONMENTAL SCIENCES - 940 East Northwest Highway,
Mount Prospect, Illinois 60056.

Areas of interest: Environmental science, environmental engineering and related technologies, especially corrosion.

Publications: JOURNAL OF ENVIRONMENTAL SCIENCES
(monthly); IES MEETINGS; PROCEEDINGS (annual). A list
of publications is available.

Information services: Answers inquiries and disseminated information on environmental sciences through exhibits, symposia and
educational programs.

THE MATERIAL HANDLING INSTITUTE, INC. - 1326 Freeport Road, Pittsburgh, Pennsylvania 15238.

> Areas of interest: Every type of industrial material handling equipment and its use, including lift trucks (internal combustion powered and electric powered), conveyors, hoists, hand lift trucks and portable elevators, industrial metal containers, monorails, package conveyors, racks, wheels and casters, dock levelers, personnel carriers, controlled mechanical storage systems, and related auxiliary equipment; safety standards; nomenclature; statistics; educational programs; industrial trade show programs.

> Publications: MHI NEWS (3 issues a year); COLLEGE INDUSTRY COMMITTEE ON MATERIAL HANDLING EDUCATION NEWS (3 issues a year); pamphlets, standards and specifications, filmstrips.

> Information services: Answers inquiries; provides reference services; rents films.

NATIONAL ASSOCIATION OF CORROSION ENGINEERS - 2400 West Loop South, Houston, Texas 77027.

> Areas of interest: Corrosion problems in water desalination, waste treatment, recycling of cooling water, and all aspects of engineering involving metals and non-metallic surfaces in corrosion forming conditions.

> Publications: CORROSION (monthly); MATERIALS PROTECTION (monthly); CORROSION ABSTRACTS (bimonthly). Books, symposia proceedings, bibliographies. A list of publications may be obtained.

> Information services: Searches of literature and duplication for fee; on-site use of library; answers inquiries.

NATIONAL ASSOCIATION OF SECONDARY MATERIAL INDUSTRIES, INC. - 330 Madison Avenue, New York, New York 10017.

> Areas of interest: Secondary materials industry throughout the United States and Canada, including the following areas of operation: nonferrous scrap metals (dealers, smelters, refiners, ingot makers); paper stock; textiles; rubber and plastics; and importers and exporters.

> Holdings: Small working collection of books and reports.

> Publications: NONFERROUS SCRAP METAL INDUSTRY; INDUSTRIAL PROFILE AND COST FACTORS IN NONFERROUS SCRAP METAL PROCESSING; THE SECONDARY MATERIAL INDUSTRIES AND ENVIRONMENTAL PROBLEMS; reports, specifications.

> Information services: Answers inquiries; provides consulting and reference services; makes referrals.

b. SPECIALIZED FIELDS OF MATERIALS TECHNOLOGY

i. Ceramics

AMERICAN CERAMIC SOCIETY, INC. - 4055 North High Street, Columbus, Ohio 43214.

Areas of interest: Ceramics; glass; refractories; porcelain enamels; cermets; composites; whitewares; structural clay products; nuclear and electronic ceramics; analytical, crystal, and colloid chemistry; solid state physics; instrumentation for high temperature reactions.

Holdings: 1,100 books; 1,200 bound volumes; an extensive file of reports, brochures, booklets, and catalogs; 270 periodicals on current subscription. The Society is a cooperating library in the ACCESS, KEY TO SOURCE LITERATURE OF THE CHEMICAL SCIENCES published by Chemical Abstracts.

Publications: JOURNAL OF THE AMERICAN CERAMIC SOCIETY--CERAMIC ABSTRACTS (monthly); THE AMERICAN CERAMIC SOCIETY BULLETIN (monthly); PHASE DIAGRAMS FOR CERAMISTS; LARGE SCALE PHASE DIAGRAMS; CERAMIC-METAL SYSTEMS AND ENAMEL BIBLIOGRAPHY AND ABSTRACTS; SYMPOSIUM ON NUCLEATION AND CRYSTALLIZATION IN GLASSES AND METALS; CERAMIC GLOSSARY; ENGINEERING PROPERTIES OF SELECTED CERAMIC MATERIALS; ALUMINA AS A CERAMIC MATERIAL; CERAMIC NUCLEAR FUELS; ELECTRONIC CERAMICS。

Information services: Answers brief inquiries; makes limited referrals; provides duplication services of noncopyrighted material for a fee; permits onsite reference.

NATIONAL CERAMIC ASSOCIATION (NCA) - P.O. Box 39, Glen Burnie, Maryland 21061.

An organization of manufacturers, distributors, dealers, teachers and ceramists.

Purpose: To promote the use of ceramic materials, to improve teaching methods; establish standards for buying, pricing and inventory; recommend standard sales policy for schools and institutions; promote standardization of entries and judging at shows; work to improve relations between the industry and raw material suppliers, transportation companies and the government. Sponsors teachers' meetings and business seminars.

Divisions: Manufacturers; Distributors; Dealers; Teachers.

Publications: CLAY CHATTER (bimonthly); BLUE BOOK (annual); HOBBY TEACHERS MANUAL.

NATIONAL INSTITUTE OF CERAMIC ENGINEERS (NICE) - 4055 North High Street, Columbus, Ohio 43214.

A professional society of ceramic engineers whose purpose is to improve the professional status of ceramic engineering; promote high standards of ceramic education and high engineering standards and practices.

Publication: NEWSLETTER (monthly).

Affiliated with: American Ceramic Society; Engineers Joint Council; Engineers' Council for Professional Development (affiliate member).

Information activities: Answers inquiries; makes referrals.

ii. Metals

ALLOY CASTING INSTITUTE - 300 Madison Avenue, New York, New York 10017.

Areas of interest: Metallurgical research in high alloy castings (ferrous castings containing alloying elements in excess of 8 per cent); chemical, physical, and mechanical properties of high alloy castings (iron-base, nickel-base, and iron-chromium-nickel types) under various thermal, corrosive, and abrasive conditions; foundry practice and techniques in castability, weldability, and machinability.

Holdings: Books, journals, reports, patents, films, photos.

Publications: Alloy Casting Bulletin (irregular); 8 Plus (quarterly); ACI Data Sheets; Standard Designations (irregular); technical papers, reprints. A technical publications list is available.

Information services: Answers inquiries; provides consulting services.

ALUMINUM ASSOCIATION - 750 Third Avenue, New York, New York 10017.

Areas of interest: Aluminum technology; alloys; standards; tensile strength and tolerances; extrusions and forgings; nomenclature of aluminum mill products.

Holdings: Books, reports, periodicals, standards and specifications, reviews, abstracts, indexes, newsletters, preprints, reprints, publication lists, bibliographies, pamphlets, press releases, clippings.

Publications: WORLD ALUMINUM ABSTRACTS (monthly); ALUMINUM IN MANUFACTURED HOUSING (seasonally); ALUMINUM HIGHWAY NEWS (irregular); RAILROAD REPORT (quarterly); BUILDING CODE NEWS (quarterly); ALUMINUM STANDARDS AND DATA (annual); ALUMINUM STATISTICAL REVIEW (annual). In addition to these serial or recurring titles, the Association has published numerous guides, manuals, standards, specifications, reference books, booklets, and pamphlets relating to aluminum. Publication lists and a directory of members are also available.

Information services: Answers inquiries; makes referrals.

AMERICAN INSTITUTE OF STEEL CONSTRUCTION, INC. - 101 Park Avenue, New York, New York 10017.

Areas of interest: Structural steel and its application to buildings and bridges.

Holdings: A file of indexed abstract cards on published and un-published, foreign and domestic, literature on structural steel technology; a small collection of books, periodicals, and reports.

Publications: MODERN STEEL CONSTRUCTION (quarterly); ENGINEERING JOURNAL (quarterly); specifications, manuals, textbooks, general and technical reports, general and technical reprints, membership list. A publications list is available.

Information services: Answers inquiries; makes referrals; provides reference, literature-searching, and abstracting and indexing services. The Institute has an information retrieval system through which it disseminates ABSTRACTS OF DOCUMENTS on structural steel application and research and a COORDINATE INDEX to the abstracts. A separate thesaurus of keywords for bridges and buildings is used for both document and photographic retrieval. The primary emphasis of the information retrieval system is on current published and unpublished material from worldwide sources. Access to this retrieval system is also pro-vided through 26 regional engineer offices located throughout the United States.

AMERICAN POWDER METALLURGY INSTITUTE - 201 East 42nd Street, New York, New York 10017.

Areas of interest: Metal powders; powder metallurgy processes and products.

Holdings: Books, monographs, reports, standards and specifica-tions, reviews, periodicals, newsletters, preprints, reprints, pub-lication lists, bibliographies, pamphlets, booklets, leaflets, brochures, press releases, clippings.

Publications: INTERNATIONAL JOURNAL OF POWDER METAL-LURGY (quarterly); POWDER METALLURGY INFORMATION BUL-LETIN (monthly); standards, proceedings. A publications list is available.

Information services: Answers inquiries; makes referrals.

AMERICAN SOCIETY FOR METALS - Metals Park, Ohio 44073.

An association dedicated to the advancement of knowledge of all aspects of metals and metallurgy.

Publications: METALS PROGRESS (monthly); ASM NEWS (month-ly); METALLURGICAL TRANSACTIONS (monthly); METALS ENGI-NEERING (quarterly); MATERIALS SCIENCE AND ENGINEER-

ING (monthly). Abstracting Services: METALS ABSTRACTS (monthly with annual cumlations); WORLD ALUMINUM AB-STRACTS (monthly, prepared tapes of abstracts are available as well as literature searches); METALFORMING DIGEST (monthly, includes abstracts, listing of books and reports in the field of metal forming and powder metallurgy). Books, Monographs and Symposia Proceedings: The society publishes selected papers from symposia. Another useful publication is the ASM Thesaurus of metallurgical terms published in 1968.

Information services: Answers inquiries, makes referrals, provides literature search services.

THE METALLURGICAL SOCIETY OF AMERICAN INSTITUTE OF MINING, METALLURGICAL, AND PETROLEUM ENGINEERS - 345 East 47th Street, New York, New York 10017.

Purpose: To set up communication systems for those interested in all phases of metallurgy and material science. To advance the science and engineering of metals and materials and to promote research pertinent to these areas. To disseminate education and knowledge in these fields through meetings and publications.

Publications: JOURNAL OF METALS (quarterly); METALLURGICAL TRANSACTIONS (monthly); OPEN HEARTH PROCEEDINGS (annual); IRONMAKING PROCEEDINGS (annual); ELECTRIC FURNACE PROCEEDINGS (annual).

Information services: Answers questions, assists with reference inquiries and makes referrals.

METAL POWDER INDUSTRIES FEDERATION - 201 East 42nd Street, New York, New York 10017.

"A trade association representing the trade, commercial, and technological interests of the metal powder producing and consuming industries."

Activities: The "Federation" acts as a clearinghouse for information on the metal powder industry both internally and for the public. Further, it promotes the advancement of research, promulgates standards and specifications, market development, and other programs.

Publications: PROGRESS IN POWDER METALLURGY (annual). Modern Developments in powder Metallurgy (a series of monographs); standards and specifications for metal powders and powder metallurgy products; P/M material standards and specifications. A complete list of publications of the Federation and its affiliates is available on request.

POWDER METALLURGY EQUIPMENT ASSOCIATION (PMEA) - 201 East 42nd Street, New York, New York 10017.

An organization of manufacturers of powder metallurgy equipment. Prepares standards for compacting presses; compiles statistics; coordinates technical exchanges for press and sintering furnace manufacturers and tool and die makers.

Publications: P/M Equipment Directory (annual); also publishes P/M Equipment Manual.

Information services: Answers inquiries, makes referrals.

iii. Plastics

THE ADHESIVE AND SEALANT COUNCIL, INC. - 1410 Higgins Road, Park Ridge, Illinois 60068.

Areas of interest: Adhesives and sealants in solid or liquid form; raw materials for adhesives and sealants, including rubber and plastics.

Publications: Specifications, reports, proceedings.

Information services: Makes referrals; permits onsite reference.

AMERICAN PAPER INSTITUTE, INC. - PLASTICS EXTRUSION COATERS GROUP, Printing-Writing Paper Division, 260 Madison Avenue, New York, New York 10016.

Areas of interest: Plastics; plastics extrusion.

Publications: STANDARDS FOR PLASTICS EXTRUSION COATING; WATER VAPOR TRANSMISSION RATE TEST PROCEDURE; SLIP; BASIS WEIGHT DETERMINATION STANDARD TEST METHOD; STANDARD MEASUREMENT OF SURFACE TREATMENT.

Information services: Answers inquiries; provides reference and consulting services; formulates standards for distribution on request.

AMERICAN SOCIETY FOR TESTING AND MATERIALS - COMMITTEE D-20 ON PLASTICS, 1916 Race Street, Philadelphia, Pennsylvania 19103.

Areas of interest: Standards for plastics; methods of testing plastics.

Publications: Parts 26 and 27 of annual BOOK OF ASTM STANDARDS; separate reprints of each standard.

Information services: Answers brief inquiries; makes referrals; permits onsite use of collection.

AMERICAN SOCIETY OF ELECTROPLATED PLASTICS, INC. - 1000 Vermont Avenue N.W., Washington, D.C. 20005.

Areas of interest: Electroplated plastics.

Publications: Specifications, reports, membership directory, monthly bulletin.

Information services: Answers inquiries; makes referrals.

CHEMICAL FABRICS AND FILM ASSOCIATION - 60 East 42nd Street, New York, New York 10017.

An organization of manufacturers of pyroxylin coated materials, supported and unsupported vinyl and urethane coated materials.

Information services: Answers inquiries; makes referrals.

NATIONAL ASSOCIATION OF PLASTIC FABRICATORS (NAPF) - 4720 Montgomery Lane, Bethesda, Maryland 20014.

Manufacturers of decorative plastic laminate used for counters, furniture, horizontal and vertical surfacing, etc.

Activities: Distributes information to members on production, sales, and management matters; sets standards of quality.

Committees: Technical; Marketing; Management.

Publishes periodic bulletins, technical reports, and specifications.

NATIONAL ASSOCIATION OF PLASTICS DISTRIBUTORS - 2217 Tribune Tower, Chicago, Illinois 60611.

An organization of suppliers and distributors in the plastics field.

Purpose: To consider common problems of management in the plastics industry.

Publications: (1) PLASTI-GRAM (bimonthly); (2) Membership Directory (annual); also publishes Plastics Purchasing Guide, and NAPD Sales Training Manual.

Information services: Answers inquiries; makes referrals.

PLASTICS INSTITUTE OF AMERICA - Stevens Institute of Technology, Hoboken, New Jersey 07030.

A research institute affiliated with Stevens Institute of Technology and other academic institutions. Has an educational function and provides information on plastics technology.

Publications: PIA REPORTER (quarterly); ANNUAL REPORT of the Institute.

PLASTICS PIPE INSTITUTE (PPI) - 250 Park Avenue, New York, New York 10017.

Manufacturers of plastics pipe and fittings and suppliers of plastic pipe raw materials. A division of the Society of the Plastics Industry whose purpose is to develop standards and to promote trade and user acceptance.

Divisions: Market; Technical.

Formerly: Thermoplastic Pipe Division of the Society of the Plastics Industry.

SOCIETY OF PLASTICS ENGINEERS, INC. - 656 West Putnam Avenue,

Greenwich, Connecticut 06830.

> Areas of interest: Plastics and polymers, their uses and properties; injection molding; blow molding; thermosetting molding; metal mold design and construction; casting and plastics tooling; forming; coloring and finishing of plastics; plastics foams; polyolefins.

> Publications: SPE JOURNAL (monthly); POLYMER ENGINEERING AND SCIENCE (bimonthly); a continuing series of technical books; papers presented at conferences.

> Information services: Answers brief inquiries, with preference given to inquiries from Society members.

THE SOCIETY OF THE PLASTICS INDUSTRY, INC. - 250 Park Avenue, New York, New York 10017.

> Areas of interest: Plastics, including machinery, mold makers, custom and proprietary molders, vinyl (film, laminators, dispersion), thermoplastics, fluorocarbons, epoxy resin formulators, expandable polystyrene molders, polyethylene film, reinforced plastics, rigid plastic containers, and cellular plastics.

> Holdings: Books, journals, reports, data, plastics publications.

> Publications: PLASTICS ENGINEERING HANDBOOK OF SPI; proceedings, manuals, bulletins, surveys, reports, booklets.

> Information services: Answers inquiries; makes referrals; provides literature-searching and reference services; permits onsite use of collection by request.

URETHANE INSTITUTE - c/o Worden & Co., Inc., 2 West 45th Street, New York, New York 10036.

> Areas of interest: Urethane foams and applications in bedding and furniture industries.

> Publications: Periodical articles, press releases for trade and consumer press.

> Information services: Answers inquiries; disseminates background information.

iv. Rubbers

ELASTIC FABRIC MANUFACTURERS INSTITUTE, INC. - P.O. Box 710, New London, Connecticut 06320; LOCATION: 105 Huntington Street, New London, Connecticut.

> Areas of interest: Elastic braid, woven elastic, braided trimmings, and power net, including set standards and procedures and test characteristics.

> Holdings: Elastic fabric specimens, physical testing equipment,

data compilations, technical reports, standards and specifications, periodicals, newsletters, reprints, pamphlets, clippings, press releases.

Publications: Data compilations, technical reports, standards and specifications, newsletter.

Information services: Answers brief inquiries; makes referrals; provides consulting and testing services. Services are primarily for Institute members, but may be provided to others, depending on the type of information sought.

NATURAL RUBBER BUREAU - 1108 16th Street N.W., Washington, D.C. 20036.

Sponsored by the Malayan Rubber Fund Board, this Bureau is one of 11 such international research information centers.

Areas of interest: Natural rubber, its production and utilization; latex; foam rubber.

Publications: NATURAL RUBBER NEWS (monthly); RUBBER DE-VELOPMENTS (quarterly); bulletins, booklets, maps, charts, photos, films.

Information services: Answers inquiries; makes referrals; provides consulting services; permits onsite reference to Bureau publications.

RUBBER MANUFACTURERS ASSOCIATION (RMA) - 444 Madison Avenue, New York, New York 10022.

An organization whose membership includes manufacturers of tires, tubes, mechanical and industrial products, footwear, sporting goods, and other rubber products.

Activities: Compiles monthly, quarterly, and annual statistics on consumption, production, and inventory of rubber and rubber products.

Committees: Environment; Governmental Relations; Industrial Relations; Natural Rubber; Public Relations; Statistics; Tax; Traffic.

Divisions: Coated Materials; Flooring; Footwear; Latex Foam; O-Rings; Heel and Sole; Industrial Rubber Products; Molded and Extruded Products; Oil Seal; Sundries; Tires.

Publications: RUBBER HIGHLIGHTS (monthly); also publishes booklets and a list of free films and other teaching aids offered by the rubber industry.

Information services: In addition to those listed above, answers inquiries; makes referrals.

RUBBER RECLAIMERS ASSOCIATION - 63 Radnor Avenue, Naugatuck, Connecticut 06770.

Convertors of scrap rubber and tires into reclaimed rubber by mechanical and chemical processes. Publishes commercial standards on scrap rubber, monthly statistics, and technical bulletins on reclaimed rubber use.

Committees: Education; Manufacturing; Scrap Rubber; Solid Waste; Technical; Traffic.

Information services: Answers inquiries; makes referrals.

RUBBER TRADE ASSOCIATION OF NEW YORK (RIA) - 11 Broadway, New York, New York 10004.

An organization of importers of crude natural rubber from the Far East, and exporters of synthetic rubber, brokers and agents, representatives of Far Eastern shippers as well as suppliers of services to the rubber trade.

Information services: Answers inquiries; makes referrals.

v. Textiles

AMERICAN ASSOCIATION FOR TEXTILE TECHNOLOGY - 11 West 42nd Street, New York, New York 10036.

Purpose: To advance the textile technology field by encouraging communications and cooperation among those involved in all its branches.

Activities: Fosters interchange of information within the industry and cooperates in educational programs related to this field.

Publications: MODERN TEXTILES (monthly).

Information services: Answers inquiries; makes referrals.

AMERICAN ASSOCIATION OF TEXTILE CHEMISTS AND COLORISTS - P.O. Box 12215, Research Triangle Park, North Carolina 22709.

Areas of interest: All aspects of use of chemicals and dyes in textile industry. Standardization, training and education and research in this and related fields.

Publications: TEXTILE CHEMIST AND COLORIST (monthly); AATCC TECHNICAL MANUAL AND MEMBERSHIP DIRECTORY (annual); PRODUCTS (annual).

Information activities: Answers inquiries; makes referrals; sponsors workshops.

AMERICAN TEXTILE MACHINERY ASSOCIATION - 1730 M Street N.W., Washington, D.C. 20036.

Areas of interest: Textile machinery; textiles; dyeing of fabrics.

Holdings: Books, periodicals, reviews, standards and specifications, newsletters, pamphlets, publication lists, press releases, clippings, photos, slides, data.

Publications: Directories.

Information services: Answers inquiries; provides consulting services; makes referrals; lends materials.

AMERICAN TEXTILE MANUFACTURERS INSTITUTE, INC. - 1501 Johnston Building, Charlotte, North Carolina 28202.

Areas of interest: Textile manufacturing.

Holdings: Economic data on textile industry.

Publications: Reports, news releases.

Information services: Answers inquiries; provides reference services.

INSTITUTE OF TEXTILE TECHNOLOGY - TEXTILE INFORMATION CENTER, P.O. Box 391, Charlottesville, Virginia 22902.

Areas of interest: Textile technology and related sciences.

Publications: TEXTILE TECHNOLOGY DIGEST (monthly journal containing abstracts of current journal articles, books, pamphlets, patents, and other published materials) with semiannual and annual author, subject, and patent number indexes and a patent concordance; TEXTILE TECHNOLOGY TERMS: AN INFORMATION RETRIEVAL THESAURUS; keyterm index (monthly); various reports, most of which are confidential to member organizations but some of which are non-confidential and available to the public; brochure describing information services; library accessions list (monthly); selected list of books recommended for a textile library.

Information services: Answers inquiries; provides consulting, computer-assisted literature-searching, selective dissemination of information, and duplication services for fee; makes interlibrary loans (and direct loans of some materials); permits onsite use of literature collection.

NORTHERN TEXTILE ASSOCIATION - 211 Congress Street, Boston, Massachusetts 02110.

Areas of interest: Textiles, including cotton, synthetic, wool, and felt; history of textiles; textile mills; textile chemistry.

Holdings: Strong library on history of textiles, textile mills, and textile chemistry.

Information services: Answers inquiries; permits onsite use of collection.

TEXTILE RESEARCH INSTITUTE - 601 Prospect Avenue, Princeton, New Jersey 08540.

Areas of interest: Textiles; polymers; fibers; physics; organic chemistry; econometrics.

Publications: TEXTILE RESEARCH JOURNAL (monthly); annual report, academic bulletin.

Information services: Answers inquiries; makes referrals; provides reference services; provides duplication services for fee; makes interlibrary loans; permits onsite reference.

3. Research Centers

AMERICAN SOCIETY FOR TESTING AND MATERIALS - 1916 Race Street, Philadelphia, Pennsylvania 19103.

Areas of interest: Standardization of specifications and methods of testing materials, including ferrous and nonferrous metals, cement, lime, gypsum, mortar, concrete, mineral aggregates, bituminous materials, soils, asbestos-cement products, masonry units, pipe and drain tile, refractories, ceramic whitewares, porcelain enamel, glass, building stone, thermal insulation, acoustical materials, paper and packaging, adhesives, wood, cellulose, casein, leather, petroleum products and lubricants, paint, naval stores, coal and coke, gaseous fuels, industrial aromatic hydrocarbons, industrial chemicals, plastics, carbon black, textiles, soap, sorptive mineral materials, organic solvents, industrial water, wax polishes, rubber, and electrical insulation; identification of phases from diffraction data, either X-ray or electron, that have been obtained from a powder; spectral absorption data and chemical structure, mass spectral data, and gas chromatographic data.

Publications: MATERIALS RESEARCH AND STANDARDS (monthly); BOOK OF ASTM STANDARDS (annual, consisting of 33 parts); ASTM INDEX TO STANDARDS (annual); ASTM PROCEEDINGS official annual record; fifty-year index and five-year indexes to ASTM technical papers and reports. A publications price list is available. ASTM is also responsible for the distribution of the POWDER DIFFRACTION DATA FILE compiled by the Joint Committee on Powder Diffraction Standards under the joint auspices of ASTM, the American Crystallographic Association, the British Institute of Physics, and the National Association of Corrosion Engineers. The POWDER DIFFRACTION DATA FILE (issued on 3x5 index cards and in book form for revised sets) is used for the identification of phases from diffraction data, either X-ray or electron, that have been obtained from a powdered crystalline material. Powder diffraction data is also available on magnetic tape, with Fortran programs for computer searching. Comprehensive indexes are available.

Information services: Answers inquiries. A standards and specifications service provides reference services and referral assistance without charge.

PURDUE UNIVERSITY
THERMOPHYSICAL PROPERTIES RESEARCH CENTER (TPRC) - 2595 Yeager
Road, West Lafayette, Indiana 47906.

> Sponsored by federal agencies, industrial organizations, and
> others. Collects, analyzes, and synthesizes information on
> thermophysical properties from world wide data bases. This
> center contains the most complete non-print collection on thermo-
> physical properties in the world.

4. International Organizations

COMMON MARKET WORKING PARTY OF THE INTERNATIONAL ASSOCIA-
TION OF USERS OF YARN OF. MAN-MADE FIBERS - 5 rue d' Anjou, 75-
Paris 8e, France.

EUROPEAN ASSOCIATION FOR THE EXCHANGE OF TECHNICAL LITERATURE
IN THE FIELD OF FERROUS METALLURGY - Case Postale, 443 Luxembourg,
Belgium.

> Purpose: To stimulate the exchange of the technical literature
> on all aspects of ferrous metallurgy, with particular reference
> to publications on the ferrous metallurgical industries of the
> USSR and other countries of the East and the Far East.

EUROPEAN COMMITTEE OF MACHINERY MANUFACTURERS FOR THE
PLASTICS AND RUBBER INDUSTRIES - 112 bd Haussmann, 75-Paris 8e,
France.

> Purpose: To promote development of the rubber and plastics
> industries.

INTERNATIONAL BUREAU FOR THE STANDARDISATION OF MAN-MADE
FIBRES - Lautengartenstrasse 12, 4052 Basel, Switzerland.

> Purpose: To establish technical rules relating to specifications
> and characteristics of various man-made fibres and to establish
> technical standards for testing purposes.

> Publications: RULES FOR STANDARDIZATION OF VARIOUS
> CATEGORIES OF MAN-MADE FIBRES.

INTERNATIONAL COPPER RESEARCH ASSOCIATION - 825 Third Avenue,
New York, New York 10022.

A nonprofit research organization supported by copper producers throughout the
world.

> Areas of interest: Chemistry and metallurgy of copper, protective
> coatings and corrosion of copper and copper alloys, copper as
> an alloy in iron and steel, copper composites in plastics and
> ceramics, etc.

> Publications: ANNUAL REPORT, and PROJECT STATUS REPORT
> (quarterly).

Information services: Answers inquiries; makes referrals.

INTERNATIONAL LEAD-ZINC ORGANIZATION - 292 Madison Avenue, New York, New York 10017.

An independent nonprofit research organization which is supported by industry.

Areas of interest: Lead and zinc, including related studies on chemistry, electrochemistry and metallurgy of these metals, their alloys, as well as production, marketing and consumption of lead and zinc products.

Publications: RESEARCH DIGEST (semiannually).

Information services: Answers inquiries; makes referrals; provides consulting service.

INTERNATIONAL RUBBER RESEARCH AND DEVELOPMENT BOARD - 19 Buckingham Street, London, WC2 N6EJ, England.

Purpose: To encourage co-operation in research and development among member units by organizing meetings, stimulating co-operation in research and development, and collating research and development programs.

Publications: Summary of Activities of Research and Development Units. Publication of research and development work is decentralized.

INTERNATIONAL RUBBER STUDY GROUP (IRSG) - Brettenham House, 5-6 Lancaster Place, London, WC2 E7ET, England.

Purpose: Foster production, consumption, and trade in both natural and synthetic rubber. Provides comprehensive statistical information on the industry.

Publications: RUBBER STATISTICAL BULLETIN (monthly); RUBBER STATISTICAL NEWS SHEET (quarterly); INTERNATIONAL RUBBER DIGEST (monthly).

INTERNATIONAL UNION OF TESTING AND RESEARCH LABORATORIES FOR MATERIALS AND STRUCTURES - 12 rue Brancion, 75-Paris 15e, France.

Purpose: To ensure the information exchange of experimental research and tests on structures and materials carried out or planned by participating laboratories.

Activities: Technical commissions, working parties; colloquia and symposia; exchange of scientific publications.

Publications: MATERIALS AND STRUCTURES (6 a year); reports of meetings. Technical Dictionary in 3 languages; Directory of Technical Committees; year book.

LATIN AMERICAN GROUP OF THE INTERNATIONAL UNION OF TESTING

AND RESEARCH LABORATORIES FOR MATERIALS AND STRUCTURES –
Libertad 1235, 3 Piso, Buenos Aires, Argentina.

> Purpose: To promote materials testing in Latin American coun-
> tries.

> Publications: CENSUS OF TESTING AND RESEARCH LABORA-
> TORIES FOR MATERIALS AND STRUCTURES IN LATIN AMERICA.

5. Other Information Sources and Programs

a. SPECIALIZED INFORMATION PROGRAMS

AMERICAN SOCIETY FOR METALS (ASM)
METALS INFORMATION SYSTEM (MIS) – Metals Park, Ohio 44075.

This organization has various specialized information services.

> 1. "Metalert" is a computer search system covering the world's
> literature in metal related fields. Individualized topics can be
> requested on a monthly subscription basis. Material is delivered
> as a computer printout and retrospective searches are also avail-
> able.

> 2. "Metadex" consists of magnetic tape data bases and contains
> complete citations of documents as well as indexing terms and
> abstract numbers from ASM's files. Contact the society for
> further details.

> 3. A new ASM Communications service is AUDIOSCOPE.
> Experts from industry, government, and university recorded live
> on ASM Audio Reports. Specially programmed roundtable sessions
> are on one hour tape cassettes. Topics include: advanced com-
> posites -- challenge to metals; embrittling effects of hydrogen
> on steel; fracture mechanics -- coupling materials to design.
> Complete list of publications and details on information retrieval
> services are available from the society.

> The Society also provides the following: Abstracting and index-
> ing; computer and manual literature searching; copying; inter-
> library loan; reference service; SDI services; state-of-the-art
> compilation.

COPPER DATA CENTER – Battelle Memorial Institute, Columbus Laboratories,
505 King Avenue, Columbus, Ohio 43201.

The Copper Data Center is sponsored by the Copper Development Association,
Inc.

> Areas of interest: Copper technology from refining of metal
> through end-use performance, including such materials as copper,
> copper-base alloys, iron and steel with copper as an alloying
> element, copper chemicals, competitive materials, and materials
> and processes used in insulated wire and cable.

Holdings: 5,000 documents.

Publications: EXTRACTS OF DOCUMENTS ON COPPER TECH-
NOLOGY (5 a year); COORDINATE INDEX TO DOCUMENTS
ON COPPER TECHNOLOGY (3 a year); and THESAURUS OF
TERMS ON COPPER TECHNOLOGY (all available only to
members of the Copper Development Association); technical
reports and data compilations (available to anyone).

Information services: Answers technical questions from anyone.
Does literature searches and provides telephone assistance for
special inquiries.

INSTITUTE OF TEXTILE TECHNOLOGY
TEXTILE INFORMATION CENTER - Route 250 West, Charlottesville, Virginia
22902.

Offers computer and manual literature searching services. SDI
programs are also available.

Computer input in form of cards and magnetic tapes can be provided on a monthly basis.

Contact the Institute for additional details.

PENNSYLVANIA STATE UNIVERSITY - Materials Research Laboratory, 1-112
Research Building, University Park, Pennsylvania 16802.

The Materials Research Laboratory, which coordinates PENNTAP's materials
science and engineering program, makes referrals to on-going programs in
materials science and engineering and conducts seminars, workshops, and short
courses in materials technology.

SOCIETY OF PLASTICS ENGINEERS - 65 Prospect Street, Stanford, Connecticut.

Offers a microprinted catalog system called "SPE Plastics Materials and Processing Catalog Library." This is made of a
master file of supplier catalog details from over 1,000 suppliers
and shows approximately 50,000 pages of catalog material.
This is issued in a 3 volume set and is updated biennially.
Contact the society for further information.

U.S. AIR FORCE
AIR FORCE MATERIALS LABORATORY
AEROSPACE MATERIALS INFORMATION CENTER - Wright-Patterson Air
Force Base, Ohio 45433.

Provides: Literature searching and SDI programs. Collects and
analyzes data. Answers inquiries and makes referrals.

U.S. ARMY
MATERIEL COMMAND
PLASTICS TECHNICAL EVALUATION CENTER (PLASTEC) - Picatinny Arsenal,
Dover, New Jersey 07801.

Offers computer and manual searches from its data banks, consisting of information on properties and applications of plastics in a variety of uses. Also provides reference service.

U.S. DEPARTMENT OF DEFENSE
DEFENSE SUPPLY AGENCY - Headquarters, Cameron Station, Alexandria, Virginia 22314.

Sponsored centers are: Chemical Propulsion Information Agency, Infrared Information and Analysis Center, Reliability Analysis Center, Metals and Ceramics Information Center, Machinability Data Center, Thermophysical Properties Information Analysis Center, and the Mechanical Properties Data Center.

These centers gather, sort, and synthesize scientific and technical information on their specific areas. Information services usually include answering of technical inquiries, referrals, specialized bibliographies, and publications.

B. LITERATURE

1. Guides and Handbooks

a. GENERAL

Adhesives 1972. BLUE BOOK NO. 2. Park Ridge, New Jersey: Noyes Corp.

"The book describes the 'on the line' adhesive products of 121 U.S. manufacturers arranged according to company names."

Aitken, D., ed. ENGINEER'S HANDBOOK OF ADHESIVES. Brighton, England: Machinery Publishing, 1972.

"A comprehensive guide to the engineering properties and applications of modern adhesives."

American Society of Tool and Manufacturing Engineers. PLASTICS TOOLING AND MANUFACTURING HANDBOOK; A REFERENCE BOOK ON THE USE OF PLASTICS AS ENGINEERING MATERIALS FOR TOOL AND WORKPIECE FABRICATION. Englewood Cliffs, New Jersey: Prentice-Hall, 1965.

ASME HANDBOOK: METALS PROPERTIES. New York: McGraw-Hill, 1954.

Bandrup, J., ed. POLYMER HANDBOOK. New York: Interscience, 1966.

Provides physical and chemical data on polymers.

BIG POTENTIAL IN RIGID ORIENTED PLASTICS. Stamford, Connecticut: Business Communications Co., 1975.

Brady, G. MATERIALS HANDBOOK; AN ENCYCLOPEDIA FOR PURCHAS-ING MANAGERS, ENGINEERS, EXECUTIVES, AND FOREMEN. 10th ed. New York: McGraw-Hill, 1971.

> Authoritative guide to general materials field.

Cagle, C. HANDBOOK OF ADHESIVE BONDING. New York: McGraw-Hill, 1973.

> Bibliographical references.

CONCISE GUIDE TO PLASTICS. 4th ed. New York: Reinhold Publishing Corp., 1976.

> "Describes the types, properties, forms and basic chemicals from which plastics are made as well as resin manufacture, processing, applications by materials and by industries, production and prices."

Cook, G. HANDBOOK OF TEXTILE FIBERS. 5th ed. Plainfield, New Jersey: Textile Book Service, 1975.

Crosby, E. PRACTICAL GUIDE TO PLASTICS APPLICATIONS. Boston: Cahners, 1972.

> "Covers thermosetting plastics; thermoplastic materials; the methods of molding, etc."

Deanin, R., ed. TOUGHNESS AND BRITTLENESS IN PLASTICS. Washington, D.C.: American Chemical Society, 1976.

EPOXY RESIN HANDBOOK. Park Ridge, New Jersey: Noyes Data Corporation, 1972.

Hall, A. THE STANDARD HANDBOOK OF TEXTILES. 8th ed. London: Butterworth, 1975.

HANDBOOK OF METAL POWDERS. Edited by A. Poster. New York: Reinhold, 1966.

HANDBOOK OF PLASTICS AND ELASTOMERS. New York: McGraw-Hill, 1975.

Housner, H. HANDBOOK OF POWDER METALLURGY. New York: Chemical Publishing Co., 1973.

Ives, G. HANDBOOK OF PLASTICS TEST METHODS. London: Butterworth, 1972.

> Test techniques applied to mechanical properties, and other aspects of plastics as materials.

Kronenthal, R., et al., eds. POLYMERS IN MEDICINE AND SURGERY. New York: Plenum Publishing Co., 1975.

Lubin, G., ed. HANDBOOK OF FIBERGLASS AND ADVANCED PLASTICS COMPOSITES. New York: Van Nostrand, 1970.

> "Extensive compilation of information on fiberglass and advanced composites."

Lynch, C. HANDBOOK OF MATERIALS SCIENCE. 3 vols. Cleveland: Chemical Rubber Co., 1975.

MATERIALS SELECTOR. Annual. Stamford, Connecticut: Materials Engineering.

Mohr, J. HANDBOOK OF TECHNOLOGY AND ENGINEERING OF RE-INFORCED PLASTICS/COMPOSITES OF THE SPI, INC. New York: Van Nostrand, 1973.

Moncrieff, R. MAN-MADE FIBERS. New York: Wiley, 1975.

Morton, M., ed. RUBBER TECHNOLOGY. New York: Van Nostrand, 1973.

> "Systematic coverage of rubber technology."

Moss, J. PROPERTIES OF ENGINEERING MATERIALS. Cleveland: Chemical Rubber Publishing Company, 1971.

Oleesky, S., et al. HANDBOOK OF TECHNOLOGY AND ENGINEERING OF REINFORCED PLASTICS. New York: Van Nostrand, 1973.

PLASTICS HANDBOOK. Edited by P. Farago. London: Product Journals Ltd., 1968.

THE PLASTICS MANUAL. 5th ed. London: Applied Plastics c/o Scientific Press Ltd., 1971.

PLASTICS MANUFACTURING HANDBOOK AND BUYER'S GUIDE. Boston: Herman Publishing Co., 1975.

POLYMER ADDITIVES. Park Ridge, New Jersey: Noyes Data Corporation, 1972.

> "Comprehensive listing of commercially available, protective additives. Gives products of 86 U.S. manufacturers arranged according to company name. Each product is carefully indexed by chemical, generic, trivial, and trade name or registered trademark in the 'one alphabet' index at the back of the book."

RARE METALS HANDBOOK. Edited by C. Hampel. Huntington, New York: R.E. Krieger Publishing Company, 1971.

Reeves, W., et al. FIRE RESISTANT TEXTILES HANDBOOK. Westport, Connecticut: Technomic, 1974.

Roff, W., comp. FIBRES, FILMS, PLASTICS, AND RUBBERS; A HANDBOOK OF COMMON POLYMERS. London: Butterworth, 1972.

Sacharow, S. HANDBOOK OF PACKAGE MATERIALS. Westport, Connecticut: Avi Publishing Co., 1976.

Society of the Plastics Industry. PLASTICS ENGINEERING HANDBOOK. 4th ed. New York: 1975.

Titow, W., ed. PENN'S PVC TECHNOLOGY. 3rd ed. Barking, England: Applied Science Publishing, 1971.

> Reference handbook which gives chemical properties trade names and commericial details.

b. TABLES AND DATA

DIFFUSION AND DEFECT DATA. 8 issues per year. Trans Tech Publications, 21330 Center Ridge Road, Rocky River, Ohio 4116.

> Contents includes diffusion and migration in solids and liquids, point defects in crystals, irradiation damage and effects, etc. Of special interest in materials research.

Glanvill, A. PLASTICS ENGINEER'S DATA BOOK. New York: Industrial Press, 1973.

Kreutzer, L. von B. GOSSAMER THREAD UNSTRUNG. New York: McGraw-Hill, 1955.

> "The more delicate properties of plastics and synthetics."

Mallows, D. STRESS ANALYSIS PROBLEMS IN SI UNITS. New York: Pergamon, 1972.

Seidell, A. SOLUBILITIES OF INORGANIC AND METAL ORGANIC COMPOUNDS. 4th ed. 2 vols. American Chemical Society. New York: Van Nostrand, 1958-1965.

Smithells, C. METALS REFERENCE BOOKS. 4th ed. 3 vols. London: Butterworths, 1969.

2. Dictionaries

a. TECHNICAL

Craig, A. DICTIONARY OF RUBBER TECHNOLOGY. London: Butterworths, 1969.

Dodd, A. DICTIONARY OF CERAMICS. London: Newnes, 1964.

Linton, G. THE MODERN TEXTILE AND APPAREL DICTIONARY. 4th ed. Plainfield, New Jersey: Textile Book Service, 1973.

Merriman, A. A DICTIONARY OF METALLURGY. London: McDonald and Evans, 1958.

Rosato, D. MARKETS FOR PLASTICS. New York: Van Nostrand Reinhold, 1969.

> A glossary of terms used in the industry appears on pages 419-40.

Simons, E. A DICTIONARY OF ALLOYS. New York: Hart, 1969.

> "Brief descriptions of a large number of alloys arranged in alphabetical order."

_____. A DICTIONARY OF FERROUS METALS. London: Muller, 1970.

TEXTILE TERMS AND DEFINITIONS. 7th ed. Manchester, England: Textile Institute, 1975.

b. FOREIGN LANGUAGE

Dawydoff, W. TECHNICAL DICTIONARY OF HIGH POLYMERS. New York: Pergamon, 1969.

> English, German, French, and Russian languages.

Dettner, H., comp. ELSEVIER'S DICTIONARY OF METAL FINISHING AND CORROSION IN FIVE LANGUAGES. New York: Elsevier Publishing Company, 1971.

Dorian, A. SIX-LANGUAGE DICTIONARY OF PLASTICS AND RUBBER TECHNOLOGY; A COMPREHENSIVE DICTIONARY IN ENGLISH, GERMAN, FRENCH, ITALIAN, SPANISH, AND DUTCH. London: Iliffe, 1965.

Holmstrom, J. TRILINGUAL DICTIONARY FOR MATERIALS AND STRUCTURES. Elmsford, New York: Pergamon Press, 1972.

Josselson, H., ed. RUSSIAN-ENGLISH PLASTICS DICTIONARY. Detroit: Wayne State University Press, 1970.

> "Contains approximately 8000 entries in plastics science and technology."

Santholzer, R. FIVE LANGUAGE DICTIONARY OF SURFACE COATINGS, PLATING PRODUCTS FINISHING, CORROSION, PLASTICS, AND RUBBER. Elmsford, New York: Pergamon Press, 1969.

3. Directories

a. GENERAL

ADHESIVES RED BOOK. Annual. New York: Adhesives Age.

BRITISH PLASTICS YEARBOOK. Annual. London: Iliffe Publishing Company.

CANADIAN PLASTIC DIRECTORY AND BUYER'S GUIDE. Annual. Canada, Southam Business Publications Ltd., 1450 Don Mills Road, Don Mills, Ontario, 1959.

CHEMICAL ENGINEERING DESKBOOK. Engineering Materials. CHEMICAL ENGINEERING, Vol. 79, December 4, 1973 issue.

EUROPEAN PLASTICS. Annual. New York: Heinman.

INDEX TO CLASSIFIED DIRECTORY AND FACILITIES. Planning guide. Annual issue in PLASTICS TECHNOLOGY.

INTERNATIONAL PLASTICS DIRECTORY. 3rd ed. 2 vols. New York: International Publications Service, 1970.

MATERIALS SELECTION. Annual. Mid-September issue MATERIALS ENGINEERING.

Metal Progress. DATA BOOK ISSUE. Mid-June, 1973. METAL PROGRESS, Vol. 104, nl.

PLASTICIZERS; GUIDEBOOK AND DIRECTORY. Park Ridge, New Jersey: Noyes Data Corp., 1972.

RUBBER RED BOOK. Annual. Palmerton Publishing Company, Inc., 101 West 31st Street, New York, New York 10001.

    Yearbook and guide to rubber industry of U.S. and Canada.

Tarnfield, C., ed. A GUIDE TO SOURCES OF INFORMATION IN THE TEXTILE INDUSTRY. 2nd ed. Manchester, England: Textile Institute, 1974.

U.S. National Bureau of Standards. 1973 CATALOG – STANDARD REFERENCE MATERIALS. (Special Publication 260). Washington, D.C.: Government Printing Office, 1973.

b. PRODUCTS AND MANUFACTURERS

i. United States and Canada

MACRAE'S BLUEBOOK. Annual. MacRae's Bluebook, Western Springs, Illinois 60558.

MATERIALS ENGINEERING -- MATERIALS. Selector issue. Annual.
Reinhold Publishing Company, 430 Park Avenue, New York, New York 10022.

>"Directory contains material important to engineering classified
>by type of material used."

PLASTIC ADDITIVES -- MARKETING GUIDE AND COMPANY DIRECTORY.
Technomic Publishing Company, Inc., 265 West State Street, Westport,
Connecticut 06880.

>"An in-depth analysis of the plastics additives business: pro-
>duction, marketing, products, and markets."

PLASTICS DIRECTORY OF CANADA. Annual. Southam Business Publications
Ltd., Ontario, Canada.

PLASTICS MATERIALS GUIDE. Annual. Engineering, Chemical and Marine
Press Ltd., London.

PLASTICS, PAINT, AND RUBBER. Periodical. BUYER'S GUIDE FOR
SOUTHERN AFRICA. Annual. Thompson Newspapers, South Africa
Ltd., Cape Town, South Africa.

PLASTICS TECHNOLOGY. Periodical. MACHINERY SELECTION GUIDE.
Annual. Bill Publications, Inc., 630 Third Avenue, New York, New York
10017.

PLASTICS WORLD DIRECTORY OF THE PLASTICS INDUSTRY. Annual. Cahners
Publishing Company, Inc., 270 St. Paul Street, Englewood, Colorado 80206.

POLYMER ADDITIVES; GUIDEBOOK AND DIRECTORY. Park Ridge, New
Jersey: Noyes Data Corporation, 1972.

>Prepared from data and information extracted and digested from
>technical data sheets of 85 United States manufacturers.

POLYPROPYLENE FIBERS - MANUFACTURING/MARKETING GUIDE. By M.
Ahmed. Technomic Publishing Company, Inc., 265 West State Street, Westport,
Connecticut 06880.

>"A unified treatment of the science and technology of polypro-
>pylene fibers, with particular orientation to manufacturing and
>marketing. It includes an extensive survey of pertinent scienti-
>fic, technical, trade information, and related patents."

PROGRESSSIVE PLASTIC BUYERS' GUIDE TO PLASTICS. Annual. Maclean-
Hunter Ltd., 481 University Avenue, Toronto 2, Ontario, Canada.

SWEET'S METALWORKING EQUIPMENT CATALOG FILE. Annual. Sweet's
Industrial Catalog Services, 330 West 42nd Street, New York, New York 10036.

THOMAS' REGISTER OF MANUFACTURERS. Annual. New York: Thomas
Publishing Company.

U.S. FOAMED PLASTICS MARKETS AND DIRECTORY. Technomic Publishing Company, Inc., 265 West State Street, Westport, Connecticut 06880. 1973.

> "The standard reference book of the rapidly-growing foamed plastics industry."

URETHANE PLASTICS AND PRODUCTS. Monthly. Technomic Publishing Company, Inc., 750 Summer Street, Stamford, Connecticut 06901.

WESTERN PLASTICS. Annual. Breskin Communications Inc., 415 North Brown Avenue, Scottsdale, Arizona 85251.

ii. International

BRITISH PLASTICS YEARBOOK. Annual. Engineering, Chemical and Marine Press Ltd., 33-40 Bowling Green Lane, London, EC1, England.

MANUAL OF RUBBER AND TIN PRODUCING COMPANIES. Zorn & Leigh-Hunt, Moor House, London Wall and Stock Exchange, London, EC2Y 5HB, England. 1971.

> Lists companies in Asia.

NEW TRADE NAMES IN THE RUBBER AND PLASTICS INDUSTRIES. Annual. Rubber and Plastics Research Association of Great Britain, Shawbury, Shrewsbury, Shropshire, England.

PLASTICS, PAINT AND RUBBER BUYER'S GUIDE. Annual. Thomson Publications, South Africa (Pty) Ltd., Trust House, Thibault Square, P.O. Box 80, Capetown, South Africa.

RUBBER DIRECTORY OF GREAT BRITAIN. Biannual. Maclaren and Sons Ltd., Davis House, 69-77 High Street, Croydon CR9 IQH, England.

TEXTILE GUIDE TO EUROPE. 2nd ed. Park Ridge, New Jersey: Noyes Data Corp., 1972.

> A directory to textile firms of 18 European countries.

c. MARKETING GUIDES AND SURVEYS

GLASS AND OTHER ADVANCED FIBERS. Special Study 78. Cleveland: Predicasts, Inc., 1973.

> Analysis of glass and advanced fibers industry, market forecasts, and general overview of field.

NONWOVEN DISPOSABLE PRODUCTS. Cleveland: Predicasts, Inc., 1973.

> "Analyzes outlook for nonwoven disposables and projects markets for medical products, diapers, wipes, uniforms, etc. (1975 and 1980)."

PAPER VS. PLASTICS. Cleveland: Predicasts, Inc., 1973.

"Revises up-to-date analysis and review of major markets where paper competes with plastics (cups, food and milk containers, shipping sacks, etc.) Based on market survey of suppliers and end-users. Projections to 1980."

PLASTIC FOAMS. Cleveland: Predicasts, Inc., 1973.

"Analysis of plastic foams by type (rigid and flexible urethanes, polyvinyl chloride, polystyrene, etc.) by 15 major end-users, with projections to 1980."

PLASTICS IN INDUSTRIAL PACKAGING. Stamford, Connecticut: Business Communications Co., 1976.

PLASTICS IN THE PAPER INDUSTRY. Stamford, Connecticut: Business Communications Co., 1974.

PLASTICS PROCESS MACHINERY & MARKETS. Cleveland: Predicasts, Inc., 1973.

"Historical and projected (to 1975, 1980, and 1985) consumption of resins by process, and machinery purchases by type."

PLASTICS TRENDS. Quarterly. Cleveland: Predicasts, Inc.

PLASTIC SYNTHETIC WOOD. Stamford, Connecticut: Business Communications Co., 1976.

REINFORCED PLASTICS. Cleveland: Predicasts, Inc., 1973.

"Analyzes the RF plastics industry. Historical and projected (1975, 1980 and 1985) data presented for 13 markets and 9 raw materials."

Technomic Research Staff. U.S. FOAMED PLASTICS MARKETS AND DIRECTORY. Stamford, Connecticut: Technomic Publishing Company, 1973.

TRANSPARENT ENGINEERING PLASTICS. Stamford, Connecticut: Business Communications Co., 1976.

WORLD PLASTICS MARKETS. Cleveland: Predicasts, Inc., 1973.

"Report analyzes supply and demand for plastics by type and by country. Historical data and projections through 1980 are given for 32 countries and 7 resin groups."

WORLD RUBBER & TIRE MARKETS. Cleveland: Predicasts, Inc., 1973.

"Report analyzes historical and projected (to 1980) consumption of rubber by country on the basis of anticipated demands for tire production and other markets. Data are also provided for production and consumption of natural and synthetic rubber for each country."

d. LABORATORIES AND RESEARCH CENTERS

EUROPEAN RESEARCH INDEX. 2nd ed. Guernsey, England: Francis Hodgson Ltd., 1969.

INDUSTRIAL RESEARCH LABORATORIES OF THE UNITED STATES. 14th ed. New York: R.R. Bowker, 1975.

RESEARCH CENTERS DIRECTORY. 5th ed. Detroit: Gale Research Company, 1975.

Updated by subscription service, NEW RESEARCH CENTERS.

4. Encyclopedias

American Fabrics. A.F. ENCYCLOPEDIA OF TEXTILES. 2nd ed. Englewood Cliffs, New Jersey: Prentice Hall, 1972.

Clauser, R. ENCYCLOPEDIA OF ENGINEERING MATERIALS AND PRO-CESSES. New York: Van Nostrand Reinhold, 1969.

THE ENCYCLOPEDIA OF BASIC MATERIALS FOR PLASTICS. Edited by H. Simonds. New York: Van Nostrand Reinhold, 1967.

ENCYCLOPEDIA OF POLYMER SCIENCE AND TECHNOLOGY; PLASTICS, RESINS, RUBBERS, FIBERS. 16 vols. Ed. board: H.F. Mark, chairman. New York: Interscience, 1964-1973.

MATERIALS AND TECHNOLOGY; A SYSTEMATIC ENCYCLOPEDIA OF THE TECHNOLOGY OF MATERIALS USED IN INDUSTRY AND COMMERCE, INCLUDING FOODSTUFFS AND FUELS. 8 vols. London: Longmans Ltd., 1968.

MODERN PLASTICS ENCYCLOPEDIA. Annual. New York: McGraw-Hill.

Osborne, A.K. AN ENCYCLOPEDIA OF THE IRON AND STEEL INDUSTRY. 2nd ed. London: Technical Press, 1967.

5. Bibliographies

a. BOOKS

American Society for Engineering Education, Engineering Libraries Division.

GUIDE TO LITERATURE OF METALS AND METALLURGY. Washington, D.C.: 1970.

AMERICAN SOCIETY FOR TESTING AND MATERIALS LIST OF PUBLICATIONS. Annual. The American Society for Testing and Materials, 1916 Race Street, Philadelphia, Pennsylvania 19103.

Grigsby, D., ed. ELECTRONIC PROPERTIES OF MATERIALS; A GUIDE TO THE LITERATURE. Volume 3, parts 1 and 2. New York: Plenum, 1971.

GUIDE TO METALLURGICAL INFORMATION. By E. Tapia. Special Libraries Association, 225 Park Avenue, South, New York, New York 10003. 1965.

Hyslop, M. A BRIEF GUIDE TO SOURCES OF METALS INFORMATION. Washington, D.C.: Information Resources Press, 1973.

REINFORCED PLASTICS. By M. Smith. Part 1: Fiberglass Reinforcement. Rept. no. NTIS/PS-76/0062. Part 2: Boron, Carbon and Other Reinforcing Materials. Rept. no. NTIS/PS-76/0064. Springfield, Virginia: National Technical Information Service, 1976.

> Bibliographies with abstracts.

Sommar, H. A BRIEF GUIDE TO SOURCES OF FIBER AND TEXTILE INFORMATION. Washington, D.C.: Information Resources Press, 1973.

White, D. HOW TO FIND OUT IN IRON AND STEEL. New York: Pergamon, 1970.

> "Information services and libraries. Literature guides, bibliographies, and lists of periodicals."

Yescombe, E. SOURCES OF INFORMATION ON THE RUBBER, PLASTICS AND THE ALLIED INDUSTRIES. Elmsford, New York: Pergamon, 1968.

b. SERIAL PUBLICATIONS

A S M BIBLIOGRAPHY SERIES. 10 issues/year. American Society for Metals, Metals Park, Ohio 44073.

Boodson, K. "The European metals industry and its literature." METALLURGIA, October/November 1963, pp. 69-73, 221-26.

British Ceramic Society. PUBLICATIONS. 8 issues/year. Shelton House, Stoke Road, Shelton, Stoke-on-Trent, England.

Japan Information Center of Science and Technology. CURRENT BIBLIOGRAPHY ON SCIENCE AND TECHNOLOGY: EARTH SCIENCES, MINING AND METALLURGY (Japan). Semimonthly. Japan Information Center of Science and Technology, 2-5-2 Nagatacho, Chiyoda-ku, Tokyo, Japan.

Text in Japanese.

Meyer, A. "Polymer science booklist." SPE JOURNAL, Vol. 26, September 1970, pp. 77-79.

c. RELATED BIBLIOGRAPHIES IN CURRENT JOURNALS

The following journals frequently contain book reviews and/or bibliographies. See the heading Journals and Serial Publications in this section for addresses.

ADHESIVES AGE

AMERICAN CERAMIC SOCIETY BULLETIN

AMERICAN CERAMIC SOCIETY JOURNAL

BRITISH CERAMIC SOCIETY PUBLICATIONS

BRITISH PLASTICS

BRITISH POLYMER JOURNAL

CANADIAN PLASTICS

COPPER ABSTRACTS

GLASS TECHNOLOGY

JOURNAL OF APPLIED POLYMER SCIENCE

JOURNAL OF CELLULAR PLASTICS

JOURNAL OF ELASTOPLASTICS

MATERIALS ENGINEERING

MATERIALS RESEARCH BULLETIN

METAL POWDER REPORT

METAL PROGRESS

MODERN PLASTICS

NONWOVEN PATENTS DIGEST

PHYSICS AND CHEMISTRY OF GLASSES

PLASTICS DESIGN AND PROCESSING

PLASTICS INDUSTRY NEWS, JAPAN

PLASTICS IN ENGINEERING

PLASTICS TECHNOLOGY

PLASTICS WORLD

POLYMER AGE

POLYMERICS

POLYMER MECHANICS

POLYMER NEWS

POLYMER SCIENCE AND TECHNOLOGY POST

POWDER METALLURGY SCIENCE AND TECHNOLOGY

RAPRA ABSTRACTS

REINFORCED PLASTICS

RUBBER AGE

RUBBER CHEMISTRY AND TECHNOLOGY

SPE JOURNAL

STAHL UND EISEN

6. Monographs, Reports, and Symposia Proceedings

ADVANCED MATERIALS: COMPOSITES AND CARBON. Symposium, Philadelphia, Pennsylvania, April 26, 1971. Sponsored by the American Ceramic Society. PROCEEDINGS. Columbus, Ohio: American Ceramic Society, 1972.

ADVANCES IN COATINGS & ADHESION. Symposium at 5th Polymer Conference Series. Polymer Institute, University of Detroit, 4001 West McNichols Road, Detroit, Michigan 48221. 1973.

ADVANCES IN PLASTICS TECHNOLOGY. 1st Annual Pacific Technical Conference and Technical Displays, Las Vegas, Nevada, September, 1975. PROCEEDINGS. Greenwich, Connecticut: Society of Plastic Engineers, 1976.

ADVANCES IN REINFORCED THERMOPLASTICS. Regional Technical Conference. Society of Plastics Engineers, 656 West Putnam Avenue, Greenwich, Connecticut 06830. 1972.

ADVANCES IN URETHANE SCIENCE & TECHNOLOGY. Symposium at 5th Annual Polymer Conference Series. Polymer Institute, University of Detroit, 4001 West McNichols Road, Detroit, Michigan 48221. 1970.

Aggarwal, S., ed. BLOCK POLYMERS. New York: Plenum, 1970.

"A collection of 23 papers presented at a symposium of the American Chemical Society, on the preparation and the applications of block polymers."

Allen, P. NATURAL RUBBER AND THE SYNTHETICS. New York: Halsted Press, Division of Wiley, 1972.

Overview of natural rubber and synthetics field.

American Chemical Society. ENGINEERING PLASTICS AND THEIR COM-MERCIAL DEVELOPMENT. American Chemical Society, Advances in Chemistry Series, 96. Washington, D.C.: 1969.

_____. EPOXY RESINS. Advances in Chemistry Series, 92. Washington, D.C.: 1970.

_____. MULTICOMPONENT POLYMER SYSTEMS. Advances in Chemistry Series, 99. Washington, D.C.: 1971.

American Chemical Society. Polymer Chemistry, Division. POLYMER NET-WORKS; STRUCTURE AND MECHANICAL PROPERTIES. Proceedings of the ACS Symposium on Highly Cross-linked Polymer Networks, Chicago, September 14-15, 1970. Edited by A.J. Chompff and S. Newman. New York: Plenum, 1971.

Argon, A., ed. PHYSICS OF STRENGTH AND PLASTICITY. Cambridge: M.I.T. Press, 1970.

Papers presented at a symposium which covered the field of physics of crystal plasticity.

Baird, R. INDUSTRIAL PLASTICS: BASIC CHEMISTRY, MAJOR RESINS, MODERN INDUSTRIAL PROCESSES. South Holland, Illinois: Goodheart-Wilcox, 1971.

Barden, E. FLOCKED MATERIALS: TECHNOLOGY AND APPLICATIONS 1972. Park Ridge, New Jersey: Noyes Data Corp., 1972.

Details compiled from recent patent literature.

Barker, S. POLYACETALS. New York: American Elsevier, 1971.

Covers polyactals field: chemistry properties and processes.

Beck, R. PLASTIC PRODUCT DESIGN. New York: Van Nostrand Reinhold, 1970.

"A guide to contemporary design in plastics, including material on the collapsible core and dual extrusion not otherwise available in book form."

Benjamin, W. PLASTIC TOOLING; TECHNIQUES AND APPLICATIONS. New York: McGraw-Hill, 1972.

"Guide to the basic techniques, materials, equipment and uses of plastic tooling."

Benning, C. PLASTIC FOAMS: THE PHYSICS AND CHEMISTRY OF PROD-UCT PERFORMANCE AND PROCESS TECHNOLOGY. 2 vols. Volume 1: Chemistry and physics of foam formation; Volume 2: Structure properties, and applications. New York: Interscience, Division of Wiley, 1969.

> "A comprehensive monograph on the technology and principles of plastic foam formation and their relation to specific products and processes."

Bishop, R. PRACTICAL POLYMERIZATION FOR POLYSTYRENE. Boston: Cahners, 1972.

> Reference tool for polystyrene processing.

Blow, C., ed. RUBBER TECHNOLOGY AND MANUFACTURE. London: Butterworth, 1971.

> Comprehensive coverage of rubber technology.

Braun, D. TECHNIQUES OF POLYMER SYNTHESIS AND CHARACTERIZATION. New York: Wiley-Interscience, 1972.

> Offers a detailed explanation of macromolecular field for practitioners and research personnel.

Bruins, P., ed. NEW POLYMERIC MATERIALS. Applied Polymer Symposia no. 11. Symposium held at Polytechnic Institute of Brooklyn, Brooklyn, New York, June 7-8, 1968. New York: Interscience, Division of Wiley, 1969.

Brydson, J. FLOW PROPERTIES OF POLYMER MELTS. New York: Van Nostrand Reinhold, 1971.

> "Provides a viewpoint of some of the problems encountered in the various aspects of polymer flow."

_____. PLASTICS MATERIALS. 2nd ed. London: Iliffe, 1969.

> Standard reference book on plastics.

Carter, M. ESSENTIAL FIBER CHEMISTRY. Fiber Science Series. New York: Dekker, 1971.

> "A general review of major fiber varieties which gives an overview of fiber chemistry with emphasis on applications in the textile industry."

Ceresa, R., ed. BLOCK AND GRAFT CO-POLYMERIZATION. New York: Wiley-Interscience, 1973.

THE CHALLENGES OF THERMOSETS. Regional Technical Conference. Greenwich, Connecticut: Society of Plastics Engineers, 1973.

Chompff, A.J., ed. POLYMER NETWORKS; STRUCTURE AND MECHANICAL PROPERTIES. New York: Plenum, 1971.

"Contains the papers presented at a symposium sponsored by the Division of Polymer Chemistry at the American Chemical Society Meeting held in September, 1970."

Conley, R. THERMAL STABILITY OF POLYMERS. New York: Dekker, 1970.

Volume 1: "Summarizes aspects of the theoretical principles which describe thermal and oxication phenomena in polymeric materials."

CORROSION 73. International Forum on Protection & Performance of Materials in Corrosive Environments. National Association of Corrosion Engineers, 2400 West Loop South, Houston, Texas 77027. 1973.

Crompton, T. CHEMICAL ANALYSIS OF ADDITIVES IN PLASTICS. New York: Pergamon, 1971.

"A compilation of world literature on the characterization, identification and determination of various types of additives in plastics."

Deanin, R. POLYMER STRUCTURE, PROPERTIES, AND APPLICATIONS. Boston: Cahners, 1972.

"Presents information on polymer synthetic field."

Doyle, E. THE DEVELOPMENT AND USE OF POLYURETHANE PRODUCTS. New York: McGraw-Hill, 1971.

Complete data on urethanes.

Ferry, J. VISCOELASTIC PROPERTIES OF POLYMERS. 2nd ed. New York: Wiley, 1970.

General review of viscoelastic polymers.

Finniston, H. STRUCTURAL CHARACTERISTICS OF MATERIALS. New York: Elsevier, 1971.

"Review articles by specialists compiling the background of experiment and theory which exists on the structural characteristics of materials."

Fitch, R., ed. POLYMER COLLOIDS. New York: Plenum, 1971.

"Presents the papers presented at an ACS symposium in 1970. Covers many of the problems concerned with polymer colloids."

14th TEXTILE CHEMISTRY AND PROCESSING CONFERENCE. Proceedings, New Orleans, Louisiana, April 1974. Washington, D.C.: Agricultural Research Services, U.S. Department of Agriculture, 1975.

Frisch, K., ed. PLASTIC FOAMS. Vol. 1, Part 1. New York: Dekker, 1972.

"Technology and applications of plastic foams."

Gait, A. PLASTICS AND SYNTHETIC RUBBERS. Chemical Industry Series. Elmsford, New York: Pergamon, 1970.

Galasso, F. HIGH MODULUS; FIBERS AND COMPOSITES. New York: Gordon & Breach, 1969.

"Covers Boron fibers, Carbon fibers, high modulus fibers and Composites."

Gill, R. CARBON FIBERS IN COMPOSITE MATERIALS. London: Butterworth, 1972.

"Introduces fibre composite materials and deals in detail with the development, uses, production and properties of carbon fibers."

Gould, R., ed. EPOXY RESINS. Washington, D.C.: American Chemical Society, 1970.

"Comprises a collection of original papers presented at a Division of Organic Coatings and Plastics Chemistry Symposium (American Chemical Society) in April 1968. Some of these are concerned with the preparation of novel epoxide resins of the aliphatic, cycloaliphatic, fluorine containing, triazine based and other types and some with newer curing agents such as fluorine containing amines, aminosiloxanes, carboxy terminated polyisobutylenes, phosphorus amides and extra-coordinate silicon to salts."

Great Britain. Ministry of Technology. ADHESION: FUNDAMENTALS AND PRACTICE. A report of an international conference held at the University of Nottingham, England, 20-22 September, 1966. New York: Gordon & Breach, 1969.

"Problems on the adhesive bonding of rubbers and plastics."

Hearle, J. STRUCTURAL MECHANICS OF FIBERS, YARNS, AND FABRICS. Vol. 1. New York: Interscience, Division of Wiley, 1969.

"An engineering approach to textile structures and their strength."

INDUSTRIAL CARBON & GRAPHITE. 3rd conference, London, England, April 14-17, 1970. PROCEEDINGS. London: Society of Chemical Industry, 1971.

INTERATOMIC POTENTIALS AND SIMULATION OF LATTICE DEFECTS. Edited by P.C. Gehlen, et al. New York: Plenum, 1972.

Johnson, K. CELLULAR PLASTICS RECENT DEVELOPMENTS. Park Ridge, New Jersey: Noyes Data Corporation, 1970.

_____. POLYCARBONATES--RECENT DEVELOPMENTS. Park Ridge, New Jersey: Noyes Data Corporation, 1970.

Review of processes and uses.

_____. POLYURETHANE COATINGS. Park Ridge, New Jersey: Noyes Data Corporation, 1972.

"Reviews 157 U.S. patents issued mostly since 1960 related to paint vehicles, wet look, highly glossy fabric coatings, micro-porous products, etc."

_____. VINYL AND ACRYLIC ADHESIVES INCLUDING PRESSURE SENSI-TIVES. Park Ridge, New Jersey: Noyes Data Corporation, 1971.

"Includes 123 processes."

Kresser, T. POLYOLEFIN PLASTICS. New York: Van Nostrand, 1970.

Reviews current knowledge on polyolefins.

McCrum, N. REVIEW OF THE SCIENCE OF FIBRE RE-INFORCED PLASTICS. London: H.M.S.O., 1972.

McCullough, R. CONCEPTS OF FIBER-RESIN COMPOSITES. New York: Dekker, 1971.

"Presents details on composite materials of polymers reinforced exclusively with fibres."

McDonald, M. FLEXIBLE FOAM LAMINATES. Park Ridge, New Jersey: Noyes Data Corporation, 1971.

"Reviews the U.S. patent literature on the technology of flexible foam laminates from 1960."

Mallinson, J., ed. CHEMICAL PLANT DESIGN WITH REINFORCED PLASTICS. New York: McGraw-Hill, 1969.

"Describes use of epoxy and polyester glass fiber laminates in chemical plant equipment where corrosion factors are important such as in vats, tanks, and piping."

Mark, H. CHEMICAL AFTERTREATMENT OF TEXTILES. New York: Wiley, 1971.

MARKETING REINFORCED PLASTICS. Conference. Plastics Institute, 11 Hobart Place, London SW1W OHL, England. 1969.

MATERIALS SCIENCE. 6th colloquium. Seattle, Washington, 1971. Proceedings issued as INTERATOMIC POTENTIALS AND SIMULATION OF LATTICE

DEFECTS. Edited by P.C. Gehlen, J.R. Beeler, and R.I. Jaffee. New York: Plenum, 1972.

MATERIALS SCIENCE AND ENGINEERING IN THE UNITED STATES. Unversity Park, Pennsylvania: Pennsylvania State University Press, 1970.

MECHANICAL PERFORMANCE AND DESIGN IN POLYMERS. Edited by O. Delatychi. New York: Wiley, 1971.

MECHANICAL PROPERTIES OF POLYMERIC FOAMS. Technomic Publishing Company, Inc. 265 West State Street, Westport, Connecticut 06380. 1973.

> "Contains design information for working with foams."

MATERIALS SCIENCE RESEARCH. Symposium on. Bhabha Atomic Research Centre, Information Service, Trombay, Bombay 85, India. 2 vols. 1972.

Meltzer, Y. URETHANE FOAMS TECHNOLOGY AND APPLICATIONS. Park Ridge, New Jersey: Noyes Data Corporation, 1971.

> "The book contains descriptions of 148 manufacturing processes of which 42 deal with application technology."

_____. WATER-SOLUBLE POLYMERS; TECHNOLOGY AND APPLICATIONS. Park Ridge, New Jersey: Noyes Data Corporation, 1972.

> Based on patent reviews.

Milby, R. PLASTICS TECHNOLOGY. New York: McGraw-Hill, 1973.

Moncrieff, R. MAN-MADE FIBRES. 5th ed. New York: Interscience, Division of Wiley, 1971.

> Useful overall treatment of this large industrial field.

Moore, D. THE FRICTION AND LUBRICATION OF ELASTOMERS. New York: Pergamon, 1972.

> "Includes: Fundamental concepts. Surface texture. Rolling and sliding. Lubrication phenomena. Squeeze films. Elastohydrodynamics. Viscoelasticity. Adhesion. Hysteresis. Abrasion and wear. Experimental methods."

Morton, M. RUBBER TECHNOLOGY. 2nd ed. New York: Van Nostrand Reinhold, 1973.

> "Features an exposition of the molecular structure of natural and synthetic rubbers and the compounding of these rubbers for vulcanization."

NATIONAL COLLOQUY ON THE FIELD OF MATERIALS. University Park, Pennsylvania: Pennsylvania State University, 1969.

NATIONAL PLASTICS CONFERENCE PROCEEDINGS. The Society of the Plastics Industry, Inc. Westport, Connecticut: Technomic Publishing Company, 1969.

"Forty-one leading authorities provide information on the technology, processes, end-uses, and marketing of plastics."

Norman, R. CONDUCTIVE RUBBER AND PLASTICS. New York: Elsevier, 1971.

"Reference tool especially concerned with static electricity."

Ogorkiewicz, R., ed. ENGINEERING PROPERTIES OF THERMOPLASTICS. New York: Interscience, Division of Wiley, 1970.

"Nature of plastics. Processing plastics. Deformation behaviour. Strength characteristics. Other physical properties."

Organization for Economic Cooperation and Development. GAPS IN TECHNOLOGY; PLASTICS. London: HMSO, 1969.

"From the Third Ministerial Meeting on Science of OECD Countries, 11 and 12 March 1968. A useful collection of statistical information on the plastics industry particularly in Europe and the USA. A useful general guide to the plastics industry in relation to economic conditions in OECD countries."

Park, W., ed. PLASTICS FILM TECHNOLOGY. New York: Van Nostrand-Reinhold, 1969.

"History of plastic films, methods of manufacture, coated plastic films, testing methods, physical properties of plastic films, plastic film laminates, etc."

Parratt, N. FIBRE REINFORCED MATERIALS TECHNOLOGY. 2 vols. New York: Van Nostrand-Reinhold, 1972.

Standard work based upon long familiarity of author with many of the processes.

Penn, W.S. PVC TECHNOLOGY. 3rd ed. New York: Interscience, Division of Wiley, 1972.

"A handbook of commercial PVC resins, commercial PVC compounds, general aspects of plasticisers, etc."

Pinner, S., comp. PLASTICS: SURFACE AND FINISH. Hartford, Connecticut: Davey, 1971.

Placek, C. ABS RESIN MANUFACTURE. Park Ridge, New Jersey: Noyes Data Corporation, 1970.

Overview of the industry.

_____. POLYSULFIDE MANUFACTURE. Park Ridge, New Jersey: Noyes Data Corporation, 1970.

"This report covers 73 processes dealing with polymers possessing the disulfide (-ss-) group."

PLASTICS: TODAY & TOMORROW. Joint Meeting. Sponsored by Chemical Marketing Research Association & Commercial Development Association, Pittsburgh, Pennsylvania. Commercial Development Association, 100 Church Street, New York, New York 10007. 1972.

PLASTICS IN BEARINGS. Symposium. Plastics Institute, 11 Hobart Place, London SW1W, OHL, England. Preprints.

PLASTICS IN BUILDING CONSTRUCTION: REALITIES & CHALLENGES. National Technical Conference. Society of Plastics Engineers, 656 West Putnam Avenue, Greenwich, Connecticut 06830. 1972.

Plastics Institute. CARBON FIBRES; PLASTICS AND POLYMERS. Conference Supplement no. 5. London: 1972.

Papers from carbon fibre science and technology conference held by the Plastics Institute in 1971.

PLASTICS PIPE SYMPOSIUM. 2nd International. British Plastics Federation, 47 Piccadilly, London W1V O DN, England. 1972.

Post, D. POLISHING COMPOSITIONS AND MATERIALS. Park Ridge, New Jersey: Noyes Data Corporation, 1972.

Based on reviews of patents.

POWDER COATINGS. Symposium at 5th Polymer Conference Series. Polymer Institute, University of Detroit, 4001 West McNichols Road, Detroit, Michigan 48221. 1973.

POWDER METALLURGY. National conference, Chicago, Illinois, April 17–19, 1972. Proceedings issued as PROGRESS IN POWDER METALLURGY, vol. 28. 1972.

POWDER METALLURGY: THE APPLICATIONS AND ECONOMICS OF POWDER METALLURGY & ITS RELATION TO COMPETITIVE PROCESSES. 3rd European Powder Metallurgy Symposium. 2 vols. Powder Metallurgy Joint Group, 17 Belgrave Square, London SW1, England. 1971.

POWDER METALLURGY CONFERENCE. Edited by S. Mocarski. Metal Powder Industries Federation, 201 East 42nd Street, New York, New York 10017. 1972.

POWDER METALLURGY FOR HIGH–PERFORMANCE APPLICATIONS. 18th Sagamore Army Materials Research Conference. Edited by J.J. Burke and V. Weiss. Syracuse University Press, P.O. Box 8, University Station, Syracuse, New York, 13210. 1972.

Preston, J., ed. HIGH TEMPERATURE RESISTANT FIBERS FROM ORGANIC POLYMERS. American Chemical Society Symposium, held at Atlantic City, New Jersey, September 9-11, 1968. New York: Interscience, Division of Wiley, 1969.

THE PROPERTIES OF LIQUID METALS. 2nd International Congress. Edited by M. Tanaka. Published by Maruzen Company, Ltd., Tokyo. United States Distributor: Feffer & Simons, 81 Union Square, New York, New York 10003. 1973.

Ranney, M. EPOXY AND URETHANE ADHESIVES. Park Ridge, New Jersey: Noyes Data Corporation, 1971.

Reviews the processes and uses.

_____. ETHYLENE-PROPYLENE-DIENE RUBBERS. Park Ridge, New Jersey: Noyes Data Corporation, 1970.

Based on patent reviews.

_____. FLAME RETARDANT POLYMERS. Park Ridge, New Jersey: Noyes Data Corporation, 1970.

"Summarizes selected process technology for the use of fire retardant additives and reactive intermediates in major polymeric plastic materials, with particular emphasis on recent technology in the areas of polyesters, polystyrene, and polyurethane foam. 144 patent-based processes."

_____. FLAME RETARDANT TEXTILES. Park Ridge, New Jersey: Noyes Data Corporation, 1970.

"Describes 177 commercial processes to produce flame retardant textiles and fabrics."

_____. FLUOROCARBON RESINS. Park Ridge, New Jersey: Noyes Data Corporation, 1971.

Overview of processes.

_____. POLYIMIDE MANUFACTURE. Park Ridge, New Jersey: Noyes Data Corporation, 1971.

Based on patent reviews.

_____. POLYURETHANE CODINGS. Park Ridge, New Jersey: Noyes Data Corporation, 1972.

Details from patent reviews.

_____. REINFORCED COMPOSITES FROM POLYESTER RESINS. Park Ridge, New Jersey: Noyes Data Corporation, 1972.

Based on patent reviews.

Reich, L.   ELEMENTS OF POLYMER DEGRADATION.   New York: McGraw-Hill, 1971.

An overall review of the field of interest to engineers.

REINFORCED PLASTICS: THE REVOLUTION OF '76.   31st Annual Conference of Reinforced Plastics/Composites Institute. PROCEEDINGS. New York: Society of the Plastics Industry, 1976.

REINFORCED PLASTICS CONFERENCE.   8th International.   British Plastics Federation, 47 Piccadilly, London W1V ODN, England. 1972.

REINFORCED PLASTICS IN AEROSPACE APPLICATIONS. Symposium.   Plastics Institute, 11 Hobart Place, London SW1W OHL, England. 1973.

RESINS & CONCRETE.   Symposium. Plastics Institute, 11 Hobart Place, London SW1W OHL, England. Preprints. 1973.

Rosato, D., ed.   ENVIRONMENTAL EFFECTS ON POLYMERIC MATERIALS. 2 vols. New York: Interscience, 1968.

Roy, R., ed.   MATERIALS SCIENCE AND ENGINEERING IN THE UNITED STATES.   Proceedings of the National Colloquy on the Field of Materials, held at the Pennsylvania State University, University Park, Pennsylvania April 14-16, 1969.   University Park: Pennsylvania State University Press, 1970.

Rubber and Plastics Association of Great Britain.   FIRE PERFORMANCE OF PLASTICS, A REVIEW.   Shrewsbury: 1972.

Offers a thorough survey of plastics in building from standpoint of fire protection.

Rubber, International Conference.   Paris, France, June 1-5, 1970.   Sponsored by Association Francaise des Ingenieurs du Caoutchouc et des Plastiques. PROCEEDINGS. 3 vols. Paris: The Conference, 1970.

Rumford, F.   CHEMICAL ENGINEERING MATERIALS.   London: Constable, 1966.

Sittig, M.   ACRYLIC AND VINYL FIBERS 1972.   Park Ridge, New Jersey: Noyes Data Corporation, 1972.

Based on patent reviews.

_____.   POLYAMIDE FIBER MANUFACTURE 1972.   Park Ridge, New Jersey: Noyes Data Corporation, 1972.

Based on recent patent literature.

_____. POLYESTER FIBER MANUFACTURE. Park Ridge, New Jersey: Noyes Data Corporation, 1971.

>Details from review of recent patents.

STRENGTH OF METALS & ALLOYS. 2nd international conference, Pacific Grove, California, August 30–September 4, 1970. PROCEEDINGS. 3 vols. Metals Park, Ohio: ASM, 1970.

SYMPOSIUM ON POLYMER SCIENCE AND ENGINEERING. New Brunswick, New Jersey, 1972. PROCEEDINGS. Edited by K.D. Pae, et al. New York: Plenum, 1972.

Tadmor, Z. ENGINEERING PRINCIPLES OF PLASTICATING EXTRUSION. New York: Van Nostrand Reinhold, 1970.

>Sourcebook filled with much fundamental knowledge on extrusion processes.

TEXTILE ENGINEERING CONFERENCE. Greenville, South Carolina, October, 1975. Preprints. New York: American Society for Mechanical Engineers, 1976.

Trotman, E. DYEING AND CHEMICAL TECHNOLOGY OF TEXTILE FIBRES. 4th ed. London: Charles Griffin, 1970.

>A standard information source for this industry.

Van Krevelen, D. PROPERTIES OF POLYMERS. New York: American Elsevier, 1972.

>Useful sourcebook for polymers information.

Venkataraman, K., ed. THE CHEMISTRY OF SYNTHETIC DYES. 6 vols. New York: Academic Press, 1972.

>Part of a series of reviews, on synthetic dyes, which provides an overall view of the state of the art.

Voorn, M., ed. IUPAC INTERNATIONAL SYMPOSIUM ON MACROMOLE-CULES. London: Butterworths, 1971.

>Presents papers from symposium on chain molecules.

Wall, L., ed. FLUOROPOLYMERS. High polymer series v.XXV. New York: Wiley–Interscience, 1972.

>Critically reviews field of fluoropolymers.

WATER STRUCTURE AT THE WATER–POLYMER INTERFACE. Symposium at 161st National meeting. Edited by H.H.G. Jellinek. 1972. Plenum Press, 227 West 17th Street, New York, New York 10011.

Wigley, D. MECHANICAL PROPERTIES OF MATERIALS AT LOW TEMPERA-TURES. New York: Plenum, 1971.

Useful for those in materials science.

YIELD, DEFORMATION & FRACTURE OF POLYMERS. Symposium. Plastics Institute, 11 Hobart Place, London SW1W, OBL, England. Preprints. 1973.

7. Abstracting and Indexing Services

a. GENERAL

APPLIED MECHANICS REVIEWS

APPLIED SCIENCE AND TECHNOLOGY INDEX

BRITISH TECHNOLOGY INDEX

CHEMICAL ABSTRACTS

ENGINEERING INDEX

For further details on these services see this topic in the Section I of this Guide.

b. SPECIFIC TO TOPIC

i. Plastics and Rubbers

Bulletin signaletique. Part 780: POLYMERES; CHIMIE ET TECHNOLOGIE. Monthly. Centre National de la Recherche Scientifique, Centre de Documentation du C.N.R.S., 15 Quai Anatole-France, Paris (7c), France.

PLASTIC INDUSTRY NOTES. Weekly. Chemical Abstracts Service, The Ohio State University, Columbus, Ohio 43210.

PLASTICS ABSTRACTS. Weekly. Plastics Investigations, 31 Canonsfield Road, Welwyn, Herts., England.

"A comprehensive abstracting service covering British patent specifications dealing with the science, technology, and applications of plastics."

PLASTRACS, Monthly. Syllabus, Inc., 10-64 Jackson Avenue, Long Island City, New York 11101.

An abstracting service which selects key articles from journals in plastics field and provides abstracts monthly.

POLYMERICS; A WORLD DIGEST OF PLASTICS AND RELATED TECHNOLOGY. Monthly. Yarsley Research Laboratories Ltd., Clayton Road, Chessington, Surrey, England.

POLYMER LITERATURE ABSTRACTS. Weekly. CIS, The University of Akron, 302 East Buchtel Avenue, Akron, Ohio 44304.

POLYMER MARKET ABSTRACTS. Monthly. Foster D. Snell, Inc., 29 West 15th Street, New York, New York.

POLYMER SCIENCE & TECHNOLOGY POST. Fortnightly. Chemical Abstracts Service, American Chemical Society, 1155 16th Street, N.W., Washington, D.C. 20036.

Predi-briefs. Abstracting service. Monthly. ADHESIVES AND SPECIALTY CHEMICALS. Predicasts, 200 University Circle Research Center, 11001 Cedar Avenue, Cleveland, Ohio 44106.

> "Covers adhesives and specialty chemicals including plastic additives, water treatment chemicals, automotive and related chemicals, catalysts and enzymes, printing inks, carbon black, plating compounds, and waxes."

_____. PLASTIC MATERIALS. Predicasts, 200 University Circle Research Center, 1001 Cedar Avenue, Cleveland, Ohio 44106.

> "Includes information on plastic resins, plastic compounding engineering plastics and emulsion polymers and also processing materials, plasticizers and plastic additives."

_____. PLASTIC PRODUCTS. Predicasts, 200 University Circle Research Center, 11001 Cedar Avenue, Cleveland, Ohio 44106.

> "Covers plastic film, extruded and molded plastics, foams and reinforced and laminated plastics. Applications include packaging, transportation, furniture, appliances, building and consumer products."

_____. PROCESS MACHINERY. Predicasts, 200 University Circle Research Center, 11001 Cedar Avenue, Cleveland, Ohio 44106.

> "Covers special industry machinery including chemical process equipment, plastic processing equipment, air and water treatment equipment (see Packet 390), power transmission equipment, instruments, process control equipment and textile, wood products, paper industry, printing, foundry, plastic and rubber products machinery. Contains contracts awarded for chemical plant construction as well as patents and licenses."

_____. RUBBER. Predicasts, 200 University Circle Research Center, 11001 Cedar Avenue, Cleveland, Ohio 44106.

> "Covers natural rubber, synthetic rubber, rubber products, and rubber chemicals."

RAPRA ABSTRACTS. Weekly. Rubber and Plastics Research Association of Great Britain, Shawbury, Shrewsbury, SY4, 4NR, England.

SEARCH: PLASTICS & RESINS DIVISION. Monthly. Compendium Publishers International Corporation, 2175 Lemoine Avenue, Fort Lee, New Jersey 07024.

SEARCH: RUBBER DIVISION. Monthly. Compendium Publishers International Corporation, 2175 Lemoine Avenue, Fort Lee, New Jersey 07024.

Weekly Government Abstracts. MATERIALS SCIENCES. Government Printing Office, Washington, D.C. 20042.

ii. Material Other Than Plastics and Rubbers

ABSTRACTS AND BOOK TITLE INDEX CARD SERVICE (ABTICS). Weekly. Iron and Steel Institute, 4 Grosvenor Gardens, London SW1, England.

ABSTRACTS OF AIR FORCE MATERIALS LABORATORY. Annual. U.S. Air Force, AF Materials Lab., Aerospace Materials Information Center, Wright Patterson Air Force Base, Ohio 45433.

American Society for Testing and Materials. FIVE-YEAR INDEX TO ASTM TECHNICAL PAPERS AND REPORTS. (Supplements the ASTM 50-year index.) Quinquennial. American Society for Testing and Materials, 1916 Race Street, Philadelphia, Pennsylvania 19103.

COPPER ABSTRACTS. Bimonthly. CIDEC (Conseil International pour le Developpement du Cuivre), 11 rue du Rhone, Case Postale, 1211-Geneva-3, Switzerland.

> Selected abstracts of recent literature on copper and copper alloys. In English.

METAL FINISHING ABSTRACTS. Bimonthly. Finishing Publications Ltd., 17 Cranmer Road, Hampton Hill, Middlesex, England.

METALS ABSTRACTS. Monthly. American Society for Metals, Metals Park, Ohio 44073.

POWDER METALLURGY SCIENCE & TECHNOLOGY. Monthly. Metal Sciences Group, Science Information Services, The Franklin Institute Research Laboratories, 20th & Parkway, Philadelphia, Pennsylvania 19103.

> Abstracts of current journal, report and patent literature.

Predi-briefs. Abstracting service. Monthly. TEXTILES & FIBERS. Predicasts, 200 University Circle Research Center, 11001 Cedar Avenue, Cleveland, Ohio 44106.

> "Contains information on synthetic and natural fibers and end-uses in apparel and industrial fabrics including tire cord. Inorganic fibers and textile chemicals are also included."

SEARCH: TEXTILES DIVISION. Monthly. Compendium Publishers International Corporation, 2175 Lemoine Avenue, Fort Lee, New Jersey 07024.

TEXTILE TECHNOLOGY DIGEST. Monthly. Institute of Textile Technology, Charlottesville, Virginia 22902.

WORLD ALUMINUM ABSTRACTS. Monthly. American Society for Metals, Metals Park, Ohio 44073.

WORLD TEXTILE ABSTRACTS. Semimonthly. Shirley Institute, Manchester M20 8RX, England.

c. RELATED ABSTRACTS IN CURRENT JOURNALS

The following journals frequently provide abstracts. See the heading Journals and Serial Publications in this section for addresses.

AMERICAN CERAMIC SOCIETY JOURNAL

APPLIED PLASTICS

BRITISH CERAMIC SOCIETY PUBLICATIONS

GLASS TECHNOLOGY

JOURNAL OF APPLIED POLYMER SCIENCE

MATERIALS ENGINEERING

METAL POWDER REPORT

METAL PROGRESS

MODERN PLASTICS

NONWOVEN PATENTS DIGEST

PHYSICS AND CHEMISTRY OF GLASSES

PLASTICS TECHNOLOGY

POLYMERICS

POLYMER MECHANICS

POWDER METALLURGY SCIENCE & TECHNOLOGY

RUBBER CHEMISTRY AND TECHNOLOGY

STAHL UND EISEN

TEXTILE TECHNOLOGY DIGEST

8. Journals and Serial Publications

a. DIRECTORIES AND LISTS

AYER'S DIRECTORY OF PUBLICATIONS. Annual. Philadelphia, Pennsylvania: Ayer Press.

STANDARD PERIODICAL DIRECTORY. Irregular. New York: Oxbridge Publishing Company.

ULRICH'S INTERNATIONAL PERIODICAL DIRECTORY. Annual. 2 vols. New York: R.R. Bowker Company.

b. JOURNAL AND SERIAL TITLES

ABSTRACTS AND BOOK TITLE INDEX CARD SERVICE (ABTICS). Weekly. Iron and Steel Institute, 4 Grosvenor Gardens, London SW1, England.

An abstracting service on card format.

ABSTRACTS OF AIR FORCE MATERIALS LABORATORY. Annual. U.S. Air Force, AF Materials Laboratory, Aerospace Materials Information Center, Wright Patterson Air Force Base, Ohio 45433.

ADDITIVES FOR POLYMERS. Monthly. Yarsley Research Laboratories Ltd., Clayton Road, Chessington, Surrey, England.

Includes details on plasticisers, stabilizers, absorbers, flame retardants, and other additives.

ADHESIVES AGE. Monthly. Palmerton Publishing Company, 101 West 31st Street, New York, New York 10001.

ADVANCES IN MATERIALS RESEARCH. 1967. John Wiley & Sons, 605 Third Avenue, New York, New York 10016.

ADVANCES IN POLYMER SCIENCE. Vol. II. New York: Springer Verlag, 1973.

ADVANCES IN URETHANE SCIENCE AND TECHNOLOGY. 1971. Technomic Publishing Company, 265 West State Street, Westport, Connecticut 06880.

American Ceramic Society. BULLETIN. Monthly. American Ceramic Society, 4055 North High Street, Columbus, Ohio 43214.

_____. JOURNAL. Monthly. American Ceramic Society, 4055 North High Street, Columbus, Ohio 43214.

American Society for Texting and Materials. ASTM PROCEEDINGS. Annual. American Society for Testing and Materials, 1916 Race Street, Philadelphia, Pennsylvania 19103.

_____. FIVE-YEAR INDEX TO ASTM TECHNICAL PAPERS AND REPORTS. (Supplements the ASTM 50-year index.) Quinquennial. American Society for Testing and Materials, 1916 Race Street, Philadelphia, Pennsylvania 19103.

ANNUAL REVIEW OF MATERIALS SCIENCE. Annual. Annual Reviews Inc., 4139 El Camino Way, Palo Alto, California 94306.

APPLIED PLASTICS. Monthly. Scientific Press Ltd., 11a Gloucester Road, London SW7, England.

Applied Polymer Symposium. PAPERS. Irregular. 1969, no. 14. Brooklyn Polytechnic Institute, Wiley-Interscience Publishers, 605 Third Avenue, New York, New York 10016.

BRITISH PLASTICS. Monthly. Chemical & Marine Press Ltd., 33-40 Bowling Green Lane, London, EC1, England.

BRITISH POLYMER JOURNAL. Bimonthly. Society of Chemical Industry, 14 Belgrave Square, London SW 1, England.

BULLETIN SIGNALETIQUE. Part 780: POLYMERS; CHIMIE ET TECHNOL-OGIE. Monthly. Centre National de la Recherche Scientifique, m., 15 Quai Anatole-France, Paris (7e), France.

CANADIAN PLASTICS MAGAZINE. Monthly. Southam Business Publications Ltd., 1450 Don Mills Road, Don Mills, Ontario, Canada.

Chemical Rubber Company. C R C CRITICAL REVIEWS IN MACROMOLECULAR SCIENCE, 1972. Quarterly. Chemical Rubber Company, 18901 Cranwood Parkway, Cleveland, Ohio 44128.

COPPER ABSTRACTS. Bimonthly. CIDEC (Conseil International pour le De-veloppement du Cuivre), 100 rue du Rhone, Case Postale, 1211-Geneva-3, Switzerland.

> Selected abstracts of recent literature on copper and copper
> alloys in English.

CURRENT INDUSTRIAL REPORTS. Monthly. Commerce Department, Social and Economic Statistics Administration, Census Bureau. Washington, D.C.: Government Printing Office.

> Includes reports on plastics and rubber.

GLASS TECHNOLOGY. Bimonthly. Society of Glass Technology, Thornton, 20 Hallam Gate Road, Sheffield S10 5BT, England.

International Reinforced Plastics Conference. PAPERS AND PROCEEDINGS. Biannual. British Plastics Federation, 47 Piccadilly, London W1V ODN, England.

JAPAN PLASTICS. Bimonthly, Kogyo Chosakai Publishing Company, Ltd., 7-14, 2-chome, Hongo, Bunkyo-ku, Tokyo, Japan.

> Text in English.

JOURNAL OF APPLIED POLYMER SCIENCE. Monthly. Interscience Publishers, Division of John Wiley & Sons Inc., 605 Third Avenue, New York, New York 10016.

JOURNAL OF CELLULAR PLASTICS. Bimonthly. Technomic Publishing Company, 265 West State Street, Westport, Connecticut 06901.

JOURNAL OF COATED FIBROUS MATERIALS. Quarterly. Technomic Publishing Company, Inc., 265 West State Street, Westport, Connecticut 06880.

> "This journal presents the technology of flexible coated and laminated materials."

JOURNAL OF COMPOSITE MATERIALS. Quarterly. Technomic Publishing Company, Inc., 265 West State Street, Westport, Connecticut 06880.

> "Provides new information on the expanding technology of multi-phase materials."

JOURNAL OF ELASTOPLASTICS. Quarterly. Technomic Publishing Company, Inc., 265 West State Street, Westport, Connecticut 06901.

JOURNAL OF TEFLON. 6/year. E. I. du Pont de Nemours & Company, Inc., Wilmington, Delaware 19898.

> Editions in English, French, German, Italian, Spanish, Japanese, and Portuguese.

MATERIALS DIGEST. Monthly. Technology Marketing Corporation, Four Edice Street, Stamford, Connecticut 06905.

MATERIALS ENGINEERING. Monthly. Reinhold Publishing Corporation, 430 Park Avenue, Stamford, Connecticut 10022.

MATERIALS PROTECTION. Monthly. National Association of Corrosion Engineers, 2400 West Loop South, Houston, Texas 77027.

MATERIALS RESEARCH BULLETIN. Monthly. Pergamon Press, Inc., Maxwell House, Fairview Park, Elmsford, New York 10523.

> "An international journal reporting research on crystal growth and materials preparation and characterization, dealing with the structure and properties of electronically, optically or mechanically

interesting solids considered and discussed specifically as a function of the preparational parameters."

MATERIALS TECHNOLOGY SURVEY. Fortnightly. Materials Technology
Consultants, Box 321, Northboro, Massachusetts 01532.

METAL FINISHING ABSTRACTS. Bimonthly. Finishing Publications Ltd.,
17 Cranmer Road, Hampton Hill, Middlesex, England.

METAL POWDER REPORT. Monthly. Powder Metallurgy Ltd., Berk Limited,
Stratford, London E 15, England.

METAL PROGRESS. Monthly. American Society for Metals, 114 East 2nd
Street, Little Rock, Arkansas 72203.

METALS ABSTRACTS. Monthly. American Society for Metals, Metals Park,
Ohio 44073.

METALS AND MATERIALS AND METALLURGICAL REVIEWS. Metals and
Metallurgy Trust, 17 Belgrave Square, London SW 1, England.

METAL SCIENCE JOURNAL. Bimonthly. Institute of Metals and the Institution of Metallurgists, 17 Belgrave Square, London SW 1, England.

MODERN MATERIALS. ADVANCES IN DEVELOPMENT AND APPLICATIONS.
1970, vol. 7. Academic Press, 111 Fifth Avenue, New York, New York
10003.

MODERN PLASTICS. Monthly. McGraw-Hill, Inc., 330 West 42nd Street,
New York, New York 10036.

MODERN PLASTICS INTERNATIONAL. Monthly. McGraw-Hill, Inc., 50
Avenue de la Gare, Lausanne, Switzerland.

International edition of MODERN PLASTICS. Text in English.

NONWOVEN PATENTS DIGEST. Bimonthly. International Executive Newsletters Company, 52 rue du Progres, 1000 Brussels, Belgium.

PAINT AND RESIN PATENTS. Monthly. R.H. Chandler Ltd., 42 Grays Inn
Road, London, WC 1, England.

PHYSICS AND CHEMISTRY OF GLASSES. Bimonthly. Society of Glass Technology, 20 Hallam Gate Road, Sheffield 10, England.

PLASTIC INDUSTRY NEWS. Monthly. Institute of Polymer Industry, CPO
Box 1176, Tokyo, Japan.

PLASTICS INDUSTRY NOTES. Weekly. Chemical Abstracts Service, Ohio
State University, Columbus, Ohio 43210.

PLASTICS ABSTRACTS. Weekly. Plastics Investigations, 31 Canonsfield Road, Welwyn, Herts., England.

"A comprehensive abstracting service covering British patent specifications dealing with the science, technology, and application of plastics."

PLASTICS AND RUBBER WEEKLY. Maclaren and Sons Ltd., Davis House, 69-77 High Street, Box 109, Croydon, Surrey, England.

PLASTICS DESIGN & PROCESSING. Monthly. Lake Publishing Corporation, 700 Peterson Road, Box 159, Libertyville, Illinois 60048.

PLASTICS INDUSTRY NOTES. Weekly. Chemical Abstracts Service, The Ohio State University, Columbus, Ohio 43210.

PLASTICS IN ENGINEERING. Quarterly. Factory Publications Ltd., Hermes House, 89 Blackfriars Road, London SE 1, England.

PLASTICS TECHNOLOGY; THE MAGAZINE OF PROCESS/PRODUCTION/MANUFACTURING ENGINEERING. Monthly. Bill Communications, Inc., 630 Third Avenue, New York, New York 10017.

PLASTICS WORLD. Monthly. Cahners Publishing Company, 270 St. Paul Street, Denver, Colorado 80206.

PLASTICS WORLD INFORMATION CARDS. Semiannual. Rogers Publishing Company, Inc., A subsidiary of Cahners Publishing Company, Inc., 270 St. Paul Street, Denver, Colorado 80206.

PLASTRACS. Monthly. Syllabus, Inc., 10-64 Jackson Avenue, Long Island City, New York 11101.

An abstracting service which selects key articles from journals in plastics field and provides abstracts monthly.

POLYMER AGE. Bimonthly. Rubber and Technical Press Ltd., Tenterden, Kent, England.

POLYMER ENGINEERING AND SCIENCE. Bimonthly. Society of Plastics Engineers, 656 West Putnam Avenue, Greenwich, Connecticut 06830.

POLYMERICS. Monthly. Yarsley Research Laboratories Ltd., Clayton Road., Chessington, Surrey, England.

"A selective digest of world plastics and related fields."

POLYMER LITERATURE ABSTRACTS. Weekly. CIS, The University of Akron, 302 East Buchtel Avenue, Akron, Ohio 44304.

POLYMER MECHANICS. Faraday Press, 84 Fifth Avenue, New York, New York 10011.

"English translation of MEKHANIKA POLIMEROV."

POLYMER NEWS. Monthly. Gordon and Breach, Science Publishers, Inc., 150 Fifth Avenue, New York, New York 10011.

POLYMER SCIENCE & TECHNOLOGY POST. Fortnightly. Chemical Abstracts Service, American Chemical Society, 1155 16th Street, N.W., Washington, D.C. 20036.

POWDER METALLURGY SCIENCE & TECHNOLOGY. Monthly. Metal Sciences Group, Science Information Services, The Franklin Institute Research Laboratories, 20th and Parkway, Philadelphia, Pennsylvania 19103.

Predi-briefs. Abstracting service. Monthly. ADHESIVES & SPECIALTY CHEMICALS. Predicasts, 200 University Circle Research Center, 11001 Cedar Avenue, Cleveland, Ohio 44106.

> "Covers adhesives and specialty chemicals including plastic additives, water treatment chemicals, automotive and related chemicals, catalysts and enzymes, printing inks, carbon black, plating compounds, and waxes."

_____. PLASTIC MATERIALS. Predicasts, 200 University Circle Research Center, 11001 Cedar Avenue, Cleveland, Ohio 44106.

> "Includes information on plastic resins, plastic compounding, engineering plastics and emulsion polymers and also processing materials, plasticizers and plastic additives."

_____. PLASTIC PRODUCTS. Predicasts, 200 University Circle Research Center, 11001 Cedar Avenue, Cleveland, Ohio 44106.

> "Covers plastic film, extruded and molded plastics, foams and reinforced and laminated plastics. Applications include packaging, transportation, furniture, appliances, building and consumer products."

_____. PROCESS MACHINERY. Predicasts, 200 University Circle Research Center, 11001 Cedar Avenue, Cleveland, Ohio 44106.

> "Covers special industry machinery including chemical process equipment, plastic processing equipment, air and water treatment equipment (see Packet 390), power transmission equipment, instruments, process control equipment and textile, wood products, paper industry, printing, foundry, plastic and rubber products machinery. Contains contracts awarded for chemical plant construction as well as patents and licenses."

_____. RUBBER. Predicasts, 200 University Circle Research Center, 11001 Cedar Avenue, Cleveland, Ohio 44106.

> "Covers natural rubber, synthetic rubber, rubber products and rubber chemicals."

_____. TEXTILES AND FIBERS. Predicasts, 200 University Circle Research Center, 11001 Cedar Avenue, Cleveland, Ohio 44106.

"Contains information on synthetic and natural fibers and end-uses in apparel and industrial fabrics including tire cord. Inorganic fibers and textile chemicals are also included."

PROGRESS IN APPLIED MATERIALS RESEARCH. Irregular. London: Heywood Books.

PROGRESS IN MATERIALS SCIENCE. Irregular. Pergamon Press, Inc., Maxwell House, Fairview Park, Elmsford, New York 10523.

PROGRESS IN POLYMER SCIENCE. Irregular. Pergamon Press, Inc., Maxwell House, Fairview Park, Elmsford, New York 10523.

QUARTERLY LITERATURE REPORTS, POLYMERS. Quarterly. Academic Press Inc., 111 Fifth Avenue, New York, New York 10003.

R A P R A ABSTRACTS. Weekly. Rubber and Plastics Research Association of Great Britain, Shawbury, Shrewsbury SY4 4NR, England.

REINFORCED PLASTICS. Monthly. Craftsman Publications, 87 Lambs Conduit Street, London WC 1, England.

RESIN NEWS. Monthly. A.S. O'Connor & Co., 30 Paradise Road, Richmond, Surrey, England.

RESIN REVIEW. Quarterly. Rohm and Haas Company, Independence Mall West, Philadelphia, Pennsylvania 19105.

REVIEWS IN MACROMOLECULAR CHEMISTRY. Irregular. New York: Dekker.

REVIEWS IN POLYMER TECHNOLOGY. Irregular. New York: Dekker.

"Provides reviews of the state of the art."

RUBBER AGE. Monthly. Palmerton Publishing Company, 101 West 31st Street, New York, New York 10001.

Communications medium for developments in rubber industry.

Rubber and plastics industry technical conference. RECORD. Annual. Institute of Electrical and Electronics Engineers, Inc., 345 East 47th Street, New York, New York 10017.

RUBBER CHEMISTRY AND TECHNOLOGY. 5 times/year. Division of Rubber Chemistry, The American Chemical Society, University of Akron, Box 123, Akron, Ohio 44329.

RUBBER JOURNAL. Monthly. Maclaren Publishers Ltd., Box 109, Davis House, 69-77 High Street, Croydon, CR9 1QH, England.

Review of new developments in the industry.

SEARCH: PLASTICS & RESINS DIVISION. Monthly. Compendium Publishers International Corporation, 2175 Lemoine Avenue, Fort Lee, New Jersey 07024.

SEARCH: RUBBER DIVISION. Monthly. Compendium Publishers International Corporation, 2175 Lemoine Avenue, Fort Lee, New Jersey 07024.

SEARCH: TEXTILES DIVISION. Monthly. Compendium Publishers International Corporation, 2175 Lemoine Avenue, Fort Lee, New Jersey 07024.

SOVIET PROGRESS IN POLYURETHANES. Irregular. Technomic Publishing Company, Inc., 365 West State Street, Westport, Connecticut 06880.

S P E JOURNAL. Society of Plastics Engineers, 656 West Putnam Avenue, Greenwich, Connecticut 06830.

S P I National Plastics Conference. PROCEEDINGS. Irregular. Technomic Publishing Company, 265 State Street, Westport, Connecticut 06880.

STAHL UND EISEN. Semimonthly. Verlag Stahleisen mbH, Breite Str. 27, Postfach 8229, 4 Duesseldorf 1, West Germany.

Contents page in English and French.

TEXTILE TECHNOLOGY DIGEST. Monthly. Institute of Textile Technology, Charlottesville, Virginia.

URETHANE PLASTICS AND PRODUCTS. Monthly. Technomic Publishing Company, Inc., 265 West State Street, Westport, Connecticut 06880.

VINYL TECHNOLOGY NEWSLETTER. Quarterly. Society of Plastic Engineers, Inc., 656 West Putnam Avenue, Greenwich, Connecticut 06830.

WORLD ALUMINUM ABSTRACTS. Monthly. American Society for Metals, Metals Park, Ohio 44073.

WORLD TEXTILE ABSTRACTS. Semimonthly. Shirley Institute, Manchester M20, 8RX, England.

ZERSTOERUNGSFREIE MATERIALPRUEFUNG. 12 Series Distributed in 6 Installments; 250-300 Abstracts per Year. Deutsche Gesellschaft fuer Zerstoerungsfreie Materialpruefung DGZfP, 6/yr. Bundesanstalt fuer Materialpruefung, Unter den Eichen 87, 1 Berlin, 45, West Germany.

9. Reviewing Services

ADVANCES IN MATERIALS RESEARCH. Irregular. New York: John Wiley & Sons.

ADVANCES IN POLYMER SCIENCE. Irregular. New York: Springer-Verlag.

ADVANCES IN URETHANE SCIENCE AND TECHNOLOGY. Irregular. Technomic Publishing Company, Westport, Connecticut 06880.

ANNUAL REVIEW OF MATERIALS SCIENCE. Annual Reviews Inc., 4139 El Camino Way, Palo Alto, California 94306.

Chemical Rubber Company. C R C CRITICAL REVIEWS IN MACROMOLECULAR SCIENCE. Quarterly. Chemical Rubber Company, 18901 Cranwood Parkway, Cleveland, Ohio 44128.

METALS AND MATERIALS AND METALLURGICAL REVIEWS. Metals and Metallurgy Trust, 17 Belgrave Square, London SW 1, England.

MODERN MATERIALS. ADVANCES IN DEVELOPMENT AND APPLICATIONS. Irregular. Academic Press, 111 Fifth Avenue, New York, New York 10003.

PROGRESS IN APPLIED MATERIALS RESEARCH. Irregular. London: Heywood Books.

PROGRESS IN MATERIALS SCIENCE. Irregular. Pergamon Press, Inc., Maxwell House, Fairview Park, Elmsford, New York 10523.

PROGRESS IN POLYMER SCIENCE. Irregular. Pergamon Press, Inc., Maxwell House, Fairview Park, Elmsford, New York 10523.

REVIEWS IN MACROMOLECULAR CHEMISTRY. Irregular. New York: Dekker.

REVIEWS IN POLYMER TECHNOLOGY. Irregular. New York: Dekker.

SOVIET PROGRESS IN POLYURETHANES. Irregular. Technomic Publishing Company, Inc., 265 West State Street, Westport, Connecticut 06880.

10. Document and Report Literature

a. LISTS

MONTHLY CATALOG. U.S. Superintendent of Documents, Government Printing Office, Washington, D.C. 20402.

U.S. GOVERNMENT REPORTS ANNOUNCEMENTS and U.S. GOVERNMENT REPORTS INDEX. Monthly. National Technical Information Service, Springfield, Virginia 22151.

b. CLEARINGHOUSES

National Technical Information Service. Springfield, Virginia 22151.

U.S. Superintendent of Documents. Government Printing Office, Washington, D.C. 20402.

11. Sources of Information on Masters and Doctoral Theses

DISSERTATION ABSTRACTS INTERNATIONAL. Section B. "Physical sciences and technology." University Microfilms, Xerox Company, 300 North Zeeb Road, Ann Arbor, Michigan 48106.

INTERNATIONAL MASTERS ABSTRACTS. Quarterly. University Microfilms, 300 North Zeeb Road, Ann Arbor, Michigan 48106.

MASTERS THESES IN THE PURE AND APPLIED SCIENCES, ACCEPTED BY COLLEGES AND UNIVERSITIES OF THE UNITED STATES. Annual. Lafayette, Inc., School of Mechanical Engineering, Purdue University (Thermophysical Properties Research Center).

12. Sources of Information on Statistics and Costs

a. STATISTICS

i. Guides and Handbooks

AMERICAN STATISTICS INDEX. Annual. Part 1: Index. Part 2: Abstracts. Congressional Information Service, 600 Montgomery Building, Washington, D.C. 20014.

> "Comprehensive guide and index to the statistical publications of the U.S. Government."

U.S. Department of Commerce, Domestic and International Business Administration. Bureau of Resources and Trade Assistance. SOURCES OF STATISTICAL DATA TEXTILES AND APPAREL. By M.V. Burke. Washington, D.C.: Government Printing Office, 1974.

ii. Current Journals Which Include Related Statistical Information

The following journals frequently contain statistics related to this field. See the heading Journals and Serial Publications in this section for addresses.

ADVANCES IN URETHANE SCIENCE AND TECHNOLOGY

CANADIAN PLASTICS

GLASS TECHNOLOGY

JAPAN PLASTICS

JOURNAL OF CELLULAR PLASTICS

JOURNAL OF ELASTOPLASTICS

PLASTICS AND RUBBER WEEKLY

PLASTICS WORLD

POLYMER AGE

POLYMERICS

POLYMER NEWS

POWDER METALLURGY SCIENCE & TECHNOLOGY

PROGRESSIVE PLASTICS

RAPRA ABSTRACTS

RUBBER AGE

RUBBER JOURNAL

SEARCH: PLASTICS & RESINS DIVISION

SEARCH: RUBBER DIVISION

SEARCH: TEXTILES DIVISION

SPE JOURNAL

STAHL UND EISEN

b. COSTS

Bishop, R. "Find Polystyrene Plant Costs." HYDROCARBON PROCESSING, Vol. 51, November 1972, pp. 137-40.

WORLD-PRODUCT COSTS. Irregular. Predicasts, Inc., 200 University Circle Research Center, 11001 Cedar Avenue, Cleveland, Ohio 44106.

> A subscription service of market forecasts by product in areas of agriculture, wood and paper, food, chemicals, polymers, primary metals, etc.

See also this subject in the Section I of this Guide.

13. Sources of Information on Standards, Specifications, and Patents

a. STANDARDS AND SPECIFICATIONS

AMERICAN NATIONAL STANDARDS INSTITUTE CATALOG. Annual. 1430 Broadway, New York, New York 10018.

> Lists standards in this subject area. Catalog and standards are available from the Institute.

American Society for Testing and Materials. ANNUAL BOOK OF A S T M STANDARDS. American Society for Testing and Materials, 1916 Race Street, Philadelphia, Pennsylvania 19103.

Multivolume work providing comprehensive testing results of various materials.

MATERIALS RESEARCH & STANDARDS. Monthly. American Society for Testing and Materials, 1916 Race Street, Philadelphia, Pennsylvania 19103.

Contents: properties, strength, and uses of various materials.

Struglia, Erasmus J. STANDARDS AND SPECIFICATIONS INFORMATION SOURCES. Detroit: Gale Research Company, 1965.

Coverage is both general and specific to materials sciences.

U.S. Department of Commerce. National Bureau of Standards. DIRECTORY OF U.S. STANDARDIZATION ACTIVITIES. Washington, D.C.: Government Printing Office, 1975.

Coverage is both general and specific to materials sciences.

WORLD INDEX OF PLASTIC STANDARDS. U.S. National Bureau of Standards. Washington, D.C.: Government Printing Office, 1971.

Specifications, standards, codes and regulations are available for individual states from the official state document printing offices and from local and municipal government agencies.

See also this subject in Section I of this Guide.

b. PATENTS AND TRADEMARKS

i. Journals and Serial Publications

Herman, B. ADHESIVES: RECENT DEVELOPMENTS. Park Ridge, New Jersey: Noyes Data Corp., 1976.

NONWOVEN PATENTS DIGEST. Bimonthly. International Executive News-letters Company, 52 rue du Progres, 1000 Brussels, Belgium.

PAINT AND RESIN PATENTS. Monthly. R.H. Chandler Ltd., 42 Grays Inn Road, London WC 1, England.

Ranney, M. COATINGS: RECENT DEVELOPMENTS. Park Ridge, New Jersey: Noyes Data Corp., 1976.

ii. Current Journal Which Include Related Patent Listings

The following journals frequently list patents. See the heading Journals and Serial Publications in this section for addresses.

ADHESIVES AGE

AMERICAN CERAMIC SOCIETY JOURNAL

APPLIED PLASTICS

BRITISH CERAMIC SOCIETY PUBLICATIONS

GLASS TECHNOLOGY

JOURNAL OF CELLULAR PLASTICS

METAL POWDER REPORT

MODERN PLASTICS

PLASTICS ABSTRACTS

PHYSICS AND CHEMISTRY OF GLASSES

PLASTICS AND RUBBER WEEKLY

PLASTICS INDUSTRY NEWS, JAPAN

PLASTICS TECHNOLOGY

POLYMERICS

POWDER METALLURGY SCIENCE & TECHNOLOGY

SPE JOURNAL

STAHL UND EISEN

TEXTILE TECHNOLOGY DIGEST

See also this subject in Section I of this Guide.

14. Sources of Information on
Translations, Meetings, and Proceedings of Symposia

U.S. Joint Publications Research Service. Publishes translations of scientific and technical materials from Eastern Europe, Russia, and China. These are listed in the Monthly Catalog of the Superintendent of Documents. Arlington, Virginia.

See also these subjects in Section I of this Guide.

15. Selected Literature Sources on
Pollution Control and Safety

a. POLLUTION CONTROL

i. Guides and Handbooks

Canton Textile Mills Inc. TREATMENT OF DENIM TEXTILE MILL WASTE-
WATERS: NEUTRALIZATION AND COLOR REMOVAL. By C.R. Froneberger,
et al. Rept. no. PB 253 478. Springfield, Virginia: National Technical
Information Service, 1976.

DISPOSAL OF PLASTICS. Yardley Research Laboratories Ltd., Clayton Road,
Chessington, Surrey, England. 1972.

"Contains abstracts of articles on a) disposal by incineration,
degradation, and land fill and b) reclamation by industry and
local government."

Mark, H., et al., eds. CHEMICAL AFTERTREATMENT OF TEXTILES. New
York: Interscience, Division of Wiley, 1971.

National Bureau of Economic Research. WATER POLLUTION CONTROL ACT
OF 1972. ECONOMIC IMPACTS. METAL FINISHING INDUSTRY. By J.
Ginn, et al. Rept. no. PB 248 803. Springfield, Virginia: National
Technical Information Service, 1975.

_____. WATER POLLUTION CONTROL ACT OF 1972. ECONOMIC IM-
PACTS. NON-FERROUS METALS INDUSTRY. By L. Lee, et al. Rept. no.
PB 248-804. Springfield, Virginia: National Technical Information Service,
1975.

POLLUTION CONTROL IN THE TEXTILE INDUSTRY. By H. Jones. Park
Ridge, New Jersey: Noyes Development Corporation, 1973.

"Based on various government surveys with practical examples
from U.S. patents."

Sittig, M. POLLUTION CONTROL IN THE PLASTICS AND RUBBER INDUS-
TRY. Park Ridge, New Jersey: Noyes Data Corp., 1975.

SOLID WASTE RECLAMATION AND RECYCLING. Part 1: Packaging and
Containers. Rept. no. NTIS/PS-76/0514. Part 2: Plastics. Rept. no.
NTIS/PS-76/0515. Part 3: Metals. Rept. no. NTIS/PS-76/0516. Part 4:
Glass. Rept. no. NTIS/PS-76/0517. Part 5: Paper. Rept. no. NTIS/
PS-76/0518. Springfield, Virginia: National Technical Information Service,
1976.

Bibliographies with abstracts. Covers 1964-1976.

Szelard, J. RECLAIMING RUBBER AND OTHER POLYMERS. Park Ridge, New Jersey: Noyes Development Corporation, 1973.

> Recycling of rubber and polymer products.

THE TEXTILE INDUSTRY AND THE ENVIRONMENT. Research Triangle Park, North Carolina: American Association of Textile Chemists and Colorists, 1973.

U.S. Environmental Protection Agency. MAKING POLYETHYLENE MORE DISPOSABLE. Solid Waste Management SW Series. By Irene Kiefer. Washington, D.C.: Government Printing Office, 1973.

> Condensation of feasibility study of disposal of polyethylene plastic waste prepared for Federal solid waste management program.

_____. NEW CHEMICAL CONCEPTS FOR UTILIZATION OF WASTE PLASTICS (analytical investigation, with bibliography). By M.E. Banks, W.D. Lusk, and R.S. Ottinger. Washington, D.C.: Government Printing Office, 1971.

_____. RUBBER REUSE AND SOLID WASTE MANAGEMENT (with bibiography). By R.J. Pettigrew, et al. Washington, D.C.: Government Printing Office, 1971.

U.S. National Industrial Pollution Control Council. TEXTILE INDUSTRY AND POLLUTION CONTROL, SUB-COUNCIL REPORT. Washington, D.C.: Government Printing Office, 1973.

b. SAFETY

i. Guides and Handbooks

Kuryla, W., ed. FLAME RETARDANCY AND POLYMENE MATERIALS: SCIENCE AND TECHNOLOGY. New York: Dekker, 1973.

LEGAL GUIDE TO OCCUPATION SAFETY AND HEALTH. Monthly. Clark Boardman Ltd., 435 Hudson Street, New York, New York 10014.

> Comment, interpretation and advice on new legal developments in occupation safety and health.

Peck, Theodore. OCCUPATIONAL SAFETY AND HEALTH: A GUIDE TO INFORMATION SOURCES. Detroit: Gale Research Company, 1974.

Rosato, D. PLASTICS INDUSTRY SAFETY HANDBOOK. Boston: Cahners, 1973.

TEXTILE SECTION SAFETY NEWSLETTER. Monthly. National Safety Council, 425 North Michigan Avenue, Chicago, Illinois 60611.

ii. Abstracting and Indexing Services

C I S ABSTRACTS. 8 issues per year. International Occupational Safety and Health Information Centre (CIS), International Labour Office, 1211 Geneva 22, Switzerland.

> Formerly OCCUPATIONAL SAFETY AND HEALTH ABSTRACTS.
> Available in card format. Editions in English, French, German, Italian, Rumanian, Russian, and Spanish.

Safety standards are listed in: AMERICAN NATIONAL STANDARDS INSTITUTE CATALOG. Annual. New York: The Institute.

U.S. Occupational Safety and Health Administration. SAFETY AND HEALTH GUIDELINES FOR GENERAL INDUSTRY. Rept. no. PB239 210 SDS. Springfield, Virginia: National Technical Information Service, 1975.

16. Management Information Sources

See this subject in Section I of this Guide.

17. Sources of Information on Laws and Legislation in This Field

a. FEDERAL

CODE OF FEDERAL REGULATIONS. Government Printing Office.

> Continuing series issued throughout the year. Available on a subscription basis from Government Printing Office.

FEDERAL REGISTER. Daily publication listing new regulations and revisions. Updates Code of Federal Regulations. Available on a subscription basis from Government Printing Office.

Library of Congress. DIGEST OF PUBLIC GENERAL BILLS AND REGULATIONS. Monthly. Available on a subscription basis from Government Printing Office, Washington, D.C. 20402.

UNITED STATES CODE, 1970 edition and supplements. Irregular. Washington, D.C.: Government Printing Office.

STATE

Library of Congress. MONTHLY CHECKLIST OF STATE PUBLICATIONS. Available on a subscription basis from Government Printing Office.

b. LOCAL AND MUNICIPAL

Contact local government offices for codes, ordinances and regulations.

All of these information source materials listed here are usually available in the reference department of large public libraries.

Section VI

NUCLEAR ENGINEERING

Section VI

NUCLEAR ENGINEERING

A. ORGANIZATIONS AND ASSOCIATIONS

1. Governmental Agencies

a. FEDERAL

CONGRESS

Joint Committee on Atomic Energy - U.S. Capitol, Room H-403, Washington, D.C. 20510.

Areas of interest: All matters pertaining to the military and civilian development, use, and control of atomic energy.

Publications: Hearing records and committee reports.

Information services: Answers inquiries; makes referrals; provides information and support in atomic energy matters to the Joint committee members, Congress, and the public.

DEPARTMENT OF HEALTH, EDUCATION AND WELFARE.

PUBLIC HEALTH SERVICE

Bureau of Radiological Health - 12720 Twinbrook Parkway, Rockville, Maryland 20852.

Areas of interest: Radiological health; X-rays and radioisotopes in medicine and industry; health hazards associated with exposure to ionizing radiation; X-radiations associated with consumer electronic products; radioactivity; measurement of environmental radiation levels; nuclear medicine; public health aspects of nuclear energy; nuclear power plants; radiation protection; radioactive waste disposal; radioactive materials; transportation of radioactive materials; programs in radiobiological effects; environmental surveillance and control; standards and intelligence; assistance to states in compliance and control; training and manpower development in the radiological health field; medical and occupational radiation hazards; X-ray exposure control and research grants.

Holdings: About 5,000 books, 6,000 bound journals, 5,000 technical reports, and 400 technical medical journals; microfiche files. The Bureau maintains a coordinate index and a KWIC index to its areas of interest.

Publications: The Bureau publishes its findings in RADIOLOGI-CAL HEALTH DATA AND REPORTS (a monthly publication), Public Health Service numbered reports, appropriate scientific journals, and technical report series for Bureau divisions, offices, and laboratories.

Information services: Answers inquiries; provides information to state health offices, federal and state agencies, scientific organizations, and industries; provides reference services; makes interlibrary loans; permits onsite use of collection by qualified researchers.

DEPARTMENT OF TRANSPORTATION

MATERIALS TRANSPORTATION BUREAU - 2100 Second Street S.W., Washington, D.C. 20590.

Purpose: This bureau was formed in 1974 to coordinate the Department of Transportation's overall responsibilities for transportation of hazardous materials and pipeline safety.

Areas of interest: These include transport of hazardous materials, regulation and safety, and related research.

Publications: Annual reports on transport of hazardous materials and pipeline safety, contributes to compilation of federal standards, and issues monthly newsletters.

Information and services: Answers inquiries and makes referrals to appropriate agencies.

ENERGY RESEARCH AND DEVELOPMENT ADMINISTRATION - 20 Massachusetts Avenue N.W., Washington, D.C. 20545.

The Energy Research and Development Administration was established by the Energy Reorganization Act of 1974 (replaces former Atomic Energy Commission).

Purpose: The purpose of the Energy Research and Development Administration (ERDA) is to reorganize and consolidate federal activities relating to research and development on the various sources of energy in order to develop and increase the efficiency and reliability of the use of all energy sources to meet the needs of present and future generations, to increase the productivity of the national economy and strengthen its position in regard to international trade, to make the nation self-sufficient in energy, to advance the goals of restoring, protecting and enhancing environmental quality, and to assure public health and safety.

Areas of interest: Coal research, nuclear energy, energy research centers, wind power, solar heating and cooling, geothermal energy and development, demonstration of commercial feasibility and practical applications of extraction conversion, storage, transmission, and utilization of energy forms.

Publications: Research and development reports, standards, critical review series, standard safety analysis reports, nuclear science abstracts.

Information services: Answers inquiries, makes referrals, maintains a speakers' bureau, produces and distributes films through its Office of Public Affairs. A public document room is available at 20 Massachusetts Avenue N.W., Washington, D.C. 20545.

Energy Research and Development Administration has the following energy research centers.

Bartlesville Energy Research Center - P.O. Box 1398, Bartlesville, Oklahoma 74003.

Grand Forks Energy Research Center - Box 8213, University Station, Grand Forks, North Dakota 58202.

Laramie Energy Research Center - P.O. Box 3395, University Station, Laramie, Wyoming 82070.

Morgantown Energy Research Center - P.O. Box 880, Morgantown, West Virginia 26505.

Pittsburgh Energy Research Center - 4800 Forbes Avenue, Pittsburgh, Pennsylvania 15213.

ENVIRONMENTAL PROTECTION AGENCY

OFFICE OF RADIATION PROGRAMS - 5600 Fishers Lane, Rockville, Maryland 20852.

Areas of interest: Radiological health; health hazards associated with exposure to ionizing radiation; radioactivity; measurement of environmental radiation levels; public health aspects of nuclear energy; nuclear power plants; radiation protection; radioactive waste disposal; radioactive materials; transportation of radioactive materials; programs in radiobiological effects; environmental surveillance and control; standards and intelligence; assistance to states in compliance and control; training and man-power development in the radiological health field; research grants.

Publications: The Office of Radiation Programs publishes its findings in RADIATION HEALTH DATA AND REPORTS (a monthly publication), appropriate scientific journals, and technical report series for Office of Radiation Programs' divisions and laboratories.

Information services: Answers inquiries; provides information to state health offices, federal and state agencies, scientific organizations, and industries; makes referrals to other sources of information.

NUCLEAR REGULATORY COMMISSION - 1717 H Street N.W., Washington, D.C. 20555.

This agency was established as an independent regulatory agency under provisions of the Energy Reorganization Act of 1974 to assume regulatory functions formerly assigned to the Atomic Energy Commission.

"The Nuclear Regulatory Commission's purpose is to assure that the civilian uses of nuclear materials and facilities are conducted in a manner consistent with the public health and safety, environmental quality, national security, and the antitrust laws. The major share of the Commission effort is focused on the use of nuclear energy to generate electric power."

Areas of interest: Licensing and regulation of nuclear facilities, plants, sites and handling and disposal of nuclear materials. Research in nuclear safeguards.

Publications: U.S. Nuclear Regulatory Commission publications, January - December 1975, accessions list; regulatory guides, draft environmental statements, safety information, and materials related to licensing, contracting, nuclear facilities, and operations.

Information activities: Answers inquiries, makes referrals to appropriate agencies, issues details on licensing and contracts, has speakers' bureau. Maintains a public document room at 1717 H Street N.W., Washington, D.C. 20555.

b. STATE AND MUNICIPAL

See this topic in Section I of this Guide.

2. Professional Societies, Trade Associations, and Other Scientific-Technological Related Organizations

AMERICAN NUCLEAR SOCIETY - 244 East Ogden Avenue, Hinsdale, Illinois 60521.

Purpose: The advancement of knowledge in nuclear energy and related fields. Cooperation with government agencies and educational institutions and other scientific societies who have similar areas of concern is a major function. The Society provides ICONS-Information Center on Nuclear Standards. Local sources of information on nuclear subjects are listed in PUBLIC INFORMATION ON NUCLEAR ENERGY SERVICE, a directory which is obtainable from the society upon request.

Publications: NUCLEAR NEWS (monthly, reviews developments and provides news from the field of nuclear science, includes annually a BUYERS' GUIDE listing 260 products and services with details on suppliers' names and addresses); NUCLEAR SCIENCE AND ENGINEERING (monthly, articles on research and investigation in nuclear energy); NUCLEAR TECHNOLOGY (monthly, technical papers and more generally applications in nuclear energy).

Papers presented at symposia and conferences appear in summary form in the TRANSACTIONS (biannually) and full proceedings of special conferences in SPECIAL PROCEEDINGS.

In addition, the American Nuclear Society cooperates with the U.S. Atomic Energy Commission in preparing and publishing monographs, critical review articles in AEC journals and in compiling directories.

Information services: Answers inquiries; makes referrals.

ATOMIC INDUSTRIAL FORUM, INC. - 850 Third Avenue, New York, New York 10022.

Areas of interest: Nuclear insurance and indemnity; radiation and radioisotopes, nuclear space applications; atomic energy legislation; health and safety; nuclear standards; AEC licensing and regulations; uranium mining and milling; nuclear reactor fuel ownership and financing; industry-government relations; civilian nuclear reactor politics and progress; public relations; Plowshare applications; nuclear merchant ships; uranium enrichment; safeguarding of special nuclear materials; transportation of nuclear materials.

Publications: NUCLEAR INDUSTRY (monthly); irregular publications (a price list is available).

Information services: Makes interlibrary loans; permits onsite reference by members.

NATIONAL COUNCIL ON RADIATION PROTECTION AND MEASUREMENTS (NCRP) - 4201 Connecticut Avenue N.W., Suite 402, Washington, D.C. 20008.

Areas of interest: Radiation protection and measurement, including basic radiation protection criteria and standards; maximum permissible dose equivalents for the population and radiation workers; maximum concentrations of radionuclides that may be inhaled or ingested by humans; maximum permissible body burdens; safe handling of radioactive materials; radiation monitoring methods and instruments; incineration of radioactive wastes; shielding for particle accelerators; radioactive nucleic acids and precursors; specification of radium substitutes, etc.

Publications: NCRP Reports; NCRP Statements; NCRP NEWS (published by NCRP Publications, P.O. Box 4867, Washington, D.C.).

Information services: Answers inquiries; makes referrals.

SOUTHERN INTERSTATE NUCLEAR BOARD - 800 Peachtree Street N.E., Suite 664, Atlanta, Georgia 30308.

Areas of interest: Applications of nuclear energy, space, and related sciences in industry, agriculture, medicine, commerce, and research in the 17-state Southern Region; radiation processing of chemicals, textile materials, irradiated wood-plastics, and coatings; preservation of foods by irradiation; nuclear-electric power generation; nuclear-related manufacturing and processing; radioisotopes; radioactive waste burial; etc.

Publications: Annual report; monthly newsletter; brochures, conference proceedings, research reports, press releases.

Information services: Answers inquiries; provides advisory and counseling services; assists development groups and technical services agencies; makes referrals, conducts briefings, seminars, conferences, and research studies; permits onsite use of collection.

3. Research Centers

BATELLE MEMORIAL INSTITUTE - Radiation Effects Information Center - Columbus Laboratories, 505 King Avenue, Columbus, Ohio 43201.

Areas of interest: Effects of space radiation, radiation from nuclear propulsion and auxiliary power systems in aerospace vehicles, and nuclear weapons on nuclear-powered and other aerospace vehicles, materials, devices, and systems (excluding biological systems).

Information services: Answers technical inquiries concerning current research and development projects and radiation effects data; provides engineering and reference services; prepares bibliographies and compilations. Pertinent literature is available for onsite use. Services are primarily for government agencies and their contractors, but may be made available to others as the need arises.

BROOKHAVEN NATIONAL LABORATORY - Associated Universities, Inc. - Upton, Long Island, New York 11973.

Areas of interest: Nuclear physics; nuclear chemistry; solid state physics; nuclear medicine; biology; nuclear engineering; health physics; electronics and instrumentation; metallurgy.

Holdings: 85,000 books and bound journals; 1,500 unbound journals; 100,000 AEC technical reports.

Publications: WEEKLY SELECTED READING LIST; annual report, technical reports.

Information services: Answers inquiries; makes interlibrary loans; permits onsite use of literature collection.

PHYSICAL DATA GROUP - Lawrence Radiation Laboratory, P.O. Box 808, Livermore, California 94551.

Areas of interest: Neutron and photon cross-section data required for weapons, weapon effects, shielding, and reactor calculations (experimental and evaluated data); static and dynamic parameters of critical assemblies and bulk experiments relating to neutron multiplication and transport; development of advanced processing codes for producing constants required for neutronic and photonic codes, making use of latest generation computational facilities.

Publications: State-of-the-art reviews, data compilations.

Information services: Answers inquiries, primarily for government agencies and their contractors, and for others as workload permits.

SANDIA LABORATORIES - Office of Industrial Cooperation--3416 - Albuquerque, New Mexico 87115.

Areas of interest: Innovations in science and technology in which industry may be interested; atomic energy technology; transfer of technology.

Publications: Industrial briefs (business-oriented summaries describing machines, apparatus, instruments, materials, devices, processes, and techniques developed by AEC contractors); reports.

Information services: Answers inquiries; provides consulting services; provides documentation ordered from permuted title index and other miscellaneous information releases.

UNIVERSITY OF CALIFORNIA - Lawrence Radiation Laboratory - Building 62, Room 207, Berkeley, California 97720.

Areas of interest: Inorganic materials; nuclear engineering;

metallurgy; ceramics; physical and inorganic chemistry; electro-chemistry; lasers and solid state physics.

Publications: Annual report; proceedings of symposium (held every 2 years); brochures detailing work of Division; preprints of journal articles.

Information services: Answers inquiries; permits onsite use of literature collection.

4. International Organizations

EURATOM - 200 rue de la Loi, 1040 Brussels, Belgium.

Purpose: To coordinate the research, development, and commer-cialization of nuclear power among member states and agencies. To promote standardization of safety, instrumentation, information exchange and related matters in the use of nuclear power.

Publications: BULLETIN OF THE EUROPEAN COMMUNITIES (monthly); JOURNAL OFFICIEL; EURO SPECTRA (quarterly). Reports (of the Commission); Information service material.

EUROPEAN ATOMIC ENERGY SOCIETY (EAES) - M. Osterreichische, Studien-gesellschaft fur Atomenergi, Lenaugasse 10, Vienna 8, Austria.

Purpose: To foster co-operation in atomic energy research and in its peaceful applications.

Publications: Symposia proceedings (biennially).

EUROPEAN ATOMIC FORUM - 26 rue de Clichy, 75-Paris 9e, France.

Purpose: To further information exchange on atomic research and development.

Publications: Congress proceedings.

EUROPEAN NUCLEAR ENERGY AGENCY - 38 bd Suchet, 75-Paris 16e, France.

Purpose: The development of systems whereby the utility of nuclear power for humane and constructive purposes may be fostered and advanced.

Programs: The programs of this agency are diverse and are car-ried out by a series of committees, by international meetings, support of research centers and libraries.

Publications: Annual reports, conference proceedings, results of specialized studies and joint research undertakings.

INTERNATIONAL AGENCY FOR ATOMIC ENERGY - 11 Kärntnerring, P.O. Box 590, 1011, Vienna, Austria.

Purpose: To advance the peaceful uses of atomic energy and to encourage research and development of this force for betterment of mankind.

Programs: Furtherance of use of nuclear power as a source of electricity. Training projects and support of research. Specialized information service; International Nuclear Information System (INIS).

Publications: Directories, bibliographies, statistical compendia, periodicals, etc.

INTERNATIONAL COMMISSION ON RADIATION UNITS AND MEASUREMENTS - 7910 Woodmont Avenue, Suite 1016, Washington, D.C. 20014.

Purpose: To develop international standards regarding quantities and units of radiation and radioactivity.

Program: Collection and evaluation of latest data pertinent to radiation measurement and dosimetry.

Publications: ICRU reports.

INTERNATIONAL COMMISSION ON RADIOLOGICAL PROTECTION - Clifton Avenue, Sutton, Surrey, United Kingdom.

Purpose: To advance the field of radiation protection.

Program: Publishes recommendations on radiation safety and protection.

5. Other Information Sources and Programs

a. SPECIALIZED INFORMATION SOURCES

Automated information retrieval services are available from National Technical Information Service, Springfield, Virginia 22151. Refer to their 1976 publication: A DIRECTORY OF COMPUTERIZED DATA FILES SOFTWARE, AND RELATED TECHNICAL REPORTS.

B. LITERATURE

1. Guides and Handbooks

Etherington, H. NUCLEAR ENGINEERING HANDBOOK. New York: McGraw-Hill Publishing Co., 1958.

"Comprehensive coverage of nuclear engineering."

Hagerton, J. ATOMIC ENERGY DESKBOOK. New York: Van Nostrand Reinhold, 1963.

REACTOR HANDBOOK. 4 vols. New York: John Wiley Publishing Co., 1964.

SOURCEBOOK ON ATOMIC ENERGY. By S. Glasstone. New York: Van Nostrand Reinhold Publishing Co., 1959.

U.S. Atomic Energy Commission. AEC LICENSING GUIDE: MEDICAL PROGRAMS, GUIDE FOR PREPARATION OF APPLICATIONS FOR MEDICAL USE OF RADIOISOTOPES. April 1972. Government Printing Office, Washington, D.C. 20545.

U.S. Atomic Energy Commission, Division of Technical Information. REACTOR HANDBOOK. 5 vols. Oak Ridge, Tennessee: 1960-64.

a. TABLES, DATA, AND SCIENTIFIC UNITS

COMPUTING METHODS IN REACTOR PHYSICS. Edited by H. Greenspan. New York: Gordon & Breach, 1968.

International Atomic Energy Agency. NUCLEAR DATA FOR REACTORS. Proceedings of the Second International Conference on Nuclear Data for Reactors, held by the International Atomic Energy Agency in Helsinki, 15-19 June, 1970. 2 vols. New York: Unipub, 1970.

Kunz, W. NUCLEAR TABLES. New York: Pergamon Press, 1968.

U.S. Atomic Energy Commission. NUCLEAR DATA TABLES. Washington, D.C.: National Academy of Sciences – National Research Council, 1959.

2. Dictionaries

a. TECHNICAL

Cooke, N. ELECTRONICS AND NUCLEONICS DICTIONARY. 3rd ed. New York: McGraw-Hill, 1966.

ELSEVIER'S DICTIONARY OF NUCLEAR SCIENCE AND TECHNOLOGY. New York: American Elsevier Co., 1970.

Hughs, L. DICTIONARY OF ELECTRONICS AND NUCLEONICS. New York: Barnes and Noble, 1970.

Markus, J. ELECTRONICS AND NUCLEONICS DICTIONARY. 3rd ed. New York: McGraw-Hill, 1967.

National Research Council Conference on Glossary of Terms in Nuclear Science and Technology. A GLOSSARY OF TERMS IN NUCLEAR SCIENCE AND TECHNOLOGY. New York: American Society of Mechanical Engineers, 1957.

Rau, H. DICTIONARY OF NUCLEAR PHYSICS AND NUCLEAR CHEMISTRY. 2nd ed. New York: Van Nostrand Reinhold, 1965.

b. FOREIGN LANGUAGE

Carpovich, Eugene A. RUSSIAN–ENGLISH ATOMIC DICTIONARY. 2nd rev. ed. New York: Technical Dictionaries Co., 1959.

Consultants Bureau Enterprises. RUSSIAN–ENGLISH GLOSSARY OF NUCLEAR PHYSICS AND ENGINEERING. New York: 1957.

ELSEVIER'S DICTIONARY OF NUCLEAR SCIENCE AND TECHNOLOGY. 2nd rev. ed. New York: American Elsevier Co., 1970.

Sube, R. DICTIONARY OF NUCLEAR PHYSICS AND TECHNOLOGY. New York: Pergamon, 1962.

> In English, French, German, and Russian.

3. Directories

a. GENERAL

Anthony, L. SOURCES OF INFORMATION ON ATOMIC ENERGY. New York: Pergamon, 1966.

DIRECTORY OF FEDERAL AGENCIES ENGAGED IN ENERGY RELATED AC-TIVITIES. Washington, D.C.: Government Printing Office, 1975.

INFORMATION CENTERS AND SERVICES DIRECTORY. Edited by J. Lewin. New York: Chicorel Library Publishing Corp., 1972.

> Compiled by staff of the Atomic Energy Commission. Describes data banks and services in the United States.

International Atomic Energy Agency. INIS REFERENCE SERIES. 15 vols. New York: Unipub, Inc., 1969-74.

_____. RESEARCH, TEST, AND EXPERIMENTAL REACTORS. Vol. 8. New York: Unipub, Inc., 1970.

> "Data - including diagrams, statistics, bibliographies - on new reactors in 16 nations."

The Mitre Corporation. A SURVEY OF FUEL AND ENERGY INFORMATION SOURCES. By D. Bobo, et al. Langly, Virginia: 1970.

U.S. Atomic Energy Commission. NUCLEAR REACTORS, BUILT, BEING BUILT, OR PLANNED, IN UNITED STATES AS OF DECEMBER 31, 1971. Prepared by Office of Assistant General Manager for Energy and Development, Washington, D.C. 20545. Springfield, Virginia: National Technical Information Service.

b. BIOGRAPHY

See also this subject in the section Chemical Engineering - General.

c. PRODUCTS AND MANUFACTURERS

MacRAE's BLUEBOOK. Annual. MacRae's Bluebook, Western Springs, Illinois 60558.

NUCLEAR NEWS ANNUAL BUYER'S GUIDE. American Nuclear Society, 244 East Ogden Avenue, Hinsdale, Illinois 60521.

THOMAS' REGISTER OF MANUFACTURERS. Annual. New York: Thomas Publishing Co.

d. LABORATORIES AND RESEARCH CENTERS

EUROPEAN RESEARCH INDEX. 2nd ed. Guernsey, Channel Islands, England: F. Hodgson Ltd., 1969.

INDUSTRIAL RESEARCH LABORATORIES OF THE UNITED STATES. 13th ed. New York: R.R. Bowker, 1970.

RESEARCH CENTERS DIRECTORY. 5th ed. Detroit: Gale Research Co., 1975.

    Updated by subscription service, NEW RESEARCH CENTERS.

4. Encyclopedias

See this subject in the section Chemical Engineering - General.

5. Bibliographies

a. SPECIFIC TO TOPIC

Anthony, L. SOURCES FOR INFORMATION ON ATOMIC ENERGY. Elmsford, New York: Pergamon Press, 1966.

Atomic Energy of Canada Ltd. LIST OF PUBLICATIONS (March 1971 to April 1972). Chalk River, Ontario, Canada.

Bene, G. NUCLEAR PHYSICS AND ATOMIC ENERGY. Glossarium Interpretum No. 2. New York: American Elsevier, 1960.

International Atomic Energy Agency. CONFERENCE PROCEEDINGS SERIES.

Irregular. Unipub Inc., Box 433, New York, New York 10018.

"Identifies conferences, congresses, meetings, symposia, and training courses relating to nuclear science and technology."

_____. HEAVY-WATER REACTORS. New York: Unipub, 1970.

Bibliography with over 2500 references.

_____. LIST OF BIBLIOGRAPHIES ON NUCLEAR ENERGY. Irregular. International Atomic Energy Agency, INIS Section of the Division, U.S. Address: National Agency for International Publications Inc., 317 East 34th Street, New York, New York 10016.

_____. TECHNICAL DIRECTORIES. Irregular. New York: Unipub, Inc.

_____. TECHNICAL REPORT SERIES. Irregular. New York: Unipub, Inc.

Oak Ridge National Laboratory, Tennessee. URANIUM TAILINGS BIBLIOGRAPHY. By C.F. Holoway, et al. Rept. no. ORNL 5109. Springfield, Virginia: National Technical Information Service, 1976.

U.S. Atomic Energy Commission. BIBLIOGRAPHIES OF ATOMIC ENERGY LITERATURE. U.S. Atomic Energy Commission, Division of Technical Information Extension, Box 62, Oak Ridge, Tennessee 37830. 1958.

_____. BIBLIOGRAPHIES OF INTEREST TO THE ATOMIC ENERGY PROGRAM. Compiled by James M. Jacobs, et al. Oak Ridge, Tennessee: Division of Technical Information, 1962.

_____. SELECTED ANNOTATED BIBLIOGRAPHY OF CIVIL, INDUSTRIAL, AND SCIENTIFIC USES FOR NUCLEAR EXPLOSIONS. Compiled by Robert C. Kelly. Springfield, Virginia: National Technical Information Service, 1972.

U.S. Congress. Joint Committee on Atomic Energy. CURRENT MEMBERSHIP OF THE JOINT COMMITTEE ON ATOMIC ENERGY, CONGRESS OF THE U.S.: JOINT COMMITTEE ON ATOMIC ENERGY MEMBERSHIP PUBLICATIONS, AND OTHER PERTINENT INFORMATION THROUGH THE 94TH CONGRESS 1ST SESSION, MARCH 1976. Washington, D.C.: Government Printing Office, 1976.

U.S. National Technical Information Service. AEC-NASA TECH BRIEFS WHICH INCLUDE DETAILS ON MANUFACTURING PROCESSES. Irregular. Springfield, Virginia 22151.

Wiren, H. GUIDE TO LITERATURE ON NUCLEAR ENGINEERING. American Society for Engineering Education, Suite 400, 1 Dupont Circle, Washington, D.C., 20036. 1970.

b. RELATED BIBLIOGRAPHIES IN CURRENT JOURNALS

The following journals frequently contain book reviews and/or bibliographies. See the heading Journals and Serial Publications in this section for addresses.

ATOM (ENGLAND)

ATOMIC ENERGY IN AUSTRALIA

ATOM-INFORMATIONEN

ATOM NEWS

ATOMIC ENERGY LAW JOURNAL

ATOMIC ENERGY REVIEW

ATOMIC ENERGY SOCIETY OF JAPAN JOURNAL

ATOMKI KOZLEMENYEK

BRITISH NUCLEAR SOCIETY JOURNAL

CANADIAN RESEARCH AND DEVELOPMENT

ENERGIA NUCLEARE

ENERGIE NUCLEAIRE

ENERGY CONVERSION

ENERGY DEVELOPMENTS

EURONUCLEAR

INDUSTRIES ATOMIQUES

INFORMATION BULLETIN ON ISOTOPIC GENERATORS

INIS ATOMINDEX

INTERNATIONAL JOURNAL OF APPLIED RADIATION AND ISOTOPES

ISOTOPES AND RADIATION TECHNOLOGY

JOURNAL OF NUCLEAR ENERGY

JOURNAL OF NUCLEAR MATERIALS

NUCLEAR CANADA

NUCLEAR DATA

NUCLEAR ENERGY

NUCLEAR ENGINEERING AND DESIGN

NUCLEAR ENGINEERING INTERNATIONAL

NUCLEAR FUSION

NUCLEAR INSTRUMENTS AND METHODS

NUCLEAR PHYSICS

NUCLEAR SCIENCE AND ENGINEERING

NUCLEAR STANDARDS NEWS

NUCLEAR TECHNOLOGY

NUCLEUS

RADIOISOTOPES

SOVIET ATOMIC ENERGY

TRANSATOM BULLETIN

6. Monographs, Reports, and Symposia Proceedings

American Nuclear Society. Annual meeting transactions. American Nuclear Society, 244 East Ogden Avenue, Hinsdale, Illinois 60521.

ANALYTICAL CHEMISTRY OF NUCLEAR FUELS. IAEA. New York: Unipub, 1973.

> "The proceedings of a Vienna Panel, this publication examines methods of determining major components of nuclear fuels, and assesses fuel burn-up. Contents include: analytical methods for determining uranium, plutonium, and thorium in nuclear fuels; advantages and limitations of mass spectrometry for measuring isotopic distributions of elements, and application to nuclear fuel burn-up; post-irradiation isotopic analysis of reactor fuels."

Ayres, J. DECONTAMINATION OF NUCLEAR REACTORS AND EQUIPMENT. New York: Ronald Press, 1970.

> General information on decontamination operations, particularly nuclear reactors."

Bell, G. NUCLEAR REACTOR THEORY. New York: Van Nostrand-Reinhold, 1970.

Benedict, M. NUCLEAR CHEMICAL ENGINEERING. New York: McGraw-Hill, 1957.

Berry, W. CORROSION IN NUCLEAR APPLICATIONS. New York: Wiley, 1971.

> "Has three major sections, dealing respectively with water-, liquid metal- and gas-cooled reactors."

Boland, J. NUCLEAR REACTOR INSTRUMENTATION (IN-CORE). Prepared under the direction of the American Nuclear Society for the Division of Technical Information, United States Atomic Energy Commission. New York: Gordon & Breach, 1970.

> "Radiation effects, radiation effects on electrical circuits, pressure measurements, temperature instrumentation, neutron and gamma flux measurements, instruments for reactor transient tests. A

monograph primarily for scientists and engineers involved in the
applications of nuclear energy."

Brookhaven National Laboratory. ENERGY FACILITY LOCATION: A RE-
GIONAL VIEWPOINT. By P. Meier. Rept. no. BNL 20435. Springfield,
Virginia: National Technical Information Service, 1976.

Cohen, P. WATER COOLANT TECHNOLOGY OF POWER REACTORS. New
York: Gordon & Breach, 1970.

"A compilation of the elements of knowledge and experience in
the chemical and physical processes relevant to water reactors,
and an interpretative summary of the specialized literature."

CONCRETE FOR NUCLEAR REACTORS. International Seminar. Sponsored by
American Concrete Institute, Berlin, Germany. ACI Special Publication no. 34.
3 vols. Edited by Clyde E. Kesler. Detroit: American Concrete Institute,
1972.

CONSTRUCTIVE USES OF ATOMIC ENERGY. International Conference.
American Nuclear Society, 244 East Ogden Avenue, Hinsdale, Illinois 60521.
1968.

Davey, W. TECHNIQUES IN FAST REACTOR CRITICAL EXPERIMENTS 1970.
New York: Gordon & Breach, 1970.

ELEMENTARY PARTICLES, AMSTERDAM INTERNATIONAL CONFERENCE.
Sponsored by European Organization for Nuclear Research (CERN), Amsterdam,
Netherlands. Edited by A.G. Tenner & M.J.G. Veltman. 1972. Publisher:
North-Holland Publishing Company, Amsterdam. U.S. Distributor: American
Elsevier Publishing Co., 52 Vanderbilt Avenue, New York, New York 10017.

ENERGY CONVERSION ENGINEERING CONFERENCE, 8TH. American Chemical Society, 1155 16th St. N.W., Washington, D.C. 20036. 1973.

Flagg, J., ed. CHEMICAL PROCESSING OF REACTOR FUELS. Nuclear Science and Technology Series, vol. 1. New York: Academic Press, 1971.

Glower, D. EXPERIMENTAL REACTOR ANALYSIS & RADIATION MEASURE-
MENTS. Nuclear Engineering Series. New York: McGraw-Hill, 1965.

Hansen, R., ed. SEISMIC DESIGN FOR NUCLEAR POWER PLANTS. Cambridge: M.I.T. Press, 1970.

"Geological and seismological factors influencing the assessment
of a seismic threat to nuclear reactors."

INNOVATIVE APPLICATIONS OF RADIATION. Conference sponsored by
Atomic Energy Commission and Southern Interstate Nuclear Board, Dallas, Texas.
Springfield, Virginia: National Technical Information Service, 1971.

International Atomic Energy Agency. ADVANCED AND HIGH-TEMPERATURE GAS-COOLED REACTORS. Proceedings of a Symposium on Advanced and High-Temperature Gas-Cooled Reactors Held by the International Atomic Energy Agency in Julich, 21-25, October 1968. New York: Unipub, 1969.

_____. FUTURE OF NUCLEAR STRUCTURE STUDIES. Proceedings of a Panel on the Future of Nuclear Structure Studies Held in Dubna, 1-3 July 1968. New York: Unipub, 1969.

_____. LARGE RADIATION SOURCES FOR INDUSTRIAL PROCESSES. New York: Unipub, 1970.

"Proceedings of a symposium on the utilization of large radiation sources and accelerators in industrial processing held by the IAEA in Munich, August 18-22, 1969. Papers describe developments in radiation-induced processes.

_____. NUCLEAR DATA FOR REACTORS: SECOND INTERNATIONAL CON-FERENCE. 2 vols. New York: Unipub, 1970.

_____. NUCLEAR DESALINATION. Proceedings of a Symposium on Nuclear Desalination Held by the International Atomic Energy Agency in Madrid, 18-22 November 1968. New York: Elsevier, 1969.

_____. NUCLEAR POWER PLANT CONTROL AND INSTRUMENTATION. Vienna: 1971.

Proceedings of a working group meeting in Vienna.

_____. NUCLEAR TECHNIQUES FOR STUDYING PESTICIDE RESIDUE PROB-LEMS. Proceedings of a Panel on Isotopic Tracer and Radioactivation Techniques for Studying Residue Problems with Particular Reference to Food Entering International Trade, Organized by the Joint FAO/IAEA Division of Atomic Energy in Food and Agriculture, and Held in Vienna, 16-20 December, 1968. New York: Unipub, 1971.

_____. PEACEFUL USES OF NUCLEAR EXPLOSIONS. New York: Unipub, 1970.

_____. PERFORMANCE OF NUCLEAR POWER REACTOR COMPONENTS. Proceedings of a Symposium on Performance of Nuclear Power Reactor Components Held by the International Atomic Energy Agency in Prague, 10-14 November, 1969. New York: Unipub, 1970.

_____. SMALL AND MEDIUM-SIZE POWER REACTORS. Proceedings of a Panel on Small and Medium Size-Power Reactors, Held in Vienna, 24-28 June 1968. New York: Unipub, 1969.

_____. THEORY OF NUCLEAR STRUCTURES. Trieste lecture 1969. Vienna: 1970.

_____. Nuclear Reactors Technical Committee. SYMPOSIUM, PEACEFUL

NUCLEAR EXPLOSIONS. January 20-24, 1975, Vienna. Proceedings. New York: Unipub, 1975.

INTERNATIONAL CONFERENCE ON THE PEACEFUL USES OF ATOMIC ENERGY. Geneva Proceedings. New York: United Nations, 1971.

Kumarovskii, A. SHIELDING MATERIALS FOR NUCLEAR REACTORS. Irregular. New York: Pergamon.

Lish, K. NUCLEAR POWER PLANT SYSTEMS AND EQUIPMENT. New York: Industrial Press, 1972.

> Good information source on this specialty field.

Moore, R., ed. NUCLEAR POWER. Beverton, Oregon: International Scholarly Book Services, 1972.

> Overview of nuclear reactors and nuclear energy. Cites details on particular locations.

NEW METHODS AND TECHNIQUES IN NUCLEAR MATERIALS SCIENCE. Diffraction analysis conference. Edited by I.F. Ferguson. 1971. Publisher: Her Majesty's Stationery Office, London. U.S. Distributor: Pendragon House, 899 Broadway Avenue, Redwood City, California 94063.

NUCLEAR ENERGY CENTERS AND AGRO-INDUSTRIAL COMPLEXES. (TRS 140). IAEA. New York: Unipub, 1973.

> "Agro-industrial complexes are today being used to develop arid areas of the world. This publication considers the status and potential of these complexes and covers: low cost energy; desalination of sea water; power-intensive industries; desalted water in agriculture; economic and social considerations."

NUCLEAR POWER FOR TOMORROW. Conference. American Nuclear Society, 244 East Ogden Avenue, Hinsdale, Illinois 60521. 1972.

NUCLEAR POWER PLANT CONTROL & INSTRUMENTATION. 1st Working Group Meeting. Unipub, Inc., Box 433, New York, New York 10016. 1972.

NUCLEAR REACTORS. International Conference. Canadian Nuclear Association, 11 Adelaide St. West, Toronto 1, Canada. 1972.

NUCLEAR SOLUTIONS TO WORLD ENERGY PROBLEMS. International Conference and Winter Meeting. Edited by Ruth Farmakes. 1972. American Nuclear Society, 244 East Ogden Avenue, Hinsdale, Illinois.

NUCLEAR TECHNICIAN MANPOWER SYMPOSIUM. Springfield, Virginia: National Technical Information Service, 1972.

NUCLEAR TECHNIQUES IN THE BASIC METAL INDUSTRIES, SYMPOSIUM. New York: Unipub, Inc., 1973.

PLUTONIUM UTILIZATION. Conference on Continuing Education in Engineering. 3rd Annual Conference on Nuclear Power and Environmental Assessment. Berkeley, California, September 1975. Proceedings. Berkeley: University Extension, University of California at Berkeley, 1975.

POWER PLANT DYNAMICS, CONTROL AND TESTING SYMPOSIUM. 2nd conference, Knoxville, Tennessee, September 1975. Proceedings. Knoxville: Nuclear Engineering Department, University of Tennessee, 1976.

REACTOR MATERIALS PERFORMANCE. Topical Meeting. American Nuclear Society, 344 East Ogden Avenue, Hinsdale, Illinois 60521. 1972.

Simmons, J. RADIATION DAMAGE IN GRAPHITE. International Series of Monographs on Nuclear Energy, vol. 102. New York: Pergamon, 1965.

Simnad, M. FUEL ELEMENT EXPERIENCE IN NUCLEAR POWER REACTORS. New York: Gordon & Breach, 1971.

> "Reviews the experience gained on the use of fuel elements in power reactors."

_____. MATERIALS AND FUELS FOR HIGH-TEMPERATURE NUCLEAR APPLICATIONS. Cambridge, Massachusetts: MIT Press, 1964.

SODIUM-WATER REACTIONS. Specialists' Meeting, U.S.S.R., 1972. Springfield, Virginia: National Technical Information Service, 1972.

STRUCTURAL MECHANICS IN REACTOR TECHNOLOGY. 1st International Conference, Berlin, September 20-24, 1971. Proceedings issued in NUCLEAR ENGINEERING & DESIGN, vol. 19, no. 2 (1972) and vol. 20, no. 1 (1972).

SYMPOSIUM ON THERMODYNAMICS WITH EMPHASIS ON NUCLEAR MATERIALS AND ATOMIC TRANSPORT IN SOLIDS. Vienna, 1965. THERMODYNAMICS: PROCEEDINGS. Vienna: International Atomic Energy Agency, 1966.

U.S. Atomic Energy Commission. ANALYSIS OF ELEMENTAL BORON (with lists of references). By Morris W. Lerner. Springfield, Virginia: National Technical Information Service, 1970.

_____. CIVILIAN NUCLEAR POWER: REACTOR FUEL CYCLE COSTS FOR NUCLEAR POWER EVALUATION. Washington, D.C.: Government Printing Office, December 1971.

_____. FORECAST OF GROWTH OF NUCLEAR POWER. Washington, D.C.: Division of Operations Analysis and Forecasting, Government Printing Office, January, 1971.

_____. FUNDAMENTAL NUCLEAR ENERGY RESEARCH, 1971. Supplemental Report to Annual Report to Congress for 1971. Washington, D.C.: Government Printing Office, January 1972.

_____. MAJOR ACTIVITIES IN ATOMIC ENERGY PROGRAMS, JANUARY-DECEMBER 1971. Washington, D.C.: Government Printing Office, January 1972.

_____. NUCLEAR INDUSTRY. Division of Industrial Participation. Washington, D.C.: Government Printing Office, 1970.

_____. NUCLEAR INDUSTRY. Division of Industrial Participation. Washington, D.C.: Government Printing Office, 1971.

_____. NUCLEAR INDUSTRY FUEL SUPPLY SURVEY. April 1972. Office of Assistant Director for Raw Materials, Division of Production and Materials Management. Springfield, Virginia: National Technical Information Service, 1973.

_____. NUCLEAR POWER 1973-2000. Forecasting Branch, Office of Planning and Analysis. Washington, D.C.: Government Printing Office, 1972.

_____. 3RD INTERNATIONAL SYMPOSIUM ON RESEARCH MATERIALS FOR NUCLEAR MEASUREMENTS. October 5-8, 1971. Oak Ridge National Laboratory. Springfield, Virginia: National Technical Information Service, 1972.

U.S. Congress. NUCLEAR POWER AND RELATED ENERGY PROBLEMS, 1968-70. Washington, D.C.: Government Printing Office, 1971.

U.S. Congress. House Interior and Insular Affairs Committee. ENERGY DEMAND STUDIES, ANALYSIS AND APPRAISAL, 92D CONGRESS. Washington, D.C.: Government Printing Office, 1972.

U.S. Department of Health, Education, and Welfare. Radiological Health Bureau. RADIATION SAFETY AND PROTECTION IN INDUSTRIAL APPLICATIONS. Proceedings of Symposium August 16-18, 1972, Washington, D.C. (with lists of references). Edited by H.F. Klein. Washington, D.C.: Government Printing Office, October 1972.

U.S. Federal Energy Administration Project Independence. PROJECT INDEPENDENCE BLUEPRINT TASK FORCE REPORT. Rept. no. PB 248 499. Springfield, Virginia: National Technical Information Service, 1975.

Whiting, R. RADIATION CHEMICAL PROCESSING. Park Ridge, New Jersey: Noyes Data Corp., 1969.

WORKING GROUP MEETING ON NUCLEAR POWER PLANT CONTROL AND INSTRUMENTATION. Vienna, 1971. PROCEEDINGS. Organized by the International Atomic Energy Agency. New York: Unipub, 1972.

7. Abstracting and Indexing Services

a. GENERAL

APPLIED SCIENCE AND TECHNOLOGY INDEX

BRITISH TECHNOLOGY INDEX

CHEMICAL ABSTRACTS

ENGINEERING INDEX

See this topic in the Section I of this Guide for additional details.

b. SPECIFIC TO TOPIC

BULLETIN SIGNALETIQUE. Part 150: "Physique, Chimie et Technologie Nucleaires." Monthly. Centre de Documentation du CNRS, 15 Quai Anatole-France, Paris (7e), France.

ENERGY ABSTRACTS FOR POLICY ANALYSIS. Monthly. Government Printing Office, Washington, D.C. 20402.

ERDA (ENERGY RESEARCH AND DEVELOPMENT ADMINISTRATION) ABSTRACTS. Monthly. Government Printing Office, Washington, D.C. 20402.

EURO-ABSTRACTS; AN ABSTRACTS JOURNAL. Monthly. Commission of the European Communities, Poste 1003, Luxembourg.

INDEX DE LA LITTERATURE NUCLEAIRE FRANCAISE. Monthly. Commissariat a l'Energie Atomique, B.L. no. 2, 91 Gif-Sur-Yvette, France.

> Text in French; summaries in English.

International Atomic Energy Agency. Vienna. Semimonthly. LIST OF BIBLI-OGRAPHIES ON NUCLEAR ENERGY.

> Published semimonthly in English, with titles in original language.

NUCLEAR SCIENCE ABSTRACTS. Government Printing Office, Washington, D.C. 20402.

> Ceased publication with volume 30, 1976.

PREDI-BRIEFS. Abstracting service. Monthly. ENERGY. Predicasts, 200 University Circle Research Center, 11001 Cedar Avenue, Cleveland, Ohio 44106.

> "Contains information on conventional fuels - coal, crude oil, natural gas and nuclear energy - and newer technologies such as solar energy, geothermal energy, synthetic natural gas and fuel cells. Low-sulfur fuels and other pollution aspects briefed."

SELECTED ABSTRACTS OF NON-U.S. LITERATURE ON PRODUCTION AND

INDUSTRIAL USES OF RADIOISOTOPES. Quarterly. Oak Ridge National Laboratory, Box X, Oak Ridge, Tennessee 37830.

U.S. Atomic Energy Commission. Semimonthly. NUCLEAR SCIENCE AB-STRACTS. Washington, GPO. Annual cumulated index.

c. RELATED ABSTRACTS IN CURRENT JOURNALS

The following journals frequently provide abstracts. See the heading Journals and Serial Publications in this section for addresses.

ATOMIC ENERGY REVIEW

ATOMIC ENERGY SOCIETY OF JAPAN JOURNAL

ENERGIA NUCLEARE

ENERGIE NUCLEAIRE

ENERGY CONVERSION

JOURNAL OF NUCLEAR ENERGY

JOURNAL OF NUCLEAR SCIENCE AND TECHNOLOGY (JAPAN)

NUCLEAR FUSION

NUCLEAR SCIENCE AND ENGINEERING

NUCLEUS

RADIOISOTOPES

8. Journals and Serial Publications

a. DIRECTORIES AND LISTS

AYER'S DIRECTORY OF PUBLICATIONS. Annual. Philadelphia, Pennsylvania: Ayer Press.

STANDARD PERIODICAL DIRECTORY. Irregular. New York: Oxbridge Publishing Co.

ULRICH'S INTERNATIONAL PERIODICAL DIRECTORY. 2 vols. Annual. New York: R.R. Bowker Co.

b. JOURNAL AND SERIAL TITLES

ADVANCED ENERGY CONVERSION. Irregular. Pergamon Press, 41-01 21st Street, Long Island City, New York 11101.

ADVANCES IN NUCLEAR PHYSICS. Irregular. New York: Plenum Publishing.

ADVANCES IN NUCLEAR SCIENCE AND TECHNOLOGY. Edited by E.J. Henley and J. Lewins. New York: Academic, 1972.

A E C L REVIEW. Monthly. Atomic Energy of Canada Ltd., 275 Slater Street, Ottawa, Canada.

ANNUAL REVIEW OF NUCLEAR SCIENCE. Annual Reviews Inc., 4139 El Camino Way, Palo Alto, California 94306.

ATEN BULLETIN D'INFORMATION. Bimonthly. Association Technique pour l'Energie Nucleaire, 26 rue de Clichy, 75-Paris (9e), France.

ATOM (ENG). Monthly. United Kingdom Atomic Energy Authority, Public Relations Branch, 11 Charles II St., London, SW 1, England.

ATOM (U.S.). Monthly. Los Alamos Scientific Laboratory, University of California, Box 1663, Los Alamos, New Mexico 87544.

ATOMIC ENERGY IN AUSTRALIA. Quarterly. Australian Atomic Energy Commission, 45 Beach Street, Coogee, NSW 2034, Australia.

ATOMIC ENERGY LAW JOURNAL. Quarterly. Hanover Lamont Corp., 89 Beach Street, Boston, Massachusetts 02111.

ATOMIC ENERGY REVIEW. Quarterly. International Atomic Energy Agency, Kaertner Ring 11, Vienna 1, Austria.

    Supplements. Text in English.

Atomic Energy Society of Japan. JOURNAL/NIHON GENSHIRYOKU GAKKAI SHI. Monthly. Japan Atomic Energy Research Institute, 1-13 Shimbashi, Minato-Ku, Tokyo, Japan.

    Text in Japanese. Summaries in English.

ATOM-INFORMATIONEN. Monthly. Deutsches Atomforum e.V., Allianzplatz Haus X, 53 Bonn, West Germany.

ATOMKI KOZLEMENYEK. Quarterly. Akademia A ommag Kutato Intezete, Bem ter 18-c, Debrecen, Hungary.

    Supplements. Text in English and Hungarian.

ATOM NEWS. U.K. Atomic Energy Authority, 11 Charles Second Street, London, SW 1, England.

BRITISH NUCLEAR ENERGY SOCIETY JOURNAL. Quarterly. British Nuclear Energy Society, 1-7 Great George Street, Westminster, London, SW 1, England.

BULLETIN SIGNALETIQUE. Part 150: "Physique, Chimie et Technologie Nucleaires." Monthly. Centre de Documentation du CNRS, 15 Quai Anatole-France, Paris (7e), France.

Canada. Dominion Bureau of Statistics. SERVICE BULLETIN: ENERGY STATISTICS. Annual. Douglas F. Heney, Dominion Bureau of Statistics, Tunney's Pasture, Ottawa 3, Ontario, Canada.

CANADIAN RESEARCH & DEVELOPMENT. Bimonthly. Maclean-Hunter Ltd., 481 University Avenue, Toronto 101, Ontario, Canada.

COMMENTS ON NUCLEAR & PARTICLE PHYSICS. Bimonthly. Gordon & Breach Science Publishers Ltd., 150 Fifth Avenue, New York, New York 10011.

CONSTRUCTION STATUS OF NUCLEAR POWER PLANTS. Monthly. National Technical Information Service, Springfield, Virginia 22151.

ENERGIA NUCLEARE. Monthly. Centro Informazioni Studi Esperienze, Casella Postale 3986, Milan, Italy.

    Text in English and Italian.

ENERGIE NUCLEAIRE. Monthly. Societe de Productions Documentaires, 80, Route de Saint-Cloud, 92 - Rueil - Malmaison, France.

ENERGY ABSTRACTS FOR POLICY ANALYSIS. Monthly. Government Printing Office, Washington, D.C. 20402.

ENERGY CONVERSION. Quarterly. Pergamon Press, Maxwell House, Fairview Park, Elmsford, New York 10523.

    Text in English, French, and German.

ENERGY DEVELOPMENTS; ATOMIC, PETROLEUM, & OTHERS. International Review Service, 15 Washington Place, New York, New York 10003.

ENERGY INFORMATION REPORTED TO CONGRESS. Quarterly. Government Printing Office, Washington, D.C. 20402.

ERDA (ENERGY RESEARCH AND DEVELOPMENT ADMINISTRATION). ABSTRACTS. Monthly. Government Printing Office, Washington, D.C. 20402.

EURO-ABSTRACTS: AN ABSTRACTS JOURNAL. Monthly. Commission of the European Communities, Handlsblatt GmbH, Postfach 1102, D 4 Dusseldorf 1, West Germany.

EURONUCLEAR. Monthly. Morgan Brothers Publishing Ltd., 28 Essex Street, London, WC 2, England.

EURO-SPECTRA. Quarterly. European Community Information Service, 2100 M Street, N.W., Suite 707, Washington, D.C. 20037.

INDEX DE LA LITTERATURE NUCLEAIRE FRANCAISE. Monthly. Commissariat a l'Energie Atomique, B.I. No. 2, 91 Gif-Sur-Yvette, France.

Text in French; summaries in English.

INDUSTRIES ATOMIQUES; REVUE INTERNATIONALE POUR LES APPLICATIONS PACIFIQUES DE L'ENERGIE NUCLEAIRE. Monthly. Editions Sadesi, 1225 Chene-Bourg, Geneva, Switzerland.

Text in French.

INFORMATION BULLETIN ON ISOTOPIC GENERATORS. Semiannual. European Nuclear Energy Agency, 38 Boulevard Suchet, 75 Paris (16e), France.

Text in English & French.

INIS ATOMINDEX. 8/yr. International Nuclear Information System, International Atomic Energy Agency, Kaerntner Ring 11, Box 590, A-1011, Vienna, Austria.

International Atomic Energy Agency. LIST OF BIBLIOGRAPHIES ON NUCLEAR ENERGY. Three issues per year. P.O. Box 590, A-1011, Vienna, Austria.

Published in English, with titles in original languages.

INTERNATIONAL ATOMIC ENERGY AGENCY BULLETIN. Quarterly. International Atomic Energy Agency. U.S. Address: National Agency for International Publications, Inc., 317 East 34th Street, New York, New York 10016.

INTERNATIONAL JOURNAL FOR RADIATION PHYSICS AND CHEMISTRY. Quarterly. Pergamon Press, Maxwell House, Fairview Park, Elmsford, New York 10523.

INTERNATIONAL JOURNAL OF APPLIED RADIATION AND ISOTOPES. Monthly. Pergamon Press, Journals Department, Maxwell House, Fairview Park, Elmsford, New York 10523.

ISOTOPE. Quarterly. South African Atomic Energy Board, Private Bag 256, Pretoria, South Africa.

Text in Afrikaans and English.

ISOTOPES AND RADIATION TECHNOLOGY. Quarterly. U.S. Superintendent of Documents, Government Printing Office, Washington, D.C. 20402.

IZOTOPTECHNIKA/ISOTOPE TECHNOLOGY. Monthly. MTA Izotopintezete, Hungarian Academy of Sciences, Institute of Isotopes, Budapest, Hungary.

Text and summaries in English and Russian.

JOURNAL OF NUCLEAR MATERIALS. Monthly. North Holland Publishing Co., Box 211, Amsterdam, Netherlands.

JOURNAL OF NUCLEAR SCIENCE AND TECHNOLOGY. Monthly. Atomic Energy Society of Japan, c/o Japan Atomic Energy Research Institute, 1-13 Shimbashi, Minato-ku, Tokyo, Japan.

JOURNAL OF PHYSICS G: NUCLEAR PHYSICS. 9 issues per year. Institute of Physics and Physical Society, 47 Belgrave Square, London, SW 1, England.

NEWSLETTER ON ISOTOPIC GENERATORS AND BATTERIES. Quarterly. European Nuclear Energy Agency, 38 bd. Suchet, 75 Paris (6e), France.

NUCLEAR CANADA/CANADA NUCLEAIRE. Monthly. Canadian Nuclear Association, 11 Adelaide Street, W., Toronto 1, Ontario, Canada.

NUCLEAR DATA SHEETS. Monthly. Academic Press Inc., 111 Fifth Avenue, New York, New York 10003.

NUCLEAR DATA TABLES. Monthly. Academic Press, 111 Fifth Avenue, New York, New York 10003.

NUCLEAR ENGINEERING AND DESIGN. Monthly. North-Holland Publishing Co., Box 211, Amsterdam C, Netherlands.

NUCLEAR ENGINEERING INTERNATIONAL. Monthly. IPC Electrical-Electronic Press Ltd., Dorset House, Stamford Street, London, SE 1, England.

NUCLEAR FUSION. 6/year. International Atomic Energy Agency, Kaerntner Ring 11, Vienna 1, Austria.

NUCLEAR INDIA. Monthly. Jagannath Sharma, Publications Officer, Department of Atomic Energy, Chatrapati Shivaji Maharaj Marg, Bombay 1, India.

NUCLEAR INSTRUMENTS AND METHODS. 7/year. North-Holland Publishing Co., Box 211, Amsterdam C, Netherlands.

NUCLEAR NEWS. Monthly. Nuclear Society, 244 East Ogden Avenue, Hinsdale, Illinois 60521.

NUCLEAR PHYSICS. Monthly. North-Holland Publishing Co., Box 211, Amsterdam, Netherlands.

NUCLEAR POWER NEWSLETTER. Monthly. American Public Power Association, 2600 Virginia Avenue, N.W., Washington, D.C. 20037.

NUCLEAR SAFETY. Bimonthly. U.S. Atomic Energy Commission, Superintendent of Documents, Government Printing Office, Washington, D.C. 20402.

NUCLEAR SCIENCE AND ENGINEERING. Monthly. American Nuclear Society, 244 East Ogden Avenue, Hinsdale, Illinois 60521.

NUCLEAR SCIENCE JOURNAL. Quarterly. Atomic Energy Council, 1-1 Lane 20, Sin Y. Road, Section 1, Taipei, Taiwan (Formosa) Republic of China.

NUCLEAR TECHNOLOGY; APPLICATIONS FOR NUCLEAR SCIENCE, NU-CLEAR ENGINEERING AND RELATED ARTS. Monthly. American Nuclear Society, 244 East Ogden Avenue, Hinsdale, Illinois 60521.

NUCLEONICS WEEK (incorporates NUCLEONICS). Weekly. McGraw-Hill Publishing Co., 1221 Avenue of the Americas, New York, New York 10020.

Reporting news and developments on nuclear industry.

NUCLEUS (West Pakistan). Quarterly. Pakistan Atomic Energy Commission, Box 3112, Karachi 29, West Pakistan.

NUKLEONIKA. Monthly. Nuclear Energy Information Center, Polish Government for Use of Nuclear Energy, Palace of Culture and Science, Warsaw, Poland.

Text and summaries in English, French, Polish, and Russian.

Oak Ridge Associated Universities. NEWSLETTER. Monthly. (Formerly: Oak Ridge Institute of Nuclear Studies. NEWSLETTER) Information Services Department, Oak Ridge Associated Universities, Inc., Box 117, Oak Ridge, Tennessee 37830.

OCCUPATIONAL SAFETY AND HEALTH ABSTRACTS. Monthly. International Occupational Safety and Health Information Centre (CIS), International Labour Office, 1211 Geneva 22, Switzerland.

PREDI-BRIEFS. Abstracting service. Monthly. ENERGY. Predicasts, 200 University Circle Research Center, 11001 Cedar Avenue, Cleveland, Ohio 44106.

"Contains information on conventional fuels -- coal, crude oil, natural gas and nuclear energy -- and newer technologies such as solar energy, geothermal energy, synthetic natural gas and fuel cells. Low-sulfur fuels and other pollution aspects briefed."

PROGRESS IN NUCLEAR ENERGY. Irregular. New York: Pergamon Press.

RADIOISOTOPES. Monthly. Japan Radioisotope Association, 28-45 Honko-magome 2-chome, Bunkyo-ku, Tokyo, Japan.

Text in English and Japanese; summaries in English.

RADIOPROTECTION. Quarterly. Societe Francaise de Radioprotection, Dunod Editeur, 92 rue Bonaparte, 75 Paris (6e), France.

Summaries in English and French.

REACTOR TECHNOLOGY. Quarterly. U.S. Superintendent of Documents, Government Printing Office, Washington, D.C. 20402.

SELECTED ABSTRACTS OF NON-U.S. LITERATURE ON PRODUCTION AND INDUSTRIAL USES OF RADIOISOTOPES. Quarterly. Oak Ridge National Laboratory, Box X, Oak Ridge, Tennessee 37830.

SOVIET ATOMIC ENERGY. Monthly. Academy of Sciences of the USSR, Consultants Bureau, 227 West 17th Street, New York, New York 10011.

SOVIET JOURNAL OF NUCLEAR PHYSICS. Monthly. American Institute of Physics, 335 East 45th Street, New York, New York 10017.

   English translation of YADERNAYA FIZIKA.

TRANSACTIONS. Biannual. American Nuclear Society, 244 Ogden Street, Hinsdale, Illinois 60521.

   "Summaries of all papers presented at the Annual and Winter meetings of the American Nuclear Society."

TRANSATOM BULLETIN: A MONTHLY GUIDE TO NUCLEAR LITERATURE IN TRANSLATION. Monthly. EURATOM - CID, Agence et Messageries de la Presse, SA, 1 rue de la Petite Ile, Brussels 7, Belgium.

U.S. Energy Research and Development Administration. NUCLEAR SCIENCE ABSTRACTS. Semimonthly, with annual cumulated index. Government Printing Office, Washington, D.C. 20402.

U.S. Nuclear Regulatory Commission. NUCLEAR REGULATORY COMMISSION ISSUANCES; OPINIONS AND DECISIONS OF THE NUCLEAR REGULATORY COMMISSION WITH SELECTED ORDERS. Washington, D.C.: Government Printing Office, 1976.

WASHINGTON ATOMIC ENERGY REPORT AND GUIDELETTER. Fortnightly. Box 389, Washington, D.C. 20044.

9. Reviewing Services

ADVANCES IN NUCLEAR PHYSICS. Irregular. New York: Plenum Publishing.

ADVANCES IN NUCLEAR SCIENCE AND TECHNOLOGY. Irregular. Edited by Ernest J. Henley and Jeffery Lewins. New York: Academic.

ANNUAL REVIEW OF NUCLEAR SCIENCE. Annual Reviews Inc., 4139 El Camino Way, Palo Alto, California 94306.

ISOTOPES AND RADIATION TECHNOLOGY. Quarterly. Division of Technical Information, Division of Isotopes Development, United States Atomic Energy Commission. Superintendent of Documents, Government Printing Office, Washington, D.C. 20402.

PROGRESS IN NUCLEAR ENERGY. Irregular. New York: Pergamon Press.

10. Document and Report Literature

a. LISTS

AEC- NASA TECHNICAL BRIEFS. Irregular. Washington, D.C.: Government Printing Office.

MONTHLY CATALOG. U.S. Superintendent of Documents, Government Printing Office, Washington, D.C. 20402.

U.S. GOVERNMENT REPORTS ANNOUNCEMENTS and U.S. GOVERNMENT REPORTS INDEX. Monthly. National Technical Information Service, Springfield, Virginia 22151.

b. CLEARINGHOUSES

National Energy Information Center. 1404 12th Street, N.W. Washington, D.C. 20461.

U.S. SUPERINTENDENT OF DOCUMENTS. Government Printing Office, Washington, D.C. 20402.

11. Sources of Information on Masters and Doctoral Theses

DISSERTATION ABSTRACTS INTERNATIONAL. Section B: "Physical Sciences and Technology." Monthly. University Microfilms, Xerox Co., 300 North Zeeb Road, Ann Arbor, Michigan 48106.

INTERNATIONAL MASTERS ABSTRACTS. Quarterly. University Microfilms, 300 North Zeeb Road, Ann Arbor, Michigan 48106.

MASTERS THESES IN THE PURE AND APPLIED SCIENCES, ACCEPTED BY COLLEGES AND UNIVERSITIES OF THE UNITED STATES. Annual. Lafayette, Indiana: School of Mechanical Engineering, Purdue University, Thermophysical Properties Research Center.

12. Sources of Information on Statistics and Costs

a. STATISTICS

i. Guides and Handbooks

AMERICAN STATISTICS INDEX. Annual. Part 1: Index. Part 2: Abstracts. Congressional Information Service, 600 Montgomery Building, Washington, D.C. 20014.

"Comprehensive guide and index to the statistical publications of the U.S. Government."

Canada. Dominion Bureau of Statistics. SERVICE BULLETIN. ENERGY STATISTICS. Annual. Douglas F. Heney, Dominion Bureau of Statistics, Tunney's Pasture, Ottawa 3, Ontario, Canada.

ii. Current Journals
Which Include Related Statistical Information

The following journals frequently contain statistics related to this field. See the heading Journals and Serial Publications in this Section for addresses.

ATOMIC ENERGY IN AUSTRALIA

ATOM - INFORMATIONEN

CANADIAN RESEARCH AND DEVELOPMENT

INTERNATIONAL ATOMIC ENERGY AGENCY. POWER RESEARCH REACTORS IN MEMBER STATES

NUCLEAR CANADA

SOVIET JOURNAL OF NUCLEAR PHYSICS

b. COSTS

International Atomic Energy Agency. NUCLEAR ENERGY COSTS AND ECONOMIC DEVELOPMENT. Proceedings of a Symposium on Nuclear Energy Costs and Economic Development. International Atomic Energy Agency in Istanbul, 20-24 October 1969. New York: Unipub, 1970.

13. Sources of Information on Standards, Specifications, and Patents

a. STANDARDS AND SPECIFICATIONS

AMERICAN NATIONAL STANDARDS INSTITUTE. CATALOG. Annual. 1430 Broadway, New York, New York 10018.

American Nuclear Society. STANDARDS. Hinsdale, Illinois: 1975.

COMPILATION OF NUCLEAR STANDARDS. 8th ed. Part I: "United States Activities," compiled by J. Paul Blakely, ORNL - NSIC - 98. Part II: "National (excluding U.S.) and International Activities," J. Paul Blakely, Chairman, ORNL-NSIC-102. Oak Ridge National Laboratory, National Safety Information Center, Oak Ridge, Tennessee 37830. 1971.

EXPERIENCE & TRENDS IN NUCLEAR LAW, INTER-REGIONAL TRAINING COURSE ON THE LEGAL ASPECTS OF NUCLEAR ENERGY. New York: Unipub, 1972.

NUCLEAR STANDARDS NEWS. Monthly. American Nuclear Society, 244 East Ogden Avenue, Hinsdale, Illinois 60521.

Struglia, Erasmus J. STANDARDS AND SPECIFICATIONS INFORMATION SOURCES. Detroit: Gale Research Company, 1965.

Coverage is both general and specific to nuclear engineering.

U.S. Department of Commerce. National Bureau of Standards. DIRECTORY OF U.S. STANDARDIZATION ACTIVITIES. Washington, D.C.: Government Printing Office, 1975.

Coverage is both general and specific to nuclear engineering.

Specifications, standards, codes, and regulations are available for individual states from the official state document printing offices and from local and municipal government agencies.

b. PATENTS

i. Guides and Handbooks

U.S. Atomic Energy Commission. PATENT LITERATURE ON PROCESS RADIA-TION AND IRRADIATOR DESIGN. Part I: United States patents 1950–68 (with list of references). By R.E. Greene. Springfield, Virginia: National Technical Information Service, 1972.

ii. Current Journals Which Include Patent Listings

The following journals frequently list patents. See the heading Journals and Serial Publications in this section for addresses.

EURO ABSTRACTS

INDUSTRIES ATOMIQUES

INFORMATION BULLETIN ON ISOTOPIC GENERATORS

NUCLEAR TECHNOLOGY

14. Sources of Information on
Translations, Meetings, and Proceedings of Symposia

TRANSATOM BULLETIN. Monthly. EURATOM-CID Agence et Messageries de la Presse, SA, 1 rue de la Petite Ile, Brussels 7, Belgium.

U.S. JOINT PUBLICATIONS RESEARCH SERVICE. Publishes translations of scientific and technical materials from Eastern Europe, Russia, and China. These are listed in the MONTHLY CATALOG of the Superintendent of Documents. Arlington, Virginia.

For sources of information on translations, meetings, and papers from conventions, see this subject in Section I of this Guide.

15. Selected Literature Sources on Pollution Control and Safety

a. POLLUTION CONTROL

ENVIRONMENTAL SURVEILLANCE IN THE VICINITY OF NUCLEAR FACILITIES. Symposium, Augusta, Georgia, January 24-26, 1968. Proceedings available from: Springfield, Illinois: C.C. Thomas, 1968.

International Atomic Energy Agency. ENVIRONMENTAL ASPECTS OF NUCLEAR POWER STATIONS. New York: 1971.

> "Proceedings of a symposium, New York, August 10-14, 1970 organized by IAEA in cooperation with USAEC. Major topics were nuclear power as an energy source; standards for the control of effluents; effluent control and monitoring; considerations affecting steam power station site selection and benefit-risk assessments."

_____. NUCLEAR TECHNIQUES IN ENVIRONMENTAL POLLUTION. New York: Unipub, 1971.

> Proceedings of a symposium on Use of Nuclear Techniques in the Measurement and Control of Environmental Pollution, held by the International Atomic Energy Agency in Salzburg, 26-30 October, 1970.

Linder, P. AIR FILTERS FOR USE AT NUCLEAR FACILITIES. Technical Reports Series no. 122. New York: Unipub, 1970.

NUCLEAR POWER AND THE ENVIRONMENT. IAEA. New York: Unipub, 1973.

> "Here is a useful timely book for the nonspecialist on environmental aspects of nuclear power stations. Prepared by IAEA and the World Health Organization (WHO), the book considers: role of atomic energy in meeting future power needs; radiation protection standards; safe handling of radioactive materials; plant construction, transmission lines, thermal discharges; public health considerations. Photographs, diagrams, and a bibliography enhance the value of this nontechnical look at energy and the environment."

Parker, F. PHYSICAL AND ENGINEERING ASPECTS OF THERMAL POLLUTION. London: Butterworth, 1970.

> Offers practical guidance on design of cooling water systems.

POLLUTION CONTROL CONFERENCE. Sponsored by U.S. Atomic Energy Commission, Oak Ridge, Tennessee. National Technical Information Service, Springfield, Virginia 22151. 1973.

RADIOACTIVE CONTAMINATION OF THE MARINE ENVIRONMENT. Symposium on the Interaction of Radioactive Contaminants with the Constituents of the Marine Environment. New York: Unipub, 1973.

RESEARCH AND DEVELOPMENT STUDIES IN ENVIRONMENTAL POLLUTION AND IN REACTOR COOLING SYSTEMS. 2 Symposia, Cleveland, Ohio, May 5-6, 1969 and Washington, D.C., November 19, 1969. Proceedings issued as: AIChE CHEMICAL ENGINEERING PROGRESS SYMPOSIA SERIES, no. 104, vol. 66 (1970).

THERMAL DISCHARGE: ENGINEERING & ECOLOGY. Thermo-fluids Conference. Edited by E.O. Storr. 1972. Institution of Engineers, Australia, 157 Gloucester Street, Sydney, N.S.W., 2000, Australia.

U.S. Environmental Protection Agency. RADIOACTIVE WASTE DISCHARGES TO ENVIRONMENT FROM NUCLEAR POWER FACILITIES ADDENDUM - 1 to BRH/DER 60-2 (with list of references). By J.E. Logsdon. Radiation Programs Office. Springfield, Virginia: National Technical Information Service, October 1971.

U.S. Nuclear Regulatory Commission. Office of Management Information and Program Control. RADIOACTIVE MATERIALS RELEASED FROM NUCLEAR POWER PLANTS, 1974. Rept. no. PB 254296. Springfield, Virginia: National Technical Information Service, 1976.

University of Connecticut. Institute of Water Resources. UNCONVENTIONAL NUCLEAR POWER PLANT SITING OPTIONS AND WATER RESOURCE IMPACTS. By C. Stern. Rept. no. PB248-460, Springfield, Virginia: National Technical Information Service, 1975.

USE OF LOCAL MINERALS IN THE TREATMENT OF RADIOACTIVE WASTE. New York: Unipub, 1972.

b. SAFETY

i. Guides and Handbooks

Fitzgerald, J. APPLIED RADIATION PROTECTION AND CONTROL. 2 vols. New York: Gordon, 1970.

Graham, J. FAST REACTOR SAFETY. Nuclear Science & Technology Series. New York: Academic, 1971.

International Atomic Energy Agency. GUIDELINES FOR THE LAYOUT AND CONTENTS OF SAFETY REPORTS FOR STATIONARY NUCLEAR POWER PLANTS. Safety Series no. 34. New York: Unipub, 1970.

_____. MANUAL ON SAFETY ASPECTS OF THE DESIGN AND EQUIPMENT OF HOT LABORATORIES. New York: Unipub, 1969.

_____. RADIATION DAMAGE IN REACTOR MATERIALS. 2 vols. New York: Unipub, 1969.

"Proceedings of the symposium on Radiation Damage in Reactor Materials, held by the International Atomic Energy Agency in Vienna, 2-6 June 1969. Contains the full text of 57 papers (6 presented in French and 4 in Russian), their discussions and a panel discussion. Covers the fundamental radiation damage mechanisms which are technologically important in terms of materials for advanced reactor systems."

_____. RADIATION SAFETY IN HOT FACILITIES. New York: Unipub, 1970.

"Proceedings of a symposium on radiation safety problems in the design and operation of hot facilities, organized by the International Atomic Energy Agency and held at Saclay, 13-17 October 1969."

_____. SAFEGUARDS TECHNIQUES. New York: Unipub, 1970.

"Proceedings of a symposium on progress in safeguards techniques organized by the International Atomic Energy Agency and held in Karlsruhe, 6-10 July 1970."

_____. SAFE OPERATION OF NUCLEAR POWER PLANTS. New York: Unipub, 1969.

Peck, Theodore. OCCUPATIONAL SAFETY AND HEALTH: A GUIDE TO INFORMATION SOURCES. Detroit: Gale Research Co., 1974.

"Physics of Nuclear Materials Safeguards," NUCLEAR TECHNOLOGY, vol. 14, no. 1, April 1972, pp. 1-112.

Proceedings of a symposium held at the annual meeting of the American Nuclear Society.

PREVENTING NUCLEAR THEFT: GUIDELINES FOR INDUSTRY & GOVERNMENT. Symposium on Implementing Nuclear Safeguards. Praeger Publishers, Inc., 111 Fourth Avenue, New York, New York 10003. 1971.

RADIATION PROTECTION STANDARDS: QUO VADIS. 6th Annual Midyear Topical Symposium on Radiation Protection Standards. Health Physics Society, P.O. Box 564, Richland, Washington 99352. 1971.

Robertson, J. GUIDE TO RADIATION PROTECTION. New York: Halstead, 1976.

Russell, C.R. REACTOR SAFEGUARDS. International Series of Monographs on Nuclear Energy. Series 11: Reactor Operational Problems, vol. 1. New York: Pergamon, 1963.

SAFEGUARDS, THE PRESS AND PUBLIC. Panel discussion at the 16th Annual Meeting. Institute of Nuclear Materials, New Orleans, Louisiana. 1975.

> Available as 16mm film or videotape cassette on loan basis from Dr. G. Robert Keepin, Institute of Nuclear Materials Management. Los Alamos Scientific Laboratory, Los Alamos, New Mexico 87545.

Safety standards are listed in: CATALOG. Annual. New York: American National Standards Institute.

Thompson, T. TECHNOLOGY OF NUCLEAR REACTOR SAFETY. Cambridge, Massachusetts: MIT Press, 1973.

U.S. Atomic Energy Commission. REACTOR SAFETY, LITERATURE SEARCH. Compiled by H.D. Raleigh. Springfield, Virginia: National Technical Information Service, June 1972.

_____. REVIEWS TO INSURE SAFETY OF NUCLEAR OPERATIONS. Nevada Operations Office. Springfield, Virginia: National Technical Information Service, 1971.

U.S. Congress Joint Committee on Atomic Energy. INVESTIGATION OF CHARGES RELATING TO NUCLEAR REACTOR SAFETY. 94th Congress, 2nd session. Washington, D.C.: Government Printing Office, 1976.

_____. NUCLEAR POWERPLANT SITING AND LICENSING LEGISLATION: HEARINGS BEFORE THE COMMITTEE. 94th Congress, 1st session. Washington, D.C.: Government Printing Office, 1976.

U.S. Department of Labor. Labor Statistics Bureau. BLS report. WORK INJURIES IN ATOMIC ENERGY, 1970, SURVEY OF PRIVATELY OWNED ESTABLISHMENTS, REPORT TO ATOMIC ENERGY COMMISSION. Washington, D.C.: Government Printing Office, 1972.

U.S. Nuclear Regulatory Commission. Office of Special Studies. NUCLEAR ENERGY CENTER SITE SURVEY--1975. Part I: Summary and Conclusions, . Rept. no. PB248 612. Part II: The U.S. Electric Power System and the Potential Role of Nuclear Energy Centers, Rept. no. PB248 613. Part III: Technical Considerations, Rept. no. PB248 614. Part IV: Practical Issues of Implementation, Rept. no. PB248 615. Part V: Resource Availability and Site Screening, Rept. no. PB248 616. Appendix A: United States Map Coarse Screening Results, Rept. no. PB248 617. Springfield, Virginia: National Technical Information Service, 1976.

ii. Abstracting and Indexing Services

EURO ABSTRACTS. Monthly. Luxembourg: Office for Official Publications of the European Communities.

OCCUPATIONAL SAFETY AND HEALTH ABSTRACTS. Monthly. International Occupational Safety and Health Information Center (CIS), International Labour Office, 1211 Geneva 22, Switzerland.

> Available in card format.

iii. Journals and Serial Publications

LEGAL GUIDE TO OCCUPATIONAL SAFETY AND HEALTH. Monthly. Clark Boardman Ltd., 435 Hudson Street, New York, New York 10014.

> Comment, interpretation, and advice on new legal developments in occupation safety and health.

NUCLEAR REGULATORY COMMISSION ISSUANCES. Monthly. Nuclear Regulatory Commission, Washington, D.C. 20555.

See also this subject in Section I of this Guide.

16. Management Information Sources

See this topic in Section I of this Guide.

17. Sources of Information on Laws and Legislation in This Field

a. FEDERAL

CODE OF FEDERAL REGULATIONS. Continuing series issued throughout the year. Available on a subscription basis from Government Printing Office, Washington, D.C. 20402.

FEDERAL REGISTER. Daily publication listing new regulations and revisions. Updates CODE OF FEDERAL REGULATIONS. Available on a subscription basis from Government Printing Office, Washington, D.C. 20402.

Library of Congress. DIGEST OF PUBLIC GENERAL BILLS AND REGULATIONS. Biweekly. Available on a subscription basis.

NUCLEAR REGULATORY COMMISSION ISSUANCES. Monthly. Nuclear Regulatory Commission, Washington, D.C. 20555.

UNITED STATES CODE. Washington, D.C.: Government Printing Office.

b. STATE

Library of Congress. MONTHLY CHECKLIST OF STATE PUBLICATIONS.

Available on a subscription basis from Government Printing Office, Washington, D.C. 20402.

c. LOCAL AND MUNICIPAL

Contact local government offices for codes, ordinances, and regulations.

All of these information source materials listed here are usually available in the reference department of large public libraries.

Specifications, standards, codes and regulations are available for individual states from the official state document printing offices and from local and municipal government agencies.

Section VII

PAPER AND PULP ENGINEERING

Section VII

PAPER AND PULP ENGINEERING

A. ORGANIZATIONS AND ASSOCIATIONS

1. Governmental Agencies

a. FEDERAL

The descriptive material provided in Section 1, Chemical Engineering --
General, in Part A on governmental information programs has a bearing on
pulp and paper engineering. Agencies listed there are actively involved in
matters relating to the different chemical industries and could be considered
as information sources for pulp and paper engineering processes as well.

DEPARTMENT OF AGRICULTURE - Division of Forest Products and Engineering
Research - Forest Service, 12th Street and Independence Avenue, S.W.,
Washington, D.C. 20250.

> Areas of interest: Forest products utilization; timber and wood
> quality; processes and product development; wood chemistry and
> fiber products; wood engineering; forest engineering, including
> timber harvesting systems.

> Publications: Technical reports, handbooks, monographs.

DEPARTMENT OF COMMERCE - National Bureau of Standards - Office of
Engineering Standards Services, Product Standards Section, Washington, D.C.
20234.

"The Products Section cooperates with and assists groups of producers, distributors, users and others interested in the development of voluntary standards for
various products."

Paper and pulp product standards have been developed through this program.

b. STATE AND MUNICIPAL

See this topic in Section I of this Guide.

2. Professional Societies, Trade Associations, and Other Scientific-Technological Related Organizations

AMERICAN FOREST INSTITUTE - 1835 K Street N.W., Washington, D.C. 20006.

Areas of interest: Forest management; conservation, including fire prevention, insects, and diseases; watersheds; forest industry lands; tree farms; logging; sawmills and pulp mills and their products; organic chemicals produced by wood distillation and hydrolysis plants; plywood and veneer mills.

Holdings: Booklets, slides, photos, maps, charts, forestry statistics, national tree farm records, reports.

Publications: BIBLIOGRAPHY OF TEACHING AIDS; news journal, proceedings, reports, state-of-the-art reviews.

Information services: Provides reference and consulting services; makes referrals; makes evaluation of school textbooks; lends films and slides to educational institutions, industry, authors, and publishers; permits onsite reference to forestry statistics and national tree farm records at the discretion of the Institute.

AMERICAN PAPER INSTITUTE, INC. - 260 Madison Avenue, New York, New York 10016.

Areas of interest: Pulp, paper, and paperboard industry, including such aspects as manufacturing, imports, international trade, economics, marketing services, governmental relations, public relations, financial management, industrial relations, recruitment of manpower, and transportation.

Publications: MONTHLY STATISTICAL SUMMARY; LABOR AND OPERATING REPORT (monthly and annual); CAPITAL AND INCOME SURVEY (quarterly and annual); PULP, PAPER AND PAPERBOARD CAPACITY SURVEY (annual); STATISTICS OF PAPER (with annual supplements); API REPORT; weekly production ratio report; weekly news release; OF PAPER AND OPPORTUNITY; GROW WITH AN EXCITING BUSINESS; HOW YOU CAN MAKE PAPER; HOW PAPER SERVES AMERICA; fact sheet.

Information services: Answers inquiries; makes referrals; permits onsite use of collection.

AMERICAN PULPWOOD ASSOCIATION - 605 Third Avenue, New York, New York 10016.

Areas of interest: Pulpwood, including technology, safety and training, statistics, forest management, legislation, harvesting, logging, transportation, storage, utilization, and conservation.

Publications: THE PULPWOOD ANNUAL; APA QUARTERLY; technical releases; papers, legislative bulletins, statistical reviews, training guides, film directory, pamphlets. A price list is available.

450

Information services: Answers inquiries; provides consulting services to members and subscribers.

NATIONAL COUNCIL OF THE PAPER INDUSTRY FOR AIR AND STREAM IMPROVEMENT, INC. - 103 Park Avenue, New York, New York 10017.

Areas of interest: Environmental control and pollution abatement by the pulp, paper, and paperboard industries, focused on and including air, water, and solids handling and disposal.

Publications: MONTHLY BULLETIN; REGULATORY REVIEW; TECHNICAL BULLETIN (irregular); annual report to members.

Information services: Answers inquiries.

NATIONAL FOREST PRODUCTS ASSOCIATION - 1619 Massachusetts Avenue N.W., Washington, D.C. 20036.

Areas of interest: Light frame and heavy timber construction; building code requirements; government specifications; fire insurance ratings; wood applications in building construction.

Publications: FOREST PRODUCTS NEWSLETTER (weekly); LUMBER AND WOOD PRODUCTS LITERATURE; technical reports; specifications; wood construction data.

Information services: Answers inquiries; provides consulting and reference services; permits onsite use of collection.

NORTHWEST PULP AND PAPER ASSOCIATION - 2633 Eastlake Avenue East, Seattle, Washington 98102.

Areas of interest: Pulp and paper industry of Oregon and Washington, including research, equipment, economics, and legislation; effect of pulp and paper industry operations on air and water in the Pacific Northwest.

Information services: Answers inquiries; makes referrals; provides bibliographic services.

PAPER INDUSTRY MANAGEMENT ASSOCIATION - 2570 Devon Avenue, Des Plaines, Illinois 60018.

Areas of interest: Management development; management techniques; training; pulp; paper; pulping; bleaching; paper stock preparation; web forming; mechanical water removal; paper drying; paper finishing; paper coating; paper converting; stream improvement; waste disposal; papermaking technology; industrial engineering; maintenance; production; marketing.

Publications: AMERICAN PAPER INDUSTRY (monthly); PAPER AND PULP MILL CATALOG AND ENGINEERING HANDBOOK (annual); PIMA DIRECTORY (annual).

Information services: Answers inquiries; prepares in-depth reviews of selected processes and management functions; organizes professional and training seminars.

SOCIETY OF AMERICAN FORESTERS - 1010 16th Street N.W., Washington, D.C. 20036.

Areas of interest: Forestry education; forest economics; management, engineering policy, products, and recreation; forest fires; forest entomology and pathology; wildlife and range management; watershed management; silviculture; history of professional forestry.

Holdings: Small reference collection of books, journals, and reports.

Publications: JOURNAL OF FORESTRY (monthly); FOREST SCIENCE (quarterly); proceedings, monographs, bulletins.

Information services: General inquiries are answered by the staff; technical inquiries are referred to member specialists. Onsite use of the literature collection is permitted.

SODA PULP MANUFACTURERS ASSOCIATION - 200 Park Avenue, New York, New York 10017.

An organization of pulp and paper companies manufacturing wood pulp by the soda process.

Affiliated with: American Paper Institute.

Information services: Answers inquiries; makes referrals.

SOUTHERN FOREST PRODUCTS ASSOCIATION - P.O. Box 52468, New Orleans, Louisiana 70150; LOCATION: 3525 North Causeway Boulevard, Metairie, Louisiana.

Areas of interest: Transportation, safety, production, trade, and research in the uses of Southern Pine lumber.

Publications: Technical literature on lumber uses.

Information services: Answers inquiries; provides consultation on proper specification of lumber for various uses; makes referrals. Services are available to architects, engineers, and other specifiers and users.

TECHNICAL ASSOCIATION OF THE PULP AND PAPER INDUSTRY - 360 Lexington Avenue, New York, New York 10017.

Areas of Interest: Pulp manufacture; paper and paperboard manufacture; coating and graphic arts; engineering; testing; research and development; corrugated containers; air and water pollution.

Publications: TAPPI (monthly); BIBLIOGRAPHY OF PAPERMAKING AND U.S. PATENTS (annual).

Information services: Answers inquiries; makes referrals.

3. Research Centers

THE INSTITUTE OF PAPER CHEMISTRY - P.O. Box 1048, Appleton, Wisconsin 54911; LOCATION: 1043 East South River Street, Appleton, Wisconsin.

Areas of interest: Paper chemistry and technology in the broadest sense; pulping; bleaching; refining; paper and paperboard converting; wood chemistry; cellulose and carbohydrate chemistry; lignin chemistry; fiber microscopy; plastics, textiles, and packaging applications; graphic arts; starch and other paper finish components; and instrumentation and testing.

Publications: THE ABSTRACT BULLETIN OF THE INSTITUTE OF PAPER CHEMISTRY (monthly); KEYWOOD SUPPLEMENT TO THE ABSTRACT BULLETIN in printed and computer tape form (monthly); SCAN; technical reports, a bibliographic series, academic catalog, microcard patent service, conference proceedings, informational pieces.

Information services: Answers inquiries; provides consulting, reference, literature-searching, bibliographic, translating, and duplication services; lends doctoral theses; permits onsite reference; conducts market research surveys within the pulp, paper, and related industries.

4. International Organizations

EUROPEAN CONFEDERATION OF THE PULP, PAPER AND FIBREBOARD INDUSTRY - 14 rue Decrayer, 1050 Brussels, Belgium.

EUROPEAN FEDERATION OF PARTICLE BOARD ASSOCIATIONS - Wilhelmstrasse 25, Postfach 23, 63 Giessen 1, Germany.

Aims: To look after the interests of members and facilitate relations between them; solve relevant problems of common concern; promote initiatives for the wider use of particle boards; submit unanimous decisions of member associations to international organizations and governments.

Publications: Annual reports, technical documents.

5. Other Information Sources and Programs

a. SPECIALIZED INFORMATION SOURCES

See Institute of Paper Chemistry entry in Part 3, Research Centers, of this Section.

B. LITERATURE

1. Guides and Handbooks

Britt, K., ed. HANDBOOK OF PULP AND PAPER TECHNOLOGY. 2nd ed. New York: Van Nostrand Reinhold, 1970.

> Covers recent advances in the paper and pulp industries.

Forbes, R. FORESTRY HANDBOOK. Edited for the Society of American Foresters. New York: Ronald, 1955.

> "Working methods and techniques, formulas, tables, converting factors, and related data most commonly used in forestry in the United States and Canada."

HANDBOOK OF THE NORTHERN WOOD INDUSTRIES, 1974-75: TIMBER, WOOD PULP, PAPER. 11th ed. New York: International Publishers Series, 1974.

Higham, R. A HANDBOOK OF PAPERBOARD AND BOARD; ITS MANUFAC- TURING TECHNOLOGY, CONVERSION, AND USAGE. London: Business Books, 1970. Order from International Publishing Service, New York.

PRACTICAL COMPUTER APPLICATIONS FOR THE PULP AND PAPER IN- DUSTRY. San Francisco: Miller Freeman Publications, 1975.

PULP AND PAPER INDUSTRY CORROSION PROBLEMS. International Semi- nar on Pulp and Paper Industry Corrosion Problems, 1974. Houston, Texas: National Association of Corrosion Engineers, 1975.

Rendle, B. WORLD TIMBERS. 3 vols. Toronto: University of Toronto Press, 1969-70.

Technical Association of the Pulp and Paper Industry. Pressing and Drying Committee. PAPER MACHINE WET PRESS MANUAL; A PROJECT OF THE PRESSING AND DRYING COMMITTEE. Edited by J.F. Atkins. Atlanta, Georgia: Atlanta Technical Association of the Pulp and Paper Industry, 1972.

_____. YEARBOOK. Annual. 360 Lexington Avenue, New York, New York 10017.

Weiner, J., et al. ENERGY PRODUCTION AND CONSUMPTION IN THE PAPER INDUSTRY. Appleton, Wisconsin: Institute of Paper Chemistry, 1974.

2. Dictionaries

a. TECHNICAL

Beekman, W. ELSEVIER'S WOOD DICTIONARY. 3 vols. New York: Elsevier, 1968.

THE DICTIONARY OF PAPER, INCLUDING PULPS, BOARDS, PAPER PROPERTIES AND RELATED PAPERMAKING TERMS. New York: American Paper and Pulp Association, 1940.

THESAURUS OF PULP AND PAPER TERMS. Pointe Claire, Canada: Pulp and Paper Research Institute of Canada, 1971.

b. FOREIGN LANGUAGE

Weck, J. DICTIONARY OF FORESTRY IN FIVE LANGUAGES: GERMAN, ENGLISH, FRENCH, SPANISH, RUSSIAN. Amsterdam and New York: Elsevier, 1966.

3. Directories

a. PRODUCTS AND MANUFACTURERS

APR - PAPIER - RUNDSCHAU. Annual. P. Keppler Verlag Kg, Industriestr. 2, D-6056 Heusenstamm, West Germany.

Directory of German and European paper industry.

INTERNATIONAL PULP AND PAPER DIRECTORY. Biennial. San Francisco: Miller Freeman Publishers.

THE KLINE GUIDE TO PAPER AND PULP INDUSTRY. 2nd ed. Fairfield, New Jersey: C.H. Kline & Co., 1973.

LOCKWOOD'S DIRECTORY OF THE PAPER AND ALLIED TRADES. Annual. Lockwood Publishing Co., 551 Fifth Avenue, New York, New York 10017.

PAPER AND PULP MILL CATALOG AND ENGINEERING HANDBOOK. Annual. Paper Industry Management Association, 2570 Devon Avenue, Des Plaines, Illinois 60018.

PAPER TECHNOLOGY YEARBOOK. Annual. London: British Paper and Board Makers Association.

PHILLIPS' PAPER TRADE DIRECTORY OF THE WORLD. Annual. S.C. Phillips and Co., Ltd., 50-51 Fetter Lane, London EC 4, England.

PIMA PAPER AND PULP MILL CATALOG. Paper Industry Management Association, 2570 Devon Avenue, Des Plaines, Illinois 60018.

POST'S PULP AND PAPER DIRECTORY, THE AUTHORITATIVE INDUSTRY GUIDE TO NORTH AMERICAN PULP AND PAPER MILLS - EQUIPMENT, GRADES, PRODUCTION, AND PERSONNEL. Annual. San Francisco: Miller Freeman Publishers.

PULP AND PAPER INDUSTRY. ANNUAL. Washington, D.C.: Organization for Economic Cooperation and Development.

PULP AND PAPER MAGAZINE OF CANADA'S REFERENCE MANUAL AND BUYERS' GUIDE. Annual. Southam Business Publications Ltd., Westmount, Quebec.

Contains buyers guide and technical flow sheets.

TAPPI YEARBOOK. Annual. Technical Association Pulp & Paper Industry, 360 Lexington Avenue, New York, New York 10017.

THOMAS' REGISTER OF MANUFACTURERS. Annual. New York: Thomas Publishing Co.

b. LABORATORIES AND RESEARCH CENTERS

EUROPEAN RESEARCH INDEX. 2nd ed. Guernsey, Channel Islands, England: F. Hodgson Ltd., 1969.

INDUSTRIAL RESEARCH LABORATORIES OF THE UNITED STATES. 13th ed. New York: R.R. Bowker, 1970.

RESEARCH CENTERS DIRECTORY. 5th ed. Detroit: Gale Research Co., 1975.

Updated by subscription service, NEW RESEARCH CENTERS.

4. Encyclopedias

See this topic in Section I of this Guide.

5. Bibliographies

a. SPECIFIC TO TOPIC

INSTITUTE OF PAPER CHEMISTRY BIBLIOGRAPHIC SERIES. Irregular. The Institute of Paper Chemistry, Box 1048, Appleton, Wisconsin 54910.

PAPERMAKING, CONVERTING, ALLIED SCIENCE AND TECHNOLOGY; BIBLIOGRAPHY AND PATENTS, 1968-1970. Atlanta: Technical Association of the Pulp and Paper Industry, 1972.

b. RELATED BIBLIOGRAPHIES IN CURRENT JOURNALS

The following journals frequently contain book reviews and/or bibliographies. See the heading Journals and Serial Publications in this section for addresses.

CANADIAN PULP AND PAPER INDUSTRY

INDIAN PULP AND PAPER

INTERNATIONAL PAPER BOARD INDUSTRY

JAPANESE TECHNICAL ASSOCIATION OF THE PULP AND PAPER INDUSTRY

PAPER AGE

PAPER-MAKER (ENGLAND)

PAPER TECHNOLOGY

PAPETIER DE FRANCE

PAPIER

PAPIER CARTON ET CELLULOSE

PAPIER UND DRUCK

PAPIRIPAR

PRINTING ABSTRACTS

PULP AND PAPER INTERNATIONAL

PULP AND PAPER MAGAZINE OF CANADA

TAPPI

WOCHENBLATT FUER PAPIERFABRIKATION

WORLD'S PAPER TRADE REVIEW

ZELLSTOFF UND PAPIER

6. Monographs, Reports, and Symposia Proceedings

a. GENERAL

Britt, K. HANDBOOK OF PULP AND PAPER TECHNOLOGY. 2nd ed. New York: Van Nostrand Reinhold, 1970.

THE EFFECT OF GROWTH ACCELERATION ON THE PROPERTIES OF WOOD. Symposium sponsored by Forest Products Laboratory, & API/TAPPI Research Liaison Committee, Madison, Wisconsin. Forest Products Laboratory, Forest Service, Madison, Wisconsin. 1972.

FOREST INDUSTRIES MARKETING. 7th Annual Conference, Eugene, Oregon, June 1970. Sponsored by University of Oregon Forest Industries Management Center. Proceedings issued as: SUCCESSFUL MARKETING AT HOME & ABROAD, Eugene, Oregon: University of Oregon Forest Industries Management Center, 1970.

Johnson, K. SYNTHETIC PAPER FROM SYNTHETIC FIBERS. Park Ridge, New Jersey: Noyes Data Corporation, 1971.

"This book provides a summary of the United States patent literature through 1970, relating to 74 processes for producing synthetic papers on conventional papermaking machinery."

McDonald, M. PAPER RECYCLING AND THE USE OF CHEMICALS. Park Ridge, New Jersey: Noyes Data Corp., 1971.

Based on United States patents.

PAPERMAKERS CONFERENCE, 1ST. Technical Association of the Pulp & Paper Co., 1 Dunwoody Park, Atlanta, Georgia 30341.

PAPERMAKING SYSTEMS AND THEIR CONTROL. 2 vols. London: Benn Publishing Co., 1970.

PULP & PAPER INDUSTRY TECHNICAL CONFERENCE, 20. Institute of Electrical & Electronics Engineers, 345 East 47th Street, New York, New York 10017. 1974.

PULP & PAPER INSTRUMENTATION SYMPOSIUM.. 13th International. Edited by R.W. Webster. Instrument Society of America, 400 Stanwix Street, Pittsburgh, Pennsylvania 15222. 1972.

PULP AND PAPER MANUFACTURE. 2nd ed. 3 vols. Edited by J. Stevenson. New York: McGraw-Hill, 1969.

"Deals with all processes and activities related to the manufacture of paper and paper goods."

Ranney, M. PAPER COATINGS BASED ON POLYMERS. Park Ridge, New Jersey: Noyes Data Corp., 1971.

Review of patent literature.

_____. SYNTHETIC PAPER FROM SYNTHETIC FIBERS. Park Ridge, New Jersey: Noyes Data Corp., 1971.

Compiled from details in recent patents.

RECENT ADVANCES IN TESTING AND PAPER SYNTHETICS. Joint Paper Synthetics Conference, Boston, Massachusetts, October, 1974. Proceedings. Atlanta: Technical Association of the Pulp and Paper Industry, 1974.

Roth, L. CONSTITUTION AND PULPING OF ASPEN AND POPLAR WOODS. Supplement 2 available from Institute of Paper Chemistry, Appleton, Wisconsin. 1972.

Scrimgeour, J., ed. INSTRUMENTATION IN THE PULP AND PAPER INDUSTRY. Irregular. Instrument Society of America, 400 Stanwix Street, Pittsburgh, Pennsylvania 15222.

THEORY AND DESIGN OF WOOD & FIBER COMPOSITE MATERIALS, Conference. Syracuse University Press, Box 8, University Station, Syracuse, New York 13210. 1972.

TRANSPORT AND HANDLING IN THE PULP AND PAPER INDUSTRY. 1st International Symposium, Rotterdam, Netherlands. Proceedings. San Francisco: Miller Freeman Co., 1975.

28TH ALKALINE PULPING CONFERENCE, SEATTLE, WASHINGTON. September, 1974. Proceedings. Atlanta: Technical Association of the Pulp and Paper Industry, 1976.

U.S. Department of Commerce. Washington, D.C. 20230.

> This agency regularly prepares industrial reports on particle board and pulp. For specific details see current and cumulative issues of MONTHLY CATALOG, Government Printing Office, Washington, D.C. 20402.

Ward, K., Jr. CHEMICAL MODIFICATION OF PAPERMAKING FIBERS. Fiber Science Series, vol. 4. New York: Dekker, 1972.

Weiner, J. COATING EQUIPMENT AND PROCESSES. Appleton, Wisconsin: The Institute of Paper Chemistry, 1970.

Wenzl, H. THE CHEMICAL TECHNOLOGY OF WOOD. Translated from German by F.E. Brauns and D.A. Brauns. New York: Academic, 1970.

> "Reviews the latest developments in research, techniques, and technical processes in the field of chemical-technological utilization of wood."

WOOD PRODUCTS CLINIC. 27th Annual Northwest. 1972. Northwest Wood Products Clinic, c/o Engineering Extension Service, Washington State University, Pullman, Washington 99163.

b. MARKET SURVEYS AND STUDIES

PAPER AND PULP DEMAND TO 1985. Predicasts, Inc., 200 University Circle Research Center, 11001 Cedar Avenue, Cleveland, Ohio 44106. 1974.

> "Projections of demand for 1975, 1980, and 1985 for 50 grades of pulp, paper and board. Includes detailed analysis and graphs for major grades."

WORLD PAPER AND BOARD. Predicasts, Inc., 200 University Circle Research Center, 11001 Cedar Avenue, Cleveland, Ohio 44106. 1974.

> "Historical data and projections to 1975 and 1980 for newsprint, printing papers, and other paper and board for 50 countries. World and regional supply-demand factors analyzed."

7. Abstracting and Indexing Services

a. GENERAL

APPLIED SCIENCE AND TECHNOLOGY INDEX

BRITISH TECHNOLOGY INDEX

CHEMICAL ABSTRACTS

ENGINEERING INDEX

See Section I of this Guide for further details on these services.

b. SPECIFIC TO TOPIC

FORESTRY ABSTRACTS. Quarterly. Compiled from world literature. Farnham Royal, England: Commonwealth Agricultural Bureaux. Annual index.

INSTITUTE OF PAPER CHEMISTRY ABSTRACT BULLETIN. Monthly. The Institute of Paper Chemistry, Box 1048, Appleton, Wisconsin 54911.

> Abstracts of world's scientific-technical literature on paper, pulp, wood science, and technology, and allied disciplines.

Muszaki lapszemle. FAIPAR, PAPIR-ES NYOMDAIPAR/TECHNICAL ABSTRACTS. WOODWORKING INDUSTRY. PAPER- & PRINTING INDUSTRY. Monthly. Orszagos Muszaki Konyvtar es Dokumentacios Kozpont, Reviczky u.6, Budapest, Hungary.

PREDI-BRIEFS. Abstracting service. Monthly. PULP AND PAPER. Predicasts, 200 University Circle Research Center, 11001 Cedar Avenue, Cleveland, Ohio 44106.

> "Covers woodpulp and related products, paper, paperboard and allied products, hardbound and particleboard."

PRINTING ABSTRACTS (England). Monthly. Randalls Road, Leatherhead, Surrey, England.

SEARCH: PULP AND PAPER DIVISION. Monthly. Compendium Publishers International Corp., 2175 Lemoine Avenue, Fort Lee, New Jersey 07024.

WOOD PRODUCTS INDUSTRY ABSTRACT BULLETIN. 6/year. Engineering Extension Service, College of Engineering, Washington State University, Pullman, Washington 99163.

c. RELATED ABSTRACTS IN CURRENT JOURNALS

The following journals frequently contain abstracts of related literature. See the heading Journals and Serial Publications in this section for addresses.

AUXILIAIRE

CANADIAN PULP AND PAPER INDUSTRY

JAPANESE TECHNICAL ASSOCIATION OF THE PULP AND PAPER INDUSTRY

PAPER-MAKER (ENGLAND)

PAPER TECHNOLOGY

PAPIER CARTON ET CELLULOSE

PAPIER UND DRUCK

PAPIRIPAR

WOCHENBLATT FUER PAPIERFABRIKATION

ZELLSTOFF UND PAPIER

8. Journals and Serial Publications

a. DIRECTORIES AND LISTS

AYER'S DIRECTORY OF PUBLICATION. Annual. Philadelphia, Pennsylvania: Ayer Press.

STANDARD PERIODICAL DIRECTORY. Irregular. New York: Oxbridge Publishing Co.

ULRICH'S INTERNATIONAL PERIODICAL DIRECTORY. Annual. 2 vols. New York: R.R. Bowker Co.

b. JOURNAL AND SERIAL TITLES

ABSTRACT BULLETIN OF THE INSTITUTE OF PAPER CHEMISTRY. Monthly. Box 1048, Appleton, Wisconsin 54911.

> Abstracts of world's scientific-technical literature on paper, pulp, wood science and technology, and allied disciplines.

AMERICAN PAPER INDUSTRY. Monthly. Paper Industry Management Association, 2570 Devon Avenue, Des Plaines, Illinois 60018.

AMERICAN PAPER INSTITUTE MONTHLY STATISTICAL SUMMARY. Monthly. American Paper Institute, 260 Madison Avenue, New York, New York 10016.

AMERICAN PAPER MERCHANT. Monthly. Peacock Business Press, Inc., 200 South Prospect Avenue, Park Ridge, Illinois 60068.

ANNUAL REVIEW OF LIGNIN CHEMISTRY. Institute of Paper Chemistry, Box 1048, Appleton, Wisconsin 54911.

> Supplement to the ABSTRACT BULLETIN.

AUXILIARE; DES FABRICANTS DE CARTONNAGES, TRANSFORMATEURS DE PAPIER. Monthly. Industries et Arts Graphiques, Editions Technorama, 6 rue de Bizerte, Paris (17e), France.

Canadian Pulp and Paper Association. PULP AND PAPER REPORT. Irregular. Canadian Pulp and Paper Association, 2300 Sun Life Building, Montreal 110, Quebec, Canada.

Canadian Pulp and Paper Association. Technical Section. ENGINEERING DATA SHEETS. Annual. Canadian Pulp and Paper Association, Technical Section, 2300 Sun Life Building, Montreal 110, Quebec, Canada.

_____. PROCEEDINGS. Annual. Technical Section, Canadian Pulp and Paper Association, 2300 Sun Life Building, Montreal 110, Quebec, Canada.

_____. USEFUL METHODS. Annual. Canadian Pulp and Paper Association, Technical Section, 2300 Sun Life Building, Montreal 110, Quebec, Canada.

CANADIAN PULP AND PAPER INDUSTRY. Monthly. Maclean-Hunter Publishing Co., Ltd., 481 University Avenue, Toronto, Ontario, Canada.

CHEM 26/PAPER PROCESSING. Monthly. A. Stewart Hale Publishing Co., 1 Bank Street, Stamford, Connecticut 06901.

FOREST INDUSTRIES. Monthly. 500 Howard Street, San Francisco, California 94105.

Covers forestry, logging, lumber, plywood, pulpwood industries.

FORESTRY ABSTRACTS. Quarterly with annual index. Farnham Royal, England: Commonwealth Agricultural Bureaux.

Compiled from world literature.

INDIAN PULP AND PAPER. Monthly. 14 Radha Bazar Lane, Calcutta 1, India.

INTERNATIONAL PAPER BOARD INDUSTRY. Monthly. Bottler and Packer Ltd., Binsted House, Devonshire Close, Devonshire Street, London, England.

JAPANESE TECHNICAL ASSOCIATION OF THE PULP AND PAPER INDUSTRY JOURNAL. Monthly. Japanese Technical Association of the Pulp and Paper Industry, 9-11-3 chome, Chuo-ku, Tokyo, Japan.

Text in Japanese; summaries in English.

MODERN CONVERTER; THE BUSINESSPAPER OF FOIL, FILM AND PAPER CONVERSION. Semimonthly. Harcourt Brace Jovanovich Publications, 757 Third Avenue, New York, New York 10017.

Muszaki lapszemle. FAIPAR, PAPIR-ES NYOMDAIPAR/TECHNICAL AB-STRACTS. WOODWORKING INDUSTRY. PAPER- & PRINTING INDUSTRY. Monthly. Orszagos Muszaki Konyvtar es Dokumentacios Kozpont, Reviczky u.6, Budapest, Hungary.

PAPER AGE. Monthly. Walden-Mott Corp., 466 Kinderkamack Road, Oradell, New Jersey 07649.

PAPER AND PACKAGING BULLETIN. Quarterly. Lincoln House, 60 East 42nd Street, New York, New York 10017.

PAPER AND TWINE JOURNAL. Monthly. 1860 Broadway, New York, New York 10023.

The national magazine for merchants and salesmen.

PAPER FACTS AND FIGURES. Bimonthly. Corinthian Press Ltd., Elm House, 10-16 Elm Street, London, WC 1, England.

PAPER-MAKER. Monthly. Phillips and Company, Ltd., 50 Fetter Lane, London EC 4, England.

PAPER SALES. Monthly. Harcourt Brace Jovanovich, 1 East First Street, Duluth, Minnesota 55802.

PAPER TECHNOLOGY. Bimonthly. Technical Section, British Paper and Board Makers Association, Plough Place, Fetter Lane, London, EC 4, England.

PAPER TRADE JOURNAL. Weekly. Lockwood Trade Journal Co., 551 Fifth Avenue, New York, New York 10017.

PAPETIER DE FRANCE; REVUE PROFESSIONNELLE DES PAPETIERSDETAIL-LANTS ET SPECIALISTES DU STYLO. Monthly. 20 bis avenue des 2 routes, Avignon, France.

PAPIER CARTON ET CELLULOSE. 7 issues/year. Compagnie Francaise d'Editions, 40 rue du Colisee, Paris (8e), France.

Text in English, French, German, and Spanish.

PAPIER UND DRUCK; FACHZEITSCHRIFT FUER TYPOGRAFIE, POLYGRA-FISCHE TECHNIK UND PAPIERVERARBEITUNG. Monthly. VEB Fachbuch-verlag, Karl-Heine-Str. 16, 7031 Leipzig, East Germany.

Text in German; index in English and Russian.

PAPIER; ZEITSCHRIFT FUER DIE ERZEUGUNG VON HOLZTOFF, ZELLSTOFF, PAPIER UND PAPPE, CHEMISCHE TECHNOLOGIE DER CELLULOSE. Monthly. Eduard Roether Verlag, Berliner Allee 56, 6100 Darmstadt, West Germany. Supplement: PAPIERGESCHICHTE.

Summaries in English and French.

PAPIRIPAR. Bimonthly. Szabadsag ter 17, Budapest V., Hungary.

Text in Hungarian; summaries in German and Russian.

PREDI-BRIEFS. Abstracting service. Monthly. PULP AND PAPER. Predicasts, 200 University Circle Research Center, 11001 Cedar Avenue, Cleveland, Ohio 44106.

Covers woodpulp and related products, paper, paperboard and allied products, hardboard and particle-board.

PRINTING ABSTRACTS. Monthly. Randalls Road, Leatherhead, Surrey, England.

PULP AND PAPER. Monthly. Miller Freeman Publications, 400 Howard Street, San Francisco, California 94105.

PULP AND PAPER CANADA. Monthly. National Business Publications, 310 Victoria Avenue, Westmount, Quebec, H3Z 2M9 Canada.

Communication source for Canada's pulp and paper industry.

PULP AND PAPER INDUSTRY. Technical Conference. RECORD. Annual. Institute of Electrical and Electronics Engineers, Inc., 345 East 47th Street, New York, New York 10017.

PULP AND PAPER INTERNATIONAL. Monthly. Miller Freeman Publications, 123A Chaussee de Charleroi, 1060 Brussels, Belgium.

Pulp and Paper Research Institute of Canada. ANNUAL REPORT. Pulp and Paper Research Institute of Canada, 570 St. John's Road, Pointe Claire, Quebec, H9R 3JR Canada.

_____. CONTRIBUTIONS. Irregular. Pulp and Paper Research Institute of Canada, 570 St. John's Boulevard, Pointe Claire, Quebec, H9R 3JR Canada.

_____. TECHNICAL REPORTS. Irregular. Pulp and Paper Research Institute of Canada, 570 St. John's Boulevard, Pointe Claire, Quebec, H9R 3JR Canada.

PULP, PAPER AND BOARD QUARTERLY INDUSTRY REPORT. Quarterly. Bureau of Domestic Commerce, U.S. Department of Commerce, Washington, D.C. 20230.

REVUE EUROPEENNE DES PAPIERS CARTONS-COMPLEXES; REVUE PROFESIONNELLE D'IMPRESSION, TRANSFORMATION COMMERCE DU PAPIER. Semimonthly. Editions Tructidor, 20 bis. av. des 2 routes, Avignon 84, France.

SEARCH: PULP AND PAPER DIVISION. Monthly. Compendium Publishers International Corp., 2175 Lemoine Avenue, Fort Lee, New Jersey 07024.

SOUTHERN PULP AND PAPER MANUFACTURER. Monthly. Ernest H. Abernethy Publishing Co., Inc., 75 Third Street, N.W., Atlanta, Georgia 30308.

TAPPI. Monthly. Technical Association of the Pulp and Paper Industry, 360 Lexington Avenue, New York, New York 10017.

TREND (CANADA). 3 issues/year. Pulp and Paper Research Institute of Canada, 570 St. Johns Boulevard, Pointe Claire, Quebec, H9R 3JR Canada.

U.S. Department of Commerce. CURRENT INDUSTRIAL REPORTS. Monthly. Social and Economic Statistics Administration, Census Bureau, Government Printing Office, Washington, D.C. 20402.

    Includes reports on paper and pulp industry.

_____. PAPER, SUB-COUNCIL REPORT. Irregular. Government Printing Office, Washington, D.C. 20402.

_____. PULP, PAPER, AND BOARD, QUARTERLY INDUSTRY REPORT. Government Printing Office, Washington, D.C. 20402.

WOCHENBLATT FUER PAPIERFABRIKATION; FACHZEITSCHRIFT FUER DIE PAPIER-, PAPPEN- U. PAPIERSTOFF-INDUSTRIE. (Papiermacher-Berufsgenossenschaft Mainz). Semimonthly. Guentter-Staib Verlag, 795 Biberach an der Riss, Wuerttemberg, West Germany.

WOOD PRODUCTS INDUSTRY ABSTRACT BULLETIN. 6 times/year. Engineering Extension Service, College of Engineering, Washington State University, Pullman, Washington 99163.

WORLD'S PAPER TRADE REVIEW; FOR THE PULP, PAPER, CONVERTING AND ASSOCIATED ENGINEERING INDUSTRIES. Weekly. Stonehill and Gillis Ltd., Lyon Tower, High Street, Colliers Wood, London SW 19, England.

YOSHI RYUTSU TOKEI GEPPO. Monthly. Ministry of International Trade and Industry, Japan Paper Trade Association, Kami-Parupu Kaikan Building, Ginza 3-9-11, Chuoku, Tokyo, Japan.

ZELLSTOFF UND PAPIER. Monthly. Fachbuchverlag, Karl-Heine-Str. 16, 7031 Leipzig, East Germany.

    Text in German; summaries in English and Russian.

9. Reviewing Services

ANNUAL REVIEW OF LIGNIN CHEMISTRY. Supplement to the ABSTRACT BULLETIN. Institute of Paper Chemistry, Box 1048, Appleton, Wisconsin 54911.

10. Document and Report Literature

a. LISTS

MONTHLY CATALOG. U.S. Superintendent of Documents, Government Printing Office, Washington, D.C. 20402.

U.S. GOVERNMENT REPORTS ANNOUNCEMENTS and U.S. GOVERNMENT REPORTS INDEX. National Technical Information Service, Springfield, Virginia 22151.

b. CLEARINGHOUSES

Institute of Paper Chemistry. Box 1048, Appleton, Wisconsin 54911.

National Technical Information Service. Springfield, Virginia 22151.

U.S. Superintendent of Documents. Government Printing Office, Washington, D.C. 20402.

11. Sources of Information on Masters and Doctoral Theses

DISSERTATION ABSTRACTS INTERNATIONAL. Section B: "Physical Sciences and Technology." University Microfilms, Xerox Co., 300 North Zeeb Road, Ann Arbor, Michigan 48106.

INTERNATIONAL MASTERS ABSTRACTS. Quarterly. University Microfilms, 300 North Zeeb Road, Ann Arbor, Michigan 48106.

MASTERS THESES IN THE PURE AND APPLIED SCIENCES, ACCEPTED BY COLLEGES AND UNIVERSITIES OF THE UNITED STATES. Annual. Lafayette, Indiana: School of Mechanical Engineering, Purdue University, Thermophysical Properties Research Center.

The journal TAPPI regularly lists doctoral theses in this field.

12. Sources of Information on Statistics and Costs

a. STATISTICS

i. Journals and Serial Publications

American Paper Institute, Inc. MONTHLY STATISTICAL SUMMARY. Monthly. American Paper Institute, 260 Madison Avenue, New York, New York 10016.

AMERICAN STATISTICS INDEX. Annual. Part 1: Index. Part 2: Abstracts. Congressional Information Service, 600 Montgomery Building, Washington, D.C. 20014.

> Comprehensive guide and index to the statistical publications of the United States government.

McClenahan, W. "Consumption of Paper Stock by United States in 1969 and 1970." TAPPI, vol. 55, November 1972, pp. 1605–8.

PAPER FACTS AND FIGURES. Bimonthly. Corinthian Press Ltd., Elm House, 10-16 Elm Street, London, WC 1, England.

POSTS PULP AND PAPER DIRECTORY. Annual. 500 Howard Street, San Francisco, California 94105.

  Contains buyers guide, statistics and guide to the industry.

PULP, PAPER AND BOARD QUARTERLY INDUSTRY REPORT. Quarterly. Bureau of Domestic Commerce, United States Department of Commerce, Washington, D.C. 20230.

ii. Current Journals
Which Include Related Statistical Information

The following journals frequently contain statistics related to this field. See the heading Journals and Serial Publications in this section for addresses.

AMERICAN PAPER MERCHANT

AUXILIAIRE

CANADIAN PULP AND PAPER INDUSTRY

PAPER BULLETIN

PAPER-MAKER (ENGLAND)

PAPIER

PAPIER CARTON ET CELLULOSE

PULP AND PAPER

PULP AND PAPER INTERNATIONAL

PULP PAPER MAGAZINE OF CANADA

SEARCH: PULP AND PAPER DIVISION

U.S. Department of Commerce. CURRENT INDUSTRIAL REPORTS

_____. PULP, PAPER, AND BOARD QUARTERLY INDUSTRY REPORT

WOCHENBLATT FUER PAPIERFABRIKATION

WORLD'S PAPER TRADE REVIEW

b. COSTS

Perkins, J. "Cost Reduction in Sulfate Pulp Bleach Plant Design." TAPPI, vol. 55, October 1973, pp. 1494-97.

Rich, S. "Pricing Strategy Considerations for Pulp and Paper Producers." TAPPI, vol. 53, July 1970, pp. 1282-85.

Wills, C. "Applying Increment Costs in Evaluating Projects." TAPPI, vol. 54, February 1971, pp. 212-14.

WORLD-PRODUCT COSTS. Quarterly. Predicasts, Inc., 200 University Circle Research Center, 11001 Cedar Avenue, Cleveland, Ohio 44108.

> A subscription service of market forecasts by product in areas of agriculture, wood and paper, food, polymers, chemicals, primary metals, etc.

13. Sources of Information on Standards, Specifications, and Patents

a. STANDARDS AND SPECIFICATIONS

AMERICAN NATIONAL STANDARDS INSTITUTE CATALOG. Annual. The Institute, 1430 Broadway, New York, New York 10018.

> Lists standards in this subject area. Catalog and standards are available from the Institute.

Struglia, Erasmus J. STANDARDS AND SPECIFICATIONS INFORMATION SOURCES. Detroit: Gale Research Co., 1965.

> Coverage is both general and specific to paper and pulp engineering.

U.S. Department of Commerce. National Bureau of Standards. DIRECTORY OF U.S. STANDARDIZATION ACTIVITIES. Washington, D.C.: Government Printing Office, 1975.

> Coverage is both general and specific to paper and pulp engineering.

Specifications, standards, codes and regulations are available for individual states from the official state document printing offices and from local and municipal government agencies.

b. PATENTS AND TRADEMARKS

Gutcho, M. SYNTHETIC PAPER FROM FIBERS AND FILMS. Park Ridge, New Jersey: Noyes Data Corp., 1975.

Halpern, M. PULP MILL PROCESSES. Park Ridge, New Jersey: Noyes Data Corp., 1975.

_____. SYNTHETIC PAPER FROM FIBERS AND FILMS. Park Ridge, New Jersey: Noyes Data Corp., 1976.

PAPERMAKING, CONVERTING, ALLIED SCIENCE AND TECHNOLOGY; BIBLI-

OGRAPHY AND PATENTS, 1968-70. Atlanta: Technical Association of the Pulp and Paper Industry, 1972.

PULP ERA. Bimonthly. Pulp Era Press, 413 Ottokee Street, Wanson, Ohio 43567.

Patents in pulp and paper making.

See also this subject in the section Chemical Engineering - General.

The following journals frequently list patents. See the heading Journals and Serial Publications in this section for addresses.

AUXILIAIRE

CANADIAN PULP AND PAPER INDUSTRY

MODERN CONVERTER

PAPER-MAKER (ENGLAND)

PAPETIER DE FRANCE

PAPIER UND DRUCK

PAPIER

TAPPI

WOCHENBLATT FUER PAPIERFABRIKATION

WORLD'S PAPER TRADE REVIEW

ZELLSTOFF UND PAPIER

14. Sources of Information on
Translations, Meetings, and Proceedings of Symposia

U.S. Joint Publications Research Service. Arlington, Virginia. Publishes translations of scientific and technical materials from Eastern Europe, Russia, and China. These are listed in the MONTHLY CATALOG of the Superintendent of Documents.

See also these subjects in Section I of this Guide.

15. Selected Literature Sources on
Pollution Control and Safety

a. POLLUTION CONTROL

i. Air Pollution

AIR POLLUTION IN THE PULP AND PAPER INDUSTRY. 1969. SUPPLEMENT I. 1970. SUPPLEMENT II. 1972. Irregular. Institute of Paper Chemistry, Box 1048, Appleton, Wisconsin 54911.

ii. Water Pollution

ENVIRONMENTAL CONFERENCE, 9th. TAPPI, 1 Dunwoody Park, Atlanta, Georgia 80341. 1972.

Environmental Science and Engineering. FIELD SURVEILLANCE AND ENFORCE-MENT GUIDE: WOOD PULPING INDUSTRY. Rept. no. 246 753. Springfield, Virginia: National Technical Information Service, 1975.

Gramenski, E. "Problems and Potentials in Paper Recycling." PROCEEDINGS OF THE NATIONAL MATERIALS CONSERVATION SYMPOSIUM ON RESOURCE RECOVERY AND UTILIZATION. Special Technical Publication, vol. 592, pp. 132-39. Symposium held 29 April - 1 May 1974, Gaithersburg, Maryland. Philadelphia, Pennsylvania: American Society for Testing and Materials, 1975.

Jones, H. POLLUTION CONTROL AND CHEMICAL RECOVERY IN THE PULP AND PAPER INDUSTRY. Park Ridge, New Jersey: Noyes Data Corp., 1973.

> "In this book are condensed vital data from government sources of information that are scattered and difficult to pull together. Important processes are interpreted and explained by examples from 54 United States patents."

McDonald, M. PAPER RECYCLING AND THE USE OF CHEMICALS. Park Ridge, New Jersey: Noyes Data Corp., 1971.

> "A guide to the chemical processes in the recycling of paper which is taken from the 1952-1971 U.S. patent literature relating to paper recycling and the use of chemicals."

National Bureau of Economic Research. WATER POLLUTION CONTROL ACT OF 1972. ECONOMIC IMPACTS. PULP AND PAPER INDUSTRY. By R. Leone, et al. Rept. no. PB 248 801. Springfield, Virginia: National Technical Information Service, 1975.

REVIEW OF THE LITERATURE ON PULP AND PAPER WASTES. Irregular. National Council of the Paper Industry for Air and Stream Improvement, Inc., 260 Madison Avenue, New York, New York 10016.

SULFITE PULPING & RECOVERY CONFERENCE, INTERNATIONAL. TAPPI, 1 Dunwoody Park, Atlanta, Georgia 30341. 1972.

TOXIC ORGANIC MATERIALS IN INTEGRATED MILL EFFLUENTS. Appleton, Wisconsin: Institute of Paper Chemistry, 1970.

> Special annotated bibliography.

U.S. Atomic Energy Commission. ROLE OF PAPER MILL ADDITIVES AS POTENTIAL STREAM POLLUTANTS, DEVELOPMENT OF NUCLEAR ANALYTICAL TECHNIQUES, ANNUAL REPORT, JUNE 1970 - JUNE 1, 1971. By R. Milton Chatters. Springfield, Virginia: National Technical Information Service, 1971.

> Includes bibliography.

U.S. Environmental Protection Agency. AERATED LAGOON TREATMENT OF SULFITE PULPING EFFLUENTS. By H.R. Amberg. Washington, D.C.: Government Printing Office, 1971.

Includes list of references.

_____. COLOR REMOVAL FROM KRAFT PULPING EFFLUENT BY LIME ADDITION. Washington, D.C.: Government Printing Office, 1971.

Includes list of references.

_____. COLOR REMOVAL FROM KRAFT PULP MILL EFFLUENTS BY MASSIVE LIME TREATMENT. By J.L. Oswalt. Washington, D.C.: Government Printing Office, 1973.

Includes list of references.

_____. COMBINED TREATMENT OF MUNICIPAL KRAFT LINERBOARD, AND FIBERBOARD MANUFACTURING WASTES. Washington, D.C.: Government Printing Office,

Includes list of references.

_____. JOINT TREATMENT OF MUNICIPAL SEWAGE AND PULP MILL EFFLUENTS. By K.G. Voelkel. Washington, D.C.: Government Printing Office, 1972.

Includes a list of references.

_____. KRAFT PULPING EFFLUENT TREATMENT AND REFUSE, STATE OF ART. By W.G. Timpe. Washington, D.C.: Government Printing Office, 1973.

Includes list of references.

_____. MULTI-SYSTEM BIOLOGICAL TREATMENT OF BLEACHED KRAFT EFFLUENTS. Washington, D.C.: Government Printing Office, 1971.

Includes list of references.

_____. RECYCLE OF PAPERMILL WASTE WATERS AND APPLICATION TO REVERSE OSMOSIS. By D.C. Morris. Washington, D.C.: Government Printing Office, 1972.

_____. REVERSE OSMOSIS CONCENTRATION OF DILUTE PULP AND PAPER EFFLUENTS. By A.J. Wiley. Washington, D.C.: Government Printing Office, 1972.

Includes list of references.

_____. ROLE OF PAPER MILL ADDITIVES AS POTENTIAL STREAM POLLUTANTS, DEVELOPMENT OF NUCLEAR ANALYTICAL TECHNIQUES, ANNUAL REPORT, JUNE 1970-JUNE 1, 1971. By R.M. Chatters. Springfield, Virginia: National Technical Information Service, 1971.

Includes bibliography.

_____. SLIME GROWTH EVALUATION OF TREATED PULP MILL WASTE. By A.W. Anderson. Washington, D.C.: Government Printing Office, 1971.

Includes list of references.

_____. SLUDGE MATERIAL RECOVERY SYSTEM FOR MANUFACTURERS OF PIGMENTED PAPERS. By R.P. Labrecque. Washington, D.C.: Government Printing Office, 1971.

_____. STATE OF THE ART REVIEW OF PULP AND PAPER WASTE TREATMENT. By H. Gehm. Washington, D.C.: Government Printing Office, 1973.

Includes list of references.

_____. TREATMENT OF SELECTED INTERNAL KRAFT MILL WASTES IN COOLING TOWER. By J.A. McAlister. Washington, D.C.: Government Printing Office, 1971.

Includes list of references.

_____. WASTE TREATMENT RESEARCH NEWSLETTER, PAPER AND FOREST INDUSTRIES, FOOD PROCESSING. Washington, D.C.: Government Printing Office, 1971.

Water pollution control research series. Includes bibliographies.

U.S. National Industrial Pollution Control Council. PAPER AND WOOD PACKAGING IN SOLID WASTE, SUB-COUNCIL REPORT. Washington, D.C.: Government Printing Office, 1972.

WATER AND AIR POLLUTION NEWSLETTER; FOR THE PAPER AND PULP INDUSTRY. Monthly. International Executive Newsletters Co., 52 rue du Progres, 1000 Brussels, Belgium and 35 West Elm Street, Littleton, New Hampshire 03561.

WATER POLLUTION IN THE PULP AND PAPER INDUSTRY. Irregular series. 251. I. General and In-plant Modifications; 252. II. Testing Methods; 253. III. Stream Response; 254. IV. Treatment Methods. A. Physical Operations; 255. IV. Treatment Methods. B. Biological Systems; 256. IV. Treatment Methods. C. Chemical Processes, General; Observations, and Reports. Institute of Paper Chemistry, P.O. Box 1048, Appleton, Wisconsin 54911.

b. SAFETY

AMERICAN NATIONAL STANDARDS INSTITUTE CATALOG. Annual. New York: American National Standards Institute.

LEGAL GUIDE TO OCCUPATIONAL SAFETY AND HEALTH. Monthly. Clark Boardman Ltd., 435 Hudson Street, New York, New York 10014.

Comment, interpretation and advice on new legal developments in occupational safety and health.

Peck, Theodore. OCCUPATIONAL SAFETY AND HEALTH: A GUIDE TO INFORMATION SOURCES. Detroit: Gale Research Co., 1974.

PULP AND PAPER SECTION SAFETY NEWSLETTER. Monthly. National Safety Council, 425 North Michigan Avenue, Chicago, Illinois 60611.

Safety code for paper and pulp mills.

16. Management Information Sources

See this topic in Section I of this Guide.

17. Sources of Information on
Laws and Legislation in This Field

a. FEDERAL

CODE OF FEDERAL REGULATIONS. Continuing series issues throughout the year. Available on a subscription basis from Government Printing Office, Washington, D.C. 20402.

FEDERAL REGISTER. Daily publication listing new regulations and revisions. Updates CODE OF FEDERAL REGULATIONS. Available on a subscription basis from Government Printing Office, Washington, D.C. 20402.

Library of Congress. DIGEST OF PUBLIC GENERAL BILLS AND REGULATIONS. Available on subscription basis from Government Printing Office, Washington, D.C. 20402.

UNITED STATES CODE, 1970 edition and supplements. Government Printing Office, Washington, D.C. 20402.

b. STATE

Library of Congress. MONTHLY CHECKLIST OF STATE PUBLICATIONS. Available on a subscription basis from Government Printing Office, Washington, D.C. 20402.

c. LOCAL AND MUNICIPAL

Contact local government offices for codes, ordinances, and regulations.

All of these information source materials listed here are usually available in the reference department of large public libraries.

PETROLEUM ENGINEERING

A. ORGANIZATIONS AND ASSOCIATIONS

1. Governmental Agencies

a. FEDERAL

DEPARTMENT OF THE INTERIOR

A. Bureau of Mines - Washington, D.C. 20240.

Objectives: The Bureau of Mines conducts research and administers regulatory programs necessary for performance of the governmental function to stimulate the private sector toward the production of an appropriate and substantial share of the national mineral and fuel needs in a manner that best protects the public interest. Specifically, concern is directed toward the satisfaction of current and emerging needs; the real cost of such achievements; the assessment of related social-economic factors; minimization of occupational hazards to workers; reduction of wastes; and assurance that mineral raw materials are supplied and mineral-based products are used and disposed of without objectionable social and environmental cost. To accomplish these objectives, the Bureau performs research, provides information to the public, conducts inquiries, inspects mines, and administers laws pertinent to the extraction, processing, use, reuse, and disposal of minerals and mineral fuels.

Functions include surveillance and evaluations of the industrial and commercial outlook for minerals and fuel deposits; studies to determine the relationship of mineral supply, demand and technology to the national and world economy; studies and projects concerning the relationship of the mineral industries to environmental problems; collection, evaluation, and publication of mineral industry statistics; and conducting engineering studies regarding effective mining practices. Also included are research programs concerning extraction, processing, use, and disposal of minerals and mineral fuels; helium production; and research on mine health and safety.

1.   Office of Mineral Information - U.S. Bureau of Mines, 18th and C Streets N.W., Washington, D.C.   20240.

> Areas of interest: Mining; minerals; coal; petroleum; environmental problems linked to extraction, processing, and use of minerals and fuels; mine safety; natural gas; helium; metallurgy; statistics and economics of mineral resources and fuels.

> Publications: MINERALS YEARBOOK (annual, in four volumes); BULLETINS; INFORMATION CIRCULARS; REPORTS OF INVESTIGATIONS; MINERAL INDUSTRY SURVEYS (weekly, monthly, quarterly, and annual commodity reports); NEW PUBLICATIONS-- BUREAU OF MINES (monthly list, with annual and five-year cumulation); listings of journal articles by Bureau specialists are included.

> Information services: The Office's Division of Public Information answers inquiries and handles distribution of information to public news media through press releases, copies of speeches, and special publications. The Office arranges for the publication of reports produced by all the Bureau's divisions. The Division of Production and Distribution, 4800 Forbes Avenue, Pittsburgh, Pennsylvania, distributes single copies of the Bureau's free publications and also circulates on a free-loan basis motion pictures produced by the Bureau in cooperation with industry sponsors. A catalog of films is available from the Washington office. Bureau publications for sale may be purchased from the Superintendent of Documents, Washington, D.C.   20402.

2.   Bartlesville Petroleum Research Center - U.S. Bureau of Mines, P.O. Box 1398, Bartlesville, Oklahoma 74003.

> Areas of interest: Chemistry and composition of crude oil, natural gas, and related materials; production, transportation, storage, processing, and utilization of crude oil, natural gas, and their products and byproducts; thermodynamic properties of petroleum compounds; nuclear energy applications for petroleum production; composition of oilfield brines; composition of vehicular emissions and their effect on air pollution.

> Holdings: Books, journals, reports, patents, maps, microfilms, some in foreign languages; extensive file of crude-oil analyses from domestic and foreign sources; file of oilfield brine analyses; well-log file of Oklahoma oilfields.   The Center is a depository for government publications on petroleum.

> Publications: Reports of studies by the Center are included in the BUREAU OF MINES MONTHLY LIST OF PUBLICATIONS (cumulated annually).

> Information services: Answers inquiries; provides literature-searching services; makes interlibrary loans; permits onsite reference.

3.   Laramie Energy Research Center - U.S. Bureau of Mines, P.O. Box 3395, University Station, Laramie, Wyoming 82070.

Areas of interest: Chemistry of petroleum; natural gas; asphalt; shale oil; properties and occurrence of oil shale; oil-shale conversion technology; production of petroleum and natural gas.

Holdings: Maps; extensive files of crude oil and oil-field brine analyses; oil-shale cores and samples.

4. Morgantown Energy Research Center - U.S. Bureau of Mines, P.O. Box 880, Morgantown, West Virginia 26505; LOCATION: Collins Ferry Road, Morgantown, West Virginia.

Areas of interest: Coal: Coal and coal waste analysis; fly ash utilization; desulfurization of coal; stack gas analysis and purification; nuclear analytical methods; pneumatic transportation of coal; combustion; coal conversion to fuel and high-Btu gas; liquid fuels from coal and tar. Petroleum: Secondary recovery of petroleum; petroleum production stimulation by thermal and fluid injection methods; reservoir performance prediction; subsurface fracture evaluation.

Holdings: 10,000 books; 250 periodical titles; 1,000 reports; Office of Coal Research publications; crude oil and oilfield brine analyses; cores and logs from selected Eastern oilfields.

Publications: Reports on research by the Center appear in the Bureau of Mines publications series (Office of Mineral Information), technical journals, and reference media.

Information services: Answers inquiries; provides reference and duplication services; makes referrals; makes interlibrary loans; permits onsite use of collection.

5. Pittsburgh Energy Research Center - U.S. Bureau of Mines, 4800 Forbes Avenue, Pittsburgh, Pennsylvania 15213.

Areas of interest: Coal preparation; coal utilization, including combustion, carbonization, and conversion to synthetic fuels and chemicals; conversion of coal to electrical energy; chemical structure and physical properties of coal; coal analysis; removal of air pollutants from flue gases.

Publications: Reports of studies by the Center are included in the Bureau's series of publications listed under Office of Mineral Information.

Information services: Answers inquiries; makes referrals.

6. San Francisco Petroleum Laboratory - U.S. Bureau of Mines, 1429 Appraisers Building, San Francisco, California 94111.

Areas of interest: Extractive techniques for petroleum and natural gas; prediction of reservoir performance; petroleum production stimulation; fluid flow problems; chemistry of oilfield waters.

Publications: Reports of studies by the Laboratory are included in the Bureau's series of publications listed under Office of Mineral Information.

Information services: Answers inquiries; provides technical
assistance.

B. OFFICE OF HEARINGS AND APPEALS - U.S. Department of the Interior,
Washington, D.C. 20240.

The Office of Hearings and Appeals is responsible for departmental
quasijudicial and related functions. Hearing examiners and five
formal boards of appeal render decisions in cases pertaining to
contract disputes; Indian probate matters; public and acquired
lands and their resources; submerged offshore lands of the Outer
Continental Shelf; mine health and safety; oil import quotas; and
enforcement of the importation and transportation of rare and en-
dangered species.

Contact the Office for additional details on services.

C. OFFICE OF OIL AND GAS - U.S. Department of the Interior, Washington, D.C. 20240.

The Office of Oil and Gas (OOG) was established May 6, 1946,
by the Secretary of the Interior in response to Presidential letter
of May 3, 1946. Additional responsibilities have been assigned
under authority of the Defense Production Act and the Voluntary
Agreement relating to foreign petroleum supply. On October 22,
1971, OOG was reorganized by the Secretary of the Interior to
include the duties of the former Oil Import Administration.

The Office of Oil and Gas serves as a focal point for leadership
and information on petroleum matters in the federal government,
and the principal channel of communication between the federal
government, the petroleum industry, the oil producing states,
and the public. The Office maintains the capability to respond
effectively to emergencies affecting the nation's supply of oil
and gas, and, in the interests of national security, administers
restrictions on the importation of crude oil, unfinished petroleum
oils, and finished petroleum products, including residual fuel oil
to be used as fuel.

The Office of Oil and Gas consists of a headquarters staff and
field representatives stationed in Olney, Maryland; Denton,
Texas; Battle Creek, Michigan; Denver, Colorado; Santa Rosa,
California; and Houston, Texas.

The Office provides information and services to government
agencies with respect to physical and economic factors affecting
the petroleum and gas industries, both in the United States and
abroad. It recommends programs and policies to improve the
position and capability of the oil and gas industries, as required
by the Mining and Minerals Policy Act of 1970 (84 Stat. 1876;
30 U.S.C. 21a).

The Office conducts a continuing study of the effects of oil and
gas production, transportation, manufacturing, and consumption on
the environment, and reviews for accuracy and completeness the

Environmental Impact Statements which relate to oil and gas.

D. OECD PETROLEUM ADVISORY COMMITTEE (PAC) - Office of Oil and Gas, U.S. Department of the Interior, 18th and C Streets N.W., Washington, D.C. 20240.

Committee provides the U.S. Department of the Interior, the U.S. Department of State, and the U.S. Delegate to the Organization for Economic Cooperation and Development (OECD) Oil Committee with the views of industry with respect to matters which may be considered in the OECD relating to oil and energy. Members also contribute to continuing OECD studies on oil supply and demand prospects.

DEPARTMENT OF TRANSPORTATION - Materials Transportation Bureau, Washington, D.C. 20590.

This agency is concerned with transport and handling of hazardous materials and pipeline safety. It has two offices--Office of Hazardous Materials Operations and Office of Pipeline Safety Operations--to provide regulatory service and safety information.

FEDERAL POWER COMMISSION

OFFICE OF PUBLIC INFORMATION - 441 G Street N.W., Washington, D.C. 20426.

Areas of interest: Generation, transmission, and distribution of electricity; production and transportation of natural gas; energy sources; pooling systems (interconnection and coordination of facilities); charges for interstate transmission and sale of electricity and natural gas; conservation and development of natural resources as related to licensing of power projects and natural gas production; construction of natural gas pipelines and non-federal hydroelectric projects; import and export of natural gas and export of electric power.

Holdings: Periodic reports of electric utility companies and of natural gas companies subject to FPC jurisdiction; proposals, expansion plans, proceedings on applications, rate and accounting data, maps, charts.

Publications: NEWS DIGEST (daily); technical, statistical, and special reports; maps, news releases. A booklet, "A Guide to FPC Public Information," and a publications list are available.

Information services: Answers inquiries; makes referrals; supplies copies of records and tapes of data on cost basis; makes files available for public inspection. Some informational services are also available at the following branch offices:

Bureau of Power Regional Office
Federal Power Commission
730 Peachtree Building, Room 500
Atlanta, Georgia 30308

Bureau of Power Regional Office
Federal Power Commission
610 South Canal Street
Chicago, Illinois 60607

Bureau of Power Regional Office
Federal Power Commission
819 Taylor Street
Fort Worth, Texas 76102

Bureau of Power Regional Office
Federal Power Commission
26 Federal Plaza, 22d Floor
New York, New York 10007

Bureau of Power Regional Office
Federal Power Commission
555 Battery Street
San Francisco, California 94111

b. STATE AND MUNICIPAL

See Section I of this Guide.

2. Professional Societies, Trade Associations, and Other Scientific-Technological Related Organizations

AMERICAN ASSOCIATION OF OILWELL DRILLING CONTRACTORS - 211 North Ervay, Room 505, Dallas, Texas 75201.

Purpose: To maintain communication among oilwell drilling contractors and to act as an exchange facility for information about activities related to oilwell drilling.

Activities: Compilation of statistics, safety instruction and educational programs for schools.

Publication: DRILLING CONTRACTOR (bimonthly); safety and technical manuals; drilling reports and training materials for courses.

Information services: Answers inquiries; makes referrals.

AMERICAN ASSOCIATION OF PETROLEUM GEOLOGISTS, INC. - Box 979, Tulsa, Oklahoma 74101.

Purpose: To advance the science of geology especially as it relates to petroleum and natural gas and to promote exploration, experimentation, and research in these and corresponding areas.

Activities: Sponsors research in geothermal conditions, mineralogical strata, deep well drilling techniques, and in similar programs of interest to its membership.

Publications: BULLETIN OF THE AMERICAN ASSOCIATION OF PETROLEUM GEOLOGISTS (monthly).

Information services: Answers inquiries; makes referrals.

AMERICAN INSTITUTE OF MINING, METALLURGICAL AND PETROLEUM ENGINEERS, INC. - United Engineering Center, 345 East 47th Street, New York, New York 10017.

Areas of interest: Mining, metallurgical, and petroleum engineering; coal; industrial minerals; minerals benefication; extractive metallurgy; physical metallurgy; geology; engineering economics and education in the fields of mining, metallurgy, and petroleum.

Publications: JOURNAL OF METALS (monthly); MINING ENGINEERING (monthly); JOURNAL OF PETROLEUM TECHNOLOGY (monthly); SOCIETY OF PETROLEUM ENGINEERS JOURNAL (quarterly); newsletter, bound volumes of transactions, proceedings, books, reprints.

Information services: Answers brief inquiries; makes referrals.

AMERICAN PETROLEUM INSTITUTE - 1801 K Street N.W., Washington, D.C. 20006.

Producers, refiners, marketers and transporters of petroleum and allied products.

Purpose: To promote communication among industry, government agencies, and researchers involved in the development or use of petroleum products.

Activities: Sponsors research on petroleum, compiles statistics, and fosters the development of standards and codes and safe practices in petroleum industries.

Publications: STATISTICAL BULLETIN (weekly); PETROLEUM TODAY (quarterly); and manuals, codes, standards, and other materials. A list of publications is available.

Information services: Answers inquiries; makes referrals; acts as a clearinghouse for data and information on the petroleum industry.

AMERICAN PETROLEUM REFINERS ASSOCIATION - 1110 Ring Building, Washington, D.C. 20036.

Petroleum refiners processing less than 30,000 barrels per day.

Purpose: To provide a communications link for smaller petroleum producers.

Information services: Answers inquiries; makes referrals.

AMERICAN SOCIETY OF LUBRICATION ENGINEERS – 838 Busse Highway, Park Ridge, Illinois 60068.

> Areas of interest: Lubrication practices; lubricant reconditioning and disposal; properties of lubricants, seals, and packings; bearings and bearing lubrication; hydraulics and hydraulic machinery; lubrication equipment.

> Publications: LUBRICATION ENGINEERING (monthly); ASLE TRANSACTIONS (quarterly).

> Information services: Makes referrals.

ASSOCIATION OF OIL PIPE LINES – RCA Building, Suite 1208, 1725 K Street N.W., Washington, D.C. 20006.

> Areas of interest: Oil pipeline construction; transportation of crude oil, petroleum, gasoline, kerosene, jet fuel, and natural gas liquids.

> Publications: SHIFTS IN PETROLEUM TRANSPORTATION and TABLES OF STATISTICS ON OIL TRANSPORTATION (both annual); data.

> Information services: Answers inquiries; makes referrals; provides reference and consulting services.

ASSOCIATION OF OILWELL SERVICING CONTRACTORS – 1700 Davis Building, Dallas, Texas 75202.

Oilwell servicing contractors, manufacturers, and distributors of equipment; others related to the industry.

> Areas of interest: Research and development, business methods, and management for oilwell contracting, information compilation in this area and safe practices.

> Publications: AOSC NEWSLETTER (monthly); WELL SERVICING (bimonthly); AOSC Directory of Officers and Members (annual).

> Information services: Answers inquiries; makes referrals.

CANADIAN GAS ASSOCIATION – 55 Scarsdale Road, Don Mills, Ontario, Canada.

An independent, nonprofit membership organization serving 600 company and individual members representing natural gas industry in Canada.

> Areas of interest: Principal field of research: heat exchanger design, domestic water heaters, gas utilization, combustion, heat transfer and accessories.

> Publications: Annual and quarterly reports.

> Information services: Answers inquiries; makes referrals. Testing, research and development and other services are available on a contract basis through an affiliated testing laboratory Cantest Ltd. Contact the association for additional details.

INDEPENDENT PETROLEUM ASSOCIATION OF AMERICA - 1101 16th Street N.W., Washington, D.C. 20036.

An association of independent oil and gas operators, land and royalty owners, and others (suppliers, drilling contractors, bankers, oil attorneys, trucking contractors, rig building contractors) interested in the production of crude and lubricating oil, gasoline, and natural gas.

Areas of interest: Production of petroleum and gas, oil reserves and sources, marketing and distribution of petroleum products, etc. Activities of this organization are carried on through a series of committees which include the following areas: cost, public land, import policies, natural gas, oil recovery, shale oil and related matters.

Information services: Answers inquiries; makes referrals.

INDEPENDENT REFINERS ASSOCIATION OF AMERICA - 1801 K Street N.W., Suite 1101, Washington, D.C. 20006.

Areas of interest: Petroleum production, refining, equipment, chemicals and byproducts of industry.

Activities: Acts as clearinghouse for information on work of independent refiners.

Publication: IRAA BULLETIN (irregular).

Information services: Answers inquiries; makes referrals.

INSTITUTE OF GAS TECHNOLOGY - TECHNICAL INFORMATION SERVICES; Institute of Gas Technology, 3424 South State Street, Chicago, Illinois 60616.

Areas of interest: Natural and manufactured gas production, transmission, storage, distribution, and utilization; training of engineers in gas technology; properties of gases; analytical methods and instruments; flow of gases, including the use of radioactive tracers in measuring flow in pipelines; combustion; gasification reactions; natural gas fuel cells; internal coatings for pipes; liquefied natural gas; air conditioning; environmental science; total energy; economic studies.

Publications: GAS ABSTRACTS (monthly); GAS SCOPE (bimonthly); RESEARCH BULLETINS; TECHNICAL REPORTS; annual report.

Information services: Answers inquiries; makes referrals; provides reference services; makes interlibrary loans; provides literature- and patent-searching, bibliographic, and duplication services for fee; provides MASTIR (Microfilmed Abstract System for Technical Information Retrieval) information service; permits onsite reference.

INTERSTATE OIL COMPACT COMMISSION - P.O. Box 53127, Oklahoma City, Oklahoma 73105; LOCATION: 900 Northeast 23d Street, Oklahoma City, Oklahoma.

Areas of interest: Oil and gas conservation, including conservation activities in the various states; engineering, including production methods, conservation practices, and fundamental geology;
research in well spacing, underground storage, gas-oil ratio,
bottom-hole pressure, and production histories; legal aspects of
repressuring and recycling; public lands, including oil and gas
development in the wilderness area; regulatory practices, including
rules and uniform reporting forms; secondary recovery and pressure
maintenance.

Publications: COMPACT COMMENTS (monthly); LEGAL REPORT
(annual); STRIPPER WELL SURVEY (annual); COMPACT BULLETIN
(semiannual); COMMITTEE BULLETIN (semiannual); motion picture,
directory, special publications.

Information services: Answers inquiries; makes referrals; permits
use of collection.

NATIONAL LP-GAS ASSOCIATION - 79 West Monroe Street, Chicago,
Illinois 60603.

Areas of interest: Liquefied petroleum and propane and butane
gas, including appliances, equipment, storage, handling, transportation, safety standards, statistics, education, advertising and
promotion, and research and development of uses for LP-gas.

Publications: NLPGA TIMES; reports, standards and specifications.

Information services: Answers inquiries; provides reference and
consulting services.

NATIONAL LUBRICATING GREASE INSTITUTE - 4635 Wyandotte Street,
Kansas City, Missouri 64112.

Companies manufacturing or selling all types of lubricating greases; suppliers
of such companies; and technical and educational organizations.

Activities: Promotes research and testing for the development of
better lubricating greases and better grease lubrication engineering service to industry; collects and disseminates technical data;
conducts forums.

Publication: NLGI SPOKESMAN (monthly).

Information services: Answers inquiries; makes referrals.

NATIONAL PETROLEUM COUNCIL - 1625 K Street N.W., Suite 601,
Washington, D.C. 20006.

An organization which acts in an advisory capacity to the Secretary of the
Interior by representing the interests and problems of the petroleum industry.
Conducts studies relating to petroleum production.

Information services: Answers inquiries; makes referrals. Contact
the council for details on assistance with more involved information requests.

NATIONAL PETROLEUM REFINERS ASSOCIATION - 1725 DeSales Street N.W., Suite 802, Washington, D.C. 20036.

Areas of interest: Petroleum production, petrochemical development and problems of refining industry.

Committees of this organization are: asphalt, computer, industrial relations, accident prevention, manufacturing, petrochemical, fuels and lubricants, insurance.

Information services: Answers inquiries; makes referrals.

NATURAL GAS PROCESSORS ASSOCIATION - 808 Home Federal Building, Tulsa, Oklahoma 74103.

Areas of interest: Natural gas liquids manufacturing, including specifications and product utilization; thermodynamic research.

Publications: Periodicals, reports, books, bibliographies, specifications, proceedings. A list of publications is available.

Information services: Answers inquiries; provides consulting services. Services are provided free to some users and at cost to others.

NATURAL GAS PROCESSORS SUPPLIERS ASSOCIATION - 808 Home Federal Building, Tulsa, Oklahoma 74103.

Manufacturers and wholesalers of natural gas processing equipment and supplies.

Activities: Provides technical data to plant operators and engineers.

Publications: Publishes ENGINEERING DATA BOOK (revised every 5-8 years).

Information services: Answers inquiries; makes referrals.

PETROLEUM EQUIPMENT CONTRACTORS ASSOCIATION - 474 Schiller Street, Elizabeth, New Jersey 07206.

An association of firms contracting for the service of installation of service station equipment, gasoline bulk plant equipment, and related lines. Membership concentrated in Pennsylvania, Delaware, and New Jersey.

Purpose: To improve the service and distribution, installation and maintenance techniques in oil equipment installation field.

Information services: Answers inquiries; makes referrals.

PETROLEUM EQUIPMENT INSTITUTE - Dowell Building, Tulsa, Oklahoma 74114.

An organization whose membership consists of distributors and manufacturers of equipment used in service stations, bulk plants, and other petroleum marketing operations.

Publications: PEI NEWSLETTER (semimonthly); PETROLEUM EQUIP-

MENT DIRECTORY (annual).

Information services: Answers inquiries; makes referrals.

PETROLEUM EQUIPMENT SUPPLIERS ASSOCIATION - 1703 First City National Bank Building, Houston, Texas 77002.

Manufacturers of oil field drilling and production equipment; oil field supply stores and specialty service firms.

Publication: NEWSLETTER (biweekly).

Information services: Answers inquiries; makes referrals.

PETROLEUM INDUSTRY RESEARCH FOUNDATION - 60 East 42d Street, New York, New York 10017.

Purpose: To conduct economic research and disseminate information concerning the oil industry. Publishes books, booklets, special reports on oil industry subjects.

SOCIETY OF PETROLEUM ENGINEERS OF AIME - 6200 North Central Expressway, Dallas, Texas 75206.

Purpose: Promote the high standards of engineering in this field, sponsor research and foster development of standards for training and production in this industry.

Areas of interest: Drilling and rock mechanics, environmental quality, fluid mechanics, oil recovery, gas technology, drilling, geology, petroleum production and by-product utilization, etc.

Activities: Symposia, career education programs, research projects, continuing education and audiovisual programs.

Publications: JOURNAL OF PETROLEUM TECHNOLOGY (monthly); SPE JOURNAL (quarterly); TRANSACTIONS (annual); MEMBERSHIP DIRECTORY (annual).

Information services: Answers inquiries; makes referrals.

SOCIETY OF PROFESSIONAL WELL LOG ANALYSTS, INC. - 13507 Tosca Lane, Houston, Texas 77024.

An association whose purpose is to advance the science of well logging techniques by various means and to maintain standards for this type of endeavor. The organization attempts to collect, record, and to disseminate data and information in this field.

Publications: THE LOG ANALYST (bimonthly); LOGGING SYMPOSIUM TRANSACTIONS (annual).

Information services: Answers inquiries; makes referrals.

3. Research Centers

See organizations listed in Part 2 of this Section.

4. International Organizations

ARAB ORGANIZATION OF PETROLEUM EXPORTING COUNTRIES - P.O. Box 20501, Kuwait.

ECONOMIC RESEARCH COMMITTEE OF THE GAS INDUSTRY - 4 Avenue Palmerston, 1040 Brussels, Belgium.

FEDERATION OF THE OIL INDUSTRY OF THE EEC - 1 rue de Spa, 1040 Brussels, Belgium.

INTERNATIONAL COLLOQUIUM ON GAS MARKETING - 23 rue Philibert Delorme, 75-Paris 17e, France.

INTERNATIONAL GAS UNION - Institution of Gas Engineers, 17 Grosvenor, England.

ORGANIZATION OF THE PETROLEUM EXPORTING COUNTRIES - 1010 Vienna, Austria.

PERMANENT COUNCIL OF THE WORLD PETROLEUM CONGRESS - Institute of Petroleum, 61 New Cavendish Street, London, W1, England.

5. Other Information Sources and Programs

a. SPECIALIZED INFORMATION SERVICES

AMERICAN PETROLEUM INSTITUTE - Central Abstracting and Index Service, 1271 Avenue of the Americas, New York, New York 10020.

1. SELECTIVE DISSEMINATION OF INFORMATION SERVICE: Based on search of chemical abstracts, this weekly service provides a printout of pertinent items matching users profile of interests. Users may adjust profiles as needs change.

2. RETROSPECTIVE SEARCHES: This includes patents as well as technical literature. Through this outlet users have access to over 125,000 journal articles and approximately 65,000 patents.

3. ABSTRACTING AND INDEXING: Publications in the area of petroleum industries is also a program of this institution and is described in previous parts of this section of the Guide.

Contact the Institute for prices and details on above listed services.

ILLINOIS INSTITUTE OF TECHNOLOGY - Institute of Gas Technology, Technical Information Services, 3424 South State Street, Chicago, Illinois 60616.

Microfilmed Abstract System for Technical Information Retrieval (MASTIR).

Subject coverage: Gas technology in general. Specific topics include liquified natural gas, safety, gas production from various sources, processing, management and engineering economics of gas production.

Service provides: Bibliographies, abstracts of wide variety of literature on microfilm, literature searches, patent searches, state-of-the-art and SDI service.

Contact the institute for additional information.

NATURAL GAS PROCESSORS ASSOCIATION - 808 Home Federal Building, Tulsa, Oklahoma 74103.

1.   NGPA/OSU COMPUTER PROGRAM FOR LIQUID DENSITY: This is "a new program to predict the density of liquid mixtures of known composition." Densities for both saturated and sub-cooled mixtures can be calculated. Saturated densities are cal-culated by a modified Rackett procedure and adjusted to subcooled conditions through a modification of the Tact equation. The pro-gram is restricted to hydrocarbons and a limited number of non-hydrocarbons. For additional details contact the Natural Gas Processors Association.

2.   MARK V COMPUTER PROGRAM: To predict physical and thermodynamic properties of natural gas and condensate systems in: processing plant units; expanders and compressors; pipelines at flowing conditions; field separators; reservoir depletion studies. "Mark V computes the fraction of the feed mixture existing as a vapor and as a liquid at the given pressure and temperature and the composition of co-existing phases and a complete set of thermodynamically consistent properties for each phase and for the combined two phase mixture."

For further information on Mark V Computer Program, contact the Natural Gas Processors Association.

THE UNIVERSITY OF TULSA - Information Services Department, 1136 North Lewis Avenue, Tulsa, Oklahoma 74110.

Technical Information Searching Service includes areas of explora-tion, production and transportation of crude oil and natural gas. Provides comprehensive coverage of world literature and patents, abstracts of pertinent articles, and bibliographies. Data base consists of chemical abstracts, Engineering Index Geoscience and Geophysical abstracts, and other major indexing and abstracting publications as well as a specialized file of over 400,000 abstract cards covering an extensive time period.

Contact the agency for additional information on searching services.

B. LITERATURE

1. Guides and Handbooks

American Society for Testing and Materials. Committee D-2 on Petroleum Products and Lubricants. MANUAL ON HYDROCARBON ANALYSIS. 2nd ed. Special Technical Publication, no. 332A. Philadelphia: 1968.

Bell, H. PETROLEUM TRANSPORTATION HANDBOOK. New York: McGraw-Hill, 1963.

> Concerned with moving petroleum and its products from oilfield to refinery to market; with special attention to characteristics problems.

Bland, W. PETROLEUM PROCESSING HANDBOOK. New York: McGraw-Hill, 1967.

> "Experts in many areas of processing have contributed to this comprehensive sourcebook on the manufacturing function of the petroleum industry. Covers the basic types of processes, equipment, materials, controls and instrumentation, construction, maintenance, safety, and offsite facilities and utilities."

EQUIPMENT DESIGN HANDBOOK FOR REFINERIES AND CHEMICAL PLANTS. Houston: Gulf Publishing Co., 1971.

FINAL REPORT, OIL AND GAS RESOURCES, RESERVES AND PRODUCTIVE CAPACITIES, SUBMITTED IN COMPLIANCE WITH PUBLIC LAW 93-275. 2 vols. Washington, D.C.: U.S. Federal Energy Administration, 1975.

Frick, T. PETROLEUM PRODUCTION HANDBOOK. New York: McGraw-Hill, 1962.

Guthrie, V. PETROLEUM PRODUCTS HANDBOOK. New York: McGraw-Hill, 1960.

> "Information and data on selection, handling, and use of commercial petroleum and natural gas products."

International Association of Drilling Contractors. DRILLING MANUAL. 8th ed. Dallas: 1970.

> "Practical reference data on all pertinent phases of rig operation for maximum hole penetration efficiency, arranged to give quick answers to many questions."

_____. PRIMER OF OILWELL DRILLING. Dallas: 1970.

> "A 95-page, illustrated, comprehensive, non-technical explanation of drilling practices, procedures and terminology."

_____. PRIMER OF OILWELL SERVICE AND WORKOVER. Dallas: 1969.

"An illustrated manual explaining work required in maintaining and repairing producing oil wells. Covers all phases, including mechanical maintenance, major repairs, recompletion and work-over. Ten-page glossary of terms."

_____. ROTARY DRILLING HANDBOOK. Dallas.

"Sixth edition, covering new developments in the equipment and technology of drilling. Illustrations include nomographic charts and performance curves."

International Research and Technology Corp. END USES OF PETROLEUM PRODUCTS IN THE U.S. 1965-1975. Vol. 1: Sources, methods and results. Rept. no. PB246-393. Vol. 2: Tabulation of results. Rept. no. PB246 394. Springfield, Virginia: National Technical Information Service, 1975.

Katz, D. HANDBOOK OF NATURAL GAS ENGINEERING. Chemical Engineering Series. New York: McGraw-Hill, 1959.

"Design calculations and charts for equipment and methods required in production of natural gas."

Noel, H. PETROLEUM REFINERY MANUAL. New York: Van Nostrand-Reinhold, 1959.

U.S. Federal Energy Administration Project Independence. PROJECT INDEPENDENCE. A SUMMARY. Rept. no. PB 248 491. Springfield, Virginia: National Technical Information Service.

A survey of energy needs and environmental and economic aspects of petroleum, natural gas, and other energy industries.

U.S. Geological Survey. WORLDWIDE SEARCH FOR PETROLEUM OFFSHORE, STATUS REPORT FOR QUARTER OF A CENTURY, 1947-1972. Washington, D.C.: Government Printing Office, 1974.

U.S. National Bureau of Standards. HYDROCARBONS FOR FUEL - 75 YEARS OF MATERIALS RESEARCH AT N35 FINAL REPORT. Rept. no. PB 253 665. By George Armstrong. Springfield, Virginia: National Technical Information Service, 1976.

Zaba, J. PRACTICAL PETROLEUM ENGINEERS HANDBOOK. Houston: Gulf Publishing Co., 1970.

a. TABLES, DATA, AND SCIENTIFIC UNITS

American Society for Testing and Materials. ASTM-IP PETROLEUM MEASUREMENT TABLES. Philadelphia.

Gerolde, S. UNIVERSAL CONVERSION FACTORS. Tulsa: The Petroleum Publishing Co., 1971.

"Over 12,000 conversions from one system of units to another."

Institute of Petroleum. PETROLEUM MEASUREMENT. London: 1969.

U.S. Department of Interior. Bureau of Mines. REVIEW OF CURVE-FITTING METHOD OF LEAST SQUARES AS APPLIED TO PETROLEUM ENGINEERING. Washington, D.C.: Government Printing Office, 1970.

2. Dictionaries

a. TECHNICAL

American Petroleum Institute. GLOSSARY OF TERMS USED IN PETROLEUM REFINING. New York: 1970.

Boone, L. THE PETROLEUM DICTIONARY. Norman: University of Oklahoma Press, 1952.

"Gives definitions and sources of about 6000 terms used in the oil industry."

Crook, L. OIL TERMS. New York: International Publications Service, 1976.

DESK AND DERRICK STANDARD OIL ABBREVIATOR. Tulsa: The Petroleum Publishing Co., 1972.

"More than 4,000 oil and gas industry abbreviations, also symbols."

Gilpin, A. DICTIONARY OF FUEL TECHNOLOGY. New York: Philosophical Library, 1970.

Ketchian, S., ed. TECHNICAL PETROLEUM DICTIONARY OF WELL-LOGGING DRILLING AND PRODUCTION TERMS. New York: Gordon, 1965.

b. FOREIGN LANGUAGE

Cagnacci-Schwicker, A. INTERNATIONAL DICTIONARY OF METALLURGY, MINERALOGY, GEOLOGY, AND THE MINING AND OIL INDUSTRIES: ENGLISH, FRENCH, GERMAN, ITALIAN. New York: McGraw-Hill, 1970.

Contains more than 2,000 synonyms.

ELSEVIER'S OILFIELD DICTIONARY IN FIVE LANGUAGES. New York: Elsevier, 1965.

International Gas Union. ELSEVIER'S DICTIONARY OF THE GAS INDUSTRY IN SEVEN LANGUAGES. New York: Elsevier, 1961.

493

_____. SUPPLEMENT TO THE DICTIONARY OF THE GAS INDUSTRY IN SEVEN LANGUAGES. New York: Elsevier, 1973.

Persch, F. PETROLEUM AND NATURAL GAS DICTIONARY: GERMAN-ENGLISH AND ENGLISH-GERMAN. New York: International Publications Service, 1973.

3. Directories

a. GENERAL

i. United States and Canada

Bobo, D. A SURVEY OF FUEL AND ENERGY INFORMATION SOURCES. Langley, California, The Mitre Corp. Rept. no. PB197 386. Springfield, Virginia: National Technical Information Service, November 1970.

BROWN'S DIRECTORY OF NORTH AMERICAN GAS COMPANIES. Annual. Duluth, Minnesota: Harcourt Brace Jovanovich.

> Sources of data concerning personnel plant and facilities, sales revenues, gas characteristics and statistical data as well as American and Canadian utility companies.

BURMASS OIL DIRECTORIES. 3 vols. Published annually in April, June, and October. Springtown, Texas: Burmass Oil Directories.

> Covers South and Southwest.

Canada. Department of Energy, Mines and Resources. MINERAL RESOURCES BR. PETROLEUM REFINERIES IN CANADA. Ottawa, Ontario: Queen's Printer, January 1970.

CANADIAN OIL REGISTER. Annual. Calgary 2, Alberta: C.O. Nickle Publications Ltd.

COMPILATION OF INDUSTRIAL AND MUNICIPAL INJECTION WELLS IN THE U.S. 2 vols. Washington, D.C.: Environmental Protection Agency, 1974.

DIRECTORY OF GAS UTILITY COMPANIES. Annual. Tulsa, Oklahoma: Midwest Oil Register.

DIRECTORY OF GEOPHYSICAL AND OIL COMPANIES WHO USE GEO-PHYSICAL SERVICE. Annual. Tulsa, Oklahoma: Midwest Oil Register.

DIRECTORY OF OIL WELL DRILLING CONTRACTORS. Annual. Tulsa, Oklahoma: Midwest Oil Register.

DIRECTORY OF OIL WELL SUPPLY COMPANIES. Annual. Tulsa, Oklahoma: Midwest Oil Register.

HANK SEALE OIL DIRECTORY: CENTRAL UNITED STATES. Annual. Oil Men's Association of America. c/o Hank Seale, ed., 1606 South Jackson Street, Amarillo, Texas 79102.

HANK SEALE OIL DIRECTORY: EASTERN UNITED STATES. Annual. Oil Men's Association of America. c/o Hank Seale, ed., 1606 South Jackson Street, Amarillo, Texas 79102.

HANK SEALE OIL DIRECTORY: LOUISIANA, MISSISSIPPI, ARKANSAS, TEXAS GULF COAST AND EAST TEXAS. Annual. Oil Men's Association of America. c/o Hank Seale, ed., 1606 South Jackson Street, Amarillo, Texas 79102.

HANK SEALE OIL DIRECTORY: TEXAS INCLUDING SOUTHEAST NEW MEXICO. Annual. Oil Men's Association of America. c/o Hank Seale, ed., 1606 South Jackson Street, Amarillo, Texas 79102.

NATURAL GAS PROCESSING PLANTS IN CANADA. Annual. Publication Distribution Office, Mineral Resources Branch, Department of Energy, Mines and Resources, Ottawa, Ontario, K1A OE4, Canada.

OFFSHORE TECHNOLOGY YEARBOOK. Biennial. Box 1589, Dallas, Texas 75221.

OIL AND GAS DIRECTORY. Annual. Houston, Texas: Oil and Gas Directory Publishing Co.

> Annual publication with worldwide coverage.

OIL DIRECTORY OF ALASKA. Annual. Tulsa, Oklahoma: Midwest Oil Register.

OIL DIRECTORY OF CANADA. Annual. Tulsa, Oklahoma: Midwest Oil Register.

OIL DIRECTORY OF HOUSTON, TEXAS. Annual. Tulsa, Oklahoma: Midwest Oil Register.

> "Contains approximately 4,700 individual names and includes all phases of the oil industry in Houston."

PETROLEUM ENGINEER INTERNATIONAL. ENERGY OPERATIONS AND EQUIPMENT FORECAST. Annual. Petroleum Engineer Publishing Co., P.O. Box 1589, Dallas, Texas 75221.

> Energy management report. Survey of 100 oil/gas companies financial operations.

PIPELINE AND PILE LINE CONTRACTORS. Annual. Tulsa, Oklahoma: Midwest Oil Register.

> "Contains approximately 8,000 individual names and is the only

directory of its kind published. It lists operating and home office personnel and covers Crude Oil, Products and Natural Gas;
shows miles of pipe line by sizes and the number of pumping
stations owned by each company."

PIPELINE AND UNDERGROUND UTILITIES CONSTRUCTION DIRECTORY. Annual. Houston, Texas: Oildom Publishing Co.

"Lists all pipeline and utility construction companies and their
personnel."

PIPELINE NEWS DIRECTORY OF PIPE LINES AND EQUIPMENT. Annual.
Houston, Texas: Oildom Publishing Co.

Lists all pipeline owning and operating companies (crude products,
natural gas, gas distribution) etc.

"Semi-annual Inventory of New Plants and Facilities; Petroleum and Energy
Products." CHEMICAL ENGINEERING, Vol. 79, October 2, 1972, pp. 74-76.

U.S. Department of Interior. Bureau of Mines. MAJOR NATURAL GAS
PIPELINES, AS OF JUNE 30, 1971. Washington, D.C.: Government Printing Office, 1971.

Colored map.

_____. NATURAL GAS PROCESSING PLANTS IN UNITED STATES. Biennial.
Bureau of Mines, Department of the Interior, Washington, D.C. 20240.

_____. PETROLEUM REFINERIES IN THE UNITED STATES. Bureau of Mines,
Publications Distribution Section, 4800 Forbes Avenue, Pittsburgh, Pennsylvania
15213. 1973.

Gives number and capacity of petroleum refineries, by P.A.D.
district and state.

U.S. Federal Power Commission. GAS SUPPLIES OF INTERSTATE PIPELINE
COMPANIES. Washington, D.C.: Government Printing Office, 1969.

_____. MAJOR NATURAL GAS PIPELINES AS OF DEC. 31, 1974. Washing-
ton, D.C.: Government Printing Office, 1975.

Colored map.

USA OIL INDUSTRY DIRECTORY. Annual. Tulsa, Oklahoma: The Petroleum
Publishing Co.

"Complete directory of U.S. oil and gas industry. Names, titles,
addresses for home office, district, division, field, and plant locations."

ii. International

Dewbury, C., ed. GAS INDUSTRY DIRECTORY AND UNDERTAKINGS OF THE WORLD. New York: British Book Centre, 1972.

EASTERN HEMISPHERE PETROLEUM. Annual. Tulsa, Oklahoma: Petroleum Publishing Co.

> "Lists over 2,100 firms, listed by country, in Europe, Africa, Middle East, Asia, and the Pacific engaged in every aspect of Eastern Hemisphere petroleum."

EUROPEAN PETROLEUM DIRECTORY. Annual. New York: International Publications Service.

EUROPEAN PETROLEUM YEARBOOK. Annual. Hamburg, German Federal Republic: Vieth Verlag.

Franco, L., et al. THE PETROLEUM INDUSTRY IN WESTERN EUROPE: A GUIDE TO INFORMATION SOURCES. New York: Garland Publishing Co., 1975.

GAS JOURNAL DIRECTORY. Annual. Walter King, Ltd., Holborn Hall, 100, Gray's Inn Road, London, WC 1, England.

INTERNATIONAL OFFSHORE RIG REGISTER. See issues of PETROLEUM EN-GINEER. Petroleum Engineer Publishing Co., Box 1589, Dallas, Texas 75221.

INTERNATIONAL PETROLEUM ANNUAL. U.S. Bureau of Mines, 4500 Forbes Avenue, Pittsburgh, Pennsylvania 05213.

INTERNATIONAL PETROLEUM REGISTER. Biannual. New York: Palmer Publications.

> A roster of approximately 30,000 oil companies.

LATIN AMERICAN PETROLEUM. Annual. Tulsa, Oklahoma: The Petroleum Publishing Co.

> "Country-by-country listing of firms, people, and government agencies which make up the Mexican, Central, and South American oil industry."

OIL AND PETROLEUM YEARBOOK. Annual. London: Walter E. Skinner.

> A worldwide directory of petroleum companies.

OIL DIRECTORY OF COMPANIES OUTSIDE THE USA AND CANADA. Annual. Tulsa, Oklahoma: Midwest Oil Register.

PETROCHEMICAL INDUSTRY OF JAPAN. Park Ridge, New Jersey: Noyes Data Corp., 1970.

PIPELINE AND GAS JOURNAL ANNUAL INTERNATIONAL E/C/S DIRECTORY.
(May issue of PIPELINE AND GAS JOURNAL.) Petroleum Engineer Publishing
Co., Box 1589, Dallas, Texas 75221.

> Covers gas processing, manufacturing, and liquid gas.

REFINING, CONSTRUCTION, PETROCHEMICAL AND NATURAL GAS PRO-
CESSING PLANTS OF THE WORLD. Annual. Tulsa, Oklahoma: Midwest
Oil Register.

> "Contains over 15,000 names. Covers home office and operating
> personnel in the world. Also shows types of refineries, capaci-
> ties, and products manufactured."

Rojana, C. CHEMICAL AND PETROCHEMICAL INDUSTRIES OF RUSSIA AND
EASTERN EUROPE, 1960-1980. New York: Praeger, 1975.

Walter, R. SKINNER'S WHO'S WHO IN WORLD OIL AND GAS. Annual.
London: Financial Times Publishing Co.

WORLDWIDE OFFSHORE CONTRACTORS. Annual. Tulsa, Oklahoma: Petroleum Publishing Co.

> "Lists offshore drilling companies, producers, and facilities; auxil-
> iary equipment firms; workboat owners/operators; geophysical com-
> panies."

WORLDWIDE PETROCHEMICAL DIRECTORY. Annual. Tulsa, Oklahoma: Petroleum Publishing Co.

> "Lists every known petrochemical plant in the world."

WORLDWIDE REFINING AND GAS PROCESSING. Annual. Tulsa, Oklahoma:
Petroleum Publishing Co.

> "Complete address listing of all crude oil and natural gas proces-
> sing companies and capacities."

b. BIOGRAPHY

WHO'S WHO IN THE GAS INDUSTRY. Annual. Benn Brothers Ltd., Hanover
House, 73-78 High Holborn, London, WC 1, England.

c. PRODUCTS AND MANUFACTURERS

i. United States

DIRECTORY OF OIL MARKETING AND WHOLESALE DISTRIBUTORS. Annual.
Tulsa, Oklahoma: Midwest Oil Register.

ENERGY MANAGEMENT REPORT/D/ANNUAL NO: ENERGY EQUIPMENT

PLANNING GUIDE. Dallas: Petroleum Engineer Publishing Co.

HYDROCARBON PROCESSING CATALOG. Annual. Houston: Gulf Publishing Co.

> "Catalog of equipment, materials, and services, codes and standards for processing crude oil, natural gas, and petrochemicals."

MACRAE'S BLUEBOOK. Annual. MacRae's Bluebook, Western Springs, Illinois 60558.

OIL MARKETING AND WHOLESALE DIRECTORY. Annual. Tulsa: Midwest Oil Register.

PETROLEUM EQUIPMENT AND SERVICES. 6/year. Petroleum Equipment and Services Co., 4300 North Central Expressway, Dallas, Texas 75204.

THOMAS' REGISTER OF MANUFACTURERS. Annual. New York: Thomas Publishing Co.

ii. International

BRITISH PETROLEUM EQUIPMENT AND SERVICES. Annual. London: Council of British Manufacturers of Petroleum Equipment.

PETROLEUM ENGINEER--WORLD WIDE NEW EQUIPMENT GUIDE. Dallas: Petroleum Engineer Publishing Co.

> Guide to companies selling oil and gas drilling and transportation equipment.

d. LABORATORIES AND RESEARCH CENTERS

EUROPEAN RESEARCH INDEX. 2nd ed. Guernsey, Channel Islands, England: F. Hodgson Ltd., 1969.

INDUSTRIAL RESEARCH LABORATORIES OF THE UNITED STATES. 13th ed. New York: R.R. Bowker, 1970.

RESEARCH CENTERS DIRECTORY. 5th ed. Detroit: Gale Research Co., 1975.

> Updated by subscription service, NEW RESEARCH CENTERS.

4. Encyclopedias

INTERNATIONAL PETROLEUM ENCYCLOPEDIA. Annual. Tulsa, Oklahoma: Petroleum Publishing Co.

Details on petroleum industry of 80 nations. Has maps and contains forecasts.

See also this subject in the section Chemical Engineering - General.

5. Bibliographies

a. GENERAL MATERIALS

American Chemical Society. Division of Chemical Literature. LITERATURE OF THE COMBUSTION OF PETROLEUM. Advances in Chemistry Series, 20. Washington, D.C.: 1958.

COAL GASIFICATION AND LIQUIFACTION TECHNOLOGY. Vol. 1: 1964-1973. Rept. no. NTIS/PS-76/0390. Vol. 2: 1974-1976. Rept. no. NTIS/PS-76/0391. By E. Harrison. Springfield, Virginia: National Technical Information Service, 1976.

Bibliographies with abstracts.

GAS AND STEAM TURBINES, GENERAL: CORROSION AND EROSION. By G. Habercom, Jr. Rept. no. NTIS/PS-76/0088. Springfield, Virginia: National Technical Information Service, 1976.

Citations from ENGINEERING INDEX.

LIQUIFIED NATURAL GAS. Quarterly. Boulder, Colorado: U.S. National Bureau of Standards, Institute for Basic Standards.

A literature survey service.

NATURAL GAS. Part 1: Supply, Demand and Utilization for 1964-76. Rept. no. NTIS/PS-76/0498. Part 2: Marine Transportation. Rept. no. NTIS/PS-76/0499. By E.J. Lehmann. Springfield, Virginia: National Technical Information Service, 1976.

Bibliographies with abstracts.

PETROLEUM LITERATURE INDEX. Annual. Amarillo, Texas: National Petroleum Bibliography.

SELECTED DOCUMENTS OF THE INTERNATIONAL PETROLEUM INDUSTRY. Annual. Organization of Petroleum Exporting Countries, 1010 Vienna, Austria.

SELECTED LIST OF U.S. BUREAU OF MINES PUBLICATIONS ON PETROLEUM AND NATURAL GAS, 1971-1975. By V. Hutchison. Rept. no. BERC/IC-75-2. Springfield, Virginia: National Technical Information Service, 1975.

Sell, G., comp. 50 YEARS OF PETROLEUM TECHNOLOGY. London: Institute of Petroleum, 1968.

Swanson, E. A CENTURY OF OIL AND GAS IN BOOKS; A DESCRIPTIVE BIBLIOGRAPHY. New York: Appleton, 1960.

> "An annotated bibliography of books and monographs published in English from the mid-19th century to August 1959."

U.S. Department of Commerce. National Bureau of Standards. SURVEY OF LITERATURE FOR LIQUIFIED NATURAL GAS. Quarterly. National Bureau of Standards, Cryogenic Data Center, Boulder, Colorado 80302.

U.S. Department of Interior. Bureau of Mines. LIST OF BUREAU OF MINES PUBLICATIONS ON OIL SHALE AND SHALE OIL, 1917-68. Washington, D.C.: Government Printing Office, 1969.

_____. SELECTED LIST OF BUREAU OF MINES PUBLICATIONS ON PETRO-LEUM AND NATURAL GAS, 1961-70. Washington, D.C.: Government Printing Office, 1972.

_____. SIMPLIFIED OIL AND GAS WELL INFORMATION RETRIEVAL SYS-TEM. By J.G. Craig, 1971. Washington, D.C.: Government Printing Office, 1971.

b. RELATED BIBLIOGRAPHIES IN CURRENT JOURNALS

The following journals frequently contain book reviews and/or bibliographies. See the heading Journals and Serial Publications in this section for addresses.

AMERICAN GAS ASSOCIATION MONTHLY

AUSTRALASIAN OIL AND GAS REVIEW

DRILLING - DCW

ENERGY MANAGEMENT REPORT

ENERGY MANAGEMENT REPORT /D/

EUROPE AND OIL

GAS IN CANADA

GAS (U.S.)

GAS WORLD

HYDROCARBON PROCESSING

INDEPENDENT PETROLEUM MONTHLY

INSTITUTE OF PETROLEUM JOURNAL

INSTITUTION OF GAS ENGINEERS JOURNAL

JOURNAL OF LUBRICATION TECHNOLOGY

JOURNAL OF PETROLEUM TECHNOLOGY

NATIONAL PETROLEUM NEWS

OIL AND GAS JOURNAL

OIL STATISTICS

PETRO/CHEM ENGINEER

PETROCHEMICAL NEWS

PETROLEUM CHEMISTRY (USSR)

PETROLEUM ENGINEER

PETROLEUM EQUIPMENT AND SERVICES

PETROLEUM REVIEW

PETROLEUM TIMES

PIPELINE AND GAS JOURNAL

PIPELINE INDUSTRY

PIPES AND PIPELINES INTERNATIONAL

PM REPORT

WORLD OIL

See also this subject in Section I of this Guide.

6. Monographs, Reports, and Symposia Proceedings

Brantly, J. HISTORY OF OIL WELL DRILLING. Houston: Gulf Publishing Co., 1972.

> "The complete story of the oil well drilling industry compiled for the first time in one volume of 1,525 pages with more than 1,700 illustrations."

Brownstein, A. U.S. PETROCHEMICALS. Tulsa: Petroleum Publishing Co., 1972.

Campbell, J. GAS CONDITIONING AND PROCESSING. 2nd ed. Tulsa: Petroleum Publishing Co., 1970.

> "Production, conditioning, transmission, and processing of natural gas and its associated liquids."

CHEMICAL & PETROLEUM INSTRUMENTATION SYMPOSIUM. 13th Annual. Published as: INSTRUMENTATION IN THE CHEMICAL AND PETROLEUM IN-DUSTRIES. Vol. 8. Edited by Irving C. Young. Instrument Society of Amer-ica, 400 Stanwix Street, Pittsburgh, Pennsylvania 15222. 1972.

Chilingar, C., ed. OIL AND GAS PRODUCTION FROM CARBONATE ROCKS. New York: Elsevier, 1972.

"Introduction to carbonate reservoir rocks. Pore geometry of carbonate rocks. Fluid flow in carbonate reservoirs. Formation evaluation. Estimation of oil and gas reserves and the production forecast in carbonate reservoirs."

CURRENT EVALUATION OF 2 1/4 CHROME 1 MOLYBDENUM STEEL IN PRESSURE VESSELS & PIPING. Symposium at 3rd Pressure Vessels and Piping Conference and 27th Petroleum Mechanical Engineering Conference. Sponsored by American Society of Mechanical Engineers and Metal Properties Council, New Orleans, Louisiana. New York: American Society of Mechanical Engineers, 1972.

DECLINING DOMESTIC RESERVES -- EFFECT ON PETROLEUM AND PETROCHEMICAL INDUSTRY. 3rd Annual Symposium of Petroleum and Petrochemical Division at 71st National Meeting. New York: American Institute of Chemical Engineers, 1972.

Defense Research Information Centre. Orpington, England, 1973. THE FATE OF OIL SPILT AT SEA (AD-763 042). Springfield, Virginia: National Technical Information Service, 1973.

"Studies effects of natural forces such as wind wave motion currents temperature and other physical, chemical, and biological factors on movement, dispersal, and destruction of treated and untreated oil at sea and in the inter-tidal zone."

DISTILLATION. 4th Annual ISA Education Symposium. Instrument Society of America, Pittsburgh, Pennsylvania 15222. 1972.

Foster Association, Washington, D.C. AN ANALYSIS OF THE REGULATORY ASPECTS OF NATURAL GAS SUPPLY (PB-219 667). Springfield, Virginia: National Technical Information Service, 1973.

"Discusses the present regulation of supply and distribution of gas in the United States and identifies possible future changes in the regulation of gas."

Goldman, G. LIQUID FUELS FROM COAL. Park Ridge, New Jersey: Noyes Data Corp., 1972.

Based on review of coal liquefaction and other patents.

Hahn, A. THE PETROCHEMICAL INDUSTRY; MARKETS AND ECONOMICS. New York: McGraw-Hill, 1966.

Analyzes the industry trends and economics of various production processes.

Harris, L. AN INTRODUCTION TO DEEPWATER FLOATING DRILLING OPERATIONS. Tulsa: Petroleum Publishing Co., 1972.

Thorough study of today's technology.

Institution of Mining and Metallurgy. MINING AND PETROLEUM TECHNOL-
OGY. Vol. 1. London: 1969.

> "Proceedings of the Ninth Commonwealth Mining and Metallurgi-
> cal Congress 1969. Petroleum papers cover a wide range of
> technology relating to the offshore production of liquid petroleum
> and natural gas. The papers deal particularly with offshore drill-
> ing, the production facilities and the construction and operation
> of the necessary pipelines."

KINETIC REACTORS IN THE CHEMICAL AND PETROLEUM INDUSTRIES, Sym-
posium at 14th Chemical and Petroleum Instrumentation Symposium and 5th ISA
Education Symposium. Instrument Society of America, 400 Stanwix Street,
Pittsburgh, Pennsylvania 15222. 1973.

McDermott, J. LIQUIFIED NATURAL GAS TECHNOLOGY. Park Ridge, New
Jersey: Noyes Data Corp., 1973.

> Covers liquefaction, storage and handling, marine transportation,
> and regasification.

National Petroleum Council. Committee on U.S. Energy Outlook. U.S. EN-
ERGY OUTLOOK: AN INITIAL APPRAISAL, 1971-85. 2 vols. Washington,
D.C.: 1971.

Nelson, W. PETROLEUM REFINERY ENGINEERING. 4th ed. New York:
McGraw-Hill, 1958.

> Discusses methods of processing petroleum, design and cost of
> refinery equipment, etc.

1972 NPRA QUESTION AND ANSWER SESSION ON REFINERY MAINTE-
NANCE. Tulsa: Petroleum Publishing Co., 1972.

> "Panel of 15 professionals gives workable answers to 254 questions
> dealing with keeping refineries on-stream at the lowest possible
> cost -- from day-to-day operations to unit turn-around procedures."

1972 NPRA QUESTION AND ANSWER SESSION ON REFINING TECHNOLOGY.
Tulsa: Petroleum Publishing Co., 1972.

> "Latest technological answers to problems in Light Oil Catalytic
> Processing; Catalytic Cracking; Refinery Support Operations; Hydro-
> gen Processing; General Refining Processing."

9TH WORLD PETROLEUM CONGRESS. Tokyo, Japan, May 1976. Proceedings. Seven vols. Barking, Essex, England: Applied Science Publishers, 1976.

NORTHERN PIPELINE RESEARCH CONFERENCE. 1st Canadian National Conference. National Research Council of Canada, Montreal Road, Ottawa, Ontario K1A OR6, Canada. 1972.

Ocean Systems, Inc. DEVELOPMENT OF A HIGH SEAS OIL RECOVERY SYS-

TEM (AD-759 522). Springfield, Virginia: National Technical Information Service, 1973.

> "A model test development program was conducted of a double weir oil recovery system for rapid recovery of oil spilled on the high seas. A summary of model test results and prototype performance projections is given."

OFFSHORE TECHNOLOGY. 4th Annual Conference, Houston, Texas, May 1-3, 1972. 2 vols. New York: American Institute of Mining, Metallurgical, and Petroleum Engineers, 1972.

ORIGIN AND REFINING OF PETROLEUM. H.G. McGrath and M.E. Charles, symposium co-chairmen. Advances in Chemistry Series 103. Washington, D.C.: American Chemical Society, 1971.

> "A symposium co-sponsored by the Division of Petroleum Chemistry of the American Chemical Society and the Canadian Society for Chemical Engineering of the Chemical Institute of Canada at the ACS-CIC joint conference, Toronto, Canada, May 25-28, 1970."

PETROLEUM AND CHEMICAL INDUSTRY. 21st Annual Conference. Piscataway, New Jersey: Institute of Electrical and Electronic Engineers, 1974.

PETROLEUM & CHEMICAL INDUSTRY CONFERENCE. 19th Annual. Institute of Electrical and Electronics Engineers, 345 East 47th Street, New York, New York 10017. 1972.

PETROLEUM CONGRESS. 8th World Petroleum Congress. United States Distributor: American Petroleum Institute, 1271 Avenue of Americas, New York, New York 10020. 1971.

> Vol. 1: General; Vol. 2: Geological & Exploration; Vol. 3: Geophysical exploration, drilling & production; Vol. 4: Manufacturing; Vol. 5: Products; Vol. 6: Transportation, conservation & general topics; Vol. 7: Index.

PETROLEUM DERIVED CARBONS. Symposium of the Divisions of Petroleum Chemistry and Industrial and Engineering Chemistry of the American Chemical Society, Philadelphia, Pennsylvania, April 1975. Proceedings, ACS Symposium Series no. 21. Washington, D.C.: American Chemical Society, 1976.

PETROLEUM PIPELINE TRANSPORTATION & STORAGE. Panel Discussion at 8th World Petroleum Congress. United States Distributor, New York: American Petroleum Institute, 1971.

PRESSURE VESSELS & PIPING CONFERENCE. 3rd Annual and Petroleum Mechanical Engineering Conference, 27th Annual. New York: American Society of Mechanical Engineers, 1972.

Ranney, M. LIQUID FUELS FROM OIL SHALE AND TAR SANDS. Park Ridge, New Jersey: Noyes Data Corp., 1972.

> Based on patent searches.

REFINING PETROLEUM FOR CHEMICALS; A SYMPOSIUM. L.J. Spillane and H.P. Leftin, symposium co-chairmen. Advances in Chemistry Series 97. Washington, D.C.: American Chemical Society, 1970.

> "Sponsored by Division of Petroleum Chemicals and Division of Industrial and Engineering Chemicals at the 158th meeting of the American Chemical Society, New York, New York, September 10-12, 1969."

Sittig, M. INDUSTRIAL GASES MANUFACTURE AND APPLICATIONS. Park Ridge, New Jersey: Noyes Data Corp., 1967.

> "This book discusses conventional cryogenic air separation and purification techniques in considerable detail."

Stobaugh, R. PETROCHEMICAL MANUFACTURING AND MARKETING GUIDE. Vol. 1: Aromatics and Derivatives. Houston: Gulf Publishing Co., 1966.

SYMPOSIUM ON ENERGY, RESOURCES, AND THE ENVIRONMENT. Mitre Corp., McLean, Virginia. Springfield, Virginia: National Technical Information Service, 1973.

> Vol. I: Introductory and context sessions (PB-219 953).
> Vol. II: Panel sessions on energy and resource issues (PB-219 954).
> Vol. III: Panel sessions on environmental, economical and institutional issues (PB-219 955).
> Vol. IV: Recapitulation of energy, resources, environmental, economic and institutional issues (PB-210 956).

THERMAL STRUCTURAL ANALYSIS PROGRAMS: A SURVEY & EVALUATION. Symposium at 3rd Pressure Vessels and Piping Conference and 27th Petroleum Mechanical Engineering Conference. American Society of Mechanical Engineers, 345 East 47th Street, New York, New York 10012. 1972.

37th MIDYEAR MEETING. API Refining Division. PROCEEDINGS. Vol. 52. 1972. American Petroleum Institute-Division of Refining, 1801 K Street, N.W. Washington, D.C. 20006.

22ND ANNUAL PETROLEUM AND CHEMICAL INDUSTRY CONFERENCE. Milwaukee, Wisconsin, September 1975. Record of Conference Papers. Piscataway, New Jersey: Institute of Electrical and Electronic Engineers, 1975.

U.S. Coast Guard. CRUDE OIL BEHAVIOR ON ARCTIC WINTER ICE. AD-754 261. Springfield, Virginia: National Technical Information Service, 1973.

> "Investigates oil spread rate, oil absorption and aging of oil on snow and ice surfaces, and effectiveness of various cleanup pro-

cedures by conducting small control oil spills on the Bering Sea in Northwestern Alaska using Prudhoe Bay crude oil."

U.S. Congress. OIL LAND LEASING ACT OF 1920 WITH AMENDMENTS AND OTHER LAWS RELATING TO MINERAL LANDS. March 2, 1919-August 12, 1968. Washington, D.C.: Government Printing Office, 1968.

U.S. Congress. Economic Joint Committee. NATURAL GAS REGULATION AND TRANS-ALASKA PIPELINE, HEARINGS. 92d Congress, 2d session, June 7-22, 1972. Washington, D.C.: Government Printing Office, 1972.

U.S. Congress. House Interior and Insular Affairs Committee. ENERGY DEMAND STUDIES ANALYSIS AND APPRAISAL. 92d Congress; September 1972. Washington, D.C.: Government Printing Office, 1972.

_____. FUEL AND ENERGY RESOURCES, 1972, HEARINGS. 92d Congress, 2d session. Pt. 1: April 10-13, 1972. Washington, D.C.: Government Printing Office, 1972.

U.S. Congress. House Science and Astronautics Committee. INVENTORY OF ENERGY RESEARCH. Prepared for Task Force on Energy of Subcommittee on Science, Research, and Development, 92d Congress, 2d session, by Oak Ridge National Laboratory with Support of National Science Foundation, March 1972. 2 vols. Washington, D.C.: Government Printing Office, 1972.

U.S. Congress. Senate. ENERGY RESEARCH POLICY ALTERNATIVES, HEARING. 92d Congress, 2d Session Pursuant to Senate Resolution 45, National Fuels and Energy Policy Study on Existing Federal Energy Research and Development Policies and Future Technological Options, June 7, 1972. Washington, D.C.: Government Printing Office, 1972.

_____. GOVERNMENTAL INTERVENTION IN THE MARKET MECHANISM. Hearings before the Subcommittee on Antitrust and Monopoly of the Committee on the Judiciary, Senate, 91st Congress, 1st session, Senate Resolution 40. Washington, D.C.: Government Printing Office, 1969-70.

_____. SOME ENVIRONMENTAL IMPLICATIONS OF NATIONAL FUELS POLICIES. Report Prepared by the Staff of the Committee on Public Works, Senate, 1970. Washington, D.C.: Government Printing Office, 1970.

"Presents findings of a study of optimization on coal gasification processes to determine those processes which hold more promise for potential economic development for the commercial production of a high-caloric-value pipeline gas."

U.S. Department of Defense. MAINTENANCE AND OPERATION OF GAS SYSTEMS. Washington, D.C.: Government Printing Office, 1970.

U.S. Department of Interior. LIQUID FUELING AND DISPENSING FACILITIES, DESIGN MANUAL 22. Rev. ed. Washington, D.C.: Government Printing Office, 1971.

U.S. Department of Interior. Bureau of Mines. CHAR OIL ENERGY DEVEL-OPMENT, PERIOD OF PERFORMANCE. THE CHAR OIL ENERGY DEVELOP-MENT PROCESS IS THE MULTISTAGE FLUIDIZED-BED PYROLYSIS OF COAL TO PRODUCE OIL, GAS AND CHAR. September 1966-February, 1970, In-terim Report 1, Process Development Unit Results and Commercial Analyses. Washington, D.C.: Government Printing Office, 1970.

_____. ENGINEERING EVALUATION OF PROJECT GASOLINE, CONSOL SYNTHETIC FUEL PROCESS. Washington, D.C.: Government Printing Of-fice, 1970.

_____. FINAL REPORT OF THE ADVISORY COMMITTEE ON PROJECT GASOLINE. National Academy of Engineering, R & D report no. 62, Period of Performance, January - October 1970. Washington, D.C.: Government Printing Office, 1970.

_____. GREEN RIVER FORMATION LITHOLOGY AND OIL SHALE CORRE-LATIONS IN THE PICEANCE CREEK BASIN, COLORADO. Washington, D.C.: Government Printing Office, 1970.

_____. HAZARDS OF SPILLAGE OF LNG INTO WATER. AD-754 498. Springfield, Virginia: National Technical Information Service, 1973.

> "Assesses the hazards of vapor explosions, and shows that explo-sions under various experimental conditions did not ignite the flammable vapor cloud and that the energy release is very mod-est relative to chemical explosions."

_____. HEAVY OIL RESERVOIRS IN ARKANSAS. Washington, D.C.: Gov-ernment Printing Office, 1969.

_____. INTERSTITIAL COMPOUNDS AS FUEL CELL CATALYSTS, THEIR PREPARATIVE TECHNIQUES AND ELECTROCHEMICAL TESTING. Washington, D.C.: Government Printing Office, 1970.

_____. NATURAL GAS LIQUIDS: A REVIEW OF THEIR ROLE IN THE PE-TROLEUM INDUSTRY, 1969. Washington, D.C.: Government Printing Of-fice, 1969.

_____. 1970 FINAL REPORT CONSOL SYNTHETIC FUEL PROCESS. Wash-ington, D.C.: Government Printing Office, 1970.

_____. 1970 FINAL REPORT PROJECT FUEL CELL. Washington, D.C.: Government Printing Office, 1970.

_____. OFFSHORE PETROLEUM STUDIES: COMPOSITION OF THE OFF-SHORE U.S. PETROLEUM INDUSTRY AND ESTIMATED COSTS OF PRODUC-ING PETROLEUM IN THE GULF OF MEXICO. XF 214 117. Washington, D.C.: Government Printing Office, 1973.

_____. OIL SHALE, POTENTIAL SOURCE OF ENERGY, 1972. Washington, D.C.: Government Printing Office, 1972.

_____. OIL TAGGING SYSTEM STUDY. 1970. Washington, D.C.: Government Printing Office, 1970.

_____. OPTIMIZATION OF COAL GASIFICATION PROCESSES 1972. 2 vols. Washington, D.C.: Government Printing Office, 1972.

_____. PIPELINE GAS FROM LIGNITE GASIFICATION, CURRENT COMMERCIAL ECONOMICS. Washington, D.C.: Government Printing Office, 1969.

_____. POTENTIAL APPLICATIONS FOR NUCLEAR EXPLOSIVES IN A SHALE-OIL INDUSTRY, 1969. Washington, D.C.: Government Printing Office, 1969.

_____. POTENTIAL OIL RECOVERY BY WATERFLOODING RESERVOIRS BEING PRODUCED BY PRIMARY METHODS, 1970. Washington, D.C.: Government Printing Office, 1970.

_____. ROLE OF PETROLEUM AND NATURAL GAS FROM THE OUTER CONTINENTAL SHELF IN NATIONAL SUPPLY OF PETROLEUM AND NATURAL GAS. Washington, D.C.: Government Printing Office, 1970.

_____. SIMPLIFIED OIL AND GAS WELL INFORMATION RETRIEVAL SYSTEM, 1971. Washington, D.C.: Government Printing Office, 1971.

_____. STORAGE STABILITY OF GASOLINE, DEVELOPMENT OF STABILITY PREDICTION METHOD AND STUDIES OF GASOLINE COMPOSITION AND COMPONENT REACTIVITY. By F.G. Schwartz. Washington, D.C.: Government Printing Office, 1972.

Includes list of references.

_____. SULFUR CONTENT OF CRUDE OILS OF THE FREE WORLD, 1967. Washington, D.C.: Government Printing Office, 1967.

_____. SUMMARY REPORT ON PROJECT GASOLINE, PERIOD SEPTEMBER 1963 - JUNE, 1969. Vol. 1. Washington, D.C.: Government Printing Office, 1970.

_____. TRANS-ALASKA PIPELINE. An analysis of the economic and security aspects of the Trans-Alaska pipeline. Washington, D.C.: Government Printing Office, 1971-72.

Vol. 1: Summary. 1971, published 1972.
Vol. 2: Supporting analyses. 1971, published 1972.
Vol. 3: Supplement, energy and policy alternatives. 1972.

_____. U.S. ENERGY NEEDS THROUGH THE YEAR 2000. By W. Dupree, et al. Washington, D.C.: Government Printing Office, 1973.

U.S. Federal Council for Science and Technology. Panel for the Committee on Energy, Research and Development Goals. EXTRACTION OF ENERGY FUELS. PB-220-328. Washington, D.C.: Government Printing Office, 1973.

> Discusses short-term needs (1972-80), intermediate-term needs (1975-85), and long-term needs (1980-2000), and gives a technical assessment of: stimulation of petroleum and natural gas production; production of oil from tar sands; development of oil shale; underground gasification of coal; oil and gas production from organic wastes; and primary extraction of coal.

U.S. NASA. COMPRESSED GAS HANDBOOK. Rev. ed. NAS 1.21:3045/2. Washington, D.C.: Government Printing Office, 1971.

U.S. Office of the President. OFFSHORE MINERAL RESOURCES, A CHALLENGE AND AN OPPORTUNITY. Washington, D.C.: Government Printing Office, 1970.

_____. OIL IMPORT QUESTION, A REPORT ON THE RELATIONSHIP OF OIL IMPORTS TO THE NATIONAL SECURITY BY THE CABINET TASK FORCE ON OIL IMPORT CONTROL. Washington, D.C.: Government Printing Office, 1970.

Van Houton Associates. ECONOMIC AND ENGINEERING ANALYSIS FOR DELIVERY OF REFINED PETROLEUM PRODUCTS. AD-759 511. Springfield, Virginia: National Technical Information Service, 1973.

> Discusses the economics of providing petroleum products to Northeastern United States consumers, using various alternatives of refinery location, through the year 2000.

World Petroleum Congresses. PROCEEDINGS. Quadrennial since 1963. Elsevier Publishing Co., Barking, Essex, England (inquire World Petroleum Congresses, 61 New Cavendish St., London W1, England).

> Proceedings published in host country.

Young, I., ed. INSTRUMENTATION IN THE CHEMICAL AND PETROLEUM INDUSTRIES. Irregular. Pittsburgh: Instrument Society of America.

> "A collection of papers concerning control instrumentation, emphasizes safety, intrinsic and otherwise, in the design, operation, and maintenance of instrumentation systems during the start-up, running, and shut-down of a process."

7. Abstracting and Indexing Services

a. GENERAL

APPLIED MECHANICS REVIEWS

APPLIED SCIENCE AND TECHNOLOGY INDEX

BRITISH TECHNOLOGY INDEX

CHEMICAL ABSTRACTS

ENGINEERING INDEX

See Section I of this Guide for additional details on these services.

b. SPECIFIC TO TOPIC

American Chemical Society. Petroleum Research Fund Advisory Board. AB-STRACTS OF RESEARCH, 1970. Petroleum research fund 1971-1972. 2 vols., Washington, D.C.: 1972.

API ABSTRACTS OF REFINING LITERATURE. Weekly. American Petroleum Institute, 1271 Avenue of the Americas, New York, New York 10020.

> Has four subject areas: petroleum refining and petrochemicals; air and water conservation; transportation and storage; petroleum substitutes.

BULLETIN OF CURRENT REFERENCES ON PETROLEUM AND GEOLOGICAL SCIENCES. Quarterly. Central Library, Oil and Natural Gas Commission, Research Training Institute, Kaulag Road, Dehra Dun, India.

Centre de Recherches de Pau. BULLETIN. Semiannual. Societe Nationale des Petroles d'Aquitaine, Centre de Recherches, av. P. Angot, 64 Pau, France.

> Text in French; summaries in English and French.

CHEMICAL ABSTRACTS - ORGANIC CHEMISTRY SECTIONS. Fortnightly. Chemical Abstracts Service, American Chemical Society, 1155 16th Street, N.W., Washington, D.C. 20036.

FUEL ABSTRACTS AND CURRENT TITLES: A SUMMARY OF WORLD LITERA-TURE ON ALL TECHNICAL AND SCIENTIFIC ASPECTS OF FUEL AND POWER. Monthly. Institute of Fuel, 18 Devonshire Street, London WN 2AV, England.

GAS ABSTRACTS. Monthly. Institute for Gas Technology, 3424 South State Street, Chicago, Illinois 60616.

INDEXES TO REFINING LITERATURE. Abstracts. Monthly. American Petro-leum Institute, 1271 Avenue of the Americas, New York, New York 10020.

> Includes three services: 1. alphabetical subject index (monthly publication) to major articles on subject; printed as a book; 2. dual dictionary – a coordinate index providing manual search capability of terms for documents having more than one descriptive subject heading; 3. magnetic tapes for computer searches.

INSTITUTE OF PETROLEUM ABSTRACTS. Quarterly. Institute of Petroleum, 61 New Cavendish Street, London, W1M 8AR, England.

PETROLEUM ABSTRACTS. Weekly. Department of Information Services, College of Petroleum, Sciences and Engineering, University of Tulsa, Tulsa, Oklahoma 74110.

Predi-briefs. ABSTRACTING SERVICE. Monthly. ENERGY. Predicasts, 200 University Circle Research Center, 11001 Cedar Avenue, Cleveland, Ohio 44106.

> "Contains information on conventional fuels -- coal, crude oil, natural gas and nuclear energy -- and newer technologies such as solar energy, geothermal energy, synthetic natural gas and fuel cells. Low-sulfur fuels and other pollution aspects briefed."

SEARCH: PETROLEUM DIVISION. Monthly. Compendium Publishers International Corp., 2175 Lemoine Avenue, Fort Lee, New Jersey 07024.

c. RELATED ABSTRACTS IN CURRENT JOURNALS

The following journals frequently provide abstracts. See the heading Journals and Serial Publications in this section for addresses.

EUROCOM PRESS INFORMATION

GAS PROCESSING/CANADA

JOURNAL OF PETROLEUM TECHNOLOGY

PETROLEUM CHEMISTRY (USSR)

PROPANE/CANADA

SOCIETY OF PETROLEUM ENGINEERS. JOURNAL.

8. Journals and Serial Publications

a. DIRECTORIES AND LISTS

AYER'S DIRECTORY OF PUBLICATIONS. Annual. Philadelphia: Ayer Press.

STANDARD PERIODICAL DIRECTORY. New York: Oxbridge Publishing Co., 1973.

ULRICH'S INTERNATIONAL PERIODICAL DIRECTORY. 2 vols. Annual. New York: R.R. Bowker Co.

b. JOURNAL AND SERIAL TITLES

ABSTRACTS OF EXPLORATION AND PRODUCTION LITERATURE AND PATENTS. Weekly. Tulsa, Oklahoma: University of Tulsa.

ADVANCES IN PETROLEUM CHEMISTRY AND REFINING. Irregular. John Wiley & Sons, Inc., 605 Third Avenue, New York, New York 10016.

ALBERTA OIL AND GAS INDUSTRY. Monthly statistics. Alberta Oil and Gas Conservation Board, 603 Sixth Avenue, S.W., Calgary 1, Alberta, Canada.

AMERICAN ASSOCIATION OF PETROLEUM GEOLOGISTS BULLETIN. Weekly. Box 979, Tulsa, Oklahoma 74101.

American Chemical Society. Petroleum Research Fund Advisory Board. AB-STRACTS OF RESEARCH, 1970. PETROLEUM RESEARCH FUND, 1971-1972. 2 vols. Washington, D.C.: 1972.

American Gas Association. Bureau of Statistics. GAS UTILITY AND PIPE-LINE INDUSTRY PROJECTIONS. Irregular. American Gas Association, Bureau of Statistics, 1515 Wilson Boulevard, Arlington, Virginia 22209.

AMERICAN GAS ASSOCIATION MONTHLY. 11 issues/year. 1515 Wilson Boulevard, Arlington, Virginia 22209.

American Petroleum Institute. ABSTRACTS OF AIR AND WATER CONSERVA-TION LITERATURE AND PATENTS. Weekly. American Petroleum Institute, 555 Madison Avenue, New York, New York 10022.

     Available in microfilm.

_____. ABSTRACTS OF PETROLEUM SUBSTITUTES LITERATURE AND PAT-ENTS. Monthly. Central Abstracting and Indexing Service, American Petroleum Institute, 555 Madison Avenue, New York, New York 10022.

     Available in microform and card format.

_____. ABSTRACTS OF REFINING PATENTS. Weekly. American Petroleum Institute, 555 Madison Avenue, New York, New York 10022.

     Available on microform or on card format.

_____. ABSTRACTS OF TRANSPORTATION AND STORAGE LITERATURE AND PATENTS. Monthly. American Petroleum Institute, 555 Madison Avenue, New York, New York 10022.

     Available in microform.

American Petroleum Institute. Department of Statistics. WEEKLY STATISTICAL BULLETIN. (Chart Supplement). American Petroleum Institute, 1101 17th Street N.W., Washington, D.C. 20036.

American Public Gas Association. MEMORANDUM BULLETINS. Irregular. American Public Gas Association, 2600 Virginia Avenue, N.W., Washington, D.C. 20037.

ANNUAL BULLETIN OF GAS STATISTICS FOR EUROPE. Annual. U.N. Economic Commission for Europe, U.N. Publications, United Nations, Room 1059, New York, New York 10017.

Text in English, French and Russian.

API ABSTRACTS OF REFINING LITERATURE. Weekly. American Petroleum Institute, 1271 Avenue of the Americas, New York, New York 10020.

Has four subject groups: petroleum refining and petrochemicals; air and water conservation; transportation and storage; petroleum substitutes.

API PATENT ALERT ABSTRACTS. Weekly. American Petroleum Institute, 1271 Avenue of the Americas, New York, New York 10020.

Reports patents in petroleum refining and petrochemical industry. Covers United States, United Kingdom, South Africa, Japan, rest of European countries.

AUSTRALASIAN OIL AND GAS REVIEW. Monthly. Tracer Petroleum and Mining Publications Pty, Ltd., Box 4318, Sydney, N.S.W. 2001, Australia.

Formerly: AUSTRALASIAN OIL AND GAS JOURNAL.

AUSTRALIAN GAS JOURNAL. Bimonthly. Australian Gas Association, 320 St. Kilda Road, Melbourne, Victoria, Australia.

BRITISH PETROLEUM EQUIPMENT AND SERVICES. Biennial. Council of British Manufacturers of Petroleum Equipment, 118 Southwark Street, London, SE1, England.

BULLETIN OF CURRENT REFERENCES ON PETROLEUM AND GEOLOGICAL SCIENCES. Quarterly. Central Library, Oil and Natural Gas Commission, Research Training Institute, Kaulag Road, Dehra Dun, India.

BUTANE PROPANE. Quarterly. Societe J.A.M., 7 Square de Chatillon, Paris (14e), France.

Summaries in English and French.

BUTANE-PROPANE NEWS. Monthly. Butane-Propane News, Inc., Box 3027, Arcadia, California 91006.

Canada. Bureau of Statistics. CRUDE PETROLEUM AND NATURAL GAS INDUSTRY. Annual. Bureau of Statistics, Ottawa, Canada.

_____. PETROLEUM REFINERIES. Annual. Bureau of Statistics, Ottawa, Canada.

_____. REFINED PETROLEUM PRODUCTS. Annual. Bureau of Statistics, Ottawa, Canada.

Canada. Dominion Bureau of Statistics. SERVICE BULLETIN: ENERGY STA-TISTICS. Weekly. Dominion Bureau of Statistics, Tunney's Pasture, Ottawa 3, Ontario, Canada.

Canadian Gas Association. STATISTICAL SUMMARY OF THE CANADIAN GAS INDUSTRY. Annual. Canadian Gas Association, 55 Scarsdale Road, Don Mills, Ontario Canada.

CANADIAN OIL REGISTER. Annual. C.O. Nickle Publications Co., Ltd., 330 Ninth Avenue, S.W., Calgary, Alberta, Canada.

CANADIAN PETROLEUM. Monthly. Business Publications Ltd., 1450 Don Mills Road, Don Mills, Ontario, Canada.

> Formerly: CANADIAN PETRO ENGINEERING.

Canadian Petroleum Association. STATISTICAL YEARBOOK. Annual. Canadian Petroleum Association, Calgary, Alberta, Canada.

Centre de Recherches de Pau. BULLETIN. Semiannual. Societe Nationale des Petroles d'Aquitaine, Centre de Recherches; av. P. Angot, 64 Pau, France.

> Text in French; summaries in English and French.

CHEMICAL ABSTRACTS - ORGANIC CHEMISTRY SECTIONS. Fortnightly. American Chemical Society, Chemical Abstracts Service, 1155 16th Street N.W., Washington, D.C. 20036.

CHEMICAL AND PETROLEUM ENGINEERING. Bimonthly. Consultants Bureau, 227 West 17th Street, New York, New York 10011.

> English translation. Formerly: SOVIET JOURNAL OF CHEMI-CAL ENGINEERING.

THE DRILLING CONTRACTOR. Bimonthly. International Association of Drilling Contractors, 211 North Ervoy Building, Suite 505, Dallas, Texas 75201.

> Official publication of the association providing details of current developments in the field.

DRILLING-D C W. Monthly. Box 19305, Dallas, Texas 75219.

> Formerly: DRILLING INTERNATIONAL.

DRILLING MAGAZINE. Monthly. Associated Publishers, Inc., 8383 Stemmons Freeway, Suite 151 Empire Central, Dallas, Texas 75247.

ENERGY DEVELOPMENTS. Monthly. International Review Service, Inc., 15 Washington Place, New York, New York 10003.

> Covers the fields of petroleum, petrochemicals, atomic energy, human environment, and pollution, space, desalination and the applications of science and technology. Focuses on interests of developing countries.

ENERGY MANAGEMENT REPORT/D/ENERGY EQUIPMENT PLANNING GUIDE. Monthly. Petroleum Engineer Publishing Co., Box 1589, Dallas, Texas 75221.

EUROCOM PRESS INFORMATION. 9/year. EUROCOM, European Fuel Merchants Union, Riponne 5, Lausanne, Switzerland.

EUROPE AND OIL; MONTHLY. Ferdinand S. Metzger, Antwerpene Street, 12, 8 Munich 23, West Germany.

In English.

FUEL ABSTRACTS AND CURRENT TITLES: A SUMMARY OF WORLD LITERATURE ON ALL TECHNICAL AND SCIENTIFIC ASPECTS OF FUEL AND POWER. Institute of Fuel, 18 Devonshire Street, London WN 2 AV, England.

GAS (U.S.); THE MAGAZINE OF THE GAS INDUSTRY. Monthly. 4151 Southwest Freeway, Suite 735, Houston, Texas 77027.

GAS ABSTRACTS. Monthly. Institute of Gas Technology, 3424 South State Street, Chicago, Illinois 60616.

GAS AGE. Monthly. Moore Publishing Co., Ojibway Building, Duluth, Minnesota 55802.

GAS IN CANADA. Monthly. Canadian Gas Industry, Box 418, 15 Yonge Street N., Richmond Hill, Ontario, Canada.

GAS INDUSTRIES. Monthly. Paul Lady & Wm. O. Dannhausen, 333 North Michigan Avenue, Chicago, Illinois 60601.

GAS IN INDUSTRY AND COMMERCE. Monthly. Walter King Ltd., 100 Gray's Inn Road, London, WC1X 8AP, England.

Formerly: GAS IN INDUSTRY.

GAS JOURNAL. Weekly. Walter King Ltd., Holborn Hall, 100 Gray's Inn Road, London, WC 1, England.

GAS MARKETING. Monthly. Walter King Ltd., Holborn Hall, 100 Gray's Inn Road, London, WC 1, England.

GAS PROCESSING/CANADA. 6/year. Sanford Evans Publishing (Alta) Ltd., Suite 1, 5512 McLeod Trail, Calgary, Alberta, Canada.

GAS SCOPE. Bimonthly. Institute of Gas Technology, 3424 South State Street, Chicago, Illinois 60616.

GAS WORLD. Weekly. Benn Brothers Ltd., Bouverie House, Fleet Street, London, EC 4, England.

Formerly: GAS WORLD and GAS AND COKE.

HISTORICAL STATISTICS OF THE GAS INDUSTRY. Irregular. American Gas Association, Bureau of Statistics, 1515 Wilson Boulevard, Arlington, Virginia 22209.

HYDROCARBON NEWS. Monthly. Petroleum Engineer Publishing Co., Box 1589, Dallas, Texas 75221.

HYDROCARBON PROCESSING. Monthly. Gulf Publishing Co., Box 2608, Houston, Texas 77001.

IMPERIAL OIL REVIEW. Bimonthly. 111 St. Clair Avenue, W., Toronto 7, Canada.

> Text in English and French.

Independent Petroleum Association of Canada. IPAC PETROLEUM NEWS. Irregular. Independent Petroleum Association of Canada, No. 501, 600-6th Avenue, Calgary 1, Alberta, Canada.

> Formerly: IPAC NEWSLETTER.

INDEPENDENT PETROLEUM MONTHLY. Monthly. 1430 South Boulder Avenue, Box 1019, Tulsa, Oklahoma 74101.

INDEXES TO REFINING LITERATURE. Abstracts. Monthly. American Petroleum Institute, 1271 Avenue of the Americas, New York, New York 10020.

> Includes three services: 1. alphabetical subject index (monthly publication) to major articles on subject, printed as book; 2. dual dictionary - a coordinate index providing manual search capability of terms for documents having more than one descriptive subject heading; 3. magnetic tapes for computer searches.

INDUSTRIAL GAS. Monthly. Harbrace Publications Inc., Harbrace Building, Duluth, Minnesota 55802.

Institute of Petroleum. ABSTRACTS. Quarterly. Institute of Petroleum, 61 New Cavendish Street, London, W1M 8AR, England.

_____. INSTITUTE OF PETROLEUM STANDARDS FOR PETROLEUM AND ITS PRODUCTS. Annual. Institute of Petroleum, 61 New Cavendish Street, London, W1, England.

INSTITUTE OF PETROLEUM JOURNAL. Bimonthly. Institute of Petroleum, 61 New Cavendish Street, London, W1M 8AR, England.

INSTITUTION OF GAS ENGINEERS JOURNAL. Monthly. Institution of Gas Engineers, 17 Grosvenor Crescent, London SW 1, England.

INTERNATIONAL OIL AND GAS YEARBOOK. American Institute of Mining, Metallurgical and Petroleum Engineers, Inc., United Engineering Center, 345 East 47th Street, New York, New York 10017.

INTERNATIONAL OIL NEWS; A FORTNIGHTLY NEWS SERVICE COVERING THE INTERNATIONAL OIL AND GAS INDUSTRY PRODUCTION, EXPLORATION, TRANSPORTATION, PROCESSING, MARKETING. Fortnightly. William F. Bland Co., World Press Center Building, 54 West 40th Street, New York, New York 10018.

Formerly: WORLD PETROLEUM REPORT.

JAPAN PETROLEUM WEEKLY. Weekly. Japan Petroleum Consultants, Box 1185, Tokyo Central, Tokyo, Japan.

JOURNAL OF CANADIAN PETROLEUM TECHNOLOGY. Quarterly. Canadian Institute of Mining and Metallurgy, 906-1117 Ste. Catherine Street W., Montreal 110, Quebec, Canada.

JOURNAL OF LUBRICATION TECHNOLOGY. Quarterly. Series F – ASME Transactions. American Society of Mechanical Engineers, United Engineering Center, 345 East 47th Street, New York, New York 10017.

JOURNAL OF PETROLEUM TECHNOLOGY. Monthly. Society of Petroleum Engineers, AIME, 6200 North Central Expressway, Dallas, Texas 75206.

LIQUIFIED NATURAL GAS. Quarterly. American Gas Association, 1515 – Wilson Boulevard, Arlington, Virginia 22209.

Liquified Petroleum Gas Association. LP-GAS MARKET FACTS. Annual. National LP-Gas Association, 79 West Monroe Street, Chicago, Illinois 60603.

Statistical handbook of the LP-Gas industry.

LIQUIFIED PETROLEUM GAS REPORT. Monthly. American Petroleum Institute, Department of Statistics, 1101-17th Street, N.W., Washington, D.C. 20036.

LOG ANALYST. Bimonthly. Society of Professional Well Log Analysts, 13507 Tosca Lane, Houston, Texas 77024.

LP-GAS. Monthly. Harcourt Brace Jovanovich Publications, One East First Street, Duluth, Minnesota 55802.

National Chemical and Petroleum Instrumentation Symposium. PROCEEDINGS. Irregular. Plenum Publishing Corp., 227 West 17th Street, New York, New York 10011.

NATIONAL OIL JOBBER. Monthly. National Oil Jobbers Council, Inc., 1701 K Street, N.W., Washington, D.C. 20006.

NATIONAL PETROLEUM NEWS. Monthly. McGraw-Hill, 1221 Avenue of the Americas, New York, New York 10020.

NATURAL GAS & L.P.G. Bimonthly. Scientific Surveys Ltd., 11a Gloucester Road, London S.W. 7, England.

NATURAL GAS LIQUIDS. Monthly. U.S. Bureau of Mines, 4800 Forbes Avenue, Pittsburgh, Pennsylvania 15213.

NLGI SPOKESMAN. Monthly. National Lubricating Grease Institute, 4635 Wyandotte Street, Kansas City, Missouri 64112.

NLPGA TIMES. Quarterly. National LP-Gas Association, 79 West Monroe Street, Chicago, Illinois 60603.

NPN BULLETIN; A WEEKLY INFORMATION SERVICE FOR IMPORTANT PEOPLE IN OIL MARKETING. Weekly. McGraw-Hill, 1221 Avenue of the Americas, New York, New York 10020.

OCCUPATIONAL SAFETY AND HEALTH ABSTRACTS. Monthly. International Occupational Safety and Health Information Centre (CIS), International Labour Office, 1211 Geneva 22, Switzerland.

    Editions in English. Available in card format.

OCEAN OIL WEEKLY REPORT. Weekly. Box 1941, Houston, Texas 77001.

    Looseleaf format.

OECD PROVISIONAL OIL STATISTICS. Quarterly. Organization for Economic Co-operation and Development, Publications Center, 1750 Pennsylvania Avenue, N.W., Suite 1207, Washington, D.C. 20006.

    Editions in English and French.

OFFSHORE. Monthly. Petroleum Publishing Co., Box 66909, Houston, Texas 77006.

OFFSHORE TECHNOLOGY. Bimonthly. Scientific Surveys (Offshore) Ltd., 11 a Gloucester Road, London, SW 7, England.

OIL. Monthly. Oil Trade Journal Publishing Corp., 218 Pan American Building, New Orleans, Louisiana 70130.

OIL & GAS DISCOVERIES. Bimonthly. John S. Herold Inc., 35 Mason Street, Greenwich, Connecticut 06830.

OIL AND GAS EQUIPMENT. Monthly. Petroleum Publishing Co., Box 1260, Tulsa, Oklahoma 74101.

OIL AND GAS INDUSTRY PURCHASING. Bimonthly. Charleson Publishing Co., Box 1105, Darian, Connecticut 06820.

OIL AND GAS INTERNATIONAL. Monthly. Oil and Gas Journal, Box 1260, Tulsa, Oklahoma 74101.

    Text in English.

OIL AND GAS JOURNAL. Weekly. Petroleum Publishing Co., 211 South Cheyenne Avenue, Tulsa, Oklahoma 74101.

OIL AND PETROCHEMICAL EQUIPMENT NEWS. Quarterly. Council of British Manufacturers of Petroleum Equipment, Irwin House, 188 Southwark Street, London, SE 1, England.

OIL AND PETROLEUM YEARBOOK. Annual. Vintry House, Queen St. Place, London, EC 4, England.

OIL DAILY; NATIONAL NEWSPAPER OF PETROLEUM. Oil Daily, Inc., 59 East Van Buren Street, Chicago, Illinois 60605.

OIL STATISTICS. Quarterly. A.K. Madan, Petroleum Information Service, Statistics Division, 22 Pusa Road, New Delhi 5, India.

Oil Technologists Association of India. JOURNAL. Semiannual. Colour Publications Pvt. Ltd., 126a Dhuruwadi, Off Dr. Nariman Road, Bombay 25DD, India.

OILWEEK; incorporating OIL IN CANADA. Weekly. Maclean-Hunter Ltd., Petro-Chem Building, 805 8th Avenue, S.W., Calgary 2, Alberta, Canada.

PATENTS INDEX. American Petroleum Institute, 1271 Avenue of the Americas, New York, New York 10020.

> Consists of 7500 patents yearly from API patent alert. Has 3 services: 1) alphabetical subject index (monthly) in book form; 2) dual dictionary for manual searching of coordinate terms; 3) magnetic tapes for computer searches.

PETRO/CHEM ENGINEER; NATURAL GASOLINE REFINING PETROCHEMICALS. Monthly. Petroleum Engineer Publishing Co., Davis Building, Box 1589, Dallas, Texas 75221.

PETROCHEMICAL INDUSTRY SERIES. Irregular. United Nations Industrial Development Organization, U.N. Publications, United Nations, Room 1059, New York, New York 10017.

PETROCHEMICAL NEWS; A WEEKLY NEWS SERVICE DEVOTED TO THE PET-ROCHEMICAL INDUSTRY. World Press Center Building, 54 West 40th Street, New York, New York 10018.

PETROLEUM ABSTRACTS. Weekly. Department of Information Services, Col-lege of Petroleum, Sciences, and Engineering, University of Tulsa, 1133 North Lewis Avenue, Tulsa, Oklahoma 74110.

> World literature and patents related to the exploration, develop-ment, and production of crude oil and natural gas.

PETROLEUM AND CHEMICAL INDUSTRY TECHNICAL CONFERENCE. Annual.

Institute of Electrical and Electronics Engineers, Inc., 345 East 47 Street, New York, New York 10017.

PETROLEUM CHEMISTRY (USSR). Quarterly. Pergamon Press, Maxwell House, Fairview Park, Elmsford, New York 10523 and Headington Hill Hall, Oxford, England.

English translation of NEFTEKHIMIYA.

PETROLEUM ENGINEER. Monthly. Petroleum Engineer Publishing Co., Box 1589, Dallas, Texas 75221.

PETROLEUM EQUIPMENT AND SERVICES. 6/year. 4300 North Central Expressway, Dallas, Texas 75206.

PETROLEUM LITERATURE INDEX. Annual. National Petroleum Bibliography, Amarillo, Texas.

PETROLEUM MIRROR. Monthly. 50 rue Stanislas Torrent, 13 Marseille (6e), France.

For the French oil and gas industry.

PETROLEUM OUTLOOK. Monthly. John S. Herold, Inc., 35 Mason Street, Greenwich, Connecticut 06830.

PETROLEUM REVIEW. Monthly. 61 New Cavendish Street, London W 1, England.

Formerly: INSTITUTE OF PETROLEUM REVIEW.

PETROLEUM STATEMENT. Monthly. U.S. Bureau of Mines, 4800 Forbes Avenue, Pittsburgh, Pennsylvania 15213.

PETROLEUM TIMES. Fortnightly. Engineering, Chemical & Marine Press Ltd., Bowling Green Lane, London EC 1, England.

PETROLEUM TODAY. Quarterly. American Petroleum Institute, Committee on Public Affairs, 1801 K Street N.W., Washington, D.C. 20006.

PIPELINE & GAS JOURNAL. Semimonthly. Petroleum Engineer Publishing Co., Box 1589, Dallas, Texas 75221.

PIPELINE AND UNDERGROUND UTILITIES CONSTRUCTION. Semimonthly. Oildom Publishing Co., 1217 Kennedy Boulevard, Bayonne, New Jersey 07002.

PIPE LINE INDUSTRY. Monthly. Gulf Publishing Co., Box 2608, Houston, Texas 77001.

Oil products, gas pipe line and gas distribution.

PIPE LINE NEWS. Monthly. Oildom Publishing Co., 1217 Kennedy Boulevard, Bayonne, New Jersey 07002.

PIPE PROGRESS. American Cast Iron Pipe Co., 2930 North 16th Street, Birmingham, Alabama 35207.

PIPES AND PIPELINES INTERNATIONAL; HOSES, TUBES, PUMPS, VALVES. Monthly. Scientific Survey Ltd., 11 a Gloucester Road, London, SW 7, England.

    Formerly: PIPES AND PIPELINES.

PLATT'S OILGRAM NEWS SERVICE. 5 issues weekly. McGraw-Hill, 1221 Avenue of the Americas, New York, New York 10020.

PM REPORT. Monthly. Petroleum Engineer Publishing Co., 800 Davis Building, Dallas, Texas 75202.

    Formerly: PETROLEUM MANAGEMENT.

Predi-briefs. ABSTRACTING SERVICE. Monthly. ENERGY. Predicasts, 200 University Circle Research Center, 11001 Cedar Avenue, Cleveland, Ohio 44106.

> "Contains information on conventional fuels -- coal, crude oil, natural gas and nuclear energy -- and newer technologies such as solar energy, geothermal energy, synthetic natural gas and fuel cells. Low-sulfur fuels and other pollution aspects briefed."

PROPANE/CANADA. Quarterly. Sanford Evans Publishing (Alta) Ltd., Suite 1, 5512 McLeod Trail, Calgary, Alberta, Canada.

RUSSIAN OIL AND GAS BULLETIN. Bimonthly. Associated Technical Services, Inc., 855 Bloomfield Avenue, Glen Ridge, New Jersey 07028.

> "Lists translated English titles of Russian research papers taken from Soviet periodical literature." Looseleaf format.

SAFETY HINTS ON DRILLING. International Association of Drilling Contractors, 211 North Ervay Building, Suite 505, Dallas, Texas 75201.

> "Eight-page monthly periodical of safety news and related material of general interest. For both field and office use."

SEARCH: PETROLEUM DIVISION. Monthly. Compendium Publishers International Corp., 2175 Lemoine Avenue, Fort Lee, New Jersey 07024.

SELECTED DOCUMENTS OF THE INTERNATIONAL PETROLEUM INDUSTRY. Annual. Organization of Petroleum Exporting Countries, Dr. Karl Lueger Ring 10, 1010 Vienna, Austria.

SOCIETY OF PETROLEUM ENGINEERS JOURNAL. Quarterly. Society of Petroleum Engineers, of AIME, 6200 North Central Expressway, Dallas, Texas 75206.

Society of Petroleum Engineers of American Institute of Mining, Metallurgical and Petroleum Engineers. TRANSACTIONS. Annual. American Institute of Mining, Metallurgical and Petroleum Engineers, Inc., Society of Petroleum Engineers, 6200 North Central Expressway, Dallas, Texas 75206.

STATISTICAL REVIEW OF THE WORLD OIL INDUSTRY. Annual. British Petroleum Co., Ltd., London, England.

TRENDS IN THE INTERNATIONAL PETROLEUM-REFINING INDUSTRY. Triennial. Ethyl Corporation Research Laboratories, 1600 West Eight Mile Road, Ferndale, Michigan 48220.

U.S. Department of Interior. INTERNATIONAL PETROLEUM ANNUAL, 1969. March 1971. Mines Bureau, 4800 Forbes Avenue, Pittsburgh, Pennsylvania 15213.

_____. PETROLEUM FORECAST. Monthly. Mines Bureau, 4800 Forbes Avenue, Pittsburgh, Pennsylvania 15213.

_____. PETROLEUM STATEMENT. Monthly. Mines Bureau, 4800 Forbes Avenue, Pittsburgh, Pennsylvania 15213.

U.S. Environmental Protection Agency. OIL POLLUTION RESEARCH NEWSLETTER. Irregular. Office of Research and Monitoring, National Environmental Research Center, Cincinnati, Edison Water Quality Research Laboratory, Environmental Protection Agency. Rockville, Maryland 20852.

UNIVERSAL NEWS; A PIPELINER'S DIGEST. Semimonthly. Universal News Publishing Co., Box 55225, Houston, Texas 77055.

WEEKLY PRODUCTION AND DRILLING STATISTICS. Alberta Oil and Gas Conservation Board, 603 Sixth Avenue SW, Calgary, Alberta, Canada.

Formerly: PRODUCTION AND DRILLING STATISTICS.

WELL SERVICING. Bimonthly. Association of Oilwell Servicing Contractors, 9106 Sovereign Row, Dallas, Texas 75247.

WORLD OIL. 14 issues/year. Gulf Publishing Co., Box 2608, Houston, Texas 77001.

WORLD PETROLEUM. Monthly. Palmer. Publications, 25 West 45th Street, New York, New York 10036.

WORLD PETROLEUM REPORT. Annual. Mona Palmer Publishing Co., 25 West 45th Street, New York, New York 10036.

9. Reviewing Services

ADVANCES IN PETROLEUM CHEMISTRY AND REFINING. John Wiley & Sons, Inc., 605 Third Avenue, New York, New York 10016. 1958-65.

PETROLEUM DERIVATIVES: CURRENT RESEARCH PAPERS. New York: Miss. Information Corp., 1976.

See also this subject in Section I of this Guide.

10. Document and Report Literature

a. LISTS

MONTHLY CATALOG. U.S. Superintendent of Documents, Government Printing Office, Washington, D.C. 20402.

U.S. GOVERNMENT REPORTS ANNOUNCEMENTS and U.S. GOVERNMENT REPORTS INDEX. National Technical Information Service, Springfield, Virginia 22151.

b. CLEARINGHOUSES

U.S. Superintendent of Documents. Government Printing Office, Washington, D.C. 20402.

National Technical Information Service. Springfield, Virginia 22151.

11. Sources of Information on Masters and Doctoral Theses

DISSERTATION ABSTRACTS INTERNATIONAL. Section B: "Physical Sciences and Technology." Monthly. University Microfilms, Xerox Co., 300 North Zeeb Road, Ann Arbor, Michigan 48106.

INTERNATIONAL MASTERS ABSTRACTS. Quarterly. University Microfilms, 300 North Zeeb Road, Ann Arbor, Michigan 48106.

MASTERS THESES IN THE PURE AND APPLIED SCIENCES, ACCEPTED BY COLLEGES AND UNIVERSITIES OF THE UNITED STATES. Annual. Thermophysical Properties Research Center, School of Mechanical Engineering, Purdue University, Lafayette, Indiana.

12. Sources of Information on Statistics and Costs

a. STATISTICS

i. Guides and Handbooks

AMERICAN STATISTICS INDEX. Annual. Part 1: Index. Part 2: Abstracts. Congressional Information Service, 600 Montgomery Building, Washington, D.C. 20014.

Comprehensive guide and index to the statistical publications of the U.S. government.

Special Libraries Association. Petroleum Section. Committee on U.S. Sources of Petroleum and Natural Gas Statistics. U.S. SOURCES OF PETROLEUM AND NATURAL GAS STATISTICS. Compiled by Margaret M. Recoq. New York: 1971.

"Indexes 231 publications which provide currently released statistical data."

U.S. Department of Interior. Bureau of Mines. ANALYSES OF NATURAL GASES, 1970. By L.E. Cardwell and L.F. Benton. Washington, D.C.: Government Printing Office, 1971.

_____. ANALYSES OF NATURAL GASES, 1969. By L.E. Cardwell and L.F. Benton. Washington, D.C.: Government Printing Office, 1970.

_____. ANALYSES OF NATURAL GASES, 1968. By L.E. Cardwell and L.F. Benton. Washington, D.C.: Government Printing Office, 1970.

_____. ANALYSES OF NATURAL GASES OF THE UNITED STATES, 1967. Washington, D.C.: Government Printing Office, 1967.

_____. MINERALS YEARBOOK. Annual. Washington, D.C.: Government Printing Office.

_____. PETROLEUM FORECAST. Monthly. Washington, D.C.: Government Printing Office.

_____. WORLD CRUDE OIL PRODUCTION REPORT. Annual. 4500 Forbes Avenue, Pittsburgh, Pennsylvania 15213.

U.S. Federal Power Commission. Bureau of Mines. NATIONAL GAS SUPPLY AND DEMAND, 1971-1990. Staff Report no. 2. Washington, D.C.: Government Printing Office, 1972.

_____. SALES BY PRODUCERS OF NATURAL GAS TO INTERSTATE PIPELINE COMPANIES, 1969. Prepared by Office of Accounting and Finance. Washington, D.C.: Government Printing Office, 1970.

_____. SALES BY PRODUCERS OF NATURAL GAS TO INTERSTATE PIPELINE COMPANIES, 1968. Prepared by Office of Accounting and Finance. Washington, D.C.: Government Printing Office, 1970.

_____. SALES BY PRODUCERS OF NATURAL GAS TO NATURAL GAS PIPELINE COMPANIES, 1967. Washington, D.C.: Government Printing Office, 1967.

_____. STATISTICS OF INTERSTATE NATURAL GAS PIPELINE COMPANIES

PURSUANT TO REQUIREMENTS OF NATURAL GAS ACT CLASSES A & B COMPANIES. Prepared by Office of Accounting and Finance. Washington, D.C.: Government Printing Office, 1971.

_____. STATISTICS OF INTERSTATE NATURAL GAS PIPELINE COMPANIES PURSUANT TO REQUIREMENTS OF NATURAL GAS ACT, CLASSES A & B COMPANIES. Prepared by Office of Accounting and Finance. Washington, D.C.: Government Printing Office, 1970.

_____. STATISTICS OF INTERSTATE NATURAL GAS PIPELINE COMPANIES PURSUANT TO REQUIREMENTS OF NATURAL GAS ACT CLASSES A & B COMPANIES COMPILED FROM ANNUAL REPORT SUBMITTED BY INTERNA- TIONAL GAS PIPELINE COMPANIES. Washington, D.C.: Government Print- ing Office, 1969.

_____. STATISTICS OF INTERSTATE NATURAL GAS PIPELINE COMPANIES, 1967; COMPILED FROM ANNUAL REPORTS SUBMITTED BY INTERSTATE GAS PIPELINE COMPANIES PURSUANT TO REQUIREMENTS OF NATURAL GAS ACT, CLASSES A AND B COMPANIES. Washington, D.C.: Government Printing Office, 1968.

_____. STATISTICS FOR INTERSTATE NATURAL GAS PIPELINE COMPANIES COMPILED FROM ANNUAL REPORTS SUBMITTED BY INTERSTATE NATURAL GAS PIPELINE COMPANIES PURSUANT TO THE REQUIREMENTS OF THE NATURAL GAS ACT, CLASSES A AND B COMPANIES, 1966. Washington, D.C.: Government Printing Office, 1966.

U.S. Geological Survey. WORLDWIDE SEARCH FOR PETROLEUM OFFSHORE, STATUS REPORT FOR QUARTER CENTURY 1947-1972. Washington, D.C.: Government Printing Office, 1974.

ii. Journals and Serial Publications

Alberta Oil and Gas Industry. MONTHLY STATISTICS. Alberta Oil and Gas Conservation Board, 603 Sixth Avenue SW, Calgary 1, Alberta, Canada.

American Gas Association. Bureau of Statistics. GAS UTILITY AND PIPELINE INDUSTRY PROJECTIONS. Irregular. American Gas Association, Bureau of Statistics, 1515 Wilson Boulevard, Arlington, Virginia 22209.

American Petroleum Institute. Department of Statistics. WEEKLY STATISTICAL BULLETIN. American Petroleum Institute, 1101 17th Street N.W., Washington, D.C. 20036.

ANNUAL BULLETIN OF GAS STATISTICS FOR EUROPE. Annual. U.N. Economic Commission for Europe. U.N. Publications, United Nations, Room 1059, New York, New York 10017.

Text in English, French, and Russian.

Canada. Bureau of Statistics. CRUDE PETROLEUM AND NATURAL GAS INDUSTRY. Annual. Bureau of Statistics, Ottawa, Canada.

_____. PETROLEUM REFINERIES. Annual. Bureau of Statistics, Ottawa, Canada.

_____. REFINED PETROLEUM PRODUCTS. Annual. Bureau of Statistics, Ottawa, Canada.

Canada. Dominion Bureau of Statistics. SERVICE BULLETIN: ENERGY STA-TISTICS. Weekly. Dominion Bureau of Statistics, Tunney's Pasture, Ottawa 3, Ontario, Canada.

Canada Gas Association. STATISTICAL SUMMARY OF THE CANADIAN GAS INDUSTRY. Annual. Canadian Gas Association, 55 Scarsdale Road, Don Mills, Ontario, Canada.

Canadian Petroleum Association. STATISTICAL YEARBOOK. Annual. Canadian Petroleum Association, Calgary, Alberta, Canada.

HISTORICAL STATISTICS OF THE GAS INDUSTRY. Irregular. American Gas Association, Bureau of Statistics, 1515 Wilson Boulevard, Arlington, Virginia 22209.

Liquified Petroleum Gas Association. LP-GAS MARKET FACTS. Annual. National LP-Gas Association, 79 West Monroe Street, Chicago, Illinois 60603.

- Statistical handbook of the LP-gas industry.

LIQUIFIED PETROLEUM GAS REPORT. Monthly. American Petroleum Insti-tute, Department of Statistics, 1101-17th Street N.W., Washington, D.C. 20036.

OECD PROVISIONAL OIL STATISTICS. Quarterly. Organization for Eco-nomic Co-Operation and Development, Publications Center, 1750 Pennsylvania Avenue N.W., Suite 1207, Washington, D.C. 20006.

Editions in English and French.

OIL STATISTICS. Quarterly. A.K. Madan, Petroleum Information Service, Statistics Division, 22 Pusa Road, New Delhi 5, India.

STATISTICAL REVIEW OF THE WORLD OIL INDUSTRY. Annual. British Petroleum Company Ltd., London, England.

WEEKLY PRODUCTION AND DRILLING STATISTICS. Alberta Oil and Gas Conservation Board, 603 Sixth Avenue, S.W., Calgary, Alberta, Canada.

Formerly: PRODUCTION AND DRILLING STATISTICS.

The following journals frequently contain statistics related to this field. See the heading Journals and Serial Publications in this section for addresses.

AMERICAN ASSOCIATION OF PETROLEUM GEOLOGISTS

AMERICAN GAS ASSOCIATION MONTHLY.

CANADIAN PETROLEUM

CHEMICAL ABSTRACTS - ORGANIC SECTIONS

ENERGY MANAGEMENT REPORT /D/

EUROCOM PRESS INFORMATION

GAS (U.S.)

GAS IN CANADA

GAS PROCESSING / CANADA

INDEPENDENT PETROLEUM MONTHLY

INSTITUTE OF PETROLEUM JOURNAL

INSTITUTION OF GAS ENGINEERS JOURNAL

INTERNATIONAL OIL AND GAS YEARBOOK

JAPAN PETROLEUM WEEKLY

LP - GAS

NATIONAL PETROLEUM NEWS

NLPGA TIMES

OIL AND GAS DISCOVERIES

OIL AND GAS INTERNATIONAL

OIL AND GAS JOURNAL

OIL AND PETROLEUM YEARBOOK

OIL DAILY

OIL TECHNOLOGISTS ASSOCIATION OF INDIA JOURNAL

OILWEEK

PETRO/CHEM ENGINEER

PETROLEUM ABSTRACTS

PETROLEUM ENGINEER

PETROLEUM EQUIPMENT AND SERVICES

PETROLEUM MIRROR

PETROLEUM OUTLOOK

PETROLEUM REVIEW

PETROLEUM TIMES

PLATT'S OILGRAM NEWS SERVICE

PM REPORT

PROPANE/CANADA

SEARCH: PETROLEUM DIVISION

U.S. DEPARTMENT OF INTERIOR
   a. INTERNATIONAL PETROLEUM ANNUAL
   b. PETROLEUM FORECAST
   c. PETROLEUM STATEMENT

WORLD PETROLEUM

b. COSTS

Dwyer, C. "Guidelines in Cost Control." PETROLEUM ENGINEER INTERNA-
TIONAL, vol. 44, August, 1972, p. 52; September, 1972, p. 74; October,
1972, p. 62.

Hahn, Albert V. THE PETROCHEMICAL INDUSTRY; MARKET AND ECONOM-
ICS. New York: McGraw-Hill, 1970.

   "An economic and market study of chemical products derived
   from petroleum fractions and by-products, and from natural gas
   constituents."

Nelson, W. GUIDE TO REFINERY OPERATING COSTS. Petroleum Publishing
Co., Tulsa, Oklahoma 74101. 1976.

   "Source of cost information for refineries, petrochemical, gas
   processing, and other processing plants.

Oak Ridge National Laboratory. ECONOMETRIC MODEL OF THE PETROLEUM
INDUSTRY. By P.L. Rice, et al. Rept. no. CONF. 751214-1. Springfield,
Virginia: National Technical Information Service, 1975.

U.S. Department of Interior. Bureau of Mines. BIBLIOGRAPHY OF INVEST-
MENT AND OPERATING COSTS FOR CHEMICAL AND PETROLEUM PLANTS.
3 vols. Washington, D.C.: Government Printing Office, 1968-70.

_____. CHANGING INVESTMENT PATTERNS OF THE U.S. PETROLEUM
INDUSTRY, 1950-68. Washington, D.C.: Government Printing Office, 1970.

_____. EFFECT OF ACCOUNTING FACTORS ON ECONOMICS OF SYN-
THETIC PIPELINE GAS. Washington, D.C.: Government Printing Office,
1970.

_____. ELECTROTHERMAL BYGAS PROCESS ESCALATED COSTS. Washington, D.C.: Government Printing Office, 1971.

_____. INFORMATION CIRCULARS. ENGINEERING COST STUDY OF DEVELOPMENT WELLS AND PROFITABILITY ANALYSIS OF CRUDE OIL PRODUCTION. By T.M. Garland and W.D. Dietzman. Washington, D.C.: Government Printing Office, 1972.

Includes bibliography.

13. Sources of Information on
Standards, Specifications, Patents, and Trademarks

a. STANDARDS AND SPECIFICATIONS

AMERICAN NATIONAL STANDARDS INSTITUTE, CATALOG. Annual. The Institute, 1430 Broadway, New York, New York 10018.

Lists standards in this subject area. Catalog and standards are available from the Institute.

Institute of Petroleum, London. INSTITUTE OF PETROLEUM STANDARDS FOR PETROLEUM AND ITS PRODUCTS. Annual. Institute of Petroleum, 61 New Cavendish Street, London, W 1, England.

Strugalia, Erasmus J. STANDARDS AND SPECIFICATIONS INFORMATION SOURCES. Detroit: Gale Research Co., 1965.

Coverage is both general and specific to petroleum engineering.

U.S. Department of Commerce. National Bureau of Standards. DIRECTORY OF U.S. STANDARDIZATION ACTIVITIES. Washington, D.C.: Government Printing Office, 1975.

Coverage is both general and specific to petroleum engineering.

Specifications, standards, codes and regulations are available for individual states from the official state document printing offices and from local and municipal government agencies.

See also this subject in Section I of this Guide.

b. PATENTS

i. Journals and Serial Publications

ABSTRACTS OF EXPLORATION AND PRODUCTION LITERATURE AND PATENTS. Weekly. University of Tulsa, 1133 North Lewis Avenue, Tulsa, Oklahoma 74110.

American Petroleum Institute. ABSTRACTS OF PETROLEUM SUBSTITUTES LITER-
ATURE AND PATENTS. Monthly. Central Abstracting and Indexing Service,
American Petroleum Institute, 555 Madison Avenue, New York, New York
10022.

_____. ABSTRACTS OF REFINING PATENTS. Weekly. American Petroleum
Institute, 555 Madison Avenue, New York, New York 10022.

_____. ABSTRACTS OF TRANSPORTATION AND STORAGE LITERATURE AND
PATENTS. Monthly. American Petroleum Institute, 555 Madison Avenue, New
York, New York 10022.

Available in microfilm.

API PATENT ALERT ABSTRACTS. Weekly. American Petroleum Institute, 1271
Avenue of the Americas, New York, New York 10020.

Reports patents in petroleum refining and petrochemical industry,
covers United States, United Kingdom, South Africa, Japan, most
of European countries.

PATENTS INDEX. Monthly. American Petroleum Institute, 1271 Avenue of
the Americas, New York, New York 10020.

Consists of 7500 patents yearly from API patent alert. Has 3 services: 1) alphabetical subject index (monthly) in book form;
2) dual dictionary for manual searching of coordinate terms;
3) magnetic tapes for computer searches.

PETROLEUM ABSTRACTS. Weekly. University of Tulsa, 1133 North Lewis
Avenue, Tulsa, Oklahoma 74110.

World literature and patents related to the exploration, development, and production of crude oil and natural gas.

PETROLEUM INTELLIGENCE. WEEKLY. Petroleum Intelligence Inc., 48 West
48th Street, New York, New York 10036.

International news of patents in petroleum field.

UNITED STATES PETROLEUM PATENTS. Annual. Noyes Development Corp.,
118 Mill Road, Park Ridge, New Jersey 07656.

ii. Current Journals
Which Include Related Patent Listings

The following journals frequently list patents. See the heading Journals and
Serial Publications in this section for addresses.

CHEMICAL ABSTRACTS - ORGANIC SECTIONS

EUROPE AND OIL

GAS JOURNAL

NLGI SPOKESMAN

OIL STATISTICS

PETRO/CHEM ENGINEER

PIPES AND PIPELINES INTERNATIONAL

14. Sources of Information on
Translations, Meetings, and Proceedings of Symposia

U.S. Joint Publications Research Service. Arlington, Virginia. Publishes translations of scientific and technical materials from Eastern Europe, Russia, and China. These are listed in the MONTHLY CATALOG of the Superintendent of Documents.

See this subject in Section I of this Guide.

15. Selected Literature Sources on
Pollution Control and Safety

a. POLLUTION CONTROL

i. Guides and Handbooks

BIODETERIORATION OF OIL SPILLS. By E. Harrison. Rept. no. NTIS/PS-76/0032. Springfield, Virginia: National Technical Information Service, 1976.

A bibliography with abstracts for 1964-1976.

THE BIOLOGICAL EFFECTS OF OIL SPILLS. By E. Harrison. Rept. no. NTIS/PS-76/0033. Springfield, Virginia: National Technical Information Service, 1976.

A bibliography with abstracts for 1964-1976.

Boech, D. OIL SPILLS AND THE MARINE ENVIRONMENT. Philadelphia: Ballinger, 1974.

ECOLOGICAL EFFECTS OF OIL POLLUTION ON LITTORAL COMMUNITIES, Symposium, Zoological Society of London, London, England, November 30 - December 1, 1970. London: Institute of Petroleum, 1971.

THE ESTABLISHMENT OF AN INTERNATIONAL COMPENSATION FUND FOR OIL POLLUTION DAMAGE. Conference. Inter-governmental Maritime Consultation Organization, 101-104 Piccadilly, London, WIV OAE, England. 1972.

THE HANDLING OF OILFIELD WATERS. Symposium. Society of Petroleum Engineers, 6200 North Central Expressway, Dallas, Texas 75206. 1972.

JOINT CONFERENCE ON PREVENTION AND CONTROL OF OIL SPILLS. Washington, D.C.: American Petroleum Institute, 1971.

Jones, H. ENVIRONMENTAL CONTROL IN THE ORGANIC AND PETRO-CHEMICAL INDUSTRIES. Park Ridge, New Jersey: Noyes Data Corp., 1971.

Miami University. (Florida) Sea Grant Institutional Program. INTERNATIONAL COOPERATION FOR THE PREVENTION OF MARINE OIL POLLUTION. Ocean and Coastal Law Program. By A. Anderson, et al. Rept. no. PB 253 290. Springfield, Virginia: National Technical Information Service, 1976.

National Bureau of Economic Research, Inc. WATER POLLUTION CONTROL ACT OF 1972. Economic Impacts, Petroleum Refining Industry. By J. Smith. Rept. no. PB 248 800. Springfield, Virginia: National Technical Information Service, 1975.

National Petroleum Council. Committee on Environmental Conservation. ENVIRONMENTAL CONSERVATION - THE OIL AND GAS INDUSTRIES. 2 vols. Washington, D.C.: 1972.

OIL ON THE SEA. Symposium on the Scientific and Engineering Aspects of Oil Pollution of the Sea. Plenum Press, 227 West 17th Street, New York, New York 10011. 1969.

OIL POLLUTION CONTROL ACT OF 1961 AND AMENDMENTS OF 1973. U.S. Code. Washington, D.C.: Government Printing Office, 1975.

OIL SPILLS AND SPILLS OF HAZARDOUS SUBSTANCES. Washington, D.C.: U.S. Environmental Protection Agency, Water Programs Operations Office, 1974.

OIL SPILL TREATING AGENTS. Industry-Government Seminar, Washington, D.C., April 1970. Proceedings. Washington, D.C.: American Petroleum Institute, 1970.

PREVENTION AND CONTROL OF OIL SPILLS. Conference, Washington, D.C., June 15-17, 1971. Sponsored by American Petroleum Institute. Washington, D.C.: American Petroleum Institute, 1971.

Sittig, M. OIL SPILLS PREVENTION AND REMOVAL HANDBOOK. Park Ridge, New Jersey: Noyes Data Corp., 1975.

Stanford Research Institute. ENERGY DEVELOPMENT: THE ENVIRONMENTAL TRADEOFFS. Vol. 1: Summary of Vols. 2-4, Rept. no. PB250 000. Vol. 2: Relative Environmental Assessment Methods to Increase Energy Production, Crude Oil, Pipeline Quality Gas, and Electricity from Western Coal, Rept. no. PB250 001. Vol. 3: Relative Environmental Ranking of Proposed Offshore Continental Shelf Areas on the Basis of Impacts of Oil Spills, Rept. no. PB250 002. Vol. 4: Background Papers, Rept. no. PB250 003. Springfield, Virginia: National Technical Information Service, 1976.

UNDERGROUND WASTE MANAGEMENT AND ARTIFICIAL RECHARGE. Underground Waste Management Symposium. American Association of Petroleum Geologists, Box 979, Tulsa, Oklahoma 74101.

U.S. Atomic Energy Commission. DEVELOPMENT OF NUCLEAR ANALYTICAL TECHNIQUES FOR OIL-SLICK IDENTIFICATION, PHASE 2A, FINAL REPORT. By H.R. Lukens. Springfield, Virginia: National Technical Information Service, 1971.

    With list of references.

U.S. Environmental Protection Agency. AIR MODULATED VACUUM OIL RECOVERY COLLECTION OF SPILLED OIL. By R.W. Sicka. Washington, D.C.: Government Printing Office, 1972.

    _____. AIR POLLUTION ASPECTS OF EMISSION SOURCES: PETROLEUM REFINERIES, BIBLIOGRAPHY WITH ABSTRACTS. Air Pollution Technical Information Center. Washington, D.C.: Government Printing Office, 1972.

    _____. CONCEPT DEVELOPMENT OF HYDRAULIC SKIMMER SYSTEM FOR THE RECOVERY OF FLOATING OIL. By J. Blacklaw. Prepared by Batelle Memorial Institute. Washington, D.C.: Government Printing Office, 1971.

    With list of references.

    _____. CONCEPT EVALUATION, RECOVERY OF FLOATING OIL USING POLYURETHANE FOAM SORBENT. By C.H. Henager. Washington, D.C.: Government Printing Office, 1972.

    With list of references.

    _____. EVALUATION OF WASTE WATERS FROM PETROLEUM AND COAL PROCESSING. By G.W. Reid. Washington, D.C.: Government Printing Office, 1973.

    With list of references.

    _____. FLOATING OIL RECOVERY DEVICE. By Physical Science Laboratory, New Mexico State University, Las Cruces, New Mexico, February 1971. Washington, D.C.: Government Printing Office, 1972.

    _____. FLUID PRODUCT PIPELINE LEAK DETECTION FROM AIRBORNE PLATFORMS. By J. Kennedy, Resources Technology Corp., Houston, Texas, December 1970. Washington, D.C.: Government Printing Office, 1972.

    _____. GELLING CRUDE OILS TO REDUCE MARINE POLLUTION FROM TANKER OIL SPILLS. Washington, D.C.: Government Printing Office, 1972.

    _____. OIL. Washington, D.C.: Government Printing Office, 1973.

    _____. OIL POLLUTION SOURCE IDENTIFICATION. By M. Lieberman. Washington, D.C.: Government Printing Office, 1974.

_____. OIL SPILLS CONTROL MANUAL FOR FIRE DEPARTMENTS. By R. Cross. Washington, D.C.: Government Printing Office, 1974.

_____. OILY WASTE DISPOSAL BY SOIL CULTIVATION PROCESS. By C.B. Kincannon. Washington, D.C.: Government Printing Office, 1973.

_____. POLYMERIC MATERIALS FOR TREATMENT AND RECOVERY OF PET-ROCHEMICAL WASTES. By Gulf South Research Institute, New Orleans, Louisiana. Washington, D.C.: Government Printing Office, 1972.

_____. PRELIMINARY INVESTIGATIONAL REQUIREMENTS, PETROCHEMICAL AND REFINERY WASTE TREATMENT FACILITIES. Washington, D.C.: Government Printing Office, 1972.

_____. RECOVERY OF FLOATING OIL ROTATING DISK TYPE SKIMMER. By Atlantic Research Systems Division, Costa Mesa, California. Washington, D.C.: Government Printing Office, 1972.

With list of references.

_____. SPILL PREVENTION TECHNIQUES FOR HAZARDOUS POLLUTING SUBSTANCES. Inventory and survey of hazardous chemical facilities in Charleston, West Virginia; Baltimore, Maryland; Texas City, Texas; and Suisun Bay-Delta Area, California. Prepared by J.L. Goodier. Washington, D.C.: Government Printing Office, 1972.

_____. SURVEY REPORTS ON ATMOSPHERIC EMISSIONS FROM THE PET-ROCHEMICAL INDUSTRY. Rept. no. EPA-450/3-73-005a, b, c and d. Washington, D.C.: 1974.

U.S. Navy. Civil Engineering Laboratory. Port Hueneme, California. MATERIALS FOR OIL SPILL CONTAINMENT BOOM. Rept. no. ADA026-139. Springfield, Virginia: National Technical Information Service, 1976.

U.S. Navy. Naval Research Laboratory. IMPACT OF PETROLEUM SPILLS ON CHEMICAL AND PHYSICAL PROPERTIES OF AIR/SEA INTERFACE. By W.D. Garrett. Springfield, Virginia: National Technical Information Service, 1972.

WATER AND AIR CONSERVATION IN THE PETROLEUM INDUSTRY. Panel Discussion at 8th World Petroleum Congress. Sponsored by Permanent Council for World Petroleum Congresses, Moscow, USSR. Published by Elsevier Publishing Co., Amsterdam. U.S. Distributor: American Petroleum Institute, 1271 Avenue of the Americas, New York, New York 10020. 1971.

WATER POLLUTION BY OIL. London: Institute of Petroleum, 1971.

ii. Journals and Serial Publications

American Petroleum Institute. ABSTRACTS OF AIR AND WATER CONSERVA-

TION LITERATURE AND PATENTS. Weekly. American Petroleum Institute, 555 Madison Avenue, New York, New York 10022.

Available in microfilm.

OCCUPATIONAL SAFETY AND HEALTH ABSTRACTS. Monthly. International Occupational Safety and Health Information Centre (CIS), International Labour Office, 1211 Geneva 22, Switzerland.

Editions in English. Available in card format.

b. SAFETY

i. Guides and Handbooks

ACCIDENT PREVENTION MANUAL. International Association of Drilling Contractors, Dallas, Texas. 1968.

A guide to recommended accident prevention methods and safe operating practices for drilling and servicing rigs, prepared by IADC Safety Committee.

AMERICAN NATIONAL STANDARDS INSTITUTE CATALOG. Annual. New York: American National Standards Institute.

Contains safety standards information.

AMF. Inc. Advanced Systems Laboratory. Galeta, California. STUDY ON CURRENT PRACTICES, TECHNOLOGIES, PROBLEMS AND RECOMMENDATIONS RELATING TO THE OVERALL SAFETY OF GAS PIPELINE DISTRIBUTION SYSTEMS. By J. Bartol, et al. Rept. no. PB251 996. Springfield, Virginia: National Technical Information Service, 1976.

GAS EVOLUTION: TANKER AND TERMINAL SAFETY. Institution's Special Publications Series, no. 1. Edited by Peter Hepple. 1971. Symposium sponsored by Institute of Petroleum, London, England. Published by: Institute of Petroleum, London. Distributed by: Applied Science Publishers, Ltd., Ripple Road, Barking, Essex, England.

GUIDES FOR FIGHTING FIRES IN AND AROUND PETROLEUM TANKS. Boston: National Fire Protection Association, 1975.

International Oil Tanker and Terminal Safety Group. INTERNATIONAL OIL TANKER AND TERMINAL SAFETY GUIDE. 2nd ed. New York: Halsted Press, 1975.

LNG (LIQUEFIED NATURAL GAS) IMPORTATION AND TERMINAL SAFETY. Conference Sponsored by National Research Council, Committee on Hazardous Materials, Boston, Massachusetts. National Academy of Sciences, 2101 Constitution Avenue NW, Washington, D.C. 20418. 1972.

National Petroleum Refiners Association. NPRA QUESTION AND ANSWER SESSION ON REFINERY MAINTENANCE. Tulsa, Oklahoma: Petroleum Publishing Co., 1972.

OUTLINE FOR DRILLING RIG SAFETY PROGRAM. International Association of Drilling Contractors, Dallas, Texas.

> "24-page manual prepared by Special IADC Safety Subcommittee, covering basic ideas for establishing and maintaining a safety program for both small and large contractors."

Page, R. PETROLEUM TANKSHIP SAFETY. New York: Heinemann, 1971.

Peck, Theodore. OCCUPATIONAL SAFETY AND HEALTH: A GUIDE TO INFORMATION SOURCES. Detroit: Gale Research Co., 1974.

> Includes petroleum industry.

THE SAFETY FACTOR AND ITS INFLUENCE ON THE OPERATING MARGIN. Instrument Society of America Chemical and Petroleum Instrument Symposium. Pittsburgh: Instrument Society of America, 1972.

U.S. Department of Interior. Bureau of Mines. DEMONSTRATION OF SAFETY PLUGGING OF OIL WELLS PENETRATING APPALACHIAN COAL MINES. By G.E. Rennick. July 1972. Coal Mine Health and Safety Research Program.

_____. LOCATING UNCHARTED OIL AND GAS WELLS, STATE OF THE ART. By F.E. Armstrong. January 1973. Coal Mine Health and Safety Research Program.

U.S. Department of Interior. Oil and Gas Office. MINIMIZING DAMAGE TO REFINERIES FROM NUCLEAR ATTACK, NATURAL AND OTHER DISASTERS. By M.M. Stephens. Washington, D.C.: Government Printing Office, 1972.

> With list of references.

ii. Journals and Serial Publications

LEGAL GUIDE TO OCCUPATIONAL SAFETY AND HEALTH. Monthly. Clark Boardman Ltd., 435 Hudson Street, New York, New York 10014.

> Comment, interpretation, and advice on new legal developments in occupation safety and health.

OCCUPATIONAL SAFETY AND HEALTH ABSTRACTS. Monthly. International Occupational Safety and Health Information Centre (CIS), International Labour Office, 1211 Geneva, 22, Switzerland.

> Editions in English, French, German, Italian, Rumanian, Russian, and Spanish. Available in card format.

16. Management Information Sources

See this topic in Section I of this Guide.

17. Sources of Information on Laws and Legislation in This Field

Huie, W. CASES AND MATERIALS ON OIL AND GAS. 2nd ed. St. Paul, Minnesota: West Publishing Co., 1972.

U.S. Department of Interior. Bureau of Mines. NATURAL GAS PIPELINE SAFETY ACT OF 1968. An Act to Authorize the Secretary of Transportation to Prescribe Safety Standards for the Transportation of Natural and Other Gas by Pipeline and for Other Purposes. Approved August 12, 1968 and amended 1974. Title 49 U.S. Code. Washington, D.C.: Government Printing Office, 1975.

_____. SUMMARY OF MINING AND PETROLEUM LAWS OF THE WORLD. Washington, D.C.: Government Printing Office, 1970-71.

ADDITIONAL INFORMATION SOURCES ON LAWS & LEGISLATION:

a. FEDERAL

CODE OF FEDERAL REGULATIONS. Continuing series issued throughout the year. Available on a subscription basis from Government Printing Office, Washington, D.C. 20042.

FEDERAL REGISTER. Daily publication listing new regulations and revisions. Updates CODE OF FEDERAL REGULATIONS. Available on subscription from Government Printing Office, Washington, D.C. 20402.

Library of Congress. DIGEST OF PUBLIC GENERAL BILLS AND REGULATIONS. Monthly. Available on a subscription basis from Government Printing Office, Washington, D.C. 20402.

b. STATE

Library of Congress. MONTHLY CHECKLIST OF STATE PUBLICATIONS. Available on a subscription basis from Government Printing Office, Washington, D.C. 20402.

Specifications, standards, codes and regulations for individual states are available from official state document printing offices in each state.

c. LOCAL AND MUNICIPAL

Contact local government offices for codes, ordinances, and regulations.

All of these information source materials listed here are usually available in the reference department of large public libraries.

ADDENDUM

ADDENDUM

LIBRARIES

The resource collections and information services of large libraries are valuable sources of data and knowledge for the entire spectrum of chemical engineering industries.

Directories and guides have been compiled which outline in some detail the scope of library collections and services throughout the world. Users of this Guide are referred to these for specific information on library programs for their particular needs.

In the following brief descriptions of library services, some of the larger library-information centers have been highlighted.

THE JOHN CRERAR LIBRARY - 35 West 33rd Street, Chicago, Illinois 60616. Tel: (312) 225-2526

A primary information center for the sciences.

> Areas of interest: Biology; botany; zoology; physiology; biochemistry; mathematics; chemistry; geology; physics (basic research aspects); cremation, international congresses and expositions; all branches of agriculture having industrial interest, e.g., food processing and preservation, farm machinery and equipment, fertilizers, and insecticides; all aspects of engineering; metallurgy; communications; transportation; medicine, including historical medicine; bacteriology, gynecology; pediatrics; leukemia.

> Holdings: Over one million volumes and pamphlets and current subscriptions to more than 13,000 periodicals and serial publications, devoted exclusively to science, technology, and medicine. The Library has a complete collection of reports from the National Aeronautics and Space Administration, a partial collection of unclassified reports of the U.S. Atomic Energy Commission and the U.S. Department of Defense, and selected reports of other U.S. government agencies.

> Publications: LEUKEMIA ABSTRACTS (monthly); TRANSLATIONS REGISTER INDEX (semimonthly); bibliographies; pamphlets describing the Library's services.

Information services: Public reference service is given in the Research Services Division, where readers may make free use of the collections. Designated by the National Library of Medicine as the Midwest Regional Medical Library, Crerar has responsibility for loan and information services to health science personnel in the six-state area of Illinois, Indiana, Iowa, Minnesota, Wisconsin, and North Dakota. Otherwise, volumes are not lent for home use except to members (membership in the Crerar Library is open to any organization or individual whose interest coincides with the objectives of the Library). The Research Information Service of the Library offers library research services on a reimbursable basis. Literature searches, prior art searches relating to patents and current awareness literature scanning are done by subject specialists and are issued in the form of abstracts, annotated bibliographies, reports, punched cards, and catalog cards. A limited number of custom translations are prepared on a per word basis. Under a grant from the National Science Foundation, the Library maintains and services a translation center with a collection of more than 150,000 translations of articles contributed by government agencies, industrial firms, and private organizations. The translations center lends copies and supplies photoduplicates of translations in its collection. Inquiries should be addressed to the National Translations Center, The John Crerar Library.

TRANSLATIONS REGISTER-INDEX lists new accessions by subject and contains a computer-produced cumulating citation index to translations available at the National Translations Center and some 30 additional sources. Subject to copyright limitations, the Library will furnish photocopies of materials in its collections (a price list for this service is available).

ENGINEERING SOCIETIES LIBRARY - 345 East 47th Street, New York, New York 10017.
Tel: (212) 752-6800

A major engineering information resource. This library serves as the information resource center for the professional Engineering Societies located in the United Engineering Center.

Areas of interest: Engineering and related physical sciences, including chemical, civil, electrical, and electronic, marine and naval, illuminating, industrial, and mechanical engineering; aerospace sciences; rockets and missiles; nuclear technology metallurgy and metallography welding; mining; shipbuilding; petroleum technology; computers; instrumentations; engineering materials; geology; geophysics; history of technology; engineering as a profession.

Holdings: 225,000 volumes; 4,000 serial titles; 10,000 maps; 60 films of research data in fluid mechanics; 20,000 unpublished papers of engineering societies; bibliographies, reports.

Publications: CLASSED SUBJECT CATALOG OF THE LIBRARY
(16 volumes, five annual supplements); annotated bibliographies
(irregular).

Information services: Provides reference services; lends books to
member societies; provides literature-searching and duplication
services for a fee; lends films. Publications cited in the Engineering Index are held by the Engineering Societies Library and
photocopies of articles are available on a fee basis.

LINDA HALL LIBRARY - 5109 Cherry Street, Kansas City, Missouri 64110.
Tel: (816) 363-4600

Provides an extensive resource collection in science, technology, and medicine.

Areas of interest: Science and technology (excluding clinical
medicine and surgery); mathematics; astronomy; physics; chemistry;
geology; the biological sciences; engineering and applied science;
history of science.

Holdings: About 500,000 pieces; over 25,000 serial titles, of
which over 12,500 are currently received, including many in
foreign languages; 6,421 reels of microfilm; 84,600 microcards;
201,500 microfiche; 75,000 standards and specifications; depository
collections of U.S. Army Map and Rand Corporation publications;
complete collection of U.S. Patents on microfilm for the period
July 1, 1946 (with cumulative indexes on microfilm); 50,000
United States and foreign geological survey and topographic maps;
comprehensive holdings of subject indexes, abstracts, and bibliographies; extensive holdings of Russian scientific and technical
books and serials and of Japanese scientific and technical serials;
a large collection of original classical works in science and technology dating back to the 15th century.

Publications: SERIALS HOLDINGS IN THE LINDA HALL
LIBRARY (October 1, 1968, with one complete revision and
one supplement annually); JAPANESE, CHINESE AND RUSSIAN
SERIALS IN THE LINDA HALL LIBRARY, 1 March 1970; BOOK
COLLECTIONS AND SERVICES OF THE LINDA HALL LIBRARY
(1967).

Information services: Permits onsite use of collection; makes
interlibrary loans; provides microfilm and photocopies, subject
to copyright regulations. A Missouri Technological Information
Center, established in the Linda Hall Library building but supported
by state and federal funds under the State Technical Services Act
and sponsored by the University of Missouri at Kansas City, provides a wide range of information services to Missouri users (services to non-Missouri users may be arranged on an individual
basis at appropriate rates).

LIBRARY OF CONGRESS - Science and Technology Division, 10 First Street
S.E., Washington, D.C. 20540.

Areas of interest: All aspects of science and technology. The Library of Congress attempts to collect definitively in all aspects of science and technology except technical agriculture and clinical medicine, which are subject specialties of the National Agricultural Library and the National Library of Medicine.

Holdings: The subject fields of science and technology are represented in the Library of Congress by more than 2 million books, nearly 20,000 current journal titles, and some 1.25 million technical reports. The Division's Science Reading Room maintains a reference collection of books in the basic sciences, including technical dictionaries, encyclopedias, and handbooks. All major abstracting and indexing journals in the physical, earth, biomedical, and social sciences are represented for purposes of current and retrospective searching. The Division also has immediate custody of an extensive collection of technical reports (some 756,000 titles), most of them in microfilm, including those issued by the Atomic Energy Commission, the National Aeronautics and Space Administration, the Department of Defense, and other government agencies. Of special interest is the Division's collection of World War II Office of Scientific Research and Development (OSRD) reports which, along with its register of unique copies in other facilities, is the most nearly complete record of the work of OSRD in existence and provides a substantially complete catalog of OSRD reports.

Publications: In addition to the publication series of the National Referral Center and the Cold Regions Bibliography Section, the Division publishes occasional bibliographies, such as NUCLEAR SCIENCE IN MAINLAND CHINA: A SELECTED BIBLIOGRAPHY (1968), and FISH PROTEIN CONCENTRATE: A COMPREHENSIVE BIBLIOGRAPHY (1970). It also compiles bibliographies and chronologies published by others, such as AIR POLLUTION PUBLICATIONS: A SELECTED BIBLIOGRAPHY WITH ABSTRACTS 1966-1968 (U.S. Public Health Service, 1969), ASTRONAUTICS AND AERONAUTICS: A CHRONOLOGY ON SCIENCE, TECHNOLOGY AND POLICY (monthly with annual cumulation, published by NASA). Most of the Division's publications are available from the Superintendent of Documents, Government Printing Office, Washington, D.C. 20402. A publications list is available.

Information services: The Division has primary responsibility for providing reference and bibliographic services on the Library's overall science and technology collections. Brief technical inquiries entailing a bibliographic response are answered without charge. Extensive literature-searching and bibliographic services are provided on a fee basis. Individuals performing their own literature searches may use the Science Reading Room on the fifth floor of the Library of Congress Annex, where microfilm readers are available for use with the technical report collection.

NATIONAL AGRICULTURAL LIBRARY - U.S. Department of Agriculture, Beltsville, Maryland 20705.

BEE CULTURE BRANCH - Agricultural Research Center, Building A, Beltsville, Maryland 20705.

DISTRICT OF COLUMBIA BRANCH - U.S. Department of Agriculture, South Building, Room 1052, 14th Street and Independence Avenue S.W., Washington, D.C. 20250.

LAW BRANCH - U.S. Department of Agriculture, South Building, Room 2016, 14th Street and Independence Avenue S.W., Washington, D.C. 20250.

Areas of interest: General agriculture; economic aspects of agriculture, including production, marketing, and transportation; botany; chemistry; animal industry; veterinary medicine; biology; rural sociology; forestry; entomology; law; food and nutrition; soils and fertilizers; all aspects of pest control and pesticides.

Holdings: 1.3 million volumes, including books and bound journals, current periodicals, reports, maps, charts, posters, and microforms. The literature is in 50 languages and includes material from 200 foreign countries. Special collections are available at the branch libraries. The Bee Culture Branch, which maintains the BEEKEEPING BIBLIOGRAPHY in card form, compiles material on beekeeping, bee diseases, plant pollination, insecticides, and floral sources of honey. The Law Branch houses material on the agricultural aspects of U.S. statute and case law; it also prepares legislative histories.

Publications: Indexes and bibliographies, including BIBLIOGRAPHY OF AGRICULTURE, a monthly index of the world literature on agriculture and related chemical and biological subjects.

Information services: Answers reference questions and makes brief literature searches without charge. Photocopying is done at cost. The holdings of the National Agricultural Library and its branches are open to the public for onsite reference, and material may be borrowed through interlibrary loans.

Automated Information System: CAIN. This system includes a data base of 43,000 journal titles from BIBLIOGRAPHY OF AGRICULTURE. Contact the agency for further details.

NATIONAL LIBRARY OF MEDICINE - 8600 Rockville Pike, Bethesda, Maryland 20014.

Areas of interest: The Library collects material exhaustively in some 40 biomedical subject areas and, to a lesser degree, in a number of related areas such as general chemistry, physics, zoology, botany, psychology, and instrumentation.

Holdings: Nearly 1.5 million books, journals, technical reports, documents, theses, pamphlets, microfilms, pictorial materials, and audiovisual materials. The History of Medicine collection numbers more than 60,000 volumes, including over 500 titles published before 1501.

Publications: INDEX MEDICUS (monthly); CUMULATED INDEX MEDICUS (annual); ABRIDGED INDEX MEDICUS (monthly); CUMU-LATED ABRIDGED INDEX MEDICUS (annual); MEDICAL SUBJECT HEADINGS (annual); LIST OF JOURNALS INDEXED IN INDEX MEDICUS (annual); NLM CURRENT CATALOG (monthly, quarterly, and annual cumulation); MONTHLY BIBLIOGRAPHY OF MEDICAL REVIEWS; TOXICITY BIBLIOGRAPHY (quarterly); SELECTED REFER-ENCES ON ENVIRONMENTAL QUALITY AS IT RELATES TO HEALTH (monthly); BIBLIOGRAPHY OF THE HISTORY OF MEDI-CINE (annual); NATIONAL MEDICAL AUDIOVISUAL CENTER CATALOG (annual); literature searches (irregular); other occa-sional publications. The Library also provides camera-ready copy for the publication of recurring bibliographies by federal agencies and professional organizations, e.g., INDEX TO DENTAL LITERA-TURE by the American Dental Association, and ENDOCRINOLOGY INDEX by the National Institute of Arthritis and Metabolic Diseases, NIH. Complete lists of titles of the recurring bibliog-raphies, literature searches, and other NLM publications are available from the Library's Office of Public Information and are included in monthly issues of INDEX MEDICUS.

Information services: NLM activities serve to supplement and support the activities of regional and other medical libraries throughout the country. Health science practitioners, educators, investigators, and students, as well as others with related needs, have access to all collections in the Library. Materials are accessible for use at the Library daily, except Sunday. The Library will lend, through other libraries only, original volumes or photocopies of journal articles not available locally. Films, audiotapes, television film recordings, and other materials from the distribution collection of the Library's National Medical Audiovisual Center are available on loan from the National Medical Audiovisual Center (Annex), Station K, Atlanta, Georgia 30324. Through its computer-based MEDLARS (Medical Literature Analysis and Retrieval System), the Library provides, on request, retrospective, one-time bibliographies (demand searches) to re-searchers, clinicians, and other health professionals. Requests for MEDLARS search services should be directed to the MEDLARS Center or Regional Medical Library serving the requester's region. A complete list of addresses for regional interlibrary loan and MEDLARS services is contained in monthly issues of INDEX MEDICUS.

NATIONAL REFERRAL CENTER - Library of Congress, Science and Technology Division, 10 First Street S.E., Washington, D.C. 20540.
Tel: (202) 426-5670 (Referral Service)
Tel: (202) 426-5680 (Registration)
Tel: (202) 426-5687 (General Inquiries)

Areas of interest: Governmental, industrial, academic, public, and private organizations and institutions, and individuals, with specialized knowledge in any area of the physical, biological, engineering, or social sciences.

Holdings: An inventory of information resources in the United States. The inventory includes broad and specific areas of interest for each organization or individual, holdings (literature and report collections, unpublished data, specimen collections, etc.), titles of representative publications, and the types of information services available. A complete information service in the sci-tech area which will track down details or provide referrals.

Publications: A DIRECTORY OF INFORMATION RESOURCES IN THE UNITED STATES: PHYSICAL SCIENCES, BIOLOGICAL SCIENCES, ENGINEERING (1965); SOCIAL SCIENCES, (1965); WATER (1966); FEDERAL GOVERNMENT WITH A SUPPLEMENT OF GOVERNMENT-SPONSORED INFORMATION RESOURCES (1967); GENERAL TOXICOLOGY (1969); PHYSICAL SCIENCES, ENGINEERING (1971, this revision of the 1965 directory excludes the biological sciences).

Information services: Provides names, addresses, telephone numbers, and brief descriptions of appropriate information resources. Functioning as an intermediary, the Center directs those who have questions concerning specific subjects to organizations or individuals with specialized knowledge of those subjects. It does not provide technical details in answer to inquiries or furnish bibliographic assistance (requests for bibliographic services should be addressed to the Science and Technology Division). Services are available without charge, by telephone, correspondence, or through personal visits.

NEW YORK PUBLIC LIBRARY - Fifth Avenue at 42nd Street, New York, New York 10018.
Tel: (212) 695-4200

Areas of interest: Aeronautics; astronomy; automobiles; biochemistry; chemistry, communications; earth sciences; electricity; electronics; engineering (all branches); industrial arts; mathematics; metallurgy; mining; navigation; paper; petroleum; physics; plastics; radio; railroads; rubber; science (history).

Holdings: Approximately 154,000 books and 365,000 bound journals.

Publications: NEW TECHNICAL BOOKS (10 issues a year).

Information services: Answers inquiries; provides duplication services through the Library's Photographic Service; permits onsite reference.

SELECTED SPECIALIZED LIBRARIES

AMERICAN GAS ASSOCIATION LIBRARY - 1515 Wilson Boulevard, Arlington, Virginia 22009.

Areas of interest: Fuel gas industry and technology and related aspects, including appliance research, sales, and promotion; industrial, commercial, and residential air conditioning and heating; incineration; liquefied natural gas; pipelines; energy conversion; engines and turbines; codes and ordinances; accident prevention and safety.

Holdings: 7,000 books; 400 current journals; 200 file drawers of pamphlets and clippings; proceedings, indexes.

Publications: PUBLICATIONS LIST (annual); A.G.A. MONTHLY; bibliographies, instruction pamphlets, manuals, data handbooks, standards, and specifications.

Information services: Answers inquiries; makes referrals; provides bibliographic and reference services; provides duplication services for fee; distributes films, slides and flip charts at published rates; lends materials; permits onsite reference.

BATTELLE MEMORIAL INSTITUTE LIBRARY - Columbus Laboratories, 505 King Avenue, Columbus, Ohio 43201.

Areas of interest: Mechanical engineering, including materials, power, structure, instrumentation, heating and cooling, fluid and solid mechanics, thermodynamics, design and development, mechanical dynamics, product and manufacturing engineering, experimental methods, and building materials; chemical engineering, including processes, chemical-metallurgical, pilot plant design, electrolytic forming and machining, devices, materials concentration, separating, and high temperature kinetics; ceramics, including nuclear applications, building materials, porcelain enamels, refractories, and glass; metallurgy, including physical, powder-and-fiber, mechanical, extractive, ferrous, nonferrous, practice, metal joining, working, technology, including fuel, textile and fiber, leather, wood, pulp and paper, lubricant, petroleum, welding, and graphic arts; energy, including materials, radiation, shielding, engineering, reactor technology, and radioisotopes; chemistry, including organic, polymer, inorganic, surface, physical, petroleum, rubber and plastics, organic coatings, and catalysts; physics, including mathematical, applied, solid state, and nuclear; biosciences, including biochemistry, biophysics, agricultural chemistry, food technology, pharmaceutical chemistry; psychology, including engineering, sensory, personnel, and human factors research.

Holdings: Over 125,000 books and bound journals, 75,000 reports, 28 drawers of microfilm and microcards, and 3,000 current journals; 11 current newspapers; patents, films, photos, maps, charts. In addition to the central library, the Institute maintains the following specialized libraries; the Slavic Library, which houses Russian and other Slavic publications, including many cover-to-cover translations of Russian technical journals; the Systems and Electronics Library, which maintains a special collection, particularly catalogs and specifications, in the fields of electronics and electrical engineering; the Economics Library, which specializes in engineering economics, market research, and research administration; and the Classified Library, which maintains the collection of classified documents.

Publications: Accessions lists.

Information services: Answers inquiries; provides technical advice on the planning of libraries and information centers; makes interlibrary loans, within certain limits; provides literature-searching and duplication services.

BROOKHAVEN NATIONAL LABORATORY - RESEARCH LIBRARY - Upton, Long Island, New York 11973.

Areas of interest: Physics, including high energy physics, neutron physics and nuclear structure, solid state physics, and low temperature physics; chemistry, including nuclear chemistry, structural and theoretical physics, inorganic chemistry, radiation and hot atom chemistry, isotope effects, analytical chemistry, and physical chemistry; biology, including genetics, general physiology, biochemistry, biophysics, molecular biology, and ecology; medical sciences, including industrial medicine, microbiology, biochemistry, physiology, medical physics, nuclear engineering, including reactor physics, chemistry and chemical engineering, metallurgy and materials science, reactor engineering, hot laboratory operations, and radiation development; instrumentation and health physics.

Publications: WEEKLY SELECTED READING LIST; ACQUISITIONS LIST (monthly).

Information services: Answers brief inquiries; provides limited reference services; makes interlibrary loans; permits onsite use of collection by appointment.

OAK RIDGE ASSOCIATED UNIVERSITIES LIBRARY - P.O. Box 117, Oak Ridge, Tennessee 37830.
Tel: (615) 483-8611

Areas of interest: Nuclear science; basic sciences as related to nuclear science.

Holdings: 23,000 books; 21,000 journals; 106,000 reports.

Information services: Permits onsite reference; makes interlibrary loans.

SOUTHWEST RESEARCH INSTITUTE LIBRARY - Location: 8500 Culebra Road, San Antonio, Texas. P.O. Drawer 28510, San Antonio, Texas 78228.

Areas of interest: Primary research field includes chemistry and chemical engineering, biochemistry, physics, electronics and electrical engineering, mechanical engineering, civil engineering, engineering mechanics, bioengineering. In addition, the Library has special facilities for research in air and water pollution and engine exhaust emissions, analytical chemistry, artificial organs, building technology, communications, corrosion, metal fatigue, nondestructive testing, ecology, electric power, explosive and impact effects, gas technology, fire research, flavors, fluid mechanics and hydrodynamics, fuels-lubricants, lubrication, gas dynamics, highway engineering, instrumentation, oceanography, metallurgy, microwaves, operations research, nuclear reactors, polymers, radiochemistry, safety engineering, waste disposal, water resources, water treatment.

Holdings: 22,000 books; 25,000 volumes of journals, with current subscriptions to 1,000 journals in English, French, Dutch, Spanish, German, Russian, Polish, Hungarian, Italian, the Scandinavian languages, Czech, Bulgarian, Yugoslavian, Arabic, Hindi, Japanese, and Turkish; 60,000 reports and documents.

Publications: Accession lists. Other divisions of the Institute issue TOMORROW THROUGH RESEARCH (bimonthly), technical reports, state-of-the-art reviews, standards and specifications, data handbooks, bibliographies, proceedings, and translations.

Information services: Answers inquiries; makes referrals; makes interlibrary loans; permits onsite use of collection. Xerographic and microfilm copying is provided for a fee.

TECHNICAL LIBRARY - Tennessee Valley Authority, 500 Union Avenue, S.W., Knoxville, Tennessee 37902.

Areas of interest: Tennessee Valley resources development; flood control; stream and air pollution control; navigation; electric power; forest management and utilization; agricultural and forest economics; mathematics and computers; fertilizers; agronomy; soil chemistry; aquatic biology; engineering (civil, electrical, mechanical, chemical, nuclear); hydraulics; hydrology; geology; architecture; meteorology.

Holdings: 70,000 books, 1,400 journal and periodical titles, 2,200 reports, 30,000 pamphlets, and 30,000 government documents, including much historical material on the Tennessee River region and development of its resources. (The Library is the official depository for most reports and studies originating in the TVA.) There is a clipping file of 400,000 items, which gives a complete day-by-day history of the TVA.

Publications: Acquisitions lists, indexes, and bibliographies, including three annual bibliographies: INDEXED BIBLIOGRAPHY OF THE TENNESSEE VALLEY AUTHORITY; TVA PROGRAM BIBLIOGRAPHY; and FLOOD DAMAGE PREVENTION, AN IN-DEXED BIBLIOGRAPHY. FERTILIZER ABSTRACTS is prepared and pub-published monthly by the Technical Library at Muscle Shoals, Alabama.

Information services: Makes interlibrary loans, permits onsite reference. Library services are also available from the two branch libraries, located as follows:

Muscle Shoals, Alabama 35661

Chattanooga, Tennessee 37401

TECHNICAL LIBRARY AND SCIENCE INFORMATION CENTER - U.S. Army
Natick Laboratories - Kansas Street, Natick, Massachusetts 01760.

Areas of interest: Chemistry; physics; biology; medicine; physiology; psychology; geography; mathematics; statistics; law; food
science and technology; textiles; clothing and personnel equipment.

Holdings: 50,000 books; 1,000 periodical titles; 50,000 reports;
abstracts and indexes in depth of literature on specific subject
areas such as terrestrial hypoxia, functional performance of clothing and personnel equipment, etc.

Publications: Bibliographies, translations of journal articles.

Information services: Answers inquiries; provides reference, literature-searching, and translation services; makes referrals; lends
materials. Library services are primarily for Natick Laboratories'
personnel, but limited access to open-literature collections is
available to others by special arrangement.

UNITED FRESH FRUIT AND VEGETABLE ASSOCIATION LIBRARY - 777 14th
Street N.W., Washington, D.C. 20005.

Areas of interest: Production and marketing of fresh fruits and
vegetables, including statistics, merchandising methods, transportation, storage and temperature, prices, containers, packaging, and
distribution; physical and chemical research on fresh fruits and
vegetables; history of fruits and vegetables; botany; nutrition,
with special reference to fruits and vegetables but covering also
the entire field in detail, by commodity, by nutrient, and by
disease entity such as beriberi, pellagra, scurvy, arteriosclerosis
and obesity.

Holdings: 64,000 reports, 800 books, 39 journal titles; data
compilations, films, photos, maps and charts, tear sheets.

Publications: UNITED FRESH OUTLOOK and POTATO DIVISION
SPUDLIGHT (weekly newsletters); SUPPLY LETTER, TOMATO
TALK, MANAGEMENT REVIEW, and TERMINAL TIMES (all
monthly); NUTRITION NOTES (approximately quarterly); FRUIT
AND VEGETABLE FACTS AND POINTERS (fact sheets on each
of 81 commodities); yearbook, technical reports, bibliography,
directory, descriptive brochure. A publications list is available.

Information services: Answers inquiries; provides consulting,
reference, and document services; makes referrals. Services are
primarily for members, but are also provided, in most cases, to
government agencies and to inquirers from universities doing
work in this field.

U.S. ARMED FORCES RADIOBIOLOGY RESEARCH INSTITUTE LIBRARY – Bethesda, Maryland 20014.

Areas of interest: Radiobiology; physiology; biochemistry; chemistry; pathology; immunology; psychology; nuclear physics; nuclear engineering; nuclear reactors; dosimetry; health physics.

Holdings: Approximately 5,000 books, 250 periodical titles, and 100,000 reports in hard copy and microfiche.

Publications: AFRRI SCIENTIFIC REPORTS; AFRRI TECHNICAL NOTES.

Information services: Answers inquiries; provides reference services; makes interlibrary loans; permits onsite use of collection. Services are available to those with a need-to-know and whose user requirements can be processed on an "as time permits" basis.

U.S. ATOMIC ENERGY COMMISSION

ARGONNE NATIONAL LABORATORY LIBRARY – 9700 South Cass Avenue, Argonne, Illinois 60439.

Areas of interest: Physical and biological sciences; mathematics; nuclear reactors and related technology.

Holdings: The library, consisting of a Central Library with eight subject-oriented branches, has an extensive collection of scientific and technical journals and books, along with a virtually complete collection of AEC unclassified reports.

Information services: The Library Services Department provides broad-scope library and technical information services to the Laboratory staff. The Central Library handles all technical processing, interlibrary loans, and contacts with outside libraries. Visitors may use the libraries during normal working hours. The publication ADDITIONS TO THE LIBRARY is available on a limited basis to government libraries. The library has a small collection of computer programs for library purposes, since many library functions are mechanized. The Information Services Department answers inquiries, handles public information functions, directs a technical exhibit program, a lecture service, and an Argonne tour service, and distributes educational literature to student requestors. The Office for Educational Affairs assists colleges and universities with regard to education in nuclear science, engineering, and related fields, assists in arranging research participation appointments for faculty members, postdoctorals, and graduate and advanced undergraduate students, maintains instructional laboratories available for use by faculty members and their students, provides educational materials, and arranges special lectures and demonstrations.

HEADQUARTERS LIBRARY – Germantown, Maryland 20767.
Tel: (301) 973-4166

BRANCH LIBRARY - BETHESDA LIBRARY - 4915 St. Elmo Avenue, Bethesda, Maryland 20014.
Tel: (301) 973-7488

Areas of interest: Nuclear science and technology; physics; chemistry; metallurgy; engineering; mathematics; biology; medicine; management; federal legislation.

Holdings: A book and journal collection of over 50,000 classified and unclassified AEC reports; a legislative collection of over 50,000 volumes.

Information services: Answers inquiries or refers inquirers to appropriate headquarters program division; makes interlibrary loans, primarily to libraries in the greater Washington areas; permits on-site reference to the unclassified collections by representatives of government agencies, their contractors, universities, and industry, provided arrangements are made in advance.

LOS ALAMOS SCIENTIFIC LABORATORY LIBRARIES - P.O. Box 1663, Los Alamos, New Mexico 87544.

Areas of interest: Theoretical physics and mathematics; experimental nuclear physics; weapons physics, design, and testing; nuclear propulsion, chemistry, and chemical engineering; explosives research; metallurgy; electronics and instrumentation; biology; medicine; industrial hygiene; reactor design and technology.

Holdings: 85,000 books; 100,000 bound periodical volumes; subscriptions to 3,400 journals and other serials; 385,000 technical reports; 400 reels of motion pictures; 5,000 reprints; 15,000 translations; patents, photos.

Publications: Technical reports, acquisition lists, translations, bibliographies.

Information services: Lends documentary materials to other organizations on a limited basis.

OAK RIDGE NATIONAL LABORATORY LIBRARY - P.O. Box X, Oak Ridge, Tennessee 37831.

Areas of interest: Nuclear science and engineering; chemistry; physical mathematics; metallurgy; biological sciences.

Holdings: 11,500 books; 95,000 journals; 320,000 reports; 20,000 translations.

Publications: Accessions list, index.

Information services: Provides reference services; makes interlibrary loans; permits onsite reference by qualified representatives of industrial or educational organizations.

U.S. FEDERAL POWER COMMISSION - LIBRARY - 441 G Street N.W., Room 2916, Washington, D.C. 20426.

Areas of interest: Electric power production and distribution; law; management; mineral resources, particularly petroleum and natural gas; public utility valuation, depreciation, and rate structures; water power; hydroelectric power projects; electric and gas engineering; atomic energy; economics; accounting; air pollution and water pollution.

Holdings: Books, journals, technical reports, legislative histories pertinent to Federal Power Commission jurisdiction.

Information services: Permits onsite reference by researchers by special permission of the librarian.

DIRECTORY ASSISTANCE IN FINDING LIBRARIES

In order to learn what libraries are located within a certain geographical area and in order to ascertain some idea of the kinds of materials they hold, reference can be made to at least three standard directories.

AMERICAN LIBRARY DIRECTORY - R.R. Bowker Co., 1180 Avenue of the Americas, New York, New York 10036. Annual.

This contains a directory listing by state, city, and village or town. Full descriptive details on all libraries academic, industrial/special, research, and public. It also has entries for library schools and for the addresses of state library associations. Therefore, one could use this directory to find if there were a science-technology library of any size within a certain location. In addition, phone numbers and names of the various department directors in a library are provided which can be helpful in getting assistance with an information problem.

DIRECTORY OF SPECIAL LIBRARIES AND INFORMATION CENTERS. 4th ed. 3 vols. Edited by Margaret Labash Young, et al. Detroit: Gale Research Co., 1977.

This covers many unusual collections and information services not covered by the AMERICAN LIBRARY DIRECTORY.

ENCYCLOPEDIA OF INFORMATION SYSTEMS AND SERVICES - Edwards Brothers, 2500 South State Street, Ann Arbor, Michigan 48104. 1974.

A guide to data banks, computerized systems and services, coordinating agencies.

The national and state library associations are also a means of finding out about libraries and their service programs. Each state has a state library association which is affiliated, but not formally a part of the American Library Association. The Special Library Association has chapters in major regions of the U.S. and many of its members are expert in information of science and technology. A newer organization, The American Association for Information

Science, is particularly concerned with automated approaches to information organization and retrieval. The other library associations already mentioned, of course, are also committed to this trend, but they differ from the American Association for Information Science in having to maintain a rather wide spectrum of library interests and projects.

The addresses of the three major associations are:

American Library Association - 50 East Huron Street, Chicago, Illinois 60611

American Society for Information Science - 2011 Eye Street, N.W., Washington, D.C. 20006

Special Libraries Association - 31 East 10th Street, New York, New York 10003

PUBLISHER'S NAMES AND ADDRESSES

APPEARING IN THIS GUIDE

PUBLISHERS' NAMES AND ADDRESSES
APPEARING IN THIS GUIDE

AIAA Technical Information Service
750 Third Avenue
New York, New York 10017

Abstracts Service
American Chemical Society
Ohio State University
Columbus, Ohio 43210

Academic Press
111 Fifth Avenue
New York, New York 10003

Addison-Wesley Publishing Co., Inc.
Reading, Massachusetts 01867

Agricultural Development Council, Inc.
630 Fifth Avenue
New York, New York 10020

Agricultural Institute of Canada
151 Slater Street
Ottawa 4, Ontario, Canada

Agricultural Pesticide Society
Research Institute
Canada Agriculture University
Sub Post Office
London 72, Ontario, Canada

Agricultural Press Ltd.
161/166 Fleet Street
London, EC4P 4AA, England

Agricultural Research Service
Publications Branch
Room 5149
South Building
Washington, D.C. 20250

Air Pollution Control Office
Box 12055
Research Triangle Park,
    North Carolina 27709

American Agricultural Economics
    Documentation Center
Room 1505 South Building
Washington, D.C. 20250

American Association for the Advance-
    ment of Science
1515 Massachusetts Avenue, N.W.
Washington, D.C. 20005

American Association of Cereal
    Chemists, Inc.
3340 Pilot Knob Road
St. Paul, Minnesota 55121

American Association of Textile
    Chemists and Colorists
P.O. Box 12215
Research Triangle Park, North Caro-
    lina 27709

American Biomedical Information Service
Box 707
Arcadia, California 91006

American Ceramic Society, Inc.
4055 North High Street
Columbus, Ohio 43214

American Chemical Society
1155 16th Street, N.W.
Washington, D.C.   20036

American Conference of Governmental
   Industrial Hygienists
Box 1937
Cincinnati, Ohio 45202

American Elsevier Publishing Co.,
   Inc.
52 Vanderbilt Avenue
New York, New York 10017

American Institute of Consulting
   Engineers
345 East 47th Street, Suite 303
New York, New York 10017

American Management Association,
   Publication Service
135 West 50th Street
New York, New York 10020

American Meat Science Association
National Livestock and Meat Board
36 South Wabash Avenue
Chicago, Illinois 60603

American Metal Market Co.
525 West 42nd Street
New York, New York 10036

American National Standards Institute
1430 Broadway
New York, New York 10018

American Petroleum Institute
1271 Avenue of the Americas
New York, New York 10020

American Potato Yearbook
834 South Avenue West
P.O. Box 398
Westfield, New Jersey 07090

The American Society for Mechanical
   Engineers
United Engineering Center
345 East 47th Street
New York, New York 10017

American Society for Metal,
   Technical and Engineering
Book Service
Metals Park, Ohio 44073

The American Society for Testing and
   Materials
1916 Race Street
Philadelphia, Pennsylvania 19103

American Society of Agricultural
   Engineers
2950 Niles Road
St. Joseph, Michigan 49085

American Society of Agronomy
677 South Segoe Road
Madison, Wisconsin 53711

American Society of Bakery Engineers
Room 1921, 2 North Riverside Plaza
Chicago, Illinois 60606

American Society of Safety Engineers
850 Busse Highway
Park Ridge, Illinois 60068

American Standards Institute
1430 Broadway
New York, New York 10018

American Tomato Yearbook
834 South Avenue, West
P.O. Box 398
Westfield, New Jersey 07090

American Water Resources Association
206 East University Avenue
Urbana, Illinois 61801

Ames Publishing Co.
One West Olney Avenue
Philadelphia, Pennsylvania 19120

Ann Arbor–Humphrey Science Publishers
Drawer 1425
600 South Wayne Road
Ann Arbor, Michigan 48106

Annual Reviews, Inc.
4239 El Camino Way
Palo Alto, California 94306

Applied Plastics Publishers
c/o Scientific Press Ltd.
11a Gloucester Road
London, SW7, England

Arlington Publishing Co.
2 North Riverside Plaza
Chicago, Illinois 60606

Edward Arnold and Co., Ltd.
25 Hill Street
London, W1X 8LL, England

Arno Press, Inc.
330 Madison Avenue
New York, New York 10017

Association for Computing Machinery
1133 Avenue of the Americas
New York, New York 10017

Association of American Pesticide
   Control Officials, Inc., Kansas
State Board of Agriculture
State Office Building
Topeka, Kansas 66612

Association of Consulting Chemists
   and Chemical Engineers, Inc.
50 East 41st Street
New York, New York 10016

Association of Official Analytical
   Chemists
Box 540
Benjamin Franklin Station
Washington, D.C.   20044

Australia Chemical Engineering
Box 250
North Sydney, N.S.W. Australia

Avi Publishing Co., Inc.
Box 831/250 East State Street
Westport, Connecticut 06880

Ayer Press
N.W. Ayer and Son, Inc.
West Washington Square
Philadelphia, Pennsylvania 19106

BPS Exhibition
6 London Street
London, W2, England

Barnes and Noble
c/o Harper & Row Publishers
10 East 53rd Street
New York, New York 10022

Beeler Publishing Co.
3030 Bridgeway
Sausalito, California 94965

W.A. Benjamin, Inc.
2725 Sand Hill Road
Menlo Park, California 94025

Benn Brothers, Ltd.
159 Fleet Street
London, EC4 2DL, England

A.M. Best Co.
Park Avenue
Morristown, New Jersey 07960

Bethune Jones
321 Sunset Avenue
Asbury Park, New Jersey 07712

Bhabha Atomic Research Centre
Information Service
Trombay, Bombay 85, India

Bill Brothers Publishing Corp.
633 Third Avenue
New York, New York 10017

Bill Publications, Inc.
630 Third Avenue
New York, New York 10017

Biochemical Society
7 Warwick Court
London, WC1R 5DP, England

Biosciences Information Service of
    Biological Abstracts
2100 Arch Street
Philadelphia, Pennsylvania 19103

Blackwell Scientific Publications, Ltd.
Osney Mead
Oxford, OX2 OEL, England

Clark Boardman, Ltd.
435 Hudson Street
New York, New York 10014

R.R. Bowker Co.
1180 Avenue of the Americas
New York, New York 10036

Breskin Communications, Inc.
415 Brown Avenue
Scottsdale, Arizona 85251

Brigham Young University Press
209 University Press Building
Provo, Utah 84602

The British Book Centre
153 East 78th Street
New York, New York 10021

British Food Manufacturing Industries
Research Association
Randalls Road
Leatherhead, Surrey, England

British Institute of Management
Management House
Park Street
London, WC2B SPT, England

British Internal Combustion
Research Institute
111/112 Buckingham Avenue
Slough, Bucks, England

British Plastics Federation
47 Piccadilly
London, W1V ODN, England

British Standards Institution
2 Park Street
London, W1A 2BS, England

Bureau of National Affairs, Inc.
1231 25th Street, N.W.
Washington, D.C. 20037

Bureau of Statistics
Ottawa, Canada

Business Communications, Inc.
2800 Euclid Avenue
Cleveland, Ohio 44115

Business Publications, Inc.
13773 North Central Expressway
Dallas, Texas 75231

Business Surveys, Ltd.
The Mead
Wellington, Surrey, England

Buttenheim Publishing Corp.
Berkshire Common
Pittsfield, Massachusetts 01201

Butterworth and Co. Publishers,
    Ltd.
88 Kingsway
London, WC2B 6AB, England

CBD Research, Ltd.
154 High Street
Beckenham, Kent, England

CCM Information Corp.
866 Third Avenue
New York, New York 10022

CRC Press
18901 Cranwood Parkway
Cleveland, Ohio 44128

Cabot Corp.
125 High Street
Boston, Massachusetts 02110

Cahners Publishing Co., Inc.
89 Franklin Street
Boston, Massachusetts 02110

Cambridge Communications Corp.
c/o Stevens House
10 South Prince Street
Lancaster, Pennsylvania 17603

Cambridge University Press
Bentley House
200 Euston Road
London, NW1 2DB, England

Canadian Department of Agriculture
Engineering Research Service
Research Branch
Ottawa 3, Ontario, Canada

Canadian Engineering Publications,
    Ltd.
46 St. Clair Avenue, East
Toronto 7, Canada

Canadian Library Association
Room 606
63 Sparks Street
Ottawa 4, Canada

Carter Thermal Engineering
Hay Mills
Birmingham, England

Catholic University of America Press
620 Michigan Avenue, N.E.
Washington, D.C. 20064

Center for Air Environment Studies
The Pennsylvania State University
226 Chemical Engineering II
University Park, Pennsylvania 16802

Center of Scientific and Technological
    Information
325 Chestnut Street
Philadelphia, Pennsylvania 19106

Central Board of Irrigation and Power
Kasturba Gandhi Marg
New Delhi, India

Centre de Documentation du C.N.R.S.
15 Quai Anatole-France
Paris (7e), France

Centre International de Documentation
    du Machinisme Agricole
Parc de Tourvoie
92 Antony
Hauts-De-Seine, France

Chapman and Hall, Ltd.
35 Red Lion Square
London, WC1R 4SG, England

Chemical Abstracts Service
American Chemical Society
Ohio State University
Columbus, Ohio 43210

Chemical and Marine Press, Ltd.
33-40 Bowling Green Lane
London, EC1, England

Chemical Daily Co., Ltd.
19-16, 3-chome
Shibaura, Minato-ku
Tokyo, Japan

Chemical Economic Services
Box 468
Princeton, New Jersey 08540

Chemical Horizons, Inc.
274 Madison Avenue
New York, New York 10016

Chemical Information Services
Stanford Research Institute
Menlo Park, California 94025

Chemical Marketing and Economic
    Division
P.O. Box 170
Brooklyn, New York 11209

Chemical Marketing Research
Association
100 Church Street
New York, New York 10007

Chemical Publishing Co., Inc.
155 West 19th Street
New York, New York 10011

Chemical Rubber Co.
18901 Cranwood Parkway
Cleveland, Ohio 44128

Chemical Society
Burlington House
London, W1V OBN, England

Chilton Book Co.
Chilton Way
Radnor, Pennsylvania 19089

Chorley and Pickersgill, Ltd.
Amberley House,
Norfold Street
Strand, London, WC2, England

Chromatic Communications, Inc.
799 Roosevelt Road
Building 4, Suite 300
Glen Ellyn, Illinois 60137

J. & A. Churchill, Ltd.
104 Gloucester Place
London, W1, England

Dean M. Clark
Board of Trade Building
141 West Jackson Boulevard
Chicago, Illinois 60604

Clissold Publishing Co.
Sun Times Building
401 North Wabash Avenue
Chicago, Illinois 60611

Colour Publications Private, Ltd.
126-A, Dhuruwadi
Off Dr. Nariman Road
Bombay 25, D D, India

Commerce Clearing House, Inc.
4025 West Peterson Avenue
Chicago, Illinois 60646

Commercial Development Association
100 Church Street
New York, New York 10007

Commercial Review, Inc.
1812 Northwest Kearney Street
Portland, Oregon 97209

Commissioner of Patents
Information Canada
Publishing Division
Ottawa, Canada

Commission Internationale Technique
de Sucrerie, General Secretary
1 Aandorenstraat
Tienen, Belgium

Committee for International Cooperation
in Information Retrieval, Among
Examining Patent Offices
32 Chemin des Colombelles
1211 Geneva 20, Switzerland

Commonwealth Agricultural Bureaux
Farnham House
Farnham Royal
Slough, SL2 3BN, Bucks, England

Commonwealth Forestry Bureau
South Parks Road
Oxford, OX1 3RB, England

Commonwealth Institute of Entomology
Publications Office
56 Queen's Gate
London, SW7 5JR, England

Commonwealth Mycological Institute
Ferry Lane
Kew, Surrey, England

Compendium Publishers International
Corp.
2175 Lemoine Avenue
Fort Lee, New Jersey 07024

Congressional Information Service
600 Montgomery Building
Washington, D.C. 20014

Constable and Co., Ltd.
10-12 Orange Street
London, WC2, England

Cornell University College of
   Agriculture
Waste Management Program
Riley Robb Hall
Cornell University
Ithaca, New York 14850

Corporate Intelligence
25 Broad Street
New York, New York 10004

Corporation des Agronomes de la
   Province de Quebec
262 Henri-Bourassa Boulevard W.
Montreal 347, Quebec, Canada

Council for Agricultural and
   Chemurgic Research
350 Fifth Avenue
New York, New York 10001

Council for Scientific and Industrial
   Research
Box 395
Pretoria, South Africa

Data Processing Management
   Association
505 Busse Highway
Park Ridge, Illinois 60068

F.A. Davis Publishing Co.
1915 Arch Street
Philadelphia, Pennsylvania 19103

Marcel Dekker, Inc.
270 Madison Avenue
New York, New York 10016

Department of the Environment
Her Majesty's Stationery Office
Atlantic House
Holborn Viaduct
London, EC1, England

Department of Trade and Industry
Warren Spring Laboratory
Gunnels Wood Road
Stevenage, Herts, SG1 2BX, England

Derwent Publications, Ltd.
Rochdale House
128 Theobalds Road
London, WC1X 8RP, England

Design Engineering Publications, Ltd.
Hermes House
89 Blackfriars Road
London, SE1, England

Diffusion Information Center
Box 2981
Cleveland, Ohio 44116

Directories Publishing Co.
R.D. 5, Box 422
Flemington, New Jersey 08822

Documentation Abstracts, Inc.
Box 8510
Philadelphia, Pennsylvania 19101

R.H. Donnelly Corp.
466 Lexington Avenue
New York, New York 10017

Economic Documentation Office
Graaf Florislaan 30a
Box 505
Hilversum, Netherlands

Edinburgh School of Agriculture
West Mains Road
Edinburgh, EH9 3JG, Scotland

Edwards Brothers, Inc.
2500 South State Street
Ann Arbor, Michigan 48104

Elsevier Publishing Co.
Box 211
Jan van Galenstraat 335
Amsterdam, Netherlands

Elsevier Scientific Publishing Co.
52 Vanderbilt Avenue
New York, New York 10017

Engineering, Chemical and Marine
    Press, Ltd.
33-39 Bowling Green Lane
London, EC1, England

Engineering Index
345 East 47th Street
New York, New York 10017

Engineering Libraries Division
The Society
Washington, D.C. 20036

Engineering Societies Library
345 East 47th Street
New York, New York 10017

Entomological Society of America
4603 Calvert Road
College Park, Maryland 20740

Environment Information Service
124 East 39th Street
New York, New York 10016

Environmental Protection Agency
555 Ridge Avenue
Cincinnati, Ohio 45213

Envoy Journals, Ltd.
67 Clerkenwell Road
London, EC1 5BP, England

Euratom
CID Agence et Messageries de la
    Presse SA
1 rue de le Petite ille
Brussels, Belgium

European Translations Centre
101 Doelenstraat
Delft, Netherlands

FAO Documentation Center
c/o Unipub, Inc.
Division of International Publications
Box 433
New York, New York 10016

F.S. Publications
Box 2458
Pulos Verdes Penninsula, California
    90274

Fancourt Technical Information
    Co., Ltd.
Box 59
St. Helier, Jersey
British Channel Islands

Farm Chemicals
37841 Euclid Avenue
Willoughby, Ohio 44094

Farm Engineering Industry
    Publications, Ltd.
64a Lansdowne Road
South Woodford
London E18, England

George F. Farrell
200 Commerce Road
Cedar Grove, New Jersey 07009

Fertilizer Institute
1015 18th Street, N.W.
Washington, D.C. 20036

Fishing News (Books), Ltd.
23 Rosemount Avenue
West Byfleet, Surrey, England

Food and Agriculture Organization
    of the United Nations
Via delle Terme di Caracalla
Rome 00100, Italy

Food Chemical News, Inc.
601 Warner Building
Washington, D.C. 20004

Food Protection and Toxicology Center
University of California
Davis, California

Food Research Institute
Stanford University
Stanford, California 94305

W.H. Freeman and Co.,
  Publishers
660 Market Street
San Francisco, California 94104

Freund Publishing House, Ltd.
Box 35010
Tel Aviv, Israel

Frozen Food Age Publishing
  Corp.
230 Park Avenue
New York, New York 10017

Gale Research Co.
Book Tower
Detroit, Michigan   48226

Godfrey Memorial Library
134 Newfield Street
Middletown, Connecticut 06457

Goodheart-Willcox Co.
123 West Taft Drive
South Holland, Illinois 60473

Gordon and Breach
  Science Publishers, Inc.
1 Park Avenue
New York, New York 10016

Gordon Publications, Inc.
20 Community Place
Morristown, New Jersey 07960

Government Printer
Bosman Street
Private Bag, 85, Pretoria
South Africa

Graphics Management Corp.
1101 16th Street, N.W.
Washington, D.C.   20036

Charles Griffin and Co., Ltd.
42 Drury Lane
London, WC2B 5RX, England

Antoni Guilleumas Brosa
Calle Parque 3, apto. 1381
Barcelona-2, Spain

Gulf Publishing Co.
Book Division
Box 2608
Houston, Texas 77001

Gusto Communications, Inc.
360 Lexington Avenue
New York, New York 10017

Hafner Publishing Co.
Subsidiary of Crowell Collier
  and Macmillan, Inc.
866 Third Avenue
New York, New York 10022

G.K. Hall and Co.
70 Lincoln Street
Boston, Massachusetts 02111

Halsted Press, Division of
  John Wiley
605 Third Avenue
New York, New York 10016

The Hamly Publishing Group
Astronaut House
Hounslow Road
Feltham, Middlesex, England

Harper & Row Publishers
49 East 53rd Street
New York, New York 10022

Hart Publishing Co., Inc.
15 West Fourth Street
New York, New York 10012

Hayden Book Co., Inc.
50 Essex Street
Rochelle Park, New Jersey 07662

Heating and Ventilating Publishers
103 Brigstock Road
Thornton Heath, Surrey, England

James H. Heinemann, Inc.,
   Publishers
19 Union Square W.
New York, New York 10003

Her Majesty's Stationery Office
   (HMSO)
Head Office
Atlantic House
Holborn Viaduct
London, EC1, England

Leonard Hill (Books), Ltd.
158 Buckingham Palace Road
London, S.W. 1, England

Hitchcock Publishing Co.
Hitchcock Building
Wheaton, Illinois 60187

Francis Hodgson, Ltd.
P.O. Box 74
Guernsey, England

Human Ecological Society
Box 146
Elsah, Illinois 62028

Human Factors Society
P.O. Box 1369
Santa Monica, California 90406

Humphrey Science Publishers, Inc.
see
Ann Arbor-Humphrey Science
   Publishers

Hungarian Central Technical Library
   and Centre for Documentation
Kultura, Box 159
Budapest 62, Hungary

Hutchinson Publishing Group, Ltd.
3 Fitzroy Square
London, NW3, England

IEEE Press
Institute of Electrical and Electronics
   Engineers, Inc.
345 East 47th Street
New York, New York 10017

IFI Plenum Data Co.
2001 Jefferson Davis Highway
Arlington, Virginia 22202

IPC Science and Technology Press,
   Ltd.
IPC House
32 High Street
Guildford
Surrey, England

Iliffe Books, Ltd.
42 Russell Square
London, England

Indian Standards Institution
Manak Bhavan
9 Bahadur Shah Zafar Marg
New Delhi 1, India

Industrial Health Foundation, Inc.
Information Services
5231 Centre Avenue
Pittsburgh, Pennsylvania 15232

Industrial Publications
A-3 Jewan Jyot, 18/20
Cowasji Patel Street
Bombay 1, India

Industrial Research Organization
Box 89
East Melbourne, Victoria 3002,
   Australia

Industrial Research Service, Inc.
90 Washington Street
Dover, New Hampshire 03820

Information Company of America
1011 Lewis Tower
225 South 15th Street
Philadelphia, Pennsylvania 19102

Information for Industry, Inc.
c/o IFI/Plenum Corp.
Subsidiary of Plenum Publishing Corp.
1000 Connecticut Avenue, N.W.
Washington, D.C. 20036

Information Resources Press
Division of Herner and Co.
2100 M Street, N.W.
Washington, D.C. 20037

Institute for Scientific Information
Publications Division
325 Chestnut Street
Philadelphia, Pennsylvania 19106

Institute National de la
   Recherche Agronomique
149, rue de Grenelle
Paris (7e), France

Institute of Chemical Engineers
16 Belgrave Square
London, SW1X 8PT, England

Institute of Electrical Engineers
(The) Savoy Place
London, WC2R OBL, England
see
London Institute of Electrical Engineers

Institute of Environmental Sciences
940 East Northwest Highway
Mt. Prospect, Illinois 60056

Institute of Food Technologists
221 North LaSalle Street
Chicago, Illinois 60601

Institute of Management Sciences
P.O. Box 6112
Providence, Rhode Island 02904

Institute of Water Pollution Control
Ledson House
53 London Road
Maidstone, Kent, England

The Institution of Agricultural
   Engineers
Penn Place, Rickmansworth
Hertfordshire, England

Institution of Chemical Engineers
15 Belgrave Square
London, England

Institution of Engineers
Australia Science House
157 Gloucester Street
Sydney 200, Australia

Institution of Fire Engineers
148 New Wald
Leicester, England

Instituto Geografico de Agostini
S.P.A.
Viale Roma 4
28100 Novara, Italy

Interdok Corporation
173 Halstead Avenue
Box 326
Harrison, New York 10528

International Association of Agricultural Librarians and Documentalists
Oxford, England

International Association of Milk,
   Food and Environmental Sanitarians,
   Inc.
P.O. Box 437
Shelbyville, Indiana 46176

International Association of Seed
   Crushers
1 Watergate
London EC4, England

International Atomic Energy Agency,
Vienna
Karntnerring 11 Box 590
A,1011 For IAEA Bulletin

U.S. address:
National Agency for International
Publications, Inc.
317 East 34th Street
New York, New York 10016

International Commission on Irrigation
and Drainage
48, Nyaya Marg
Chanakyapuri, New Delhi 110021,
India

International Executive Newsletter
Co.
52 rue du Progres
1000 Brussels, Belgium

U.S. address:
35 West Elm Street
Littleton, New Hampshire 03561

International Licensing, Ltd.
92 Cannon Lane
Pinner, Middlesex, HA5 1HT, England

International Patent Service (Interpas)
N.V., Buitenhaven 25
Ben Bosch, Netherlands

International Potash Institution
Zieglerstrasse 30
CH-3000 Berne 14
Switzerland

International Publications Service
114 East 32nd Street
New York, New York 10016

International Society of Soil Science
63 Mauritskade
Amsterdam, Netherlands

Interscience-Wiley Publishers, Inc.
605 Third Avenue
New York, New York 10016

The Interstate Printers & Publishers,
Inc.
19-27 North Jackson Street
Danville, Illinois 61832

Intertec Publishing Corp.
Subsidiary of Howard W. Sams and
Co., Inc.
1014 Wyandotte Street
Kansas City, Missouri 64105

Israel Patent Office
Ministry of Justice
Government Printer, Jerusalem

Jacobsen Publishing Co.
300 West Adams Street
Chicago, Illinois 60606

Japan Patent Center, Inc.
Box 72
Shitaya P.O. Tokyo, Japan

Japan Publications Trading Co.,
Ltd.
2-1 Sarngakucho, 1-chome
Chiyoda-ku, Tokyo

Jet Propulsion Laboratory Library
4800 Oak Grove Drive
Pasadena, California 91103

John Adams House
John Adams Street
London, WC2, England

Johnson Hill's Press, Inc.
1233 Janesville Avenue
Ft. Atkinson, Wisconsin 53538

Keith Business Library
Box 453
Ottawa, Canada

Kellogg Center for Continuing
Education
Michigan State University
East Lansing, Michigan 48823

B. Klein Publishing Co.
11 Third Street
Rye, New York 10580

R.E. Krieger Publishing Co., Inc.
Box 542
Huntington, New York 11743

Kultura
Box 149
Budapest 62, Hungary

Lakewood Publications, Inc.
731 Hennepin Avenue
Minneapolis, Minnesota 55403

Lawson Publications Ltd.
49 Clarence Street
Sydney, Australia

Lea and Febiger Publishers
600 Washington Square
Philadelphia, Pennsylvania 19106

Lead Development Association
34 Berkeley Square
London, W1, England

Library Association (Publication Office)
Store Street
London, WC1, England

London Institute of Electrical Engineers
see
Institute of Electrical Engineers

Longman Young Books, Ltd.
74 Grosvenor Street
London, W1X OAS, England

Lowry-Cocroft Abstracts
905 Elmwood Avenue
Evanston, Illinois 60202

Macdonald and Co., Ltd.
49 Poland Street
London, W1A 2LG, England

McDonald and Evans, Ltd. Publishers
8 John Street
London, WC1N 2H4, England

McGraw-Hill Book Co., Inc.
1221 Avenue of the Americas
New York, New York 10020

Machinery Publishing Co., Ltd.
Clifton House
83/117 Euston Road
London, NW1 2RE, England

Maclaren and Sons, Ltd.
Davis House
69/77 High Street
Croydon, CR9 1QH, England

Maclean-Hunter, Ltd.
481 University Avenue
Toronto, M5W 1A7, Ontario, Canada

Macmillan Publishing Co.
866 Third Avenue
New York, New York 10022

MacRae's Bluebook
Western Springs, Illinois 60558

Magazines for Industry
777 Third Avenue
New York, New York 10017

Malawi Government Printing
  Department
Box 37
Zomba, Malawi

Management Publications, Ltd.
5 Winsley Street
London, W1A 2HG, England

Manufacturing Chemists' Association,
  Inc.
1825 Connecticut Avenue, N.W.
Washington, D.C. 20009

Manufacturing Confectioner Publishing Co.
1031 South Boulevard
West Oak Park, Illinois 60302

Marquis Who's Who, Inc.
Marquis Publications Building
210 East Ohio Street
Chicago, Illinois 60611

Martindale-Hubbell Co.
One Prospect Street
Summit, New Jersey 07901

Massachusetts Institute of Technology Press (M.I.T. Press)
28 Carleton Street
Cambridge, Massachusetts 02142

C.L. Mast, Jr. and Associates
2041 Vardon Lane
Flossmoor, Illinois 60422

Meister Publishing Co.
37841 Euclid Avenue
Willoughboy, Ohio 44094

Merck and Co., Inc.
Publications Department
Rahway, New Jersey 07065

Metal Powder Industries Federation
201 East 42nd Street
New York, New York 10017

The Michigan State University Press
Box 550
East Lansing, Michigan 48823

Miller Publishing Co.
P.O. Box 67
2501 Wayzata Boulevard
Minneapolis, Minnesota 55440

Mills and Boon, Ltd.
17/19 Foley Street
London, W1A 1DR, England

Ministry of Agriculture, Fisheries and Food
H.M.S.O.
P.O. Box 659
London SE1, England

Modern Brewery Age Publishing Corp.
80 Lincoln Avenue
Stanford, Connecticut 06904

Modern Chemicals Publishing Co.
Box 810
Red Bank, New Jersey 07701

Morgan-Grampian Books, Ltd.
28 Essex Street
Strand, London, WC2, England

Frederick Muller, Ltd.
Ludgate House, 110 Fleet Street
London, EC4 2AP, England

Myers Publishing Co., Inc.
381 Park Avenue S.
New York, New York 10016

NTIS
see
National Technical Information Service

NRFEA Publications, Inc.
2340 Hampton Avenue
St. Louis, Missouri 63139

National Academy of Sciences
Committee on Fire Research Council
2101 Constitution Avenue, N.W.
Washington, D.C. 20418

National Association of Corrosion Engineers
2400 West Loop South
Houston, Texas 77027

National Fertilizer Solution Association
910 Lehmann Building
Peoria, Illinois 61602

National Institution of Agricultural
Engineering
Wrest Park
Silsoe, Bedford, England

National Livestock and Meat Board
36 South Wabash Avenue
Chicago, Illinois 60603

National Paint, Varnish and Lacquer
Association
1500 Rhode Island Avenue, N.W.
Washington, D.C. 20005

National Planning Association
1606 New Hampshire Avenue, N.W.
Washington, D.C. 20009

National Research Council
800 State National Bank Plaza
Evanston, Illinois 60201

National Safety Council
425 North Michigan Avenue
Chicago, Illinois 60611

National Standards Association, Inc.
1321 Fourteenth Street, N.W.
Washington, D.C. 20005

National Technical Information Service
Springfield, Virginia 22151

North Carolina State University
Division of Continuing Education
Box 5125
Raleigh, North Carolina 27607

Noyes Data Corp.
118 Mill Road
Park Ridge, New Jersey 07656

Occupational Hazards
614 Superior Avenue, West
Cleveland, Ohio 44113

Oceana Publications, Inc.
Dobbs Ferry, New York 10522

Ohio State University Press
Hitchcock Hall, Room 316
2070 Neil Avenue
Columbus, Ohio 43210

Oil, Paint and Drug Reporter
Now:
Chemical Marketing Reporter
Schnell Publishing Co.
100 Church Street
New York, New York 10007

Oliver and Boyd, Ltd.
Tweedal Court
14 High Street
Edinburgh, EH1 1YL, Scotland

Oregon State University
Environmental Health Sciences Center
Corvallis, Oregon 97330

Organization europeanne et Mediter-
raneene pour la Protection des Plantes
1 rue le Notre
75-Paris 16e, France

Orszagos Muszaki Konyotar es
Dokumentacios, Kozpont
Reviczky u.6
Budapest 8, Hungary

Alan Osborne and Associates
1/113 Blackheath Park
London, SE3 OHA, England

Oxbridge Publishing Co.
150 East 52nd Street
New York, New York 10022

Palmerton Publishing Co., Inc.
101 West 31st Street
New York, New York 10001

Pandex Current Indexes and Scientific
and Technical Literature
Macmillan Information
Division of Macmillan Publishing
Co.
866 Third Avenue
New York, New York 10022

Parkins Publishing Co., Ltd.
1215 Greene Avenue
Montreal 1215, Quebec, Canada

Patent and Trademark Institute of
   Canada
P.O. Box 553
Station B
Ottawa 4, Ontario, Canada

Patent Exchange, Inc.
26 Broadway
New York, New York 10004

Patent Office
25 Southampton Buildings
London, WC2A 1AY, England

Patent Office
St. Mary Cray
Orpington, Kent, BR5 3RD, England

Pennsylvania State University Press
215 Wagner Building
University Park, Pennsylvania 16802

Penton Publishing Co.
Penton Building
Cleveland, Ohio 44113

Pergamon Press
Headington Hill Hall
Oxford, OX3 0BW, England

Pergamon Press, Inc.
Maxwell House
Fairview Park
Elmsford, New York 10523

Peterson Publishing Co., Ltd.
Peterson House
Livery Street
Birmingham 3, England

Pilot Books
347 Fifth Avenue
New York, New York 10016

Plastics Institute
11 Hobart Place
London, SW1W 0HL, England

Plenum Press, Inc.
227 West 17th Street
New York, New York 10011

Political Research, Inc.
1500 Jackson Street
Continental Building
Dallas, Texas 75201

Pollution Abstracts, Inc.
P.O. Box 2369
La Jolla, California 92037

Polymer Institute
University of Detroit
4001 West McNichols Road
Detroit, Michigan 48221

Powder Metallurgy Joint Group
17 Belgrave Square
London, SW1, England

Frederick A. Praeger, Inc.
   Publishers
111 Fourth Avenue
New York, New York 10003

Predicasts, Inc.
200 University Circle Research Center
10001 Cedar Avenue
Cleveland, Ohio 44106

Prentice-Hall, Inc.
Route 9W
Englewood Cliffs, New Jersey 07632

Proceedings in Print, Inc.
Box 247
Mattapan, Massachusetts 02126

Production Publications (London), Ltd.
10-16 Elm Street
London, WC1, England

Products Journal, Ltd.
London
see Morgan-Grampian Books, Ltd.

Profit Press, Inc.
400 East 89th Street
New York, New York 10028

Publications of Canada
2813A Eglington Avenue
Scarborough 701
Ontario, Canada

Publicom, Inc.
17 Sherwood Place
Greenwich, Connecticut 06830

Purdue University Engineering
Experimental Station
Lafayette, Indiana 47907

Putnam Publishing Co.
11 East Delaware Place
Chicago, Illinois 60611

Queen's Printer
171 Slater Street
Ottawa, Canada

Refrigeration Press
Box 109
Davis House, High Street
Croydon, Surrey, England

Reinhold Publishing Corp.
430 Park Avenue
New York, New York 10022

Research and Design Institute
Box 307
Providence, Rhode Island 02901

Richard Rimbach Street
8550 Babcock Boulevard
Pittsburgh, Pennsylvania 15237

Rockefeller University Press
66th Street and York Avenue
New York, New York 10021

The Ronald Press Co.
79 Madison Avenue
New York, New York 10016

Rowman & Littlefield, Publishers
81 Adams Drive
Totowa, New Jersey 17512

Royal Australian Chemical Institute,
Headquarters: Chinies Ross House
191 Royal Parade
Parkville 3052, Australia

Royal Society for the Prevention
of Accidents, The
52 Grosvenor Gardens
London, SW1, England

Rubber and Plastics Research
Association of Great Britain
Shawbury, Shrewsbury,
Shropshire, England

Sage Hill Publishers, Inc.
111 Washington Avenue
Albany, New York 12210

Salesmen's Association of the
American Chemical Industry, Inc.
79 Madison Avenue
New York, New York 10016

Sam's Publishing Co.
3 West 57th Street
New York, New York 10019

Schnell Publishing Co., Inc.
100 Church Street
New York, New York 10007

Science Associates/International, Inc.
Publications Department
23 East 26th Street
New York, New York 10010

Scientific Surveys, Ltd.
11-a Gloucester Road
London, SW7, England

Scranton Publishing
35 East Wacker
Chicago, Illinois 60611

Charles Scribner's Sons
597 Fifth Avenue
New York, New York 10017

Select Publications, Inc.
900 Northstar Center
Minneapolis, Minnesota 55406

Service Industry Communications, Inc.
347 Madison Avenue
New York, New York 10017

The Shoe String Press, Inc.
995 Sherman Avenue
Hamden, Connecticut 06514

Siebel Publishing Co.
4049 West Peterson Avenue
Chicago, Illinois 60646

Smithsonian Institution Press
Washington, D.C. 20560

Foster D. Snell, Inc.
Hanover Road
Florham Park, New Jersey 07932

Society for Analytical Chemistry
(See Chem Society)
10 Savile Row
London, W1X 1AF, England

Society of Automotive Engineers
400 Commonwealth Drive
Warrendale, Pennsylvania 15096

Society of Plastics Engineers
656 West Putnam Avenue
Greenwich, Connecticut 06830

Society of the Chemical Industry
Publications Department
14 Belgrave Square
London, SW1, England

Society of the Plastic Industry
250 Park Avenue
New York, New York 10017

Soil Science Society of America
(See American Society of Agronomy)
677 South Segoe Road
Madison, Wisconsin 53711

Southam Business Publications, Ltd.
1450 Don Mills Road
Don Mills, Ontario, Canada

Spartan/Hayden Book Co.
50 Essex Street
Rochelle Park, New Jersey 07662

Special Libraries Association
235 Park Avenue S.
New York, New York 10003

Springer-Verlag
Heidelberger Platz 3
Berg-Wilmersdorf 1, Germany

Springer-Verlag New York, Inc.
175 Fifth Avenue
New York, New York 10010

Sweet's Industrial Catalog Services
330 West 42nd Street
New York, New York 10036

Synthetic Organic Chemical
  Manufacturing Association
261 Madison Avenue
New York, New York 10016

Syracuse University Press
P.O. Box 87
University Station
Syracuse, New York 10310

TCR Service, Inc.
140 Sylvan Avenue
Englewood Cliffs, New Jersey 07632

TTA Information Services Co.
171 Second Street
San Francisco, California 94105

TW Publishing Co.
Box 152
River Forest, Illinois 60305

Technical Information Co., Ltd.
Box 59
St. Helier, Jersey, British Channel
  Islands

Technical Press, Ltd., The
112 Westbourne Grove
London, W2 5R4, England

Techni Research Associates
Professional Center Building
Willow Grove, Pennsylvania 19090

Technology Publishing Corp.
825 South Barrington Avenue
Los Angeles, California 90049

Technomic Publishing Co., Inc.
265 West State Street
Westport, Connecticut 06880

Tennessee Valley Authority
National Fertilizer Development
 Center, Technical Library
Muscle Shoals, Alabama

Tennessee Valley Authority
Public Information Office
Knoxville, Tennessee 37902

Thomas Publishing Co.
461 Eighth Avenue
New York, New York 10001

Thompson Publications
South Africa, Ltd.
Trust House
Thibault Square, P.O. Box 80
Capetown, South Africa

Thompson Publications
Box 5601
Fresno, California 93755

Thunderbird Enterprises, Ltd.
102 College Road
Harrow, Middlesex, England

Tothill Press
161 Fleet Street
London, England

Transatlantic Publishing Corp.
North Village Green
Levittown, New York 11756

Transchem, Inc.
Chemical Translation Service
Box 669
Knoxville, Tennessee 37901

Trans Tech Publications
21330 Center Ridge Road
Cleveland, Ohio 44116

Unipub, Inc.
P.O. Box 433
New York, New York 10016

United Nations, F.A.O.
Villa delle Terme di Carcalla
Rome, Italy

The United Piece Dye Works
111 West 40th Street
New York, New York 10018

U.S. Department of Agriculture
Room 1447
South Building
Washington, D.C. 20250

U.S. Department of Agriculture
Eastern Marketing and Nutrition
 Research Division
600 East Mermaid Lane
Philadelphia, Pennsylvania 19118

U.S. Department of Commerce
Washington, D.C. 20231

U.S. Department of Defense
Alexandria, Virginia 22314

U.S. Department of Health, Education
 and Welfare
330 Independence Avenue, S.W.
Washington, D.C. 20201

U.S. Government Printing Office
Washington, D.C. 20402

U.S. International Labor Office
666 11th Street, N.W.
Washington, D.C. 20001

U.S. National Aeronautics and
    Space Administration
Scientific and Technical Information
    Division
Washington, D.C. 20546

U.S. Trademark Association
6 East 45th Street
New York, New York 10017

United Trade Press, Ltd.
9 Gough Square
Fleet Street
London, WC4, England

University Microfilms
Xerox Company
300 North Zeeb Road
Ann Arbor, Michigan 48106

University of Arizona
Engineering Experiment Station
College of Engineering
Tuscon, Arizona 85721

University of California Press
2223 Fulton Street
Berkeley, California 94720

University of Pittsburgh Press
127 North Bellefield Avenue
Pittsburgh, Pennsylvania 15260

University of Toronto Press
St. George Campus
Toronto 181
Ontario, Canada

University of Washington Press
1405 Northeast 41st Street
Seattle, Washington 98105

University of Wisconsin Press
Box 1379
Madison, Wisconsin 53701

Van Nostrand-Reinhold Co.
450 West 33rd Street
New York, New York 10001

Verlag Chemie
6940 Weinheim/Bergstrasse
Pappelallee 3, Germany

Verlag Hans Carl, K.G.
Breite Gasse 58/60
Nurnberg 1, West Germany

Viking Publications, Ltd.
Box 47062
Parklands, Transvaal, South Africa

Walker Art Center
Vineland Place
Minneapolis, Minnesota 55403

Washington National Academy of
    Science
Washington, D.C. 20418

Washington State University
College of Engineering
Research Division
Pullman, Washington 99163

Water Information Center, Inc.
14 Vanderventer Avenue
Port Washington, New York 11050

Wayne State University Press
5980 Cass
Detroit, Michigan 48202

Webb Publishing Co.
1999 Shepard Road
St. Paul, Minnesota 55116

West Publishing Co.
Kellogg Boulevard
St. Paul, Minnesota 94101

W. Arthur West Publishing Co.
35 Mason Street
Greenwood, Connecticut 06830

Western Frozen Food Institute
1467 Echo Park Avenue
Los Angeles, California 90026

Western Marketing and Nutrition
Research Division
Berkeley, California 94710

Western Retail Implement and H
Hardware Association
638 - 40th West 39th Street
Kansas City, Missouri 64111

"Where to Buy", Ltd.
John Adam House
17 John Adam Street
London, WC2N 6JH, England

Whitney Publications, Inc.
130 East 59th Street
New York, New York 10022

Wiley-Interscience Division
605 Third Avenue
New York, New York 10016

John Wiley & Sons, Inc.
605 Third Avenue
New York, New York 10016

Williams & Wilkins Co.
428 East Preston Street
Baltimore, Maryland 21202

H.W. Wilson Co.
950 University Avenue
Bronx, New York 10452

Wolfe Publishing, Ltd.
10 Earlham Street
London, England

World Health Organization
Palais des Nations
20, av. Appia
Ch-1211 Geneva 27, Switzerland

World Publishing Co.
110 East 59th Street
New York, New York 10022

Xerox Corp.
Systems and Services for Libraries
Xerox Square 007
Rochester, New York 14603

Yarsley Research Laboratories, Ltd.
Clayton Road
Chessington, Surrey, England

J.R. Young
Box 19
Anderson, South Carolina 19622

Ziff-Davis Publishing Co.
1156 15th Street, N.W.
Washington, D.C. 20005

Zinc Development Association
34 Berkeley Square
London, W1, England

Zorn and Leigh-Hunt
Moor House
London Wall and Stock Exchange
London, EC2Y 5HB, England

INDEXES

ORGANIZATIONS INDEX

Organizations, associations, and institutions referenced in this directory are listed here. Further access is provided by the comprehensive Table of Contents and the Subject Index. Alphabetization is letter by letter.

A

J

SUBJECT INDEX

This index is alphabetized letter by letter and references are to page numbers.
See the Table of Contents for specific areas of engineering.